Encyclopaedia of Mathematical Sciences
Volume 141

Low-Dimensional Topology II

Subseries Editors:
R.V. Gamkrelidze   V.A. Vassiliev

Springer
*Berlin
Heidelberg
New York
Hong Kong
London
Milan
Paris
Tokyo*

Sergei K. Lando
Alexander K. Zvonkin

Graphs on Surfaces and Their Applications

Appendix by Don B. Zagier

Springer

Sergei K. Lando
Independent University of Moscow
Higher College of Mathematics
Bolshoi Vlassievskii per. 11
121002 Moscow, Russia, and
Institute for System Research
Russian Academy of Sciences
e-mail: lando@mccme.ru

Alexander K. Zvonkin
LaBRI
Université Bordeaux I
351, cours de la Libération
33405 Talence Cedex, France
e-mail: zvonkin@labri.fr

Don B. Zagier
Max-Planck-Institut für Mathematik
Vivatsgasse 7
53111 Bonn, Germany
e-mail: zagier@mpim-bonn.mpg.de

Founding editor of the Encyclopaedia of Mathematical Sciences:
R. V. Gamkrelidze

Mathematics Subject Classification (2000): 05C10, 05C30, 12-XX, 14-XX, 14Hxx, 15A52, 20Bxx, 20F36, 30Fxx, 32G15, 57Mxx, 57M12, 57M27, 81T18, 81T40

ISSN 0938-0396
ISBN 3-540-00203-0 Springer-Verlag Berlin Heidelberg New York

This work is subject to copyright. All rights are reserved, whether the whole or part of the material is concerned, specifically the rights of translation, reprinting, reuse of illustrations, recitation, broadcasting, reproduction on microfilm or in any other way, and storage in data banks. Duplication of this publication or parts thereof is permitted only under the provisions of the German Copyright Law of September 9, 1965, in its current version, and permission for use must always be obtained from Springer-Verlag. Violations are liable for prosecution under the German Copyright Law.

Springer-Verlag is a part of Springer Science+Business Media
springeronline.com
© Springer-Verlag Berlin Heidelberg 2004
Printed in Germany

The use of general descriptive names, registered names, trademarks, etc. in this publication does not imply, even in the absence of a specific statement, that such names are exempt from the relevant protective laws and regulations and therefore free for general use.

Typeset by LE-TEX Jelonek, Schmidt & Vöckler GbR, Leipzig
Cover Design: E. Kirchner, Heidelberg, Germany
Printed on acid-free paper   46/3142 db   5 4 3 2 1 0

Alors que dans mes recherches d'avant 1970, mon attention systématiquement était dirigée vers les objets de généralité maximale, afin de dégager un language d'ensemble adéquat pour le monde de la géométrie algébrique, et que je ne m'attardais sur les courbes algébrique que dans la stricte mesure où cela s'avérait indispensable (notamment en cohomologie étale) pour développer des techniques et énoncés "pass–partout" valables en toutes dimensions et en tous lieux (j'entends, sur tous schémas de base, voire tous topos annelés de base ...), me voici donc ramené, par le truchement d'objets si simples qu'un enfant peut les connaître en jouant, aux débuts et origines de la géometrie algébrique, familiers à Riemann et à ses émules !

A. Grothendieck, Esquisse d'un programme ([129], page 8)

Whereas in my research before 1970, my attention was systematically directed towards objects of maximal generality, in order to uncover a general language adequate for the world of algebraic geometry, and I never restricted myself to algebraic curves except when strictly necessary (notably in etale cohomology), preferring to develop "pass-key" techniques and statements valid in all dimensions and every place (I mean, over all base scheme, or even based ringed topoi...), here I was brought back, via objects so simple that a child learns them while playing, to the beginnings and origins of algebraic geometry, familiar to Riemann and his followers!

A. Grothendieck, Sketch of a programme ([129], page 246)

Preface

Russian scientists have a long-standing tradition of working with high school students. The work takes many forms: in circles, special mathematical schools, olympiads, "scientifically oriented" summer camps. The subjects to discuss with students are usually chosen with great care. They must be elementary enough to be accessible to a beginner but at the same time sufficiently rich to provide a kind of an opening to the world of real research. They are often related to the researcher's field of interest.

During one such summer camp, near the ancient city of Pereslavl not far from Moscow, in the summer of 1989, we discussed in a company of friends the programs of our respective circles. To our great amazement we discovered that, while working on different subjects, we studied the same objects, namely, embedded graphs. These objects were indeed elementary, and they were a natural subject of interest for combinatorialists. But among us there were also physicists, as well as specialists in algebraic number theory, and in singularity theory. Thus began our cooperation; we tried to understand one another's works and language, which, at least for the two of us, has culminated in this book. Luckily, at the beginning we largely underestimated the width and the depth of the sea we intended to cross, otherwise we would not have dared to start. But the world we discovered is really full of wonders.

While working on the book we encountered two main difficulties which probably encounters every author trying to cover a quickly developing field. First, the domain turned out to be vaster than we initially supposed. Second, the progress of science was sometimes faster than we were able to follow it. In the course of the last decade some entirely new branches of the subject appeared "out of nowhere". Sometimes we were compelled to stop writing certain parts of the book only because at some moment one must stop anyway. The reader will notice that every chapter starts as a play with a very intriguing plot; but its end is often rather abrupt, and there is no epilogue. Well, finally, it is only natural.

An overview of the content is given in the Introduction. The authors share full responsibility for everything written in the book. All the chapters were discussed by both authors. However, we had to share the work in some way. Chapters 4 and 6 are mainly written by Sergei Lando; Chapters 1 and 2, by Alexander Zvonkin; in Chapters 3 and 5 each of us wrote certain sections and we find it unnecessary to specify this in detail.

Acknowledgments

First of all our gratitude is due to Vladimir Igorevich Arnold and to Izrail Moisseevich Gelfand: it is their inspiring influence that pushed us to work in combinatorics.

All the topics presented in this book were discussed, sometimes briefly, and sometimes at great length, with our friends and colleagues (who quite often became our teachers for certain subjects). It would be too long to indicate the particular reasons and the exact measure of our gratitude to everyone. Therefore we decided to limit ourselves to giving just a list of names: we are grateful to N. Adrianov, J. Bétréma, B. Birch, M. Bousquet-Mélou, S. Chmutov, R. Cori, J.-M. Couveignes, H. Crapo, M. Delest, P. Di Francesco, A. Dimca, S. Dulucq, S. Duzhin, T. Ekedahl, S. Gelfand, V. Goryunov, L. Granboulan, L. Habsieger, C. Itzykson, G. Jones, M. Kapranov, V. Kazakov, M. Kazaryan, S. Kerov, A. Khovanskii, Yu. Kochetkov, R. Köck, M. Kontsevich, G. Labelle, P. Leroux, V. Liskovets, A. Machì, Yu. Matiyasevich, A. Mednykh, S. Natanzon, S. Orevkov, F. Pakovitch, G. Schaeffer, L. Schneps, G. Shabat, B. Shapiro, M. Shapiro, A. Shen, M. Shubin, D. Singerman, A. Sossinsky, A. Vainshtein, V. Vassiliev, X. Viennot, J. Wolfart, M. Zaidenberg, S. Zdravkovska, J.-B. Zuber. (It is possible we have forgotten certain people but we a grateful to them nevertheless.)

Next come our students: D. Bouya, M. El Marraki, N. Hanusse, N. Magot, P. Moreau, D. Péré, E. Soboleva, J. Zipperer. The necessity to explain things to others always promotes a better understanding.

Finally, there are two persons to whom we owe special thanks: Don Zagier and Dimitri Zvonkine. Dimitri carefully read preliminary versions of the book and helped a lot in working through a number of difficult points. Professor Zagier supplied us with an enormous number of extremely useful remarks and comments at the final stage of the editing. He became almost a co-author by writing an Appendix at the end of the book. We were happy to have such collaborators.

At various stages of writing the book we had several possibilities to meet and work together. We are grateful to Laboratoire Bordelais de Recherche en Informatique (LaBRI) of the University Bordeaux I, The Independent University of Moscow (IUM), Max-Planck-Institut für Mathematik at Bonn, and Mathematisches Forschungsinstitut Oberwolfach for their hospitality. Parts of the text were used in lecture courses given by the first author at the IUM and by the second author at LaBRI.

Through these years, the first author enjoyed support from RFBR, INTAS, and NWO research grants, currently RFBR 01-01-00660, INTAS 00-259,

NWO-RFBR 047.008.005. The second author was partially supported by the European Community IHRP Program, within the Research Training Network "Algebraic Combinatorics in Europe", grant HPRN-CT-2001-00272, and by the French program "GDR Tresses" 2105 CNRS.

Moscow and Bordeaux, *Sergei Lando*
August 31, 2003 *Alexander Zvonkin*

Contents

0	**Introduction: What is This Book About**		**1**
	0.1	New Life of an Old Theory	1
	0.2	Plan of the Book	2
	0.3	What You Will Not Find in this Book	4
1	**Constellations, Coverings, and Maps**		**7**
	1.1	Constellations	7
	1.2	Ramified Coverings of the Sphere	13
		1.2.1 First Definitions	13
		1.2.2 Coverings and Fundamental Groups	15
		1.2.3 Ramified Coverings of the Sphere and Constellations	18
		1.2.4 Surfaces	22
	1.3	Maps	26
		1.3.1 Graphs Versus Maps	26
		1.3.2 Maps: Topological Definition	28
		1.3.3 Maps: Permutational Model	33
	1.4	Cartographic Groups	39
	1.5	Hypermaps	43
		1.5.1 Hypermaps and Bipartite Maps	43
		1.5.2 Trees	45
		1.5.3 Appendix: Finite Linear Groups	49
		1.5.4 Canonical Triangulation	50
	1.6	More Than Three Permutations	55
		1.6.1 Preimages of a Star or of a Polygon	56
		1.6.2 Cacti	57
		1.6.3 Preimages of a Jordan Curve	61
	1.7	Further Discussion	63
		1.7.1 Coverings of Surfaces of Higher Genera	63
		1.7.2 Ritt's Theorem	65
		1.7.3 Symmetric and Regular Constellations	68
	1.8	Review of Riemann Surfaces	70

XII   Contents

2 Dessins d'Enfants 79
    2.1 Introduction: The Belyi Theorem 79
    2.2 Plane Trees and Shabat Polynomials 80
        2.2.1 General Theory Applied to Trees 80
        2.2.2 Simple Examples 88
        2.2.3 Further Discussion 94
        2.2.4 More Advanced Examples 101
    2.3 Belyi Functions and Belyi Pairs 109
    2.4 Galois Action and Its Combinatorial Invariants 115
        2.4.1 Preliminaries 115
        2.4.2 Galois Invariants 118
        2.4.3 Two Theorems on Trees 123
    2.5 Several Facets of Belyi Functions 126
        2.5.1 A Bound of Davenport–Stothers–Zannier 126
        2.5.2 Jacobi Polynomials 131
        2.5.3 Fermat Curve 135
        2.5.4 The *abc* Conjecture 137
        2.5.5 Julia Sets 139
        2.5.6 Pell Equation for Polynomials 142
    2.6 Proof of the Belyi Theorem 146
        2.6.1 The "Only If" Part of the Belyi Theorem 146
        2.6.2 Comments to the Proof of the "Only If" Part 147
        2.6.3 The "If", or the "Obvious" Part of the Belyi Theorem . 150

3 Introduction to the Matrix Integrals Method 155
    3.1 Model Problem: One-Face Maps 155
    3.2 Gaussian Integrals 160
        3.2.1 The Gaussian Measure on the Line 160
        3.2.2 Gaussian Measures in $\mathbb{R}^k$ 162
        3.2.3 Integrals of Polynomials and the Wick Formula 163
        3.2.4 A Gaussian Measure on the Space
               of Hermitian Matrices 164
        3.2.5 Matrix Integrals and Polygon Gluings 167
        3.2.6 Computing Gaussian Integrals. Unitary Invariance .. 171
        3.2.7 Computation of the Integral for One Face Gluings .. 176
    3.3 Matrix Integrals for Multi-Faced Maps 179
        3.3.1 Feynman Diagrams 179
        3.3.2 The Matrix Integral for an Arbitrary Gluing 180
        3.3.3 Getting Rid of Disconnected Graphs 183
    3.4 Enumeration of Colored Graphs 185
        3.4.1 Two-Matrix Integrals and the Ising Model 185
        3.4.2 The Gauss Problem 188
        3.4.3 Meanders 190
        3.4.4 On Enumeration of Meanders 191
    3.5 Computation of Matrix Integrals 192

          3.5.1   Example: Computing the Volume of the Unitary Group 192
          3.5.2   Generalized Hermite Polynomials 195
          3.5.3   Planar Approximations 197
   3.6 Korteweg–de Vries (KdV) Hierarchy
        for the Universal One-Matrix Model 199
          3.6.1   Singular Behavior of Generating Functions 200
          3.6.2   The Operator of Multiplication by λ
                  in the Double Scaling Limit......................... 202
          3.6.3   The One-Matrix Model and the KdV Hierarchy 204
          3.6.4   Constructing Solutions to the KdV Hierarchy
                  from the Sato Grassmanian 206
   3.7 Physical Interpretation 210
          3.7.1   Mathematical Relations Between Physical Models 211
          3.7.2   Feynman Path Integrals and String Theory 211
          3.7.3   Quantum Field Theory Models...................... 213
          3.7.4   Other Models 214
   3.8 Appendix 215
          3.8.1   Generating Functions 215
          3.8.2   Connected and Disconnected Objects 217
          3.8.3   Logarithm of a Power Series and Wick's Formula 219

4 **Geometry of Moduli Spaces of Complex Curves** 223
   4.1 Generalities on Nodal Curves and Orbifolds 223
          4.1.1   Differentials and Nodal Curves 223
          4.1.2   Quadratic Differentials 226
          4.1.3   Orbifolds... 227
   4.2 Moduli Spaces of Complex Structures 232
   4.3 The Deligne–Mumford Compactification..................... 234
   4.4 Combinatorial Models of the Moduli Spaces of Curves........ 237
   4.5 Orbifold Euler Characteristic of the Moduli Spaces 243
   4.6 Intersection Indices on Moduli Spaces
        and the String and Dilaton Equations 249
   4.7 KdV Hierarchy and Witten's Conjecture 256
   4.8 The Kontsevich Model 257
   4.9 A Sketch of Kontsevich's Proof of Witten's Conjecture 263
          4.9.1   The Generating Function for the Kontsevich Model 263
          4.9.2   The Kontsevich Model and Intersection Theory 264
          4.9.3   The Kontsevich Model and the KdV Equation 266

5 **Meromorphic Functions and Embedded Graphs**............ 269
   5.1 The Lyashko–Looijenga Mapping
        and Rigid Classification of Generic Polynomials 270
          5.1.1   The Lyashko–Looijenga Mapping 270
          5.1.2   Construction of the LL Mapping
                  on the Space of Generic Polynomials 271

Contents

- 5.1.3 Proof of the Lyashko–Looijenga Theorem 273
- 5.2 Rigid Classification of Nongeneric Polynomials and the Geometry of the Discriminant 277
  - 5.2.1 The Discriminant in the Space of Polynomials and Its Stratification 277
  - 5.2.2 Statement of the Enumeration Theorem 279
  - 5.2.3 Primitive Strata 280
  - 5.2.4 Proof of the Enumeration Theorem 282
- 5.3 Rigid Classification of Generic Meromorphic Functions and Geometry of Moduli Spaces of Curves 288
  - 5.3.1 Statement of the Enumeration Theorem 288
  - 5.3.2 Calculations: Genus 0 and Genus 1 289
  - 5.3.3 Cones and Their Segre Classes 292
  - 5.3.4 Cones of Principal Parts 294
  - 5.3.5 Hurwitz Spaces 297
  - 5.3.6 Completed Hurwitz Spaces and Stable Mappings 299
  - 5.3.7 Extending the LL Mapping to Completed Hurwitz Spaces 300
  - 5.3.8 Computing the Top Segre Class; End of the Proof 302
- 5.4 The Braid Group Action 304
  - 5.4.1 Braid Groups 304
  - 5.4.2 Braid Group Action on Cacti: Generalities 309
  - 5.4.3 Experimental Study 312
  - 5.4.4 Primitive and Imprimitive Monodromy Groups 318
  - 5.4.5 Perspectives 325
- 5.5 Megamaps 327
  - 5.5.1 Hurwitz Spaces of Coverings with Four Ramification Points 328
  - 5.5.2 Representation of $\overline{H}$ as a Dessin d'Enfant 329
  - 5.5.3 Examples 331

6 Algebraic Structures Associated with Embedded Graphs ... 337
- 6.1 The Bialgebra of Chord Diagrams 337
  - 6.1.1 Chord Diagrams and Arc Diagrams 337
  - 6.1.2 The 4-Term Relation 339
  - 6.1.3 Multiplying Chord Diagrams 342
  - 6.1.4 A Bialgebra Structure 343
  - 6.1.5 Structure Theorem for the Bialgebra $\mathcal{M}$ 346
  - 6.1.6 Primitive Elements of the Bialgebra of Chord Diagrams 347
- 6.2 Knot Invariants and Origins of Chord Diagrams 350
  - 6.2.1 Knot Invariants and their Extension to Singular Knots . 350
  - 6.2.2 Invariants of Finite Order 353
  - 6.2.3 Deducing 1-Term and 4-Term Relations for Invariants .. 355

		6.2.4	Chord Diagrams of Singular Links 357
	6.3	Weight Systems 359	
		6.3.1	A Bialgebra Structure on the Module $\mathcal{V}$ of Vassiliev Knot Invariants........................ 359
		6.3.2	Renormalization................................... 360
		6.3.3	Weight Systems 362
		6.3.4	Vassiliev Knot Invariants and Other Knot Invariants ... 364
	6.4	Constructing Weight Systems via Intersection Graphs 367	
		6.4.1	The Intersection Graph of a Chord Diagram 367
		6.4.2	Tutte Functions for Graphs 368
		6.4.3	The 4-Bialgebra of Graphs.......................... 369
		6.4.4	The Bialgebra of Weighted Graphs 379
		6.4.5	Constructing Vassiliev Invariants from 4-Invariants 383
	6.5	Constructing Weight Systems via Lie Algebras 384	
		6.5.1	Free Associative Algebras 385
		6.5.2	Universal Enveloping Algebras of Lie Algebras 387
		6.5.3	Examples 390
	6.6	Some Other Algebras of Embedded Graphs 393	
		6.6.1	Circle Diagrams and Open Diagrams................. 393
		6.6.2	The Algebra of 3-Graphs 395
		6.6.3	The Temperley–Lieb Algebra 395

A Applications of the Representation Theory of Finite Groups (*by Don Zagier*)............................ 399
  A.1 Representation Theory of Finite Groups 399
    A.1.1 Irreducible Representations and Characters 399
    A.1.2 Examples 403
    A.1.3 Frobenius's Formula 406
  A.2 Applications 408
    A.2.1 Representations of S_n and Canonical Polynomials Associated to Partitions............................ 409
    A.2.2 Examples 415
    A.2.3 First Application: Enumeration of Polygon Gluings 416
    A.2.4 Second Application: the Goulden–Jackson Formula 418
    A.2.5 Third Application: "Mirror Symmetry" in Dimension One 423

References 429

Index 445

0
Introduction: What is This Book About

0.1 New Life of an Old Theory

The theory of *maps* (sometimes also called *embedded graphs*, or *ribbon graphs*, or *fat graphs*, or *graphs with rotation*), or, otherwise, the *topological graph theory*, is an old and well established branch of combinatorics. It may justifiably be proud of such classical results as the Euler formula (relating the number of vertices, edges, and faces of a map with the genus of the corresponding surface), and of such notably difficult modern achievements as the Four Color Theorem. But the last two decades witnessed a kind of a volcanic activity in this domain which would have been difficult to predict some 30 years ago.

The new trends do not lie in the mainstream of the preceding development of the subject. Rather, they reflect the appearance of entirely new domains of application, reaching so far as to Galois theory or to the quantum field models. It seems incredible and sometimes even improbable that "the objects so simple that a child learns them while playing" (A. Grothendieck, see the epigraph) may have so large a variety of relations. But what is written above is not just publicity. Probably it is sufficient to note that the relationships with quantum physics were discovered (and developed) by quantum physicists (see [143], [44], [146], [25] and many other papers), and those with Galois theory, by the specialists in this theory ([21], [129], [257], etc.).

On a more technical level, the vision of the main object of study also has changed. We now consider it as an entity having a kind of a triple nature. It is not merely a topological object, a graph *embedded into* (or *drawn on*) a two-dimensional surface. It is also a sequence of permutations (or, if you prefer, it "is encoded by" a sequence of permutations), which provides a relation to group theory. And it is at the same time a way of representing a ramified covering of the sphere by a compact two-dimensional manifold. Considering the sphere as the Riemann complex sphere we obtain, on the covering manifold, the structure of a Riemann surface. And Riemann surfaces rarely walk by themselves. Usually they keep company with Galois theory, with algebraic curves, moduli spaces and many other exciting subjects. It is exactly the interrelations between these three points of view that make the subject so rich.

The goal of the book is to explain the above mentioned relations and to expose the variety of new applications. The majority of the material discussed here was not presented elsewhere in a book form.

0.2 Plan of the Book

In Chapter 1 we introduce carefully, and sometimes at great length, the objects we deal with and their interrelations. These objects are: (1) constellations, that is, finite sequences of permutations; (2) ramified coverings of the sphere; (3) various species of embedded graphs (maps, hypermaps, trees, cacti, etc.); and, finally, (4) Riemann surfaces.

Chapter 2 is dedicated to a very popular subject which is called, even in the English language literature, by a French name of the theory of *dessins d'enfants* (see, for example, the collection of papers [247]). The term itself was coined by Grothendieck in his famous "Esquisse d'un programme" and means "children's drawings". Although in principle a "dessin d'enfant" means just a "map", the term is used, as a rule, in the context of the action of the absolute Galois group $\Gamma = \mathrm{Aut}(\overline{\mathbb{Q}} \,|\, \mathbb{Q})$ on maps. The existence of such an action may be deduced from the theorems known already in the 19th century, though nobody ever looked attentively in this direction. The fact that the action is faithful is a recent discovery of G. Belyi (1979) [21], and it made a profound impression on Grothendieck. The Belyi theorem relates Riemann surfaces defined over $\overline{\mathbb{Q}}$ with meromorphic functions having three critical values. Combinatorial and geometric consequences of this fact constitute the main subject of Chapter 2.

Chapter 3 is an introduction to the method of matrix integrals in map enumeration. One would naturally suppose that we speak of an application of maps to quantum fields. The idea is wrong: it works in the opposite direction! We speak of an application of quantum fields to the theory of maps. Certain models of quantum fields (namely, when a field takes values in a space of matrices) lead naturally to a presentation of the partition function of the model as the generating function for a certain kind of maps. At the same time, the maps in question may be interpreted as discrete approximations to strings; this relates the whole construction to the string theory, which is another branch of theoretical physics.

The subject of the matrix integrals has three aspects: (1) the physical foundations of the models studied; (2) the methods of calculation of matrix integrals; and (3) the interpretation of the matrix integrals in combinatorial terms, as well as the "encoding" of a combinatorial problem in terms of a matrix integral. Concerning the first aspect, we reduce its discussion to few remarks; not only we don't have enough space to develop the subject in this book, but, at least in this particular case, we must admit that we also lack a real understanding. The reader who wants to know more must address the specialized physical literature. Concerning the second aspect, we give several guidelines but don't dwell on it much. It is an exciting topic, but we would need a separate monography in order to cover it to a reasonable depth. Mainly, we concentrate here on the third aspect, thus providing an interested reader with a sort of a bridge relating the two theories (matrix integrals and map enumeration). Our feeling is that the physical literature often proposes not a bridge but a ford, which is sometimes too deep, and this may prevent a mathematically minded reader to make the most of it.

The method of matrix integrals was reinvented by mathematicians in 1986 [138] as a tool in a computation of the Euler characteristic of moduli spaces of complex algebraic curves. The paper [138] was very influential and gave a powerful incentive for an activity around this subject. The culmination point of all that was Kontsevich's proof of the Witten conjecture relating matrix models to the intersection theory on the moduli spaces [178]. An account of this range of ideas is given in Chapter 4.

In Chapter 5 we return to a general study of meromorphic functions, in order to concentrate on two topics. The first one is the Lyashko–Looijenga mapping, which relates, from yet another point of view, the enumerative questions with algebraic geometry and singularity theory. The second is the flexible classification of meromorphic functions (often called topological classification), which is related to a braid group action on constellations. The chapter finishes by an exposition of the so-called megamaps, a beautiful return of dessins d'enfants for meromorphic functions with not three but four critical values, using the above mentioned braid group action.

Finally, in Chapter 6 we explain the structure of the Hopf algebra on chord diagrams, which originates in the Vassiliev knot invariants (or "finite order invariants"). The relevance of maps here does not jump very easily to the eyes; but finally one understands that the chord diagrams are just one-vertex maps (which were by the way enumerated in the above mentioned paper [138]), and arbitrary maps play the role of the chord diagrams for links.

The book finishes with an Appendix written by Don Zagier. It may be considered as a crash-course in the representation and character theory of finite groups, with applications to the enumeration of constellations. Certain results used in the main body of the book are proved here in a very concise and elegant manner.

The difficulty level in various parts of the book is very unequal; probably we should have added "unfortunately", if it were not inevitable. Sometimes we explain in great detail rather simple things about graphs and maps; at other places we use, with a strict minimum of explanations or even without them, such things as, for example, the characters of group representations, or the absolute Galois group, and so on. The main reason is that our book is *about graphs*, and we must supply the reader with all the necessary information in order to be clear and precise. Of course, the book is also about *applications*; but, and this is the second reason, an attempt to give all the necessary prerequisites concerning groups, Riemann surfaces, Galois theory, moduli spaces, quantum fields and knots would certainly lead to a disaster. We do not pretend, by the way, to be experts in all the above mentioned fields (and if we did, the reader would probably not believe us). Nevertheless whenever possible we try our best in explaining how to work concretely, by bare hands or using a computer, with the above mentioned objects. This approach is consistent with our pedagogical credo: we do not think that a profound knowledge of a theory must necessarily precede a concrete use of the objects under study. The two processes may very well go in parallel and enrich each other.

It is clear that *every* chapter of our book could give rise to an entire book. Probably, one day such a series of books will be written. But such a project needs time. Our goal here is more modest: to give an introduction and a first account of some of the spectacular developments we witnessed these years.

0.3 What You Will Not Find in this Book

If you request the Mathematical Reviews database for a list of publications containing the subject 05C10 "Topological graph theory, imbeddings"[1], you will find more than 3000 publications. Certainly, the index 05C10 is not always assigned to papers treating, say, ramified coverings or Feynman diagrams, so one may suppose that the above list of 3000 publications is far from being complete. The conclusion is obvious: today it is absolutely unthinkable for a person (and for two persons as well) to follow the development of this branch of combinatorics. We may be practically sure that we are not aware of some important results and even entire directions of this development. Our book shows only a certain number of facets of the subject. Several books on the subject are already well-known: Ringel (1974) [241], Gross and Tucker (1987) [128], Bonnington and Little (1995) [36], as well as two first editions of the book of White [298]. It is worth noting that very recently, and almost simultaneously, several books dedicated to embedded graphs appeared: Liu [200], Mohar and Thomassen [219], White [298], Jackson and Visentin [150]. The intersection of all the above books with ours is almost inexistent.

Still, we would like to mention explicitly at least the most noticeable omissions of our book.

First of all, such a classical topic as the map coloring, which may be considered as one of the origins of map theory, is practically not touched at all. The four-color theorem [9], the coloring of maps on surfaces of higher genera [241], and many other ramifications of the subject, such as the coloring of directed graphs, are completely omitted.

Second, we don't treat either certain less classical but not less important topics, and among them there are some that we want to mention, though very

---

[1] The latter word possesses two correct spellings: 'embedding' and 'imbedding'. We use the first one.

briefly. The spectral theory of graphs is one of them. In principle it studies graphs, not maps; but, as in the case of coloring, the "genus aspect" turns out to be very important: see [60]. The theory of graph minors has grown up out of the theory of maps. More precisely, it is a very far-reaching generalization of the Pontryagin–Kuratowski characterization of planar graphs. But we don't give an account of it here; see a long series of papers by Robertson and Seymour, of which we cite only the latest one available to us at the moment of writing this introduction: [244]. There also exists a multitude of algorithmic problems concerning maps: planarity verification, random generation, least genus embedding, and so on. They are not accounted. Finally, let us mention the embedding problems for various series of graphs, such as the complete graphs, the n-dimensional cubes, and so on.

Our relationship with the map enumeration is more complicated. It would be wrong to say that we don't do it at all. In Chapter 3 we show how to enumerate certain maps by computing matrix integrals. In Chapter 5 we enumerate some classes of maps by computing the degree of the Lyashko–Looijenga mapping. In the Appendix a number of enumerative results is obtained using the character theory of finite groups. But the theory of map enumeration also possesses a wealth of more traditional combinatorial methods, such as the bijective method, the Lagrange inversion, Tutte's quadratic method, the conjugation of trees, and so on. We do not explain all this: our goal is to explore the new dimensions of map enumeration.

Among the omissions we regret the most, there are two topics that are very close to the mainstream of our exposition. The first is the finite type invariants of immersed curves. The second concerns complex dynamical systems, in particular the range of ideas around the postcritically finite systems and Thurston's obstructions.

We may but repeat what was already said before: each chapter of this book, being exposed up to a sufficient depth, deserves an entire book. In fact, there are also some publications that relate the topics of different chapters among themselves. Thus, [263] is an attempt to apply the dessins d'enfants to the quantum field models. In [164] an action of the absolute Galois group on Vassiliev's invariants is studied. Matrix integrals are used in enumerating "tangles" (certain objects related to knots) in [306].

We stop here, being unable to embrace the unembraceable. It is clear that a more appropriate title of our book should be "Certain Aspects of Graphs on Surfaces, with Few Applications".

1
Constellations, Coverings, and Maps

We start with a combinatorial–group theoretic notion of constellation. Then we pass on to topology and study ramified coverings of the sphere. Next we introduce various types of "pictures", or embedded graphs: maps, hypermaps, trees, cacti and so on, and clarify their relation to the ramified coverings and the constellations. Finally, at the end of the chapter we discuss Riemann surfaces; they are constructed as ramified coverings of the complex Riemann sphere.

The goal of this chapter is not only to introduce all these objects, but to show that they are, in a way, different manifestations of the same object. The possibility to look at it from different points of view, and to involve different intuitions, is the main source of ideas forming the foundation of this book.

1.1 Constellations

Let S_n denote the symmetric group that acts on n points. We usually multiply permutations from left to right: this is the rule adopted in all computer systems of symbolic calculations, which are now actively used in mathematical research. Accordingly, our permutations act on the right; in particular, the action by conjugation is $g^h = h^{-1}gh$, and $g^{h_1 h_2} = (g^{h_1})^{h_2}$. We write sets in curly brackets, sequences in square brackets, and group generators in angular ones.

Definition 1.1.1 (Constellation). A sequence $[g_1, g_2, \ldots, g_k]$, where $g_i \in S_n$, is called a *constellation* (or a *k-constellation*) if the following two properties are satisfied:

- the group $G = \langle g_1, g_2, \ldots, g_k \rangle$ acts transitively on the set of n points;
- the product of g_i is the identity permutation: $g_1 g_2 \ldots g_k = \text{id}$.

The integer n is called the *degree* of the constellation, and k is called its *length*.

8   1 Constellations, Coverings, and Maps

It is obvious that the constellations of length $k \leq 2$ are of no interest (for $k = 2$ they have the form $[g, g^{-1}]$ with g cyclic; for $k = 1$ the trivial constellation [id] exists only if $n = 1$). The case $k = 3$, on the other hand, is already of great interest and will be the subject of a large part of this book.

The motivation for the term "constellation" will become clear later (see Remark 1.2.24). It is amazing to see how important such a simple object as a constellation could be, and how many profound and non-trivial structures it may possess.

Definition 1.1.2 (Cartographic group). The group $G = \langle g_1, \ldots, g_k \rangle \leq S_n$ is called the *cartographic group* of the constellation $[g_1, \ldots, g_k]$.

It is important to note that the cartographic group is defined not up to isomorphism, but by its action. In other words, it is not an abstract group but a permutation group, a concrete subgroup of S_n. A cartographic group is an object associated to a constellation, and as such it may be considered as a group with a fixed set of generators. In the context of ramified coverings of the sphere (see Sec. 1.2) the group G will also be called the *monodromy group*.

According to the definition, the group G is a permutation group. In principle, nothing prevents us from considering constellations inside an arbitrary group (and probably Coxeter groups are the next candidates to study). But for the moment there is no strong motivation for such a generalization. Whatever will be the further development of the subject, in this book we will encounter only finite permutation cartographic groups.

Definition 1.1.3 (Primitive and imprimitive group). Let G be a permutation group of degree n. It is called *imprimitive* if the underlying set may be split into disjoint subsets of equal size different from 1 and n, called *blocks*, such that for any $g \in G$ the image of a block is always a block. (In other words, the action respects some non-trivial equivalence relation.) Otherwise the group is called *primitive*.

The "majority" of the permutation groups are imprimitive. For example, there exist 301 transitive permutation groups of degree 12, but only 6 of them are primitive (see [245]), including S_n and A_n. On the other hand, according to a classical result of Dixon [85], a randomly chosen set of permutations almost always generates either S_n or A_n. Therefore a primitive cartographic group different from S_n and A_n is a rare guest.

Definition 1.1.4 (Special group). We call a permutation group *special* if it is primitive and does not coincide with S_n or A_n.

See further discussion of the imprimitive case in Sec. 1.7.2.

Definition 1.1.5. Two constellations $C = [g_1, \ldots, g_k]$ and $C' = [g'_1, \ldots, g'_k]$ acting on two sets E and E' of the same cardinality are *isomorphic* if there exists a bijection $h : E \to E'$ such that $g'_i = h^{-1} g_i h$ for $i = 1, \ldots, k$. The constellations C and C' are *conjugate* if $E = E'$ and $h \in G = \langle g_1, \ldots, g_k \rangle$.

For a permutation $g \in S_n$ its *cycle structure* is the partition $\lambda \vdash n$ which consists of the lengths of the cycles of g. For example, if $n = 12$ and $g = (1, 3, 5, 7)(2, 4, 6)(8, 9, 12)$, then g contains one cycle of length 4, two cycles of length 3 and two cycles of length 1 (that is, fixed points, namely 10 and 11).

Notation 1.1.6. There exist various conventions to denote partitions. We permit ourselves the freedom of using any comprehensible notation accordingly to its convenience. Taking the partition of the above example, that is, $12 = 4 + 3 + 3 + 1 + 1$, it may be denoted as $(4, 3, 3, 1, 1)$, or as 43311, or as $43^2 1^2$. If one of the parts of a partition is a number with two or more digits, the two last methods may lead to an ambiguity. Therefore in this case we shall use only the notation of the type $(21, 1, 1, 1)$ or $(21, 1^3)$.

For the group S_n, to fix the cycle structure of a permutation means to fix its conjugacy class. For an arbitrary permutation group $H \leq S_n$, distinct conjugacy classes may have the same cycle structure.

Definition 1.1.7 (Passport). Let $C = [g_1, \ldots, g_k]$ be a constellation. The sequence $[\lambda_1, \ldots, \lambda_k]$ of partitions of n, where each λ_i is the cycle structure of the permutation g_i, $i = 1, \ldots, k$, is called the *passport* of C. If all g_i belong to a subgroup H, $G \leq H \leq S_n$, then the sequence $[K_1, \ldots, K_k]$ of the conjugacy classes of the permutations g_i in H is called the *refined passport* of C with respect to H.

According to this definition, the same constellation may have several different refined passports, depending on the choice of the group H. However, we usually omit the part of the phrase "with respect to H".

Example 1.1.8. In Example 1.1.14 below we will meet a constellation with the passport $[3^6 1^6, 2^{12}, (21, 3)]$.

Everybody who has ever worked with permutations knows that the following parameter of a permutation is very important: its degree minus the number of its cycles. For example, a permutation is even if and only if this parameter is even. In terms of partitions, if $\lambda \vdash n$, $\lambda = (d_1, \ldots, d_p)$, with all $d_i > 0$, we have this parameter equal to $n - p$.

Notation 1.1.9. If $\lambda \vdash n$ is the cycle structure of a permutation $g \in S_n$, then we denote by $v(g) = v(\lambda)$ the difference $n - p$, where p is the number of parts in λ, or the number of cycles in g. The number of cycles in g will be denoted by $c(g)$.

For example, for the partition $\lambda = 43^2 1^2 \vdash 12$ we have $v(\lambda) = 12 - 5 = 7$. If the sequence $[v(g_1), \ldots, v(g_k)]$ consists only of even numbers, then the cartographic group is even: $G \leq A_n$. In the future, the partition λ_i will play the role of the set of multiplicities of the preimages of a critical value y_i; then the corresponding number $v(\lambda_i)$ may be considered as the multiplicity of the critical value itself.

Two problems often arise in relation to constellations:

Problem 1.1.10. Having fixed a group $G \leq S_n$ and a passport $[K_1, \ldots, K_k]$, all the K_i being conjugacy classes in G, find all the constellations with the cartographic group G and with the refined passport $[K_1, \ldots, K_k]$ with respect to G.

Problem 1.1.11. Compute the number of solutions of Problem 1.1.10.

Problem 1.1.10 is in general very difficult: we know of no efficient methods of solving it. Concerning Problem 1.1.11, the following classical formula which goes back to Frobenius answers not exactly this question but a very close one, and thus provides us with a valuable and often indispensable tool in working with constellations. The proof may be found in Sec. A.1.3.

Theorem 1.1.12 (Frobenius's formula). *Let G be a finite group, and let a sequence $[K_1, \ldots, K_k]$ of (not necessarily distinct) conjugacy classes in G be given. Then the number of solutions of the equation*

$$g_1 g_2 \ldots g_k = \mathrm{id} \quad \text{such that} \quad g_i \in K_i, \quad i = 1, \ldots, k$$

is equal to

$$\frac{|K_1| \cdot \ldots \cdot |K_k|}{|G|} \sum_\chi \frac{\chi(K_1) \cdot \ldots \cdot \chi(K_k)}{\chi(\mathrm{id})^{k-2}},$$

where $|\cdot|$ denotes the size of the corresponding set, and χ ranges over all irreducible complex characters of G.

The mnemonic rule behind this formula is as follows. The number of potential candidates $[g_1, \ldots, g_k]$ for being a solution is $|K_1| \cdot \ldots \cdot |K_k|$. If the distribution of the products $g_1 g_2 \ldots g_k$ inside G were uniform, then $|K_1| \cdot \ldots \cdot |K_k|/|G|$ times we would obtain $\mathrm{id} \in G$. However, this distribution is not uniform; the sum over the characters provides the correcting term.

(Possibly you have never studied group representation theory, and don't know what an irreducible character is. Never mind: the goal of the examples given below is to show that you may get some useful information even without a profound knowledge of the subject. Another possibility is to thoroughly study Appendix A.)

Remark 1.1.13. The reason why Frobenius's formula does not reply *exactly* to the question posed in Problem 1.1.11 is that among the k-tuples $[g_1, \ldots, g_k]$ counted by it there are also those that generate not the group G but some of its subgroups $H < G$. A particular (and very nasty) case of this situation is the case when the group $H = \langle g_1, \ldots, g_k \rangle$ does not act transitively on the underlying set of n elements. Sometimes the question of the transitivity can be easily overcome: when, for example, one of the permutations is circular. In general, however, one must use a kind of an inclusion-exclusion procedure inside the lattice of subgroups of G, which may not be an easy thing to do. A complete account of this approach is contained in [215].

Example 1.1.14. Let us start with a non-trivial example and take as a group to consider the Mathieu group M_{24}. To define it properly would take us too much time and place; therefore we will not dwell on the definition here. We just ascertain that it is a very important finite group, which is a member of the famous list of 26 sporadic simple groups, and which is, according to Conway, "the most remarkable among all finite groups". In its presentation of minimal degree (which is also called the "natural representation") it is a permutation group of degree 24. Its order is equal to $N = 244823040 = 2^{10} \cdot 3^3 \cdot 5 \cdot 7 \cdot 11 \cdot 23$. The character table of M_{24} is to be found in the Atlas of Finite Groups [62], page 96. In M_{24}, there are only 26 conjugacy classes, and hence 26 irreducible characters. The columns of the table correspond to the conjugacy classes, the rows correspond to the irreducible characters, and on their intersection the value of the character on an element of the class is given.

Suppose we want to know the number of constellations with the refined passport $[3A, 2B, 21A]$ (here we use the Atlas notation of conjugacy classes). Concerning the cycle structures, the class $3A$ contains all $g \in M_{24}$ with the cycle structure $3^6 1^6$; the class $2B$, all g with the structure 2^{12}; and the class $21A$ is one of the two mutually inverse classes with the cycle structure $(21, 3)$.

What are we to do? The idea to undertake an exhaustive search of all the possible $N^3 \approx 14.7 \times 10^{24}$ constellations must certainly be avoided. Then let us use Frobenius's formula! The sum over the characters contains only 26 summands; also from time to time a character is equal to zero, and then the whole product of the three characters is equal to zero.

Table 1.1. Non-zero values of irreducible characters on conjugacy classes in M_{24}

		1080	7680	21
	id	$3A$	$2B$	$21A$
χ_1	1	1	1	1
χ_2	23	5	-1	-1
χ_8	253	10	-11	1
χ_{17}	1265	5	-15	1
χ_{19}	2024	-1	24	1

In Table 1.1 we reproduce only the part of the character table of M_{24} we really need: the four columns out of 26, which correspond to the three conjugacy classes $3A$, $2B$ and $21A$ and to the identity; and the five characters, also out of 26, that are non-zero on all the three classes. The first row gives the orders $Z(K)$ of the centralizers of the corresponding conjugacy classes K. We have $|K| = N/Z(K)$; therefore,

$$|3A| = 226688 = 2^7 \cdot 7 \cdot 11 \cdot 23,$$
$$|2B| = 31878 = 2 \cdot 3^2 \cdot 7 \cdot 11 \cdot 23,$$
$$|21A| = 11658240 = 2^{10} \cdot 3^2 \cdot 5 \cdot 11 \cdot 23.$$

Now we are able to apply the formula:

$$\frac{226688 \cdot 31878 \cdot 11658240}{244823040} \left(1 + \frac{5}{23} - \frac{110}{253} - \frac{75}{1265} - \frac{24}{2024}\right) =$$
$$2^8 \cdot 3 \cdot 7 \cdot 11^2 \cdot 23^2 \left(1 + \frac{5}{23} - \frac{10}{23} - \frac{15}{253} - \frac{3}{253}\right) = 244823040 = |M_{24}|.$$

Note that if a triple $[g_1, g_2, g_3]$ generates the whole group M_{24}, then all its conjugates $[h^{-1}g_1 h, h^{-1}g_2 h, h^{-1}g_3 h]$, $h \in M_{24}$ are pairwise distinct. Indeed, otherwise there would exist an h that would commute with g_1, g_2 and g_3 and therefore it would commute with all the elements of M_{24}. Thus, the group M_{24} would have a non-trivial center, which contradicts the fact that the group is simple. Later we will see that in our example the triples $[g_1, g_2, g_3]$ indeed generate the whole group. Therefore the fact that the number of solutions is equal to the order of the group means that in fact the solution is unique up to conjugacy. The solution will be given explicitly in Example 1.4.1.

We see that Frobenius's formula may sometimes be remarkably efficient.

Remark 1.1.15. Not all the entries of character tables are as simple as in the above example. From time to time we may encounter something like, say, $(-1 + \sqrt{-23})/2$. But anyway the final answer given by the formula should be an integer, and therefore all the square roots and other irrationalities should eventually disappear.

Exercise 1.1.16 ([5]). If the Atlas [62] is available to you, carry out similar computations for the Mathieu group M_{23} and the refined passport $[2A, 4A, 23A]$. In the natural representation of this group (of degree 23) the cycle structures of these classes are $[2^8 1^7, 4^4 2^2 1^3, 23]$.

For the group S_n, the number of classes and characters very soon becomes too big, and a "character table" in paper form becomes impossible. On the contrary, for this case there exists an elaborated combinatorial theory of characters, which permitted to develop many software packages computing them. In certain cases a study of more profound properties of characters allows one to find the result without using the software. Also there exist some *enumerative formulas* that give the desired result without having to resort to the characters at all. We do not discuss this topic in the present section, because the formulas in question often depend on certain geometric (or, rather, topological) characteristics of the constellations, such as their *genus*, and a topological interpretation of constellations will be given in the next section.

Construction 1.1.17 (Braid group action). Sometimes it would be desirable to transpose two neighboring permutations g_i and g_{i+1}, thus obtaining a new "constellation"

$$[g_1,\ldots,g_{i-1},g_{i+1},g_i,g_{i+2},\ldots,g_k].$$

Alas, usually this is impossible: the product of permutations is not commutative, and therefore the new object in general will not be a constellation (the product $\prod g_i$ will be different from id). In order to overcome this obstacle, let us introduce the following operations $\sigma_1,\ldots,\sigma_{k-1}$ that act on k-constellations:

$$\begin{aligned}\sigma_i : g_i &\mapsto g'_i = g_{i+1},\\ g_{i+1} &\mapsto g'_{i+1} = g_{i+1}^{-1}g_ig_{i+1}, \text{ and}\\ g_j &\mapsto g'_j = g_j \qquad \text{for } j\neq i, i+1.\end{aligned}$$

Now it is easily seen that the result of this operation is also a constellation, because the product $\prod g'_i = \prod g_i$ is preserved, and the groups generated by the g_i and by the g'_i coincide. The relation

$$\sigma_i\sigma_j = \sigma_j\sigma_i \quad \text{for} \quad |i-j|>1$$

is trivial.

Exercise 1.1.18. Verify the relation

$$\sigma_i\sigma_{i+1}\sigma_i = \sigma_{i+1}\sigma_i\sigma_{i+1}.$$

Thus the operations σ_i, $i=1,\ldots,k-1$, generate an action of the *braid group* B_k (see [33] or Sec. 5.4) on k-constellations. The following *invariants of the action* are obvious:

- the cartographic group G (as a concrete subgroup of S_n);
- the *unordered* refined passport of the constellation: in fact, the conjugacy classes K_i and K_{i+1} are transposed by σ_i.

If two constellations are isomorphic, then their images under σ_i are also isomorphic. Therefore, we may consider the action also on the isomorphism classes of constellations. It will play an important role in Chapter 5 because it is responsible for what we call the *flexible classification* of meromorphic functions (as opposed to their *rigid classification*).

1.2 Ramified Coverings of the Sphere

1.2.1 First Definitions

Definition 1.2.1 (Unramified covering). Let X and Y be two path connected topological spaces, and let $f: X\to Y$ be a continuous mapping. The triple (X,Y,f) is called an *unramified covering*, or simply a *covering*, of Y by X if for any $y\in Y$ there exists a neighborhood V of y such that the preimage $f^{-1}(V)\subset X$ is homeomorphic to $V\times S$, where S is a discrete set.

14   1 Constellations, Coverings, and Maps

The function f is called the *projection* from X to Y; by abuse of language it is often this function which is called a covering (when X and Y are specified by the context).

The connected components of the preimage $f^{-1}(V)$ are called the *sheets* of the covering *over* V. The preimage $f^{-1}(y)$ is called the *fiber* over y. The cardinality of S is called the *degree* of the covering and is denoted by $\deg f$. If $\deg f = n$, then the covering f is called n-sheeted, and if $n < \infty$, it is called finite-sheeted.

Two unramified coverings $f_1 : X_1 \to Y$ and $f_2 : X_2 \to Y$ are *isomorphic* if there exists a homeomorphism $u : X_1 \to X_2$ such that the following diagram is commutative:

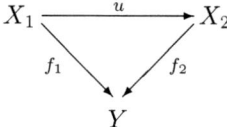

We will be mainly interested in finite-sheeted coverings.

Example 1.2.2. Let both X and Y be the unit circle S^1. For a point in S^1, take as the coordinate the angle φ measured $\mod 2\pi$. Then the mapping

$$f : \varphi \mapsto n\varphi \mod 2\pi$$

is an unramified covering of degree n. In Fig. 1.1 we show on the left n points belonging to $X = S^1$ which are the preimages of a single point in $Y = S^1$ (shown on the right).

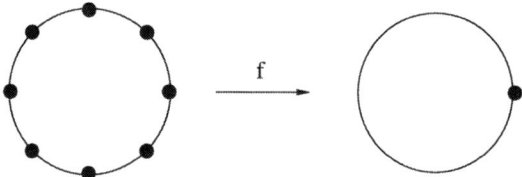

Fig. 1.1. An unramified covering of the circle by itself: $f : \varphi \mapsto n\varphi \mod 2\pi$; n different preimages of a point are shown

We may also choose as the coordinate of a point in S^1 the complex number z having $|z| = 1$; then the same function f may be expressed as $f : z \mapsto z^n$.

It is interesting to note that in this example X and Y are homeomorphic to each other. The same is obviously true for the covering of an annulus $\{(r, \varphi) | 0 \leq r_1^n < r < r_2^n\}$ by the annulus $\{(r, \varphi) | 0 \leq r_1 < r < r_2\}$ corresponding to the function

$$f : (r, \varphi) \mapsto (r^n, n\varphi),$$

where $n\varphi$, as before, is taken mod 2π. In the complex coordinate z this covering has the form $f : z \mapsto z^n$. The preimage of the horizontal segment under this covering is shown in Fig. 1.2.

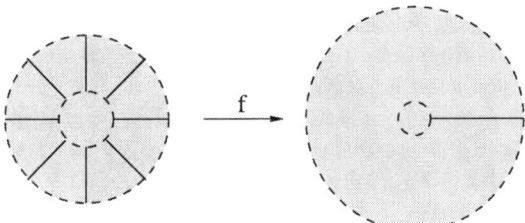

Fig. 1.2. An unramified covering of an annulus: $f : (r, \varphi) \mapsto (r^n, n\varphi)$; n different preimages of a segment are shown

When $r_1 = 0$, the annulus becomes an open disk punctured at the center. Adding the central point to both X and Y we obtain a *ramified* covering of an open disk by an open disk: the mapping remains continuous, and all the points of Y *except one* have the same number n of preimages, while the center of Y has a single preimage. In such a case we say that this preimage (the center of X) is a *critical point* and has *multiplicity* n, or *order* n, while the center of Y is a *critical value*. Quite often critical values are called *ramification points*.

Ramified coverings of the sphere S^2 will be considered a little later, when we prepare everything we need to introduce them.

Remark 1.2.3. The usage of the two terms is slightly different. If the term "covering" is used without any adjective, then it usually means "unramified covering". While speaking of a ramified covering, one must specify that it is "ramified". However, we will not always respect this rule.

1.2.2 Coverings and Fundamental Groups

Unramified coverings of a topological space Y admit a very convenient algebraic description. Namely, taken up to isomorphism, they are in bijection with the subgroups of the fundamental group of Y, taken up to conjugacy. The degree of the covering is equal to the index of the subgroup; thus, the finite-sheeted coverings are classified by the finite index subgroups of the fundamental group. An important construction involved in this correspondence is that of monodromy.

It is hardly possible to find a topology textbook that does not contain the definition of the fundamental group, and the reader may choose any of them to find all the necessary details. Nevertheless we give here the main points of this definition.

Definition 1.2.4 (Fundamental group). Let us fix a point $y_0 \in Y$. A *loop* in Y with the base point y_0 is a continuous mapping $\gamma : [0,1] \to Y$ such that $\gamma(0) = \gamma(1) = y_0$. The image $\gamma([0,1])$ is an oriented path in Y that starts at y_0 and returns to y_0; by abuse of language this path is also called a loop.

Two loops γ_0 and γ_1 are *homotopic* if there exists a continuous mapping (a homotopy) $\varphi : [0,1]^2 \to Y$ such that $\varphi(0,t) = \gamma_0(t)$, $\varphi(1,t) = \gamma_1(t)$, and $\varphi(s,0) = \varphi(s,1) = y_0$ for all s. (Intuitively, this means that the oriented path γ_0 may be continuously transformed to γ_1, without moving its ends.) Homotopy is an equivalence relation on loops.

The product $\gamma = \gamma_0\gamma_1$ of two loops γ_0 and γ_1 is the loop that follows first γ_0 and then γ_1. More formally: $\gamma(t) = \gamma_0(2t)$ for $0 \le t \le 1/2$; and $\gamma(t) = \gamma_1(2t-1)$ for $1/2 \le t \le 1$.

The homotopy classes of loops, with the above operation of a product, form a group which is called the *fundamental group* of Y with the base point y_0, and is denoted by $\pi_1(Y, y_0)$.

Exercise 1.2.5. Show that if $\gamma(t)$ and $\gamma'(t)$ are two different reparametrizations of the same path, then they are homotopic. Verify that the product of loops is well-defined, that is, if γ_0 is homotopic to γ_0', and γ_1 is homotopic to γ_1', then $\gamma_0\gamma_1$ is homotopic to $\gamma_0'\gamma_1'$. Verify the group properties for $\pi_1(Y, y_0)$.

In order not to be too formal, from now on we will not distinguish between a homotopy class of oriented loops or paths, and its concrete representative.

If Y is path connected, the groups $\pi_1(Y, y_1)$ and $\pi_1(Y, y_2)$ with two different base points are isomorphic. Indeed, take an arbitrary oriented path α from y_1 to y_2; then to any $\gamma \in \pi_1(Y, y_1)$ we may associate a loop $\alpha^{-1}\gamma\alpha \in \pi_1(Y, y_2)$ (the path $\alpha^{-1}\gamma\alpha$ goes along α^{-1} from y_2 to y_1; then along γ from y_1 back to y_1; then along α from y_1 to y_2). It is an easy exercise to show that this correspondence is an isomorphism. This isomorphism is not canonical: choosing a path β not homotopic to α we obtain another isomorphism $\gamma \mapsto \beta^{-1}\gamma\beta$. However, the freedom is not total: the latter isomorphism may be obtained by making first an *inner* automorphism of $\pi_1(Y, y_1)$, namely, the conjugation by $\beta\alpha^{-1}$, and making the initial isomorphism $\gamma \mapsto \alpha^{-1}\gamma\alpha$ after that. Thus, the fundamental group without a base point may be considered as a well defined group up to inner automorphism. We may speak of its general properties (we may say that it is free, or finite, or commutative), though we cannot speak of its concrete elements. The fundamental group of Y with a base point not specified will be denoted by $\pi_1(Y)$.

The reader is advised to compute fundamental groups for several examples, or to see them in the textbooks.

Consider a covering $f : X \to Y$. Let $E \subset X$ be the preimage of the base point: $E = f^{-1}(y_0)$. Obviously, E is in bijection with the discrete set S of Def. 1.2.4. The *monodromy* is the following action of the group $\pi_1(Y, y_0)$ on E.

Construction 1.2.6 (Monodromy). Let $\gamma \in \pi_1(Y, y_0)$ be an arbitrary element of the fundamental group. Any such γ induces a bijection $g : E \to E$.

Indeed: (i) γ is a closed oriented curve in Y; therefore $f^{-1}(\gamma)$ consists of $|S|$ oriented curves in X; (ii) γ leads from y_0 to y_0; therefore each of the curves of $f^{-1}(\gamma)$ leads from a point of E to a point of E, which gives a mapping $g : E \to E$; (iii) finally, this mapping g is invertible because γ is invertible in $\pi_1(Y, y_0)$.

It is easy to see that the correspondence $\gamma \mapsto g$ gives a group homomorphism from $\pi_1(Y, y_0)$ to the group of bijections on E: the product of the paths in $\pi_1(Y, y_0)$ corresponds to the composition of these bijections. The image G of this homomorphism is called the *monodromy group* of the covering.

The covering space X often inherits some important structures from the covered space Y. For example, if Y is oriented, we may construct the corresponding orientation of X. The process of reconstruction of a structure on X from the corresponding structure on Y is often called the *lifting*.

Construction 1.2.7 (From a covering to a subgroup). We have seen that the fundamental group $\pi_1(Y, y_0)$ acts on E. Let us fix $x_0 \in E$, and consider the subgroup M of $\pi_1(Y, y_0)$ which is the stabilizer of x_0. The choice of M obviously depends on the choice of x_0; but the stablizers of different points are conjugate. Now, the right cosets of M are in bijection with E: indeed, for $\alpha, \beta \in \pi_1(Y, y_0)$ the cosets $M\alpha$ and $M\beta$ coincide if and only if $\alpha\beta^{-1} \in M$, which is true if and only if both α and β send x_0 to the same element $x \in E$. Therefore the index of M in $\pi_1(Y, y_0)$ is equal to the cardinality of E.

Incidentally, *M is isomorphic to the fundamental group of X with the base point x_0*, because it consists of loops in $\pi_1(Y, y_0)$ whose liftings to X starting at x_0 always return to x_0.

Construction 1.2.8 (From a subgroup to a covering). In the opposite direction, let a subgroup $M \leq \pi_1(Y, y_0)$ be given. Consider the set of oriented paths in Y, with the starting point y_0 and with an arbitrary end point $y \in Y$. We will call two such paths α and β equivalent if (i) they have the common end y; and (ii) the loop $\alpha\beta^{-1}$ belongs to M. Take as the space X the set of equivalence classes of such paths. The projection $f : X \to Y$ associates to each set of equivalent paths their common end.

We leave it to the reader to verify all the properties of an unramified covering, and the fact that the choice of a subgroup M' conjugate to M leads to an isomorphic covering.

Example 1.2.9. The fundamental group of the circle, or of an annulus, is the infinite cyclic group $\mathbb{Z}$. All the finite index subgroups of this group are of the form $n\mathbb{Z}$, $n \in \mathbb{Z}_+$, with their factors being finite cyclic groups $C_n \cong \mathbb{Z}/n\mathbb{Z}$. Therefore all the possible unramified coverings of the annulus are those that we have already described in Example 1.2.2. One more subgroup contained in $\mathbb{Z}$ is $\{0\} \subset \mathbb{Z}$. It determines the infinite-sheeted covering $f : \mathbb{R} \to S^1$, $f : t \mapsto e^{it}$.

18   1 Constellations, Coverings, and Maps

We have finished our general exposition of the relations between the fundamental groups and the coverings. It was a little hasty, but it contains all the necessary information and can be worked through in order to acquire a better understanding of the subject. Let us add two more notions to our list:

Definition 1.2.10. A covering is called *regular* (or *Galois*) if the corresponding subgroup $M \triangleleft \pi_1(Y, y_0)$ is normal. In this case the monodromy group is isomorphic to the quotient group $\pi_1(Y, y_0)/M$.

A covering is called *universal* if $M = \{\mathrm{id}\}$, that is, if the covering space X is simply connected. It is, in a way, the "biggest" possible covering of Y.

1.2.3 Ramified Coverings of the Sphere and Constellations

We turn now to a particular case of *finite-sheeted coverings of a two-dimensional sphere* S^2 *punctured at* k *points* $y_1, \ldots, y_k$. Set $R = \{y_1, \ldots, y_k\}$; thus $Y = S^2 \setminus R$. We consider this sphere endowed with an orientation. This means that when we turn around any of its points, we know if we turn in the positive (that is, counter-clockwise) or negative (clockwise) direction. The fundamental group of Y is well known: it is the free group F_{k-1} with $k-1$ generators (whatever the base point y_0). However, it will be more convenient for us to choose k generators subject to one relation: in this way we will be able to associate a generator to each puncture, and thus preserve the symmetry among the punctures.

Convention 1.2.11. We construct the generators of the fundamental group of Y in the following way. We will consider the order of the punctures $y_1, \ldots, y_k$ fixed once and for all. Choose, as usual, a base point $y_0 \in Y$, and connect it with all the points y_i, $i = 1, \ldots, k$, by oriented simple non-selfintersecting curves c_i which intersect one another only at y_0. The union of the paths c_i forms a *star graph* on S^2, i.e., a tree in which one central vertex is connected to all other vertices. Now, we transform every c_i in a loop $\gamma_i \in \pi_1(Y, y_0)$ in the following way: γ_i follows c_i until it comes to a small vicinity of y_i; then it makes a complete turn around y_i in the counter-clockwise direction; and finally it returns back to y_0 along c_i^{-1}; see Fig. 1.3.

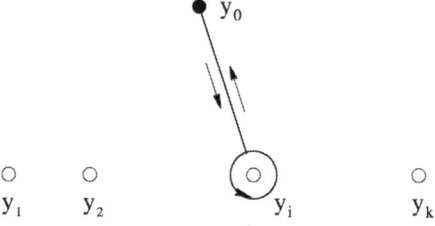

Fig. 1.3. The loop γ_i in the fundamental group $\pi_1(S^2 \setminus R, y_0)$

Finally, we impose one more condition on the loops γ_i, which is a condition on their order around y_0: *in a vicinity of y_0 the order of $\gamma_1, \gamma_2, \ldots, \gamma_k$ must be counter-clockwise*. Note that this condition still does not guarantee the unicity (up to homotopy, with the points y_0 and $y_1, \ldots, y_k$ being fixed) of such a system of loops: see Fig. 1.4.

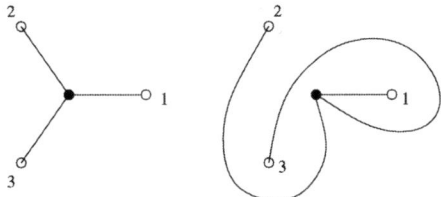

Fig. 1.4. The order of the loops $\gamma_1, \ldots, \gamma_k$ around y_0 is counter-clockwise

Definition 1.2.12 (Base star). A star graph in the target sphere with marked points, constructed as in the convention above, is called a *base star* of the covering.

With this convention, the product $\iota = \gamma_1 \ldots \gamma_k \in \pi_1(S^2 \setminus R, y_0)$ is a loop that goes around *all* the points $y_1, \ldots, y_k$; see Fig. 1.5. Taking into account that this takes place on the sphere, the above product is retractable to y_0, and thus is equal to the identity in $\pi_1(S^2 \setminus R, y_0)$.

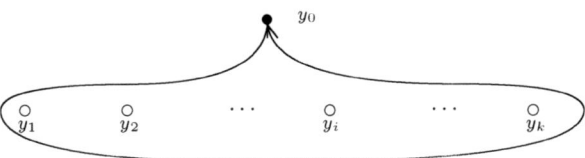

Fig. 1.5. The product $\iota = \gamma_1 \ldots \gamma_k$ goes around all the points $y_1, \ldots, y_k$, and therefore it is retractable on the sphere to y_0

We may now apply to our case the general construction of Sec. 1.2.2.

Construction 1.2.13 (Getting a constellation). (a) Let $f : X \to S^2 \setminus R$ be a finite-sheeted covering of degree n of a punctured sphere. Then to each $\gamma_i \in \pi_1(S^2 \setminus R, y_0)$ we may associate a permutation $g_i \in S_n$ acting on the set $E = f^{-1}(y_0)$.

(b) The permutations $g_1, \ldots, g_k$ generate a permutation group $G \leq S_n$, which is the monodromy group of the covering. The mapping $\gamma_i \mapsto g_i$, $i = 1, \ldots, k$, extends to a group homomorphism $\pi_1(S^2 \setminus R, y_0) \to G$.

(c) The equality $\gamma_1 \ldots \gamma_k = \mathrm{id}$ in the fundamental group of $S^2 \setminus R$ implies the equality $g_1 \ldots g_k = \mathrm{id}$ in the monodromy group G.

(d) X is path connected, therefore the permutation group G is transitive.

We summarize the above construction in the following

Proposition 1.2.14. *The sequence of permutations $[g_1, \ldots, g_k]$ is a constellation.*

What is probably more important is the fact that this construction also works in the opposite direction.

Proposition 1.2.15. *For any constellation $C = [g_1, \ldots, g_k]$ there exists an unramified covering of the punctured sphere corresponding to C.*

Proof. Let E be the underlying set of the constellation C. Consider the mapping $\pi_1(S^2 \setminus R, y_0) \to G$ taking each generator γ_i to g_i. Since the only restriction $\gamma_1 \ldots \gamma_k = \mathrm{id}$ is also valid for their images $g_1, \ldots, g_k$ in G, this mapping admits a unique extension to a group homomorphism. For a point $x \in E$ let $M_x \subset \pi_1(S^2 \setminus R, y_0)$ denote the preimage of the stabilizer of x in G. Then M_x determines a finite-sheeted covering of $S^2 \setminus R$. Since the group $G = \langle g_1, \ldots, g_k \rangle$ acts transitively on E, the covering is connected.

A final stroke, to finish with the unramified coverings of the punctured sphere:

Proposition 1.2.16. *A sequence of generators $\gamma_1, \ldots, \gamma_k \in \pi_1(S^2 \setminus R, y_0)$ being fixed, two unramified coverings of the punctured sphere $S^2 \setminus R$ are isomorphic if and only if the corresponding constellations are isomorphic.*

Proof. Indeed, an isomorphism of coverings induces an isomorphism of the corresponding constellations. The permutation $h : E \to E'$ is the restriction of the homeomorphism from Definition 1.2.1 to the fiber over y_0.

In the opposite direction, the composition of the monodromy (considered as an action of the group $\pi_1(S^2 \setminus R, y_0)$ on E) with an isomorphism $h : G \to G'$ of two constellations is a group homomorphism $\pi_1(S^2 \setminus R, y_0) \to G'$. The preimage of the stabilizer of $h(x)$ in G' coincides with M_x, and stabilizers corresponding to distinct points are conjugate.

We may at last pass on to the *ramified coverings* of the sphere. We are now going to construct from a covering of a punctured sphere a *ramified* covering of the sphere.

Construction 1.2.17 (Ramified covering). First of all let us *compactify* the punctured sphere by adding to it the missing points $y_1, \ldots, y_k$. The topological space Y thus becomes S^2; but now some of its points (namely, the points $y_1, \ldots, y_k$) do not have preimages in X. Then let us add these preimages!

For every $y_i \in Y$, $i = 1, \ldots, k$ we add as many points to X as there are cycles in the permutation g_i. We declare them the preimages of y_i, and assign to them multiplicities equal to the lengths of the corresponding cycles.

Let us fix y_i and a cycle of length d in g_i. The construction coincides "locally" with that of Example 1.2.2 (the parameter n of the example being replaced with d). Before the compactification, a small vicinity of y_i in the punctured sphere is a punctured disk $V \subset Y$; there is a punctured disk $U \subset X$ which corresponds to the chosen cycle, and which covers V with degree d. What we have done is, we added a "central point" to both annuli, thus transforming them to disks.

We preserve the old notation X, Y and $f : X \to Y$ for the new objects constructed as above.

Note that the mapping extended to the new, enlarged space X remains continuous. Note also that the multiplicity of a point $x \in X$ may be defined purely topologically: x is of multiplicity d if for any $y' \in Y$ close to $y = f(x)$, $y' \neq y$, among the n points in $f^{-1}(y')$ there exist exactly d points close to x.

The result of the above construction deserves a bit of examination. First of all, the topological space X remains a manifold. Indeed, before the point adding procedure it was a manifold (though non-compact) by Def. 1.2.1. Now, by Construction 1.2.17 the neighborhoods of the new, added points are also homeomorphic to open disks. What is even more important, the manifold X is now compact. Finally, X is orientable; it is even oriented if the sphere S^2 is. Indeed, the orientation in a neighborhood $U \subset X$ may be *lifted* from the orientation in $V = f(U)$ on the sphere, and the monodromy process does not lead to a contradiction in this definition.

Definition 1.2.18 (Ramified covering of the sphere). Let X be an orientable compact two-dimensional manifold, and let S^2 be the two-dimensional sphere. A continuous mapping $f : X \to S^2$ is called a *ramified covering* of S^2 by X if there is a finite set of points $R = \{y_1, \ldots, y_k\} \subset S^2$ such that f is obtained from an unramified covering of $S^2 \setminus R$ by Construction 1.2.17.

The degree n of the initial unramified covering is called the *degree* of the ramified covering f and is denoted by $\deg f$. The minimal set R satisfying the assumptions of the definition is the *ramification locus* of the covering. Its elements $y_1, \ldots, y_k$ are the *critical values* of f. The points $x \in f^{-1}(R)$ of multiplicity greater than one are the *critical points*. Critical values may also be called *ramification points* of the covering.

Two ramified coverings $f_1 : X_1 \to S^2$ and $f_2 : X_2 \to S^2$ are *isomorphic* if there exists an orientation preserving homeomorphism $u : X_1 \to X_2$ such that the following diagram is commutative:

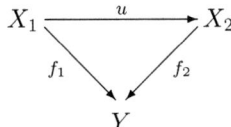

The critical values of f are those points $y \in S^2$ for which $|f^{-1}(y)| < n$. Sometimes, by abuse of language, it is convenient to add to the minimal set R certain *generic*, that is, non-critical values $y \in S^2$ (the minimal set R is *the* ramification locus, while, with some generic points being added, it becomes *a* ramification locus). Also, it may sometimes be convenient to speak of *all* the points in $f^{-1}(R)$, both critical and non-critical ones, as of critical points (thus some of them are considered as "critical points of multiplicity one").

1.2.4 Surfaces

We have already touched on two-dimensional manifolds. An accurate exposition of their theory may be found in many books on topology. The reader may look, for example, in [241] or [267]. In fact, we don't need such an exposition here, as the purely combinatorial approach developed in this book will be equivalent to it. Also, we will work mainly with *orientable* manifolds. The classification of two-dimensional orientable compact topological manifolds is well known. They are classified according to one integral parameter, their *genus*, usually denoted by the letter g. This classification is shown in Fig. 1.6.

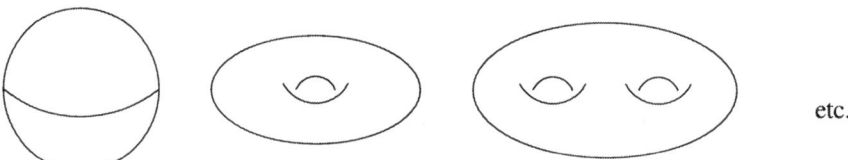

Fig. 1.6. Orientable two-dimensional manifolds of genus $g = 0$ (sphere), $g = 1$ (torus), $g = 2$, etc.

In order not to use too cumbersome terminology we introduce the following definition:

Definition 1.2.19 (Surface). A compact oriented two-dimensional topological manifold X will be called a *surface*.

Let us note that the surfaces will be considered not only orientable but oriented, that is, already endowed with a fixed orientation.

The next question to ask is, how to compute the genus of X in a ramified covering $f : X \to S^2$? The answer is given by the Riemann–Hurwitz formula. Though it is mainly used in the context of Riemann surfaces, it is purely topological in nature.

Construction 1.2.20 (Riemann–Hurwitz formula). The formula itself is rather difficult to memorize. Instead of giving the final result we prefer to explain first how it works. Though we will return to maps, and to the

Euler formula in particular, later, we will use them here. We know that for a map drawn on the sphere, with v vertices, e edges, and f faces, we have $v - e + f = 2$. Let us draw an arbitrary map on the sphere S^2, taking the set of critical values R as the set of its vertices (thus $v = k$). Now consider the lifting (that is, the preimage) of this map. It is also a map which is drawn on X; it obviously has $E = ne$ edges and $F = nf$ faces, because every edge drawn on S^2 is "repeated" n times on X, and the same is true for the faces. But this reasoning is not valid for the vertices of the map on X, because the vertices are the preimages of the critical values. Their number is the number of the points in $f^{-1}(R)$, which is equal to the total number of the cycles in the permutations g_i, $i = 1, \ldots, k$. Denoting this latter number by V we obtain

$$V - E + F = V - ne + nf = 2 - 2g.$$

Remark 1.2.21. The only information we really need in order to determine the genus of X is the constellation (in fact, even its passport suffices). Indeed, the map "below", on S^2, may be fixed once and for all. For example, we may always take as such a map a closed Jordan curve that passes through $y_1, \ldots, y_k$; then $e = k$ and $f = 2$, and everything depends on V alone, namely:

$$V - nk + 2n = 2 - 2g \quad \Longrightarrow \quad g = 1 + \frac{(k-2)n - V}{2}. \tag{1.1}$$

Any other map on S^2 which "involves" in some or another way the points $y_1, \ldots, y_k$ will give the same result. Therefore we may speak of the *genus of a constellation*.

Another form of this formula may be obtained if we recall the parameter $v(g)$ introduced in Notation 1.1.9. Taking into account that $kn - V = \sum_{i=1}^{k} v(g_k)$ we get

$$g = 1 - n + \frac{1}{2} \sum_{i=1}^{k} v(g_i), \tag{1.2}$$

or, if you like,

$$2 - 2g = 2n - \sum_{i=1}^{k} v(g_i),$$

where $2 - 2g$ is the *Euler characteristic* of the surface of genus g.

Exercise 1.2.22. Take the *star graph* on the sphere, which consists of the center y_0 and of k segments that connect the center to the vertices $y_1, \ldots, y_k$ (it is the graph we used in the monodromy construction). The map thus obtained has $k+1$ vertices, k edges, and 1 face. Apply Construction 1.2.20 to this map and compare the result to the formula (1.1). [**Hint:** The point y_0 is generic and thus gives rise to n vertices.]

Exercise 1.2.23. Show that both constellations considered in Example 1.1.14 and Exercise 1.1.16 are planar, that is, their genus is equal to zero. Therefore they define ramified coverings of the sphere by itself.

Remark 1.2.24. We may now explain the motivation behind the term *constellation*. The total preimage of the star drawn on S^2 is a *collection of stars* drawn on X and connected to each other at the critical points. A combinatorial-geometric study of this object will be one of our main tools throughout the book. Historically, the term "constellation" was first used by Jacques in [151] to denote a pair of permutations; then it was supplanted by "map" and "hypermap" (see later), and recently reanimated in the meaning we use here.

Remark 1.2.25. The genus of a two-dimensional manifold cannot be negative. This simple observation leads to certain constraints on the numbers used in Eq. (1.1). The following exercise is based on this idea.

Exercise 1.2.26 ([238]). If permutations $g_1, \ldots, g_{k-1} \in S_n$ generate a transitive subgroup of S_n, and if $v(g_1) + \ldots + v(g_{k-1}) = n - 1$, then every product $g_{i_1} \cdots g_{i_{k-1}}$, where $i_1, \ldots, i_{k-1}$ is a permutation of the indices $1, \ldots, k-1$, is a cycle of length n.

In [238] Ree remarks that he had been unable to give a direct proof of this statement even when all g_i, $i = 1, \ldots, k-1$ are cycles of length 3. Thus certain topological considerations may lead to rather non-trivial consequences for permutations.

In Chapter 5 we will study two kinds of equivalencies of ramified coverings. Unfortunately, in the literature both of them are called "topological equivalence". In order to avoid confusion, we introduce two terms: *rigid equivalence* (which means "up to isomorphism", where the isomorphism is meant that of Definition 1.2.18) and *flexible equivalence*. The latter one is defined below.

Definition 1.2.27 (Flexible equivalence of coverings). Let $f_1 : X_1 \to S^2$ and $f_2 : X_2 \to S^2$ be two ramified coverings of the sphere. They are called *flexibly equivalent* if there exist two orientation preserving homeomorphisms $u : X_1 \to X_2$ and $v : S^2 \to S^2$ taking the set of points $\{y_1, \ldots, y_k\}$ to itself such that the following diagram is commutative:

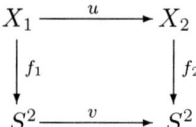

Compare this notion with that of isomorphism introduced in Definition 1.2.18. The new element here is the possibility to make a homeomorphism also on the lower level, between the two copies of S^2. What does it change?

1.2 Ramified Coverings of the Sphere

By making a homeomorphism, and even an isotopy of the sphere, we may exchange the positions of the points, say, y_i and y_{i+1}. Then the loops γ_i and γ_{i+1} will take the form shown in Fig. 1.7, and nothing will happen with the constellation.

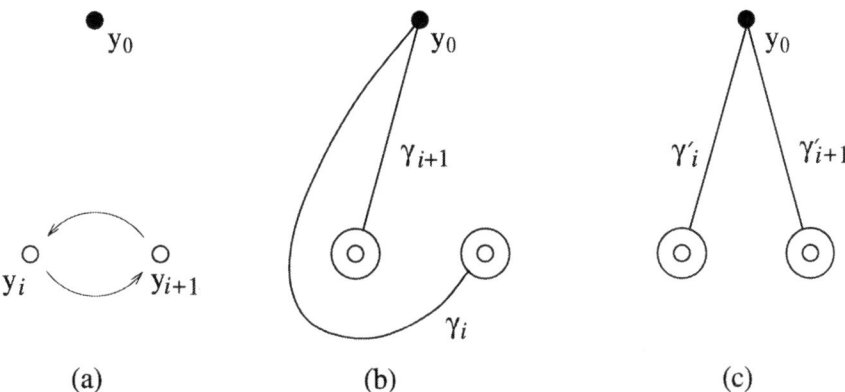

Fig. 1.7. (a) y_i and y_{i+1} exchange their places; (b) the new form of the loops γ_i and γ_{i+1}; (c) the new loops γ'_i and γ'_{i+1} that we would like to use.

But now we would probably like to change the numbering of the ramification points, declaring $y'_i = y_{i+1}$ and $y'_{i+1} = y_i$. Then the loops must also be changed. For γ_i this is simple: $\gamma'_i = \gamma_{i+1}$, while for γ_{i+1} it is more complicated. The new loop γ'_{i+1} is shown in Fig. 1.7(c), and its expression in terms of the old generators of $\pi_1(S^2 \setminus R)$ is shown in Fig. 1.8.

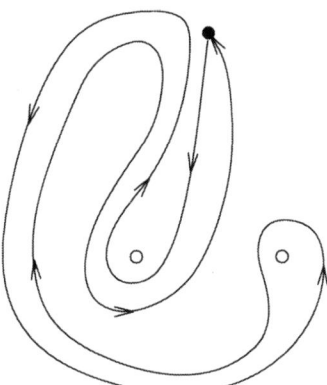

Fig. 1.8. Expression of γ'_{i+1} in terms of γ_i and γ_{i+1}.

26   1 Constellations, Coverings, and Maps

We see that $\gamma'_{i+1} = \gamma_{i+1}^{-1}\gamma_i\gamma_{i+1}$. Taking into account that the mapping $\pi_1(S^2 \setminus R) \to S_n : \gamma_i \mapsto g_i$ is a group homomorphism, we get the following transformation of the monodromy data:

$$g'_i = g_{i+1}, \quad \text{and} \quad g'_{i+1} = g_{i+1}^{-1}g_ig_{i+1},$$

while the other g_j, for $j \neq i, i+1$, don't change.

What we have found is the operation σ_i of the braid group action on the constellations introduced in Construction 1.1.17. We may conclude that this action respects the flexible classification of ramified coverings. This fact was established by Hurwitz in 1891 [144]. (He thus implicitly introduced the braid group, which was introduced explicitly by E. Artin 34 years later [14].) The opposite statement, that the flexible equivalence of ramified coverings of the sphere implies the equivalence of the corresponding constellations under the braid group action, was proved by Smilka Zdravkovska in 1970 [304]. We thus have the following theorem:

Theorem 1.2.28. *Two ramified coverings of the sphere are flexibly equivalent if and only if the corresponding constellations belong to the same orbit of the braid group action of* Construction 1.1.17.

A proof of this theorem, as well as numerous other details concerning the flexible equivalence, will be given in Sec. 5.4.

1.3 Maps

We have seen in the previous section that *maps*, that is, graphs drawn on two-dimensional manifolds, may provide a very efficient tool in solving various questions. We have even used this notion, and the Euler formula, before their proper introduction. In this section we start the accurate (and sometimes slow) presentation of all these objects and statements.

1.3.1 Graphs Versus Maps

1.3.1.1 What Do We Do When We Draw a Graph

When we work with graphs, we often draw them on a piece of paper. But we very seldom get aware of the fact that this operation of drawing adds to an "abstract" graph a supplementary structure. Loosely speaking, a graph with this additional structure is what we call a map. Look at Fig. 1.9. We see here the same graph drawn twice; but the two maps shown in this picture are different. The set of the vertices, and that of the edges, and the incidence and adjacency relations are the same for both pictures. But if we consider the *faces*, that is, the pieces into which our graph cuts the plane (or the sphere), they are different. In the left-hand side picture the faces are of "degree" 1 and

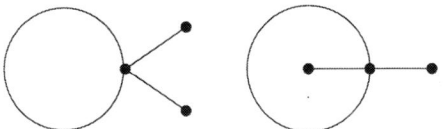

Fig. 1.9. One graph, but two maps

5, while in the right-hand side they are of "degree" 3 and 3. The notion of degree of a face will be defined below; here we just note that a graph does not have faces at all: it only has vertices and edges.

After a closer inspection of the above example we see that, combinatorially speaking, the two pictures differ in the way their edges are "arranged" around the vertex of degree 4. Namely, they have there different cyclic order. This suggests the idea to encode somehow this additional information. Here permutations are of a great help. The idea to use permutations in order to encode a map goes back to the works of Dyck and Heffter in the late 19th century [88], [140], [141], or probably even to Hamilton [132], and was revived in the 20th century by J. R. Edmonds [90]. Later on this approach was thoroughly developed by combinatorialists, see [63], [65].

We start with graphs, then pass on to two-dimensional manifolds, after that give a topological definition of a map, and then continue with its combinatorial definition using permutations.

1.3.1.2 Graphs

Except for very few explicitly stipulated cases, we will consider only connected graphs. On the other hand, we accept loops and multiple edges. No restrictions on vertices of degree 1 or 2 are imposed either. A graph will usually be denoted by a capital Γ.

Definition 1.3.1 (Graph). A *graph* $\Gamma = (V, E, I)$ is a triple consisting of a set of *vertices* V, a set of *edges* E, and an *incidence relation* I between the elements of V and the elements of E such that any edge $e \in E$ is incident either to two different vertices $v_1, v_2 \in V$, or is incident "twice" to the same vertex $v \in V$ (in the latter case the edge is called a *loop*).

If several edges are incident to the same pair of vertices v_1, v_2, then they are called *multiple edges*. A *path* in a graph is a sequence $v_0, e_1, v_1, e_2, \ldots, e_n, v_n$ (certain vertices and/or edges may be repeated) such that e_i is incident to v_{i-1} and v_i, $i = 1, \ldots, n$. We say that such a path *connects* v_0 to v_n. A path having $v_0 = v_n$ is called a *cycle*.

Definition 1.3.2. We say that a graph Γ is *connected* if any two of its vertices may be connected by a path.

Definition 1.3.3. The *degree*, or the *valency*, deg(v) of a vertex v is the number of edges incident to it (any loop incident to the vertex is counted twice).

Each edge is incident to two vertices; hence:

Proposition 1.3.4. $\sum_{v \in V} \deg(v) = 2|E|$.

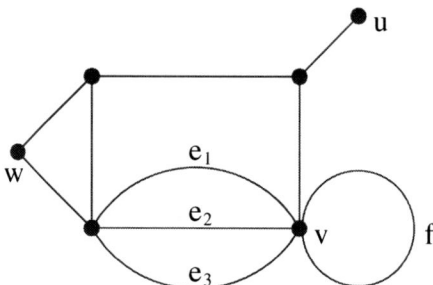

Fig. 1.10. A graph containing multiple edges e_1, e_2, e_3 and a loop f

Example 1.3.5. In the graph shown in Fig. 1.10, the edges e_1, e_2, e_3 are multiple; the edge f is a loop; the degree of the vertex v is 6; the degrees of the vertices u and w are 1 and 2 respectively; the number of edges is 10, and the sum of the degrees of the vertices is 20.

1.3.2 Maps: Topological Definition

1.3.2.1 What Is a Map?

A map is a graph drawn on a surface, or embedded into it:

Definition 1.3.6 (Topological map). A *map* M is a graph Γ embedded into a surface X (that is, considered as a subset $\Gamma \subset X$) in such a way that

- the vertices are represented as distinct points of the surface;
- the edges are represented as curves on the surface that intersect only at the vertices;
- if we cut the surface along the graph thus drawn, what remains (that is, the set $X \setminus \Gamma$) is a disjoint union of connected components, called *faces*, each homeomorphic to an open disk.

By abuse of language, we speak of the *genus of the map* M, meaning the genus of the underlying surface X (cf. Theorem 1.3.10).

Following the last definition, the maps shown in Fig. 1.9 should be regarded as drawn not on the plane but on the sphere, which is obtained from the plane by adding a point at infinity; after this operation the "outer face" becomes also homeomorphic to an open disk.

1.3.2.2 Some Comments on the Above Definition

Definition 1.3.6 seems obvious and very easy to digest. However, we must clarify certain points that may lead to confusion. First, look at Fig. 1.11. What is drawn here is not a map at all, because the "outer face" is not homeomorphic to an open disk but contains a handle instead.

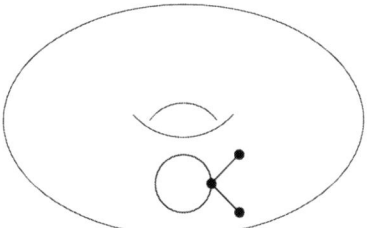

Fig. 1.11. This is not a map

It also becomes clear why we consider only connected graphs: any attempt to draw a disconnected graph on a surface will lead to a "face" not homeomorphic to a disk. However, in the rare occasions when non-connected maps will be treated, they will be considered as drawn not on a single surface but on a number of surfaces equal to the number of the graph components; see Fig. 1.12.

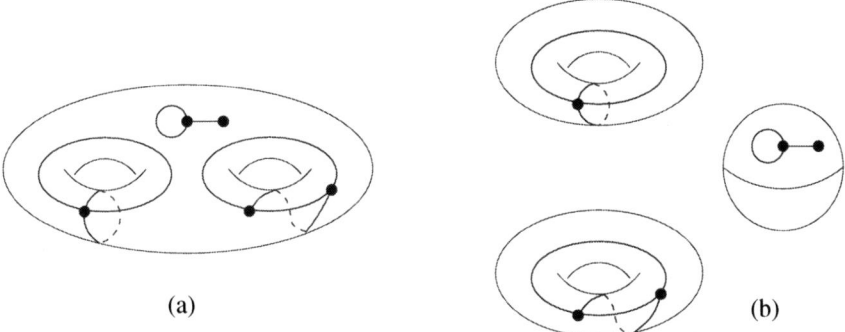

Fig. 1.12. A "disconnected map" is a disjoint union of several maps: the image (a) is wrong while the image (b) is correct

Finally, we must answer a rather tricky question: which maps we would like to consider as being isomorphic.

Definition 1.3.7. Two maps $M_1 \subset X_1$ and $M_2 \subset X_2$ are *isomorphic* if there exists an orientation preserving homeomorphism $u : X_1 \to X_2$ such that the restriction of u on Γ_1 is a graph isomorphism between Γ_1 and Γ_2.

Be careful: this seemingly harmless definition contains a trap. The difficulty is related to homeomorphisms (existing for $g \geq 1$) that realize non-trivial elements of the *mapping class group* (see [33]). The so-called *Dehn twists* form a set of generators of this group. A Dehn twist is constructed as follows: we cut the surface along a closed curve which does not bound a topological disk; then we "twist" one of the borders thus obtained through 2π; and then we glue the two things together again. The result that this operation produces on a map is shown in Fig. 1.13.

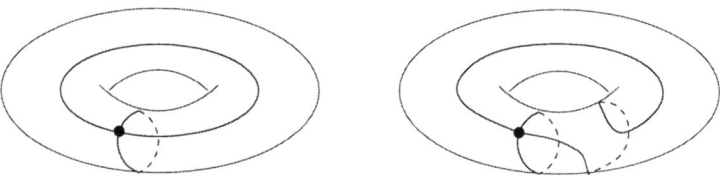

Fig. 1.13. These two pictures represent the same map

We will consider the resulting map as being the same as the initial one (or isomorphic to it), and this point of view will be coherent with the combinatorial definitions given below. Note that it is impossible to obtain one map from the other by "changing it slowly and continuously little by little", that is, by an isotopy of the surface.

Note also that a reflection symmetry of a surface does not preserve the orientation, and therefore the maps that are mirror symmetric to each other are, generally speaking, not isomorphic.

1.3.2.3 Faces and Euler's Formula

An edge is incident to a face if it belongs to the boundary of this face. If both "banks" of the edge belong to the same face, then such an edge is called an *isthmus*; we say that an isthmus is incident to the corresponding face twice.

Definition 1.3.8. The *degree*, or the *valency* $\deg(f)$ of a face f is the number of edges incident to this face (isthmuses being counted twice).

The notion of face degree is illustrated in Fig. 1.14. If we go around the boundary of a face slightly inside the face, then the number of times we pass along an edge is exactly the degree of the face.

The following proposition is similar to Proposition 1.3.4.

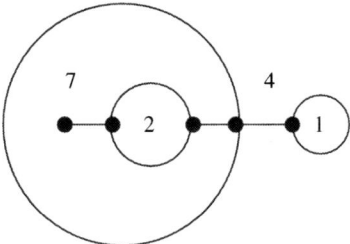

Fig. 1.14. The degrees of the faces of this map are written inside the faces

Proposition 1.3.9. $\sum_{f \in F} \deg(f) = 2|E|$, where the sum is taken over the set F of the faces of the map.

Theorem 1.3.10 (Euler characteristic). Let us associate to a map M the number
$$\chi(M) = |V| - |E| + |F|,$$
which is called its Euler characteristic. Then $\chi(M)$ does not depend on the map M itself but only on its genus g and is equal to $2 - 2g$.

For the genus zero case this theorem was already observed by Descartes, and was proved by Euler in 1752 [96]. For genera $g \geq 1$, the theorem appeared for the first time in 1812 in a paper by Simon Lhuilier [196], or, more exactly, in its "extended abstract" prepared by Gergonne, the original manuscript by Lhuilier being too long.

(The polyhedra with holes are considered in this paper as the second of three possible types of "exceptions" to the usual Euler formula: "La seconde sorte d'exception a lieu, lorsque le polyèdre est *annulaire* ; c'est-à-dire, lorsqu'étant d'ailleurs compris sous une surface unique, il a une ouverture qui le traverse de part en part." On page 186 Gergonne writes: "J'avais, depuis long-temps, remarqué les deux premières sortes d'exceptions." Therefore the formula should probably be attributed to both authors.)

A careful proof of the theorem needs some attention, but the idea is simple. We introduce operations of the kind "add/erase a vertex", or "add/erase an edge", as many of them as we are able to invent. Now we observe that any such operation preserves $\chi(M)$. For example, if we add a new vertex in the middle of an edge, then both $|V|$ and $|E|$ increase by 1, while $|F|$ does not change. What remains to be proved is the transitivity of these operations on the maps of a given genus.

Note that for non-connected maps (defined as graphs embedded into non-connected surfaces) the genus is not additive, but the Euler charactersitic is, because the numbers of vertices, edges and faces are additive.

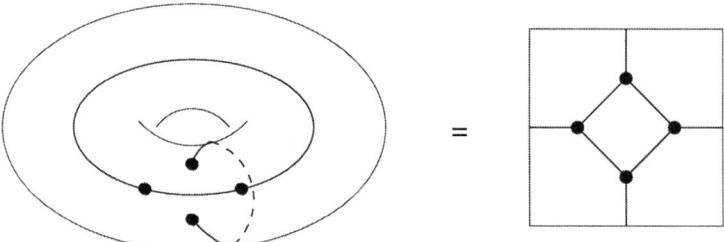

Fig. 1.15. A genus 1 embedding of the graph of the tetrahedron. In the right-hand side picture the torus is shown as a square, whose opposite sides should be identified.

1.3.2.4 Graph Embeddings

For somebody not familiar with the theory we develop here, it often comes as a refreshing surprise to learn that the graph itself not only does not determine the map (as we have already seen before), but does not even determine its genus. Let us take the graph of a tetrahedron. We have the habit to see it as the one-dimensional skeleton of a polytope; this image is very close to that of a map drawn on the sphere, which has 4 triangular faces. Thus according to Euler's formula $\chi(M) = 4 - 6 + 4 = 2 = 2 - 2g \implies g = 0$. And now look at Fig. 1.15: it is probably not immediate to recognize the same graph, but after some time we see that this graph indeed has 4 vertices and 6 edges, and every pair of vertices is joined by an edge. This time there are only two faces (one of degree 4 and the other one of degree 8), and therefore $\chi(M) = 4 - 6 + 2 = 0 = 2 - 2g \implies g = 1$.

Definition 1.3.11 (Plane vs. planar). A map of genus zero is called *plane*. A graph which can be embedded into the sphere and thus give rise to a plane map is called *planar*.

The graph of the tetrahedron is planar, but only some of its embeddings are plane.

The cube is usually considered as a map of genus 0 having 6 faces of degree 4.

Exercise 1.3.12. Embed the graph of the cube into a torus in such a way that the corresponding map would have 4 faces of degree 6.

The icosahedron is usually considered as a map of genus 0 having 12 vertices of degree 5, 30 edges, and 20 triangular faces.

Exercise 1.3.13. Embed the graph of the icosahedron into a surface of genus 4 in such a way that the corresponding map would have 12 faces of degree 5.

This exercise already exceeds our capacity to proceed by drawing pictures. We need a tool which would permit us not to draw maps but to *calculate* them. Such a tool will be introduced in Sec. 1.3.3 below. There the reader will also find a solution to the above exercise (see Example 1.3.29).

In general, the problem of finding, for a given graph, the lowest genus of a surface into which this graph can be embedded is NP-complete [278]. For particular cases, the problem may be manageable but very difficult. For example, the whole book [241] is dedicated to the proof of the following statement: for the complete graph K_n this lowest genus is equal to the number $\lceil (n-3)(n-4)/12 \rceil$ (the notation $\lceil x \rceil$ means $\min\{m \in \mathbb{Z} | m \geq x\}$; the complete graph K_n is the graph with n vertices, whose every pair of vertices is joined by an edge).

1.3.3 Maps: Permutational Model

Consider a neighborhood of a vertex of a map, see the left-hand side of Fig. 1.16. We will call the small pieces of the edges that go out of the vertex the *darts*. (In other publications they are also called "half-edges", or "1-flags".) We will also say that these darts are *incident* to the vertex. Taking into account that our surface is oriented, a cyclic order on the darts incident to each vertex is specified: we take them one after another in the counter-clockwise direction; see the right-hand side of Fig. 1.16.

Fig. 1.16. A neighborhood of a vertex

Denote the set of all the darts of edges by D; its size is twice the number of edges, $|D| = 2|E|$, that is, $|D|$ is equal to the sum of degrees of all the vertices, and also to the sum of degrees of all the faces. The collection of the described above cyclic orders on the darts corresponding to all the vertices gives us a permutation on D; we denote this permutation by σ.

There is also another permutation, which represents rather the structure of the underlying graph of the map. It shows how the darts couple to each other in order to form an edge. We denote this permutation by α: all its cycles are of length 2, or, in other words, this permutation is an involution without fixed points.

Example 1.3.14. Returning to the example of Fig. 1.9, we label the darts, for example, in the way shown in Fig. 1.17. Then α is the same for both maps,

$$\alpha = (1,2)(3,4)(5,6),$$

while σ for the map on the left is equal to

$$\sigma = (1,3,5,6)(2)(4),$$

and for the map on the right, to

$$\sigma = (1,5,3,6)(2)(4)$$

(we have shown on purpose the fixed points of σ in order to see how our rule works for vertices of degree 1).

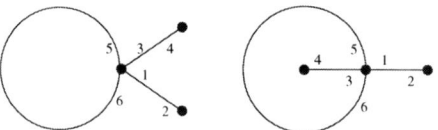

Fig. 1.17. A labelling of the darts of the maps given in Fig. 1.9

It turns out that the two permutations σ and α together permit to completely recover the structure of the map. But first we must clarify what the latter phrase means. In fact, the only thing we need to "recover" is the set of faces.

The surface is oriented; therefore we can distinguish between left and right while moving along a line. Let us set up the following convention:

Convention 1.3.15. If we start at a vertex and move along a dart, then we write the label of the corresponding dart on its left-hand bank. A dart is said to be *incident to a face* if its label is inside this face.

The faces may now be represented by a permutation φ on D, in which every face is represented as a cycle of the darts incident to it and taken in the counter-clockwise direction.

Proposition 1.3.16. *The permutation φ representing the faces of a map is computed by the formula*

$$\varphi = \alpha^{-1}\sigma^{-1},$$

where σ and α are the permutations representing the vertices and the edges of the map respectively.

The proof is obvious and may be seen in Fig. 1.18.

Example 1.3.17. For the maps of Fig. 1.17 one computes $\varphi = (1,2,6,3,4)(5)$ (for the left map), and $\varphi = (1,2,6)(3,4,5)$ (for the right one).

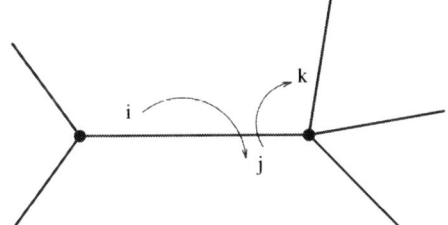

Fig. 1.18. To go around a face in the positive direction, repeat $\alpha^{-1}\sigma^{-1}$

Remark 1.3.18. (1) Historically, the Greek letters that are used to denote our three permutations are taken to resemble the first letters of the French words: σ as 'sommet' (vertex), α as 'arête' (edge), and φ like 'face' (face).

(2) One may wonder why we have written α^{-1}: is it not an involution, and therefore $\alpha^{-1} = \alpha$? In fact, this form of the formula is a preparation for the future: in a *hypermap*, see Sec. 1.5, the permutation α will be arbitrary.

(3) For the outer face of a plane map, one may have the impression that the dart labels are read in the negative direction. The impression is wrong: the face should be looked at from the "opposite side" of the sphere.

(4) The reader will easily find how to label the "two" maps of Fig. 1.13 in such a way as to get the same permutations σ, α, and φ.

Remark 1.3.19. The equality $\varphi = \alpha^{-1}\sigma^{-1}$ may be rewritten as

$$\sigma\alpha\varphi = \mathrm{id}.$$

The condition of transitivity of the group $G = \langle \sigma, \alpha, \varphi \rangle$ action on the set of darts D is equivalent to the condition of connectedness of the underlying graph. Thus, the triple $[\sigma, \alpha, \varphi]$ is a particular case of 3-constellation, in which α is an involution without fixed points.

We have seen that to any map we may associate such a 3-constellation. Now we must see that the procedure is invertible, and to any 3-constellation $[\sigma, \alpha, \varphi]$ in which α is an involution without fixed points there corresponds a map. Indeed, the data consisting of the three permutations σ, α, φ allows one to reconstruct entirely both the surface and the graph drawn on it.

Construction 1.3.20 (From a 3-constellation to a map). The usual way of constructing a two-dimensional manifold is to glue it from a number of polygons (see, for example, [267]). Therefore, our starting point will not be σ but φ.

1. Associate to each cycle of φ of length m a polygon with m sides (considered as a two-dimensional figure) and with a border oriented in the positive direction. The sides of the polygons may be identified with darts.

36   1 Constellations, Coverings, and Maps

2. Glue the sides of the polygons according to the permutation α in such a way that the orientation of the sides glued together is always opposite; this condition guarantees that the resulting surface will be oriented, see Fig. 1.19.

3. At each vertex, the darts will be automatically glued in a cyclic order according to the permutation $\sigma = \varphi^{-1}\alpha^{-1}$.

It remains to verify that in the topological space thus obtained, a neighborhood of every point is homeomorphic to an open disk in $\mathbb{R}^2$. For points inside the polygons it is true because they are parts of the plane. For points on edges it is true because we glue together exactly two polygons. Finally, for vertices it is true exactly because of the cyclic order on the "angles" of the polygons. See Fig. 1.19.

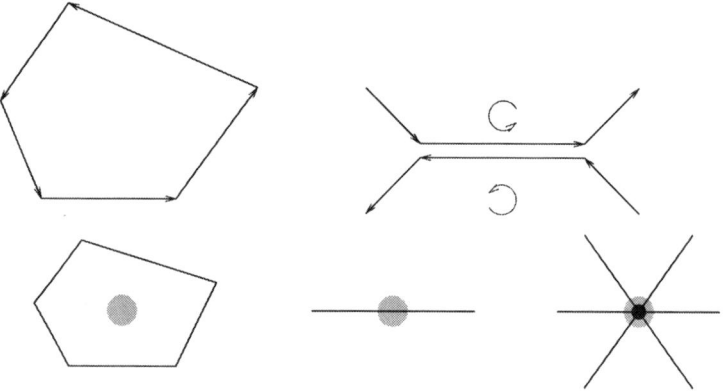

Fig. 1.19. Various stages of gluing a surface out of polygons

Remark 1.3.21. Let us return once more to the example shown in Fig. 1.11. There is no problem in labelling the darts. The permutation α is determined by the graph structure. We may even introduce σ: the cyclic order of the darts at each vertex is determined by the fact that the graph is drawn on an oriented surface. (In fact, in this example there is only one vertex of degree greater than 2, where such an order is significant.) Then why have we said that this figure does not represent a map? What exactly does not work here?

The answer is as follows. Having found α and σ, compute φ. You will get a face of degree 1 inside the loop, and an "outer face" of degree 5. Now apply the procedure of Construction 1.3.20. What you will find is not the torus but the sphere. If we insist on considering this figure as a map, we must consider it as a *plane* map. It is drawn on the sphere, and the handle of the torus stole in illegally.

Construction 1.3.22 (From a covering to a map). Let us give one more construction, which is probably even more efficient than the previous one.

According to Proposition 1.2.15, for any constellation there exists a ramified covering of the sphere determined by this constellation. Take the constellation $[\sigma, \alpha, \varphi]$, and let $y_1, y_2, y_3 \in S^2$ be the corresponding ramification points on the sphere. Then the map in question is obtained as the preimage of the following "degenerate map" on S^2, shown in Fig. 1.20: it does not have even a single edge but only a half-edge, or a dart, represented by a segment connecting y_1 to y_2.

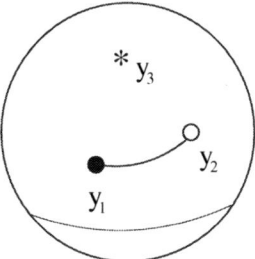

Fig. 1.20. A half-edge drawn on S^2; its preimage under the ramified covering corresponding to a constellation $[\sigma, \alpha, \varphi]$ is the map corresponding to the same constellation.

The preimages of y_2 are often called the *midpoints* of the edges, and the preimages of y_3 are often called the *centers* of the faces.

All the above considerations lead us to the following definition:

Definition 1.3.23 (Combinatorial map). A *combinatorial map* is a 3-constellation $[\sigma, \alpha, \varphi]$ in which α is an involution without fixed points. Two combinatorial maps are *isomorphic* if the corresponding constellations are isomorphic.

Remark 1.3.24. Many authors give an equivalent definition, in which they consider only the pair $[\sigma, \alpha]$ instead of the triple $[\sigma, \alpha, \varphi]$, because φ is determined by σ and α.

The correspondence between topological maps and combinatorial maps is one to many, because each topological map may be labelled in many different ways. But if two topological maps are isomorphic, then the corresponding combinatorial maps are also isomorphic.

A dictionary may easily be established between the two notions. For the vertices it looks like this:

a vertex of a map	a cycle of σ
vertex degree	length of the cycle
number of vertices	number of cycles in σ

38   1 Constellations, Coverings, and Maps

and the same for the edges and α and for the faces and φ. On a more profound level, the following definition may be given:

Definition 1.3.25 (Automorphism group). The automorphism group of a map M, denoted by $\mathrm{Aut}(M)$, is the centralizer of its cartographic group G in S_n, that is, the set of the permutations which commute with all the elements of G:
$$\mathrm{Aut}(M) = \{h \mid h^{-1}gh = g \quad \forall g \in G\}.$$

Of course, in order to verify that a permutation h belongs to this set, it is sufficient to verify that h commutes with σ and α.

Remark 1.3.26. Here the reader should make a pause and meditate a little upon the difference between the notions of isomorphism and that of automorphism. The permutation model possesses many advantages, but there is also a disadvantage with which we will often struggle in the future. Namely, it is the possibility to label the same topological object in too many ways (in the "generic" case, in $n!$ ways), and thus obtain yet more and more new triples $[\sigma, \alpha, \varphi]$. All of them are isomorphic. (In the future we will try to reduce the number of labellings by imposing various restrictions on them.) But for the notion of automorphism the corresponding difficulty does not exist. In the above definition, the triple $[\sigma, \alpha, \varphi]$ is chosen arbitrarily at first, but is fixed afterwards; therefore, a permutation commuting with it is a "true" automorphism whose existence is not due to a relabelling. For the overwhelming majority of maps the automorphism group is trivial: $\mathrm{Aut}(M) = \{\mathrm{id}\}$.

Exercise 1.3.27. Verify on examples that if a map is mirror symmetric, then the permutation expressing this mirror symmetry is not an automorphism.

Remark 1.3.28. There exists one more subtle point hidden in Definition 1.3.25. In general, the centralizer is an object associated not to a group but to a pair consisting of a group and of its subgroup (it is the centralizer of the subgroup inside a given group). A cartographic group is a permutation group, and as such is embedded into S_n. But the same "abstract" group may act on different number n of points, and thus be embedded into different S_n. For example, the group A_5 acting on 5 points has trivial centralizer, while "the same" group A_5 acting on 60 points has as a centralizer a group isomorphic to A_5 (it is the automorphism group of the icosahedron). In this latter case both groups A_5 are embedded into S_{60}, but they are embedded differently. Geometrically this becomes clear if we note that the automorphism group respects the incidence of darts to vertices, edges and faces, while the cartographic group does not respect these relations. See also the discussion in Sec. 1.7.3.

Recalling the notation $c(g)$ for the number of cycles of a permutation g (see Notation 1.1.9) we may rewrite the Euler formula:

$$\chi(M) = c(\sigma) - c(\alpha) + c(\varphi) = 2 - 2g. \qquad (1.3)$$

It is very important to understand that we may choose *any* cyclic order on the darts incident to a vertex, and we may do this independently at each vertex; the result will always be a map. The number of cyclic orderings of the darts at a vertex of degree d is $(d-1)!$. Therefore, the number of possible embeddings of a graph with vertices of degrees $d_1, \ldots, d_m$ is a priori given by the formula

$$\prod_{i=1}^{m} (d_i - 1)! \ .$$

Of course, many of these embeddings may turn out to be isomorphic.

Now at last we have a powerful tool for working with maps which does not depend on our drawing capacities.

Example 1.3.29. In order to construct a genus 4 embedding of the graph of the icosahedron (see Exercise 1.3.13), take the "ordinary" icosahedron (that is, the one of genus 0); write the corresponding permutations σ and α; and replace σ with σ^2.

Remark 1.3.30. Let us make one more remark concerning the difference between graphs and maps. The problem of graph isomorphism is one of the most difficult algorithmic problems. From the practical point of view, no polynomial algorithm has ever been found. From the theoretical point of view, the NP-completeness of this problem has not been proved either. On the contrary, the problem of map isomorphism is polynomial in time (more exactly, quadratic), and this assertion is very simple. Indeed, suppose that a putative isomorphism h sends the label 1 of the map $[\sigma, \alpha, \varphi]$ to the label i of the map $[\sigma', \alpha', \varphi']$. Then it must also send $\sigma(1)$ to $\sigma'(i)$, $\alpha(1)$ to $\alpha'(i)$, and so on. Therefore the verification of the fact that h is indeed an isomorphism requires time proportional to n. Now we must repeat this procedure for $i = 1, \ldots, n$.

1.4 Cartographic Groups

Our colleagues have often expressed to us a kind of astonishment at the fact that such a seemingly simple combinatorial structure as a pair of permutations $[\sigma, \alpha]$ possesses a rather non-trivial geometric counterpart, namely, a map, and such an interesting invariant as its genus. In fact, this is only a small part of the whole picture, because, as we shall see later, in Chapter 2, there are many other structures that are related to this pair of permutations, for example: a number field, its Galois group, a Riemann surface with its complex structure, and so on. But in this short section we propose to our reader to be "astonished in the opposite direction". Namely, isn't it astonishing that to any map we may associate a group? We mean here the cartographic group of the corresponding constellation, that is, $G = \langle \sigma, \alpha, \varphi \rangle$. Just imagine: you draw an arbitrary sketch on a piece of paper, and what you get in addition is a group. (In fact, this was our motivation in calling this group "cartographic".)

Of course, in the majority of cases the group you get is either S_n or A_n, because in general the probability of generating one of these two groups tends to 1 as $n \to \infty$, see [85], [17], [204]. But in certain (rare) cases you may have a nice surprise.

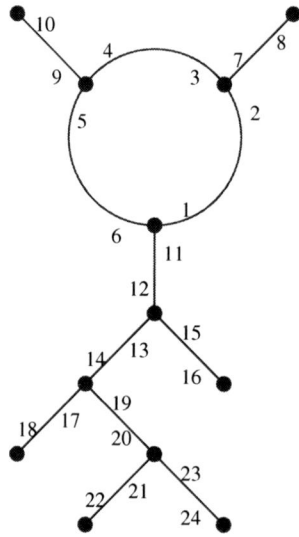

Fig. 1.21. A map with cartographic group M_{24}

Example 1.4.1. Let us consider the map shown in Fig. 1.21. We affirm that its cartographic group is the Mathieu group M_{24}.

In order to prove this assertion, we label the darts, for example in the way shown in the figure, and use the Maple package **group** (and also the fact that the group M_{24} is the only permutation group of degree 24 and of order 244823040). The computation is shown in Fig. 1.22.

The function **permgroup** generates the permutation group G of degree 24 with generators α and σ; the function **grouporder** computes the order of this group. The last but one line of the print-out shows that "the proof takes a quarter of a second".

The cycle structure of σ in this example is $3^6 1^6$; the cycle structure of α is 2^{12}; that of φ is $(21, 3)$. These three cycle structures correspond to the conjugacy classes $3A$, $2B$, and to $21A$ or to its inverse in M_{24} which we denote $(21A)^{-1}$. Because of the complete symmetry between $21A$ and its inverse we may suppose that $\varphi \in 21A$. Therefore we have found a solution of the equation

$$g_1 g_2 g_3 = \mathrm{id}, \quad g_1 \in 3A, \quad g_2 \in 2B, \quad g_3 \in 21A$$

1.4 Cartographic Groups 41

```
bixeon:/.automount/raid3/export/home/zvonkin$ maple
    |\^/|     Maple V Release 5 (Universite Bordeaux I)
._|\|   |/|_. Copyright (c) 1981-1997 by Waterloo Maple Inc. All rights
 \  MAPLE  /  reserved. Maple and Maple V are registered trademarks of
 <____ ____>  Waterloo Maple Inc.
      |       Type ? for help.
> with(group);
[DerivedS, LCS, NormalClosure, RandElement, Sylow, areconjugate, center,

    centralizer, core, cosets, cosrep, derived, groupmember, grouporder, inter,

    invperm, isabelian, isnormal, issubgroup, mulperms, normalizer, orbit,

    permrep, pres]

> alpha:=[[1,2],[3,4],[5,6],[7,8],[9,10],[11,12],[13,14],[15,16],[17,18],
> [19,20],[21,22],[23,24]];
alpha := [[1, 2], [3, 4], [5, 6], [7, 8], [9, 10], [11, 12], [13, 14],

    [15, 16], [17, 18], [19, 20], [21, 22], [23, 24]]

> sigma:=[[1,6,11],[2,7,3],[4,9,5],[12,13,15],[14,17,19],[20,21,23]];
sigma := [

    [1, 6, 11], [2, 7, 3], [4, 9, 5], [12, 13, 15], [14, 17, 19], [20, 21, 23]

    ]

> G:=permgroup(24,{alpha,sigma});
G := permgroup(24, {[[1, 2], [3, 4], [5, 6], [7, 8], [9, 10], [11, 12],

    [13, 14], [15, 16], [17, 18], [19, 20], [21, 22], [23, 24]], [

    [1, 6, 11], [2, 7, 3], [4, 9, 5], [12, 13, 15], [14, 17, 19], [20, 21, 23]

    ]})

> grouporder(G);
bytes used=1000360, alloc=917336, time=0.13
                    244823040

> quit;
bytes used=1938768, alloc=1179432, time=0.25
bixeon:/.automount/raid3/export/home/zvonkin$
```

Fig. 1.22. A computer-assisted proof that $\langle \alpha, \sigma \rangle = M_{24}$

we have looked for in Example 1.1.14 and, furthermore, we have shown that our triple of permutations generates the whole group M_{24}. Thus, this triple is unique up to conjugacy.

(We may also take the mirror image of our map. In terms of permutations this would mean taking the same α but replacing σ with σ^{-1}. Then $\alpha \in 2B$, $\sigma^{-1} \in 3A$, and the new $\varphi' = \alpha^{-1}\sigma \in (21A)^{-1}$: it is not exactly equal to $\varphi^{-1} = \sigma\alpha$ but it is conjugate to it: $\varphi' = \alpha^{-1}(\sigma\alpha)\alpha$.)

This example permits us to compare the two ways of data representation: the geometric one and the algebraic one. The geometric method is less precise but more concise. We have only two maps, namely, the above picture and its mirror image, and they represent $2 \times 24!$ triples of permutations with the cycle structures given above which generate the group M_{24}. All these triples correspond to the different ways of labelling the darts. The "economy of thought" thus obtained is very impressive. On the contrary, some information in the picture is lost: it concerns the particular choice of the embedding of M_{24} into S_{24}, and of the choice between the classes $21A$ and $(21A)^{-1}$.

One more remark is in order. We have already pointed out the difficulty of searching for constellations with a given (refined) passport. The combinatorial approach, at least in certain cases, gives us a tool: the maps with the above passport $[3^6 1^6, 2^{12}, (21, 3)]$ can be drawn by hand by attaching three trivalent trees to the three vertices on a circle. See also Exercise 1.4.4 below.

Definition 1.4.2. A plane map is called *special* if its cartographic group is special (cf. Definition 1.1.4).

Conjecture 1.4.3. (See also [131]) The set of special plane maps is finite.

For further examples see [5], [307], [3], [135].

Exercise 1.4.4. Represent the group M_{24} as the cartographic group of a map with the passport $[3^6 1^6, 2^{12}, (23, 1)]$.

Exercise 1.4.5. (a) Show that the maps in Fig. 1.23 represent the Mathieu group M_{12}. (Use the fact that it is the only group of degree 12 and of order 95040.)

(b) Find all the 18 maps with 6 edges and with the passport $[6321, 2^6, 6321]$. Verify that for 8 of them the cartographic group is M_{12}, for 6 more the group is A_{12}, and for the remaining 4 the cartographic group is imprimitive.

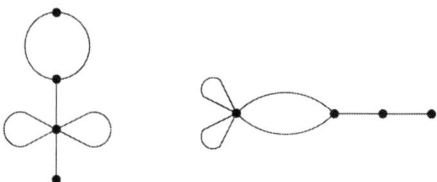

Fig. 1.23. Two maps with the cartographic group M_{12}

1.5 Hypermaps

1.5.1 Hypermaps and Bipartite Maps

Definition 1.5.1 (Hypermap). A *hypermap* is a 3-constellation.

The notions of isomorphism, automorphism, and cartographic group extend trivially to hypermaps. As before, we denote a hypermap by $[\sigma, \alpha, \varphi]$.

So, now α is not necessarily an involution without fixed points but an arbitrary permutation. What does this mean?

A graph is called *bipartite* if its vertices may be colored in two colors, say, black and white, in such a way that any edge connects vertices of different colors. A bipartite graph with one of the two possible colorings specified is called *bicolored*.

Take a map. Of course, not every graph that underlies a map is bipartite. In order to make it such, let us make the following operation of subdivision: besides the already existing vertices, which we color in black (vertices of type •), put a new white vertex (of type ○) inside each edge (see Fig. 1.24). The object we thus get is a *bicolored map*. The number of edges is multiplied by two, because the darts of the original map become the edges of the new one. The bicolored map may be described by the same triple of permutations $[\sigma, \alpha, \varphi]$. But now these permutations act not on the darts but on the edges. Each edge connects a black vertex with a white one; σ is the rotation of the edges around black vertices, and α, around the white ones. Concerning the faces, if we want to be coherent with the previous definitions, then we must declare an edge to be incident to a face if, moving inside the face along the edge in the positive direction, we move from its black end to the white one.

Fig. 1.24. We put a white vertex inside each edge

This image still bears an imprint of the fact that it was produced from a map: all its white vertices are of degree 2. If we get rid of this last restriction, then we get a hypermap: see Fig. 1.25. Therefore, the following definition is equivalent to the previous one.

Definition 1.5.2 (Hypermap). A *hypermap* is a map whose vertices are colored in black and white in such a way that each edge connects two vertices of different colors.

Thus, according to our personal preferences, we may consider hypermaps either as a generalization of maps (3-constellations with an arbitrary α), or

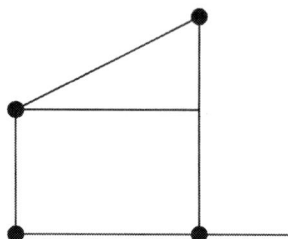

Fig. 1.25. An example of a plane hypermap; the passport is equal to $[3^2 2^2, 32^3 1, 532]$

as their particular case (bipartite maps with a fixed coloring). Whatever the choice, the permutations σ, α, and φ act now on the set of the edges E. We keep writing the labels on the left side of the edges while moving along the edges from black to white.

Historically, hypermaps were first invented as a generalization of maps, see [63], and the equivalence of the two approaches was remarked in [291]. Mathematically, a hypermap is a much more fundamental structure than that of a map, and the fact that we got rid of the rather artificial condition of α being an involution without fixed points is only one manifestation of this phenomenon. The main argument for this fundamental nature of hypermaps will be developed in Chapter 2, where we will see that they naturally correspond to Belyi functions.

Proposition 1.5.3. *Let H be a hypermap with n edges. Then its Euler characteristic is computed by the formula*

$$\chi(H) = c(\sigma) + c(\alpha) + c(\varphi) - n = 2 - 2g. \qquad (1.4)$$

Proof. The number of vertices (black and white together) is equal to $c(\sigma) + c(\alpha)$. The number of edges is equal to n. The number of faces is equal to $c(\varphi)$.

Exercise 1.5.4. Show that for the case in which a hypermap is the result of a subdivision of a map, formulas (1.3) and (1.4) coincide.

Exercise 1.5.5. Compare the above formula (1.4) with the Riemann–Hurwitz formula (1.1); note that for hypermaps we have $k = 3$.

Construction 1.5.6 (From a covering to a hypermap). It is now almost obvious that every hypermap is the preimage, via the covering determined by $[\sigma, \alpha, \varphi]$, of the *elementary hypermap* shown in Fig. 1.26. This is indeed the simplest hypermap possible: it contains one edge, one black and one white vertex (both of degree 1), and one face of degree 1. In the language of permutations it is described by the triple of permutations, all acting on the set containing a single point, and all equal to the identity.

Fig. 1.26. The elementary hypermap

1.5.2 Trees

A simple and important example of a hypermap is a bicolored plane tree. A *tree* is a connected graph without cycles. Every tree is planar, but we remind our reader that the word "plane" means that a particular embedding is chosen. Every tree is bipartite. Choosing one of the two possible colorings we obtain a particular case of a hypermap, namely, of genus 0 and with a single face, see Fig. 1.27. Quite often we will omit the tiresome adjectives "plane" and "bicolored".

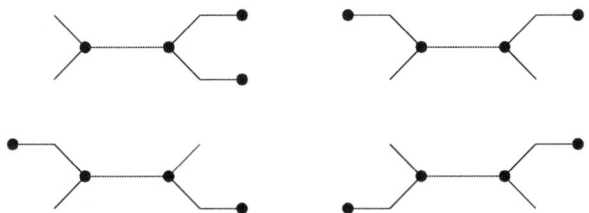
Fig. 1.27. All the bicolored plane trees with the passport $[3^2 1^2, 2^3 1^2, 8]$

A trivial induction shows that a tree with n edges has $n+1$ vertices. Therefore, if p of them are black and q are white, then we have

$$p + q = n + 1,$$

which imposes the corresponding condition on the number of cycles in σ and α and, what is the same, on the number of parts in the partitions belonging to the passport. The reader may compare the latter equality to the planarity condition (1.4). A triple of partitions of n of the form $[\lambda, \mu, n]$, for which the total number of parts in λ and μ is equal to $n+1$, is called a *valuable tree passport*.

Trees are a convenient object of study because, among many other advantages, there exist enumerative formulas for them. Before presenting them, we introduce some notions that make sense not only for trees but also for arbitrary hypermaps (and in fact for arbitrary constellations). Every enumerative combinatorialist knows that there are always some troubles with symmetric objects, and therefore one must "kill out" the symmetry. Hence, the following definition:

Definition 1.5.7 (Rooted hypermap). A *rooted hypermap* is a hypermap with a distinguished edge $e \in E$; the distinguished edge is called the *root*. While labelling a hypermap in order to represent it as a 3-constellation, we always label the root by 1.

Exercise 1.5.8. Show that the automorphism group of a rooted hypermap is always trivial. [**Hint:** By definition, an automorphism h must send 1 to 1. Then use the fact that h commutes with σ and α, and the transitivity of their action.]

Let us return to trees. Our next step is to reduce the number of their possible labellings. To do that we use the fact that φ is a circular permutation, and fix its value. Denote $c = (1, 2, 3, \ldots, n)$. For practical work with trees, it is more convenient to take φ equal not to c but to c^{-1}, thus having $\sigma\alpha = c$. Note that if we consider a rooted tree, that is, if the edge number 1 is already chosen, and if in addition $\sigma\alpha = c$, then the labels of all the edges become fixed; there remains no more ambiguity in edge labelling. We call such a rooted tree *canonically labelled*.

Exercise 1.5.9. Show that the automorphism group of a (non-rooted) plane tree is always cyclic. [**Hint:** An automorphism must commute with c.]

Remark 1.5.10. Certain trees are symmetric with respect to a midpoint of an edge; they always have a passport of the type $[\lambda, \lambda, n]$ (with n odd) and the symmetry takes white vertices to black ones and vice versa. But our model "does not see" such a symmetry, because it cannot be represented by a permutation commuting with c. Therefore, we consider these trees as asymmetric. An automorphism of a tree is always a rotation around a symmetry center which is necessarily a vertex. Nevertheless, there remains a subtle point. For trees with a passport of the type $[\lambda, \lambda, n]$ we may try to change colors: black to white, and white to black. Then a tree that does not possess any symmetry at all will give us two different trees, while for a tree which is "symmetric" with respect to a midpoint of an edge this operation will produce the same tree. See Exercise 1.5.17 below.

Notation 1.5.11. Let $\lambda = (\lambda_1, \ldots, \lambda_p) \vdash n$ be a partition of n with p parts. Represent it in the form $\lambda = 1^{d_1} 2^{d_2} \ldots n^{d_n}$, where d_i is the number of parts equal to i,

$$\sum_{i=1}^{n} d_i = p, \quad \sum_{i=1}^{n} i d_i = n.$$

Set

$$N(\lambda) = \frac{(p-1)!}{d_1! d_2! \ldots d_n!} = \frac{1}{p} \binom{p}{d_1 \; d_2 \; \ldots \; d_n}.$$

Note that $N(\lambda)$ is not necessarily an integer. The following formula, as well as its generalization to cacti (see Theorem 1.6.6) is proved in Goulden and Jackson (1992) [114]; it will also be proved by another method in Chapter 5. For trees, an essentially equivalent formula was already found in [280].

Theorem 1.5.12. *Given a valuable tree passport* $[\lambda, \mu, n]$, *the number of isomorphism classes of rooted canonically labelled trees with this passport is equal to*
$$nN(\lambda)N(\mu).$$

Taking into account that the above number is always positive we get the following

Corollary 1.5.13. *For any valuable tree passport there exists at least one tree.*

Exercise 1.5.14. Prove Corollary 1.5.13 by induction, by successively cutting off the leaves of the tree.

Any edge of a tree T may be chosen as the root. If T has no automorphisms, this gives rise to n pairwise distinct rooted trees. If it does have a non-trivial automorphism group, then the number of the corresponding rooted trees is equal to $n/|\mathrm{Aut}(T)|$. Therefore, dividing the previous formula by n, we obtain the following equivalent formulation of the Goulden–Jackson theorem:

Theorem 1.5.15. *Given a valuable tree passport* $\pi = [\lambda, \mu, n]$, *we have*
$$\sum_T \frac{1}{|\mathrm{Aut}(T)|} = N(\lambda)N(\mu), \tag{1.5}$$
where the sum is taken over all non-isomorphic non-rooted trees T with the passport π.

We will quite often encounter enumerative formulas where the objects are not counted one by one but a weight is assigned to each object, and this weight is equal to $1/|\mathrm{Aut}|$, where the denominator means the order of the automorphism group of the object. Formulas of this kind are often called *mass-formulas*[1]. Sometimes they are more convenient for practical use.

Example 1.5.16. (1) Applying formula (1.5) to the passport $[3^21^2, 2^31^2, 8]$ we get 3 as the result. At the same time we may see in Fig. 1.27 that there are 4 trees with this passport. The reason is that two of them are symmetric, with the symmetry of order 2; therefore the left-hand side of the formula is equal to $1 + 1 + \frac{1}{2} + \frac{1}{2}$. Of course, much more complicated examples of that sort may be invented. In [184] certain methods are given to count separately the trees with a given passport and with different orders of symmetry.

(2) Let us take the passport $[2^81^7, 4^42^21^3, 23]$. The number 23 is prime, therefore no symmetry is possible, and the formula (1.5) gives directly the number of different (that is, non isomorphic) trees with this passport. This number is
$$\frac{14!}{8!7!} \times \frac{8!}{4!2!3!} = 60060.$$

---

[1] The first mass-formula was proposed by H. J. S. Smith in 1867. Mass-formulas are also called Siegel–Minkowski formulas.

Exercise 1.5.17. For the passport $[321^2, 321^2, 7]$ the formula (1.5) gives 9 trees (the number 7 being a prime, there are no non-trivial automorphisms). Draw these 9 trees and verify that the counting of trees that have and that do not have a symmetry with respect to a midpoint of an edge was carried out correctly; cf. Remark 1.5.10.

Example 1.5.18. The passport $\pi = [2^8 1^7, 4^4 2^2 1^3, 23]$ treated above has already been encountered before: in Exercise 1.1.16. The partition $2^8 1^7$ is the cycle structure of the class $2A$ in M_{23}, the partition $4^4 2^2 1^3$, that of the class $4A$, and 23, that of the two mutually inverse classes $23A$ and $23A^{-1}$. Therefore, the result obtained in Exercise 1.1.16 (two constellations) gives us a hope that among 60060 trees with the passport π there may exist $2 \times 2 = 4$ trees with the cartographic group M_{23}. What should be done is (a) to find them, and (b) to verify that they do generate the whole group M_{23}. The reader may verify that this is indeed the case for the two trees shown in Fig. 1.28; the two other trees are their mirror images.

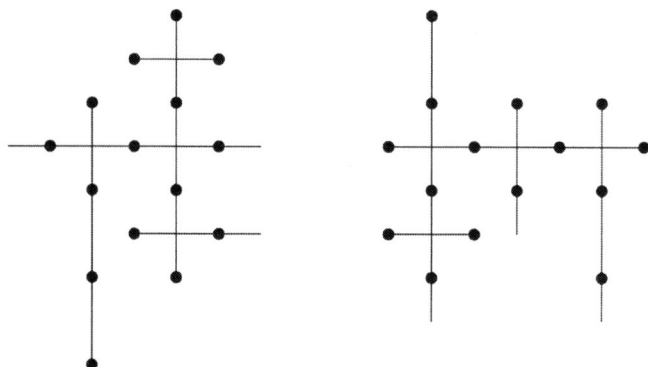

Fig. 1.28. Two of the four trees with the cartographic group M_{23}. The other two are their mirror images

Definition 1.5.19 (Special tree). A bicolored plane tree is called *special* if its cartographic group is special.

A remarkable result was obtained in [6]: there one may find the *complete list of all special trees*. Previously it was proved that their number is finite: see [221] and [3]. The above M_{23} trees belong to the list: this example has appeared in [5]. Brute force does not work here: for example, the total number of plane trees with 23 edges (without bicoloring) is 7457847082 (see [183]), but only 4 of them are special, namely, those that represent the Mathieu group M_{23}. A summary of the results on special trees is presented in Table 1.2; one may obtain 48 more such trees by exchanging the colors. See also Theorems 5.4.17 and 5.4.27.

1.5 Hypermaps

Table 1.2. Special trees: a summary (see [6])

degree	passport	group	# of special trees
6	$[41^2, 2^21^2, 6]$	$\mathrm{PGL}_2(5) \simeq S_5$	1
7	$[421, 2^21^3, 7]$	$\mathrm{PGL}_3(2) \simeq \mathrm{PSL}_2(7)$	2
7	$[3^21, 2^21^3, 7]$	$\mathrm{PGL}_3(2) \simeq \mathrm{PSL}_2(7)$	2
8	$[3^21^2, 2^31^2, 8]$	$\mathrm{PGL}_2(7)$	2
9	$[3^21^3, 2^41, 9]$	$\mathrm{P}\Gamma\mathrm{L}_2(8)$	2
9	$[3^21^3, 3^21^3, 9]$	$\mathrm{P}\Gamma\mathrm{L}_2(8)$	4
10	$[4^21^2, 2^31^4, 10]$	$\mathrm{P}\Gamma\mathrm{L}_2(9)$	1
11	$[3^31^2, 2^41^3, 11]$	$\mathrm{PSL}_2(11)$	2
11	$[4^21^3, 2^41^3, 11]$	M_{11}	2
13	$[4^22^21, 2^41^5, 13]$	$\mathrm{PGL}_3(3)$	4
13	$[3^41, 2^41^5, 13]$	$\mathrm{PGL}_3(3)$	4
13	$[6321^2, 2^41^5, 13]$	$\mathrm{PGL}_3(3)$	4
15	$[63^221, 2^41^7, 15]$	$\mathrm{PGL}_4(2) \simeq A_8$	2
15	$[4^321, 2^41^7, 15]$	$\mathrm{PGL}_4(2) \simeq A_8$	2
15	$[4^22^21^3, 2^61^3, 15]$	$\mathrm{PGL}_4(2) \simeq A_8$	2
21	$[4^421^3, 2^71^7, 21]$	$\mathrm{P}\Gamma\mathrm{L}_3(4)$	2
23	$[4^42^21^3, 2^81^7, 23]$	M_{23}	4
31	$[4^42^61^3, 2^{12}1^7, 31]$	$\mathrm{PGL}_5(2)$	6
Total			48

1.5.3 Appendix: Finite Linear Groups

The reader may find it useful to have a concise summary of definitions and notation concerning finite linear groups. Let p be a prime, and $q = p^e$, $e \geq 1$. Denote by $\mathbb{F}_q$ the finite field with q elements. The letter d will denote the dimension.

- $\mathrm{GL}_d(q)$ is the *general linear group*, that is, the group of non-degenerate linear transformations of the d-dimensional vector space $\mathbb{F}_q^d$ over $\mathbb{F}_q$, or, equivalently, the group of $d \times d$ matrices with the elements in $\mathbb{F}_q$.
- $\mathrm{AGL}_d(q)$ is the *affine linear group*, that is, the group of affine transformations
$$\{x \mapsto ax + b \mid a \in \mathrm{GL}_d(q),\ b \in \mathbb{F}_q^d\}.$$
- $\mathrm{PGL}_d(q)$ is the *projective linear group*, that is, the quotient of $\mathrm{GL}_d(q)$ by the subgroup of scalar matrices $a \cdot \mathrm{Id}$, $a \in \mathbb{F}_q$, $\mathrm{Id} \in \mathrm{GL}_d(q)$.
- $\mathrm{SL}_d(q)$ is the *special linear group*, that is, the subgroup of $\mathrm{GL}_d(q)$ containing all the matrices with the determinant equal to 1.
- $\mathrm{PSL}_d(q)$ is the *projective special group*, that is, the group of projective linear transformations (i.e., elements of $\mathrm{PGL}_d(q)$) having a representative with the determinant equal to 1. These groups are often denoted $\mathrm{L}_d(q)$.

The automorphism group of the field $\mathbb{F}_q$ is a cyclic group C_e of order e (we remind that $q = p^e$), generated by the *automorphism of Frobenius* $a \mapsto a^p$ for $a \in \mathbb{F}_q$. (To verify that this is indeed an automorphism, one may use the binomial formula and show that $(a + b)^p = a^p + b^p$ in $\mathbb{F}_q$.) A *semilinear transformation* of the d-dimensional vector space over $\mathbb{F}_q$ is a transformation $f : \mathbb{F}_q^d \to \mathbb{F}_q^d$ satisfying the conditions

$$f(x + y) = f(x) + f(y), \quad f(\lambda x) = \lambda^\sigma f(x),$$

where σ is an automorphism of $\mathbb{F}_q$.

- $\Gamma L_d(q)$ is the group of all the semilinear transformations.
- $P\Gamma L_d(q)$ is the quotient of $\Gamma L_d(q)$ by the subgroup of scalar matrices.

The groups $P\Gamma L_d(q)$, $PGL_d(q)$ and $PSL_d(q)$ act on the *projective space* of dimension d over $\mathbb{F}_q$, containing the "projective points" $(x_0 : x_1 : \ldots : x_{d-1})$ (the number of points in this space is $(q^d - 1)/(q - 1)$). The group $P\Gamma L_d(q)$ is the full automorphism group of the corresponding projective geometry (that is, it is the biggest group which sends lines to lines, planes to planes, etc.).

The group $PGL_d(q)$ is a normal subgroup in $P\Gamma L_d(q)$, with the quotient group isomorphic to $\mathrm{Aut}(\mathbb{F}_q) \simeq C_e$. Thus, if q itself is prime, the two groups coincide.

The group $PSL_d(q)$ is a normal subgroup in $PGL_d(q)$, with the quotient group isomorphic to the quotient $\mathbb{F}_q^*/\{a^d \mid a \in \mathbb{F}_q^*\}$, where $\mathbb{F}_q^* = \mathbb{F}_q \setminus \{0\}$ is the cyclic multiplicative group of the non-zero elements of $\mathbb{F}_q$. The size of the latter quotient depends on number-theoretic properties of q and d; in certain cases the groups $PSL_d(q)$ and $PGL_d(q)$ may coincide as well.

A detailed definition of the five Mathieu groups M_{11}, M_{12}, M_{22}, M_{23}, and M_{24} would lead us too far. Anyway, even when we work with these groups, we use non-standard generators. Therefore, we omit the definitions and refer the reader to the group-theoretic literature.

1.5.4 Canonical Triangulation

1.5.4.1 Positive and Negative Triangles

By introducing white vertices in order to better visualize the cycles of α, we have made only one step; but it was a step in a good direction. In this section we propose to make the similar operation with φ.

Construction 1.5.20 (Canonical triangulation). Let us take an arbitrary hypermap, and put a new vertex inside each face; we will mark such a vertex by the sign $*$. Then connect it with all the vertices that lie on the border of the face: we connect it with the vertices of type $\circ$ by a dashed line, and with the vertices of type $\bullet$, by a dotted line. Now the whole surface is subdivided

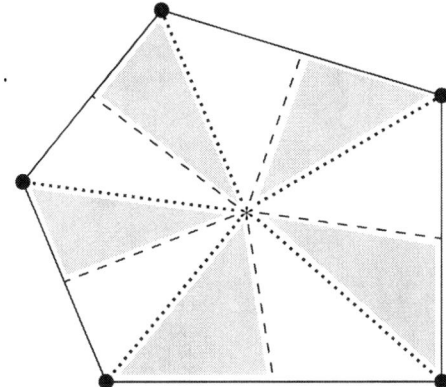

Fig. 1.29. The canonical triangulation of a face of a map

into triangles. Each triangle has three types of vertices: •, ○ and ∗; each has three types of sides: •———○, ○ − − − ∗ and ∗ · · · · · · •. Still, there are two sorts of triangles. Suppose that we look turn by turn, from the inside of a triangle, at its vertices •, ○, ∗, • (in this order!). If while doing this we turn to the positive direction, then we call the triangle *positive*. Otherwise we call it *negative*. In Fig. 1.29 the positive triangles are shaded.

The result of this construction is what is called the *canonical triangulation* of the hypermap.

The canonical triangulation of a hypermap may be rather cumbersome and therefore difficult to draw. In Fig. 1.29 we show it only for one face. The fact that the surface is oriented ensures that the "signs" (i.e., orientations) of the triangles in different faces are coherent: any positive triangle has three negative neighbors, and vice versa.

The positive triangles of the triangulation are in bijection with the edges of the hypermap. Our previous convention, which consisted in writing the edge labels to the left of the edges while moving along them from • to ○, is now equivalent to the new one:

Convention 1.5.21. We write the labels of edges of a hypermap inside the positive triangles. The same labels serve as the labels of the positive triangles themselves.

The action of the permutations σ, α, and φ becomes very transparent: they send each positive triangle to the next positive one, in the counter-clockwise direction, around the vertices of the type •, ○, and ∗ respectively.

1.5.4.2 Three Involutions

The set of *all* triangles, both positive and negative, has $2n$ elements (where n is the number of edges of the hypermap). Denote by a, b, c the following three involutions without fixed points that act on this set: a associates to each triangle its neighbor across the side of the type $*\cdots\cdots\bullet$; b does the same thing but across the side of the type $\bullet\text{———}\circ$; finally, c does the same for $\circ--- *$; see Fig. 1.30.

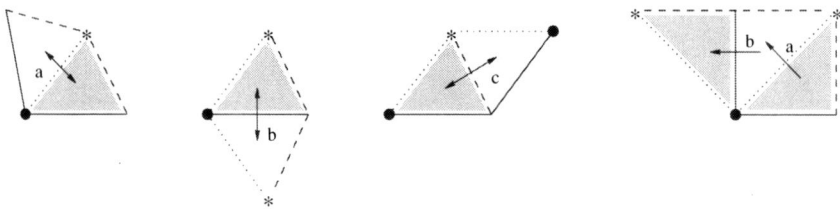

Fig. 1.30. Three involutions without fixed points that act on the set of all triangles; the right-most figure illustrates the equality $\sigma = ab$

It is easy to verify that $\sigma = ab$. If we introduce the new permutation $\bar{\sigma}$ which acts on the *negative* triangles by sending each of them to the next negative one around $\bullet$ in the counter-clockwise direction, then $\bar{\sigma} = ba$. In the same way we have $\alpha = bc$ and $\bar{\alpha} = cb$, and also $\varphi = ca$ and $\bar{\varphi} = ac$. The verification of the identity $\sigma\alpha\varphi = \text{id}$ now becomes trivial.

Remark 1.5.22. In our previous notation, the action of σ is defined only on the set of positive triangles. Now the equality $\sigma = ab$ allows us to extend the action of σ to all triangles, both positive and negative. The same is true for $\bar{\sigma} = ba$. We may then observe that $\bar{\sigma} = \sigma^{-1}$ (since a and b are involutions). And, indeed, this new extended σ turns the negative triangles in the clockwise direction.

The canonical triangulation together with the triple of involutions without fixed points a, b, c often represents a more convenient model for the study of the corresponding objects. Certain arguments of the same order will also be given in the next section.

Construction 1.5.23 (Triangulation via a covering). In order to obtain the canonical triangulation of a hypermap as the preimage via a ramified covering, we must canonically triangulate the elementary hypermap of Construction 1.5.6. The result, which we call the *elementary triangulation*, is shown in Fig. 1.31 (on the plane representing the sphere, and on the sphere itself).

1.5.4.3 Duality and Related Phenomena

Returning briefly to maps, let us recall what is the *dual map*. We put a new vertex inside each face of the original map (the "center" of the face). Then, for any edge of the original map, we draw a new edge which intersects it in

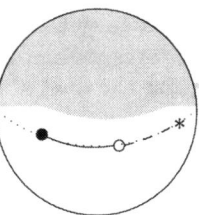

Fig. 1.31. The elementary triangulation of the sphere

its midpoint, and which connects the centers of the two faces adjacent to this original edge; see Fig. 1.32. (If these two faces coincide, the new edge thus obtained is a loop.)

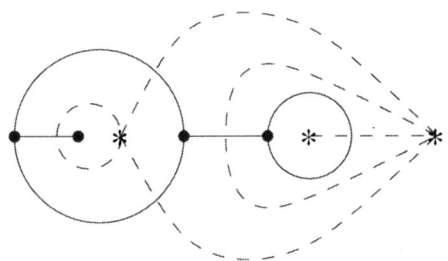

Fig. 1.32. A pair of mutually dual spherical maps

It is easy to see that the dual to the dual of a map is the original map itself. The faces of the original map are in bijection with the vertices of the dual map, and even the degrees of the new vertices are equal to the degrees of the old faces. The same may be said about the new faces and the old vertices.

In fact, this construction may also be carried out for hypermaps (see [66]; our version is slightly different). What is more, the dual hypermap may be easily seen on the canonical triangulation: it is the hypermap drawn by the dashed lines. It has two types of vertices: ∗ and ∘, while the points of type • play the role of the face centers. It is clear that if we want to draw the canonical triangulation of this new hypermap, we must preserve the same picture but only change its marking: exchange the point marks • and ∗ (the marks ∘ are preserved); exchange the line marks •———∘ and ∘ − − − ∗ (the lines of the type ∗ · · · · · · •, which connect the points • and ∗ are preserved); and finally do not forget to declare the old negative triangles positive, and vice versa.

The type of the line exchange between •———∘ and ∘ − − − ∗ shows also that we must take $b' = c$ and $c' = b$, while a is preserved. Therefore, in the language of triples of involutions without fixed points, the new object is represented by the triple (a, c, b), and this finishes our construction. *Duality is the transposition of the second and the third involutions in the triple.* The fact that the dual of dual is the original hypermap is more than clear now.

If we want to return to the permutations acting not on all the triangles but only on the triangles of the same sign, then we must take, as before, the three permutations ac, cb, and ba, and recall that they act on the triangles that previously were negative. Therefore $ac = \bar{\varphi}$, $cb = \bar{\alpha}$ and $ba = \bar{\sigma}$; thus, the dual hypermap is the constellation $[\sigma', \alpha', \varphi'] = [\bar{\varphi}, \bar{\alpha}, \bar{\sigma}] = [\varphi^{-1}, \alpha^{-1}, \sigma^{-1}]$ acting on the set of negative triangles.

Exercise 1.5.24 ([66]). Let us call *reciprocal* hypermap the result of the exchange of the black and white colors of vertices of a given hypermap.

(1) Represent the reciprocal hypermap in the form of a triple of involutions without fixed points, and in the form of a constellation.

(2) Show that the operation "dual of reciprocal of dual" gives the same result as the operation "reciprocal of dual of reciprocal".

(3) The result of the latter operation is called the *hyperdual* hypermap. What is its geometric meaning?

(4) Using the operations of duality and reciprocity construct a diagram consisting of six hypermaps, and show the relations between them.

Exercise 1.5.25. Recall the braid group action of Construction 1.1.17 on 3-constellations. Show that the operation σ_1, being applied to the $[\sigma, \alpha, \varphi]$, gives a constellation isomorphic to $[\bar{\alpha}, \bar{\sigma}, \bar{\varphi}]$, and the operation σ_2 gives a constellation isomorphic to $[\bar{\sigma}, \bar{\varphi}, \bar{\alpha}]$.

Remark 1.5.26. The above exercise shows that the action of the braid group B_3 on 3-constellations is of no interest. If 3-constellations are considered up to isomorphism, then any orbit consists of six elements (or, possibly, three, or two, or one, if some of them are isomorphic). It is only for $k > 3$ that this action gives something significant.

Remark 1.5.27. While working "in practice" with all the above permutations, involutions etc., we must not forget that the set of positive triangles and that of negative triangles are two different sets. If we label, for example, the positive triangles by $1, 2, \ldots, n$, then we must not label negative triangles by the same labels. Otherwise not only may confusion occur, but we may also overlook some interesting phenomena. For example: what does it mean to say that a map is *self-dual*? Obviously, this means that there exists an isomorphism between the map itself and its dual map. The square of this isomorphism "returns" to the original map. But it is not necessarily equal to the identity! It may be a non-trivial automorphism of the original map: see [153], [213] and Example 1.5.28 below.

The case when a map M is self-dual and also possesses non-trivial automorphisms is very interesting. Geometrically, the fact that M is self-dual means that if we draw both M and its dual in the same figure, we obtain a "double" map N which has an additional automorphism. Algebraically this means that the group $\mathrm{Aut}(N)$ is an index 2 extension of the group $\mathrm{Aut}(M)$. The list of possible pairs of such groups, for the planar case (but for the groups which include orientation reversing automorphisms) is given in [255]. This problem still awaits its extension to higher genera and to constellations with $k > 3$.

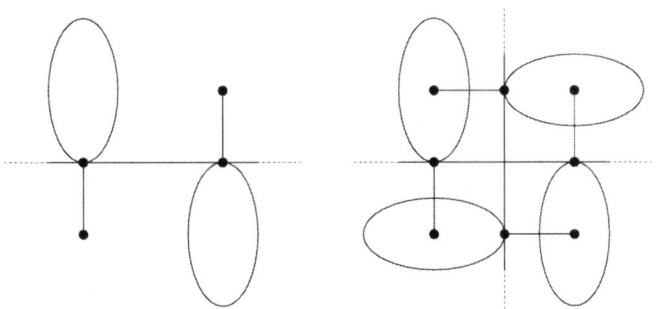

Fig. 1.33. We draw a map and its dual in the same figure

Example 1.5.28 ([254]). In Fig. 1.33 we see on the left a map which is both self-dual and symmetric, with a symmetry of order 2 (the horizontal line represents the "equator" on the sphere). Let us draw both the map itself and its dual in the same figure. The result may be seen on the right. By the way, here the midpoints of the edges acquire degree 4, and therefore we must introduce them explicitly. (One of the white vertices, which is "on the opposite side of the sphere", is not shown in the picture.) Now we see that the permutation that used to be an isomorphism between the map and its dual, becomes an automorphism of the "double", but this automorphism is of order 4.

1.6 More Than Three Permutations

Sections 1.3 and 1.5 were entirely dedicated to graphical representations of 3-constellations. Our next goal is to introduce adequate graphical images for k-constellations with $k > 3$. The reader will easily understand that there are too many possibilities, so we need a sort of guideline in order to be able to grasp the main idea. The general principle is as follows:

Principle 1.6.1 (Types of pictures). All graphical representations of k-constellations are constructed as the preimages, via the corresponding coverings, of a "very simple" picture on the sphere, which "involves" all the ramification points.

56   1 Constellations, Coverings, and Maps

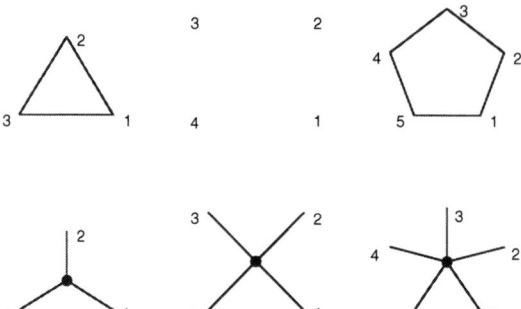

Fig. 1.34. Building blocks for drawing constellations: here $k = 4$, 5, and 6

A "principle" is neither a theorem nor an axiom: it is a vague expression of an idea, which must be clarified and specified in each particular case.

In all the illustrations given below the constellations are plane, but in the majority of cases the only reason for this is the difficulty to draw non-plane constellations.

1.6.1 Preimages of a Star or of a Polygon

Previously, maps and hypermaps were constructed as the preimages of a segment joining two of the three critical values. Now we have more than three critical values; if their number is k, then let us take $k - 1$ of them and draw either the polygon with $k - 1$ sides and with vertices at these critical values, or the star-tree with its center at a generic point and with its leaves at the critical values: see Fig. 1.34. Note that the star and the polygon can be superimposed: then the central (black) vertex of the star is placed at the center of the polygon, and its rays go to the polygon's vertices. The preimage of such a "building block" gives us one of possible graphical images of a constellation.

An example of a drawn constellation is given in Fig. 1.35. For this example we used the star-tree as a building block. One may note that all black vertices are of degree 4. This means that $k = 5$ (hence $k - 1 = 4$). The fifth critical value is placed "at infinity". Its preimages play the same role as the face centers for maps and hypermaps: there is one preimage inside each face, and its multiplicity is equal to the face degree. Note also that the four "marked" (or "colored") vertices around each black vertex always go in the same cyclic order $(1, 2, 3, 4)$. The vertices marked by 1 are the preimages of the critical value marked by 1, the same for 2, 3 and 4. The constellation is of degree 7: indeed, there are 7 black vertices. The vertices of color 1 are of degrees 3, 1, 1, 1, 1; therefore, the first partition in the passport is 31^4; the same procedure works for the other colors. It is not that easy to "see" the face degrees; it would be better to *compute* them. In order to do that, one must first mark the black vertices; then compute the permutations $g_1, \ldots, g_4$ (g_i shows how

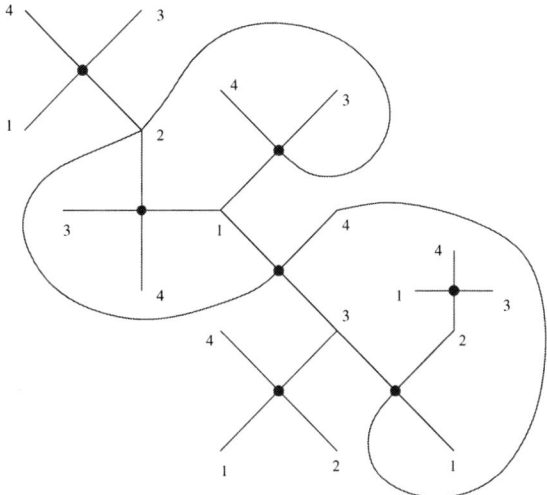

Fig. 1.35. A constellation of degree $n = 7$ (while $k = 5$)

the black vertices are rotated around each vertex of color i); and then compute g_5 in such a way as to have the product $g_1 \ldots g_5 = \mathrm{id}$. Then one will find that the face degrees are 3, 2, 1, 1. An explicit correspondence between pictures and permutations will be illustrated below for the case of cacti which are a particular case of plane constellations.

1.6.2 Cacti

Cacti will be important objects in Chapter 5, where we will study, among other things, the problem of their enumeration (or, equivalently, rigid classification) and of their flexible classification (or, equivalently, the braid group action on cacti).

A cactus is a plane constellation in which one of the permutations is cyclic and fixed once and for all. Therefore, it is convenient to change slightly our previous convention concerning the usage of the letter k:

Convention 1.6.2. While working with cacti, we consider the constellations of lenght $k + 1$ instead of k, the $(k + 1)$-st permutation being fixed.

Definition 1.6.3 (Cactus). A *cactus* is a plane constellation $[g_1, \ldots, g_k, c^{-1}]$, where c is the fixed cyclic permutation $c = (1, 2, \ldots, n)$; thus $g_1 \ldots g_{k-1} g_k = c$. Two cacti are *isomorphic* if they are isomorphic as constellations, with the additional condition that the product $g_1 \ldots g_k = c$ is fixed.

Example 1.6.4. In Fig. 1.36 the same cactus is represented in two graphical forms. If we superimpose both images, then each black vertex will be

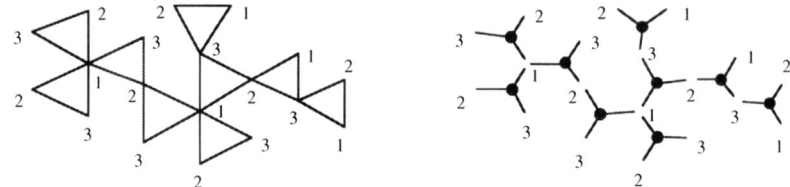

Fig. 1.36. Two graphical images of the same cactus

inside the corresponding triangle, while the "colored" vertices of both figures will coincide. In this example, the degree $n = 9$ (there are 9 triangles, and, equivalently, 9 stars), while $k = 3$ because of the new convention above.

We will use one of the two ways of drawing cacti, whichever will fit the best. For $k = 2$, the reader is advised to compare the "tree-like" presentation given here with that of Sec. 1.5.2: they differ in some insignificant details. By the way, the condition of the cyclic permutation c being fixed also corresponds to the notion of canonical labelling of trees, cf. Sec. 1.5.2.

In Fig. 1.37 the triangles of the above cactus are marked by numbers from 1 to 9. The corresponding permutations which show how the triangles are rotated around the vertices of color 1, 2, 3 are as follows:

$$g_1 = (1,5,6)(2,3,4)(7)(8)(9),$$
$$g_2 = (1,8)(2,5)(3)(4)(6)(7)(9),$$
$$g_3 = (1,9)(2)(3)(4)(5)(6)(7,8).$$

We have chosen the marking in such a way as to have $g_1 g_2 g_3 = (1, 2, \ldots, 9)$. The passport of this cactus is $[3^2 1^3, 2^2 1^5, 2^2 1^5, 9]$. If it is clear from the context that we consider a cactus, the same passport may be written as $[3^2 1^3, 2^2 1^5, 2^2 1^5]$, the last entry being omitted.

In the same way as for trees, we say that a cactus is *rooted* if one of the polygons is distinguished and marked by 1. An equivalence class of isomorphic rooted cacti is called a *non-rooted* cactus. An isomorphism of cacti, or, equivalently, an automorphism of a non-rooted cactus, must preserve the cyclic

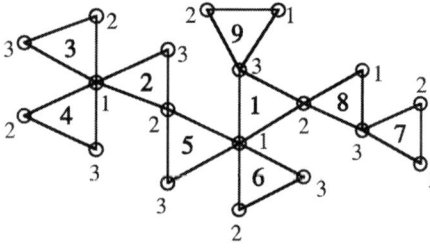

Fig. 1.37. Marking of the triangles of the above cactus

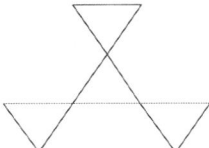

Fig. 1.38. This cactus is not symmetric

permutation $c = (1, 2, \ldots, n)$, and is therefore a power of c. Thus the isomorphism group of a non-rooted cactus is a cyclic group generated by a power of c. Geometrically, this group is that of rotations around a vertex which serves as a center. We also say that a cactus having a non-trivial automorphism group is *symmetric*, and the *order* of its symmetry is the order of the automorphism group. On the contrary, a rooted cactus does not have non-trivial automorphisms, since any automorphism must send the distinguished polygon to itself (and the label 1 to 1).

Remark 1.6.5. Even this seemingly trivial notion of cactus symmetry may have traps. For example, the cactus shown in Fig. 1.38 is *not* symmetric. Indeed, a cyclic permutation of order 4 cannot have a symmetry of order 3. Geometrically, only a vertex, and not a center of a polygon, has a right to be a symmetry center. (If we represent a cactus as a tree by gluing stars, then we must say: only a non-black vertex may be a symmetry center.)

According to Definition 1.6.3 a cactus is a plane figure. This imposes a condition on the number of its vertices. Let us take a "polygonal" representation of a cactus, and let us forget all the colors and consider the resulting picture as a plane map. Then this map has kn edges and $n+1$ faces (n polygons plus the outer face). Denoting by V the number of vertices, we get $V - kn + n + 1 = 2$, hence
$$V = (k-1)n + 1.$$
Recall that V is also the total number of cycles in the permutations $g_1, \ldots, g_k$, or, equivalently, the total number of parts in the partitions $\lambda_1, \ldots, \lambda_k$ of the passport $\pi = [\lambda_1, \ldots, \lambda_k, n]$. Thus the planarity condition may also be represented by the equality
$$\sum_{i=1}^{k} v(\lambda_i) = n - 1,$$
where $v(\lambda)$, as before, denotes $n - \#(\text{parts in } \lambda)$. We call a passport satisfying this condition *valuable cactus passport*, or *polynomial passport* (the reasons of calling it polynomial will become clear later). When it is clear from the context that we discuss cacti, we will speak simply of *valuable* passports.

A feature that makes cacti very convenient objects of study is the existence of a formula that enumerates them for any given valuable passport (Goulden and Jackson 1992 [114]; see also Sec. A.2.4).

Theorem 1.6.6. *Given a valuable passport* $\pi = [\lambda_1, \ldots, \lambda_k]$ *of degree* n, *the number of rooted cacti with this passport is equal to*

$$n^{k-1} \prod_{i=1}^{k} N(\lambda_i), \qquad (1.6)$$

where we have used Notation 1.5.11 *for* $N(\lambda)$. *Equivalently,*

$$\sum_{C} \frac{1}{|\mathrm{Aut}(C)|} = n^{k-2} \prod_{i=1}^{k} N(\lambda_i), \qquad (1.7)$$

where the sum is taken over all non-isomorphic non-rooted cacti C *with the passport* π.

If the number on the right-hand side of (1.6) is not an integer, then one may be sure that symmetric cacti do exist. However, a seemingly innocent integral result may also conceal their existence.

Example 1.6.7. For the passport $\pi = [2^6 1^6, 2^6 1^6, 6 1^{12}, 18]$ formula (1.7) gives 106722. In fact, for this passport there are 106668 asymmetric cacti, 99 cacti having symmetry of order 2, 12 cacti having symmetry of order 3, and 3 cacti having symmetry of order 6. Being assembled by the mass-formula, these numbers give indeed

$$106668 + \frac{1}{2} \times 99 + \frac{1}{3} \times 12 + \frac{1}{6} \times 3 = 106722.$$

In [35] a method is given to count separately cacti of different symmetry orders.

Remark 1.6.8. Having a valuable passport, we may add to it the trivial partition 1^n; the new passport remains valuable, and Eq. (1.6) will give the same result. Indeed, n^{k-1} will be replaced with n^k, but $N(1^n) = (n-1)!/n! = 1/n$. In the same way, if by chance the passport contains the partition 1^n, it may be eliminated without any problem.

Corollary 1.6.9. *For any valuable passport there exists at least one cactus.*

Proof. Whatever the passport, Eq. (1.6) gives a non-zero result.

This corollary was first proved by R. Thom [277] (with a minor error), and then reproved in [89] and [169]. We would like to give a purely combinatorial proof which does not use the enumerative formula. The intuitive idea consists in "cutting off a leaf"; see Fig. 1.39. A leaf, in the polygonal model, is a polygon whose all vertices except one are of degree 1. We affirm that in a valuable passport all the partitions except possibly one contain a part equal to 1. Indeed, if all the parts in a partition are ≥ 2, then the number of parts

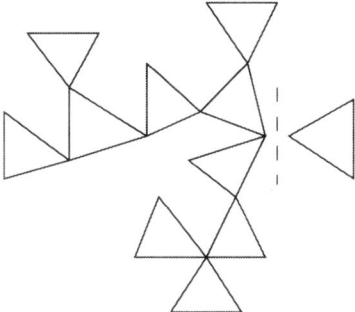

Fig. 1.39. Cutting off a leaf

is $\leq n/2$, and the existence of more than one such partition contradicts the planarity. Let us remove the parts equal to 1 from $k - 1$ partitions, and also diminish by 1 a part bigger than 1 in the remaining partition. The result is a valuable passport of degree $n - 1$. (If it contains the trivial partition 1^{n-1}, we may remove it.) Supposing by induction that there exists a cactus with this passport of degree $n - 1$, we may "glue back" the leaf that we have cut off.

This example shows how useful pictures can be. For somebody working only with permutations to find an idea of a proof would not be an easy task.

Remark 1.6.10. For general constellations, and even for maps, similar results are not valid. Imagine a plane map with n edges, and n vertices of degree 2. It is immediately obvious that such a map is unique: it is a circle with n points on it. Thus, there are two faces of degree n, and it is impossible to have a map with the same vertex degrees and with two faces of degrees, say, $n - 1$ and $n+1$. In Chapter 11 of [150] one may find a list of all impossible pairs of vertex and face partitions up to eight "edges". A simple existence criterion has not been found so far; therefore the above result for cacti should be considered as a kind of a miracle.

Exercise 1.6.11. If $k = n - 1$, the only valuable passport consists of $n - 1$ partitions of the type 21^{n-2}. In this case the formula (1.6) gives n^{n-2}, that is, the same result as in Cayley's formula enumerating labelled trees with n vertices (and without any planar structure). Find a bijection between Cayley trees and cacti with $k = n - 1$ colors.

1.6.3 Preimages of a Jordan Curve

The reader must be already tired of the numerous ways to draw constellations. Nevertheless below we discuss, though very briefly, one more such way. It is a generalization, for more than three critical values, of the canonical triangulation of hypermaps.

62   1 Constellations, Coverings, and Maps

$y_1 \quad y_2 \quad \ldots\ldots \quad y_k$

Fig. 1.40. A Jordan curve that passes through all k critical values and divides the sphere in two polygons with k sides each

Let us return to the notation k for the total number of critical values, denoted, as usual, $y_1, \ldots, y_k$. Let us draw a Jordan curve on the sphere that passes through the points $y_1, \ldots, y_k$ in this specific order. In Fig. 1.40 this Jordan curve is shown as a straight line (an equator on the sphere), but it might as well be a curve of an arbitrary form. This curve divides the sphere in two polygons with k sides each, which we call positive and negative (the positive polygon is shaded). If we turn inside the positive polygon in the positive direction, we see its vertices in the order $y_1, y_2, \ldots, y_k$, while in the negative polygon we see them in the opposite order $y_k, y_{k-1}, \ldots, y_1$.

The preimage of this picture via a covering of degree n with critical values at $y_1, \ldots, y_k$ consists of n positive and n negative polygons. All the neighbors of each positive polygon are negative, and vice versa. In the same way as for the canonical triangulation of a hypermap, we may introduce k involutions without fixed points $a_1, \ldots, a_k$: here a_i sends each polygon to its neighbor across the side marked $(i-1)$–i (the indices are taken $\mod k$). Permutations $g_1, \ldots, g_k$ act on the set of positive polygons and may be represented as $g_i = a_i a_{i+1}$.

The only problem with this representation is that "in real life" such a picture can be rather difficult to draw.

Exercise 1.6.12. Show that there exist 4 non-isomorphic (plane) constellations with the passport $\pi = [21, 21, 21, 21]$. Draw them in the form of preimages of a Jordan curve.

However, this representation may be very useful. One of the situations in which it becomes practically indispensable is the drawing of a composition of constellations. Imagine a ramified covering $h : X \to Z$ which is a composition $h : X \xrightarrow{f} Y \xrightarrow{g} Z$; here $Y = Z = S^2$. Topologically this means that on the y-plane, besides the image of Fig. 1.40, another image is drawn, which represents a constellation obtained via a ramified covering $Y \xrightarrow{g} Z$. The question is how to obtain the corresponding picture on X? In order to do that we must just take the part of the picture corresponding to g which is inside the positive polygon on the y-plane, and repeat it in each positive polygon on X, and then do the same for negative polygons (we must also take care to glue them correctly on the border). Though there remain some algorithmic problems, the general idea of the construction thus becomes clear.

1.7 Further Discussion

1.7.1 Coverings of Surfaces of Higher Genera

In this section we will briefly discuss how to construct coverings $f : X \to Y$ where Y is a surface of a genus greater than zero. We will temporarily denote the genera of different surfaces by the letters p, q, r.

First of all note that there exist non-trivial *unramified* coverings of Y. The reason is that its fundamental group $\pi_1(Y)$ is not trivial. In order to find a set of generators of this group let us take an arbitrary one-vertex and one-face map on Y. This map has $2q$ edges, where q is the genus of Y, and all of these edges are loops. We may choose the vertex of our map as the base point, orient the loops in an arbitrary way, and then they will serve as the generators of the fundamental group. There is also one relation on the generators: if we go along the border of our unique face, we obtain a closed path which can be retracted to the base point inside the face. (The usual convention is to go around the face in the positive direction.)

Example 1.7.1. In Fig. 1.41 we show two one-vertex and one-face maps of genus 2 (the thin lines show which sides of the 8-gon must be identified). These maps correspond to two different choices of generators. The relation for the first choice is $aba^{-1}b^{-1}cdc^{-1}d^{-1} = \mathrm{id}$, and for the second one, $abcda^{-1}b^{-1}c^{-1}d^{-1} = \mathrm{id}$. Of course, there exist many other possibilities. However, in practically all topology textbooks they choose our first representation as the "canonical" one. Note that our convention to glue together the oriented sides with the opposite orientations (see Construction 1.3.20 and Fig. 1.19) is coherent with the fact that the closed path goes first, say, along a generator a, and then along its inverse a^{-1}.

In order to represent an n-sheeted unramified covering of a surface Y of genus q we must construct a homomorphism of $\pi_1(Y)$ to S_n, or, in other words, give $2q$ permutations h_i, acting transitively on n sheets, each one corresponding to a generator of $\pi_1(Y)$, and which satisfy the same relation as the generators of $\pi_1(Y)$. Thus, a covering of Y is also represented by a sequence of permutations, but this sequence does not form a constellation, since its elements satisfy a relation of a different kind. For example, if we choose the left gluing of Fig. 1.41, then an unramified covering of degree n is represented by a sequence of 4 permutations $[h_1, h_2, h_3, h_4]$, $h_i \in \mathrm{S}_n$, satisfying the relation $h_1 h_2 h_1^{-1} h_2^{-1} h_3 h_4 h_3^{-1} h_4^{-1} = \mathrm{id}$.

If X and Y are surfaces of genus p and q respectively, and $f : X \to Y$ is an unramified covering of degree n, then the Euler characteristic of X is equal

64   1 Constellations, Coverings, and Maps

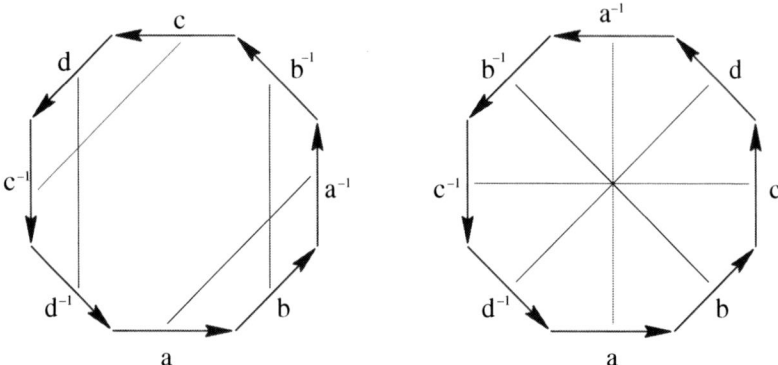

Fig. 1.41. Two choices of generators for the fundamental group of the surface of genus 2

to the Euler characteristic of Y multiplied by n:

$$2 - 2p = n(2 - 2q).$$

The proof is obvious: for any map on Y its preimage on X has n times more vertices, edges, and faces. A trivial consequence of this formula is the fact that if Y is a torus, then X is also a torus.

Now suppose that Y contains k ramification points $y_1, \ldots, y_k$. Then we must take a one-vertex and one-face *hypermap* (denote its single vertex by y_0), adding to the above map k *half-edges* which go from y_0 to y_i, $i = 1, \ldots, k$. (We suppose that the edges of the hypermap do not pass through the ramification points.) The hypermap thus constructed has $2q$ edges and k half-edges, one vertex of degree $2q + k$, and one face of the same degree. To each half-edge we associate a loop which goes first from y_0 to y_i along this half-edge, then goes counter-clockwise around y_i, and then returns along the same half-edge in the opposite direction. In this way we obtain $2q + k$ loops which generate the fundamental group of $Y \setminus \{y_1, \ldots, y_k\}$. As before, these loops satisfy one relation: we obtain it by going along the border of the face of our hypermap. Note that for every edge both the corresponding generator and its inverse appear in the relation, while every generator corresponding to a half-edge appears only once.

Now, a representation of an n-sheeted ramified covering of Y with k ramification points is a sequence of $2q + k$ permutations of degree n acting transitively on n sheets, each permutation being associated to a generator of the fundamental group, and these permutations must satisfy the same relation that satisfy the generators. It is reasonable to fix a concrete choice of the "*base hypermap*" and to give the following definition.

Definition 1.7.2 (Generalized constellation). A *generalized constellation* is a sequence of $2q+k$ permutations $[h_1, \ldots, h_{2q}, g_1, \ldots, g_k]$, $h_i, g_j \in S_n$, acting transitively on n points and satisfying the relation

$$h_1 h_2 h_1^{-1} h_2^{-1} \ldots h_{2q-1} h_{2q} h_{2q-1}^{-1} h_{2q}^{-1} g_1 g_2 \ldots g_k = \mathrm{id}. \tag{1.8}$$

The *cartographic group* of this generalized constellation is the permutation group

$$G = \langle h_1, \ldots, h_{2q}, g_1, \ldots, g_k \rangle \leq S_n.$$

Exercise 1.7.3. Draw a base hypermap which corresponds to the above definition.

Exercise 1.7.4. Find an analogue of the Riemann–Hurwitz formula for coverings of surfaces of higher genera.

1.7.2 Ritt's Theorem

Definition 1.7.5 (Decomposable covering). Let $h : X \to Z$ be a covering (ramified or not). We say that h is *decomposable* if it can be represented as

$$h : X \xrightarrow{f} Y \xrightarrow{g} Z, \tag{1.9}$$

where f and g are coverings of degrees greater than 1.

The following very important theorem was proved in 1922 by Ritt [243] for the planar case, but it remains valid also in the general case.

Theorem 1.7.6 (Ritt's theorem). *A covering is decomposable if and only if its cartographic group is imprimitive.*

Proof. In one direction the proof is obvious. Suppose that h is decomposable as in formula (1.9). Denote $\deg f = m$, $\deg g = n$ and $\deg h = mn$. Denote by z_0 a base point of h in Z; by y_j, $j = 1, \ldots, n$, the n points of the g-preimage of z_0 in Y; and by x_{ij}, $i = 1, \ldots, m$, $j = 1, \ldots, n$, the mn points of the f-preimage of the set $y_1, \ldots, y_n$. Then the n sets $f^{-1}(y_1), \ldots, f^{-1}(y_n)$ form blocks.

Suppose now that we have a covering h of degree mn with an imprimitive cartographic group having n blocks of size m. Let h be represented by a generalized constellation $[h_1, \ldots, h_{2r}, g_1, \ldots, g_k]$, where r is the genus of Z. These permutations also act on blocks, thus creating $2r + k$ permutations $h'_1, \ldots, h'_{2r}, g'_1, \ldots, g'_k$ of degree n. The new permutations obviously satisfy the relation similar to (1.8) for the genus r instead of q (since the initial permutations of degree mn satisfy it), and they obviously act transitively on n blocks (since the initial permutations act transitively). Therefore, the sequence $[h'_1, \ldots, h'_{2r}, g'_1, \ldots, g'_k]$ is a generalized constellation. It describes a ramified covering $g : Y \to Z$ (for this covering, we take the same k ramification points as for h).

It remains to construct a covering $f : X \to Y$ which will complete the composition (1.9). Let the set of points $y_1, \ldots, y_n$ be the g-preimage of the base point z_0. By construction, the "points" y_j are nothing else but blocks, each of them containing m points on X. Let us denote these sets by $B_1, \ldots, B_n$. Choose the point, say, y_1 as the base point for the covering $f : X \to Y$ we are looking for. We must construct certain permutations acting on the set B_1.

The covering $h : X \to Z$ has k ramification points on Z (k may well be equal to zero). We declare all their g-preimages in Y ramification points of f (if some of these points are in fact not "true" ramification points, the permutations corresponding to them will be identities.) Let their number be l. As in the preceding section, we prepare a one-vertex and one-face hypermap on Y, with its single vertex at y_1, with $2q$ edges if the genus of Y is q, and with l half-edges. Next we find the $2q + l$ corresponding generators of the fundamental group of $Y \setminus \{\text{the ramification points}\}$. We must associate to each such generator a permutation acting on B_1.

In order to do that we draw the g-images of each of the above $2q + l$ loops on the surface Z. Next, we represent them in terms of the generators of the fundamental group of the surface Z minus k points. Finally, we replace each generator in this representation with the corresponding permutation h_i or g_j, $i = 1, \ldots, 2r$, $j = 1, \ldots, k$. In this way we obtain $2q + l$ permutations of degree mn acting on the mn preimages of z_0 on X. But all these permutations preserve the block B_1, since they all stabilize the point y_1 on Y. Taking their restriction on B_1 we obtain the permutations we need.

Finally, we note that the relation which the above constructed permutations on B_1 must satisfy is indeed valid. It corresponds to a closed path which is retractable in Y; and the image under g of such a path is retractable in Z. Therefore the corresponding permutation is the identity.

The theorem is proved.

There exists a quadratic algorithm [16] which verifies if a given permutation group is imprimitive, and if so the algorithm finds the blocks. However, there remain many other algorithmic problems which are related to the composition and which are not yet studied, even in the planar case. In vague terms, the problem is as follows. On the "intermediate" surface Y, two pictures coexist: one of them represents the constellation corresponding to $g : Y \to Z$, while the other one represents the one-vertex and one-face base hypermap we have used in representing the covering $f : X \to Y$. What we need is a reasonable description of their mutual disposition, and an algorithm which would find this description for a given imprimitive (generalized) constellation.

The notion of *wreath product* of groups helps one to better understand the structure of imprimitive groups.

Definition 1.7.7 (Semidirect product). Let K and G be two groups, and suppose that G acts on K by automorphisms of K. Then their *semidirect product*, denoted by $K \rtimes G$, is the set

$$K \rtimes G = \{(a,b) \mid a \in K, b \in G\}$$

with the following multiplication operation:

$$(a,b) \cdot (c,d) = (a \cdot c^{b^{-1}}, bd).$$

Exercise 1.7.8. Verify the following statements.
 1. The above multiplication operation is associative.
 2. The identity element in $K \rtimes G$ is (id, id), and the inverse is

$$(a,b)^{-1} = ((a^b)^{-1}, b^{-1}).$$

 3. The subgroup $K' = \{(a, \text{id}) \mid a \in K\}$ is isomorphic to K and is normal in $K \rtimes G$. (Sometimes by abuse of language they say that K itself is a normal subgroup in $K \rtimes G$.)
 4. The subgroup $G' = \{(\text{id}, b) \mid b \in G\}$ is isomorphic to G, and its action on K' by conjugation reflects the original action of G on K, namely,

$$(\text{id}, b)^{-1}(a, \text{id})(\text{id}, b) = (a^b, \text{id}).$$

 5. The mapping $(a,b) \mapsto b$ is a surjective homomorphism from $K \rtimes G$ to G with kernel K'.

Definition 1.7.9 (Wreath product). Let F and G be permutation groups of degrees m and n respectively. Let us take $K = F^n = F \times \ldots \times F$ (n times), and let G act on K by permuting the factors. Then the *wreath product* of F and G, denoted by $F \wr G$, is the semidirect product

$$F \wr G = F^n \rtimes G.$$

It is convenient to think of the group $F \wr G$ as of the permutation group permuting mn elements of a rectangular matrix with columns of size m and rows of size n: the factors of the group F^n permute elements inside the columns, while the group G permutes the columns themselves. Obviously, such an action is imprimitive, the columns of the matrix being its blocks.

The elements of $F \wr G$ may be represented as

$$[f_1, f_2, \ldots, f_n; g], \quad f_i \in F, \quad g \in G.$$

The above element acts as follows: first each f_i permutes the elements of the ith column; then the columns themselves are permuted by g^{-1}.

The order of the group $F \wr G$ is $|F \wr G| = |F|^n \cdot |G|$.

Proposition 1.7.10. 1. *Every imprimitive group of degree mn with blocks of size m is a subgroup of $S_m \wr S_n$.*

 2. *If the cartographic group of a covering $f : X \to Y$ is F, and the cartographic group of a covering $g : Y \to Z$ is G, then the cartographic group H of their composition $h = X \to Z$ can be represented as a subgroup of the wreath product $F \wr G$.*

Our cautious formulation "can be represented as", instead of just "is", is explained by the necessity of a certain procedure of identifying points of different blocks.

1.7.3 Symmetric and Regular Constellations

Definition 1.7.11 (Automorphism group). The *automorphism group* of a constellation $C = [g_1, \ldots, g_k]$ is the group

$$A = \mathrm{Aut}(C) = \{h \in \mathrm{S}_n \mid h^{-1}g_i h = g_i, \ i = 1, \ldots, k\}.$$

In other words, A is the centralizer of the cartographic group G of C in S_n.

The reader is advised to consult once more Remark 1.3.28.

The existence of non-trivial automorphisms gives rise to a procedure which is called factorization, or quotiening, and which, taking into account the previous discussion, becomes simple:

Proposition 1.7.12. *If a constellation has a non-trivial automorphism group, then its cartographic group acts imprimitively, the orbits of the action of the automorphism group being the corresponding blocks. The same statement is true for any subgroup of the full automorphism group.*

A constellation whose darts are the above orbits of the action of a group acting by automorphisms is called a *quotient constellation*.

Exercise 1.7.13. Take the plane map corresponding to the 3-prism. Its automorphism group is the dihedral group D_6 of order 6. This group contains as subgroups the cyclic groups C_3 and C_2. Find the quotients of the 3-prism corresponding to the groups D_6, C_3 and C_2. We may note that the factorization by C_3 preserves the C_2-symmetry, while the result of the factorization by C_2 does not have any symmetry at all. What is the reason for this difference?

Lemma 1.7.14. *Let a group G of order N act on itself by multiplication on the right: an element $x \in G$ acts by sending $a \in G$ to ax. Then the centralizer of G in S_N is obtained by the action of G on itself by multiplication on the left: an element $y \in G$ acts by sending $a \in G$ to $y^{-1}a$.*

We multiply by y^{-1} on the left in order to have the action first of y_1 and then of y_2 be equal to the action of $y_1 y_2$.

Proof. The commutation of the two actions is obvious: acting first by x on the right, and then by y on the left we obtain $y^{-1}(ax)$, while acting in the opposite order we obtain $(y^{-1}a)x$.

Now suppose that a mapping $p : G \to G$ commutes with the right multiplication by x, that is, $p(ax) = p(a)x$ for any $a, x \in G$. Denoting $p(\mathrm{id}) = y^{-1}$ we obtain $p(a) = p(\mathrm{id} \cdot a) = p(\mathrm{id})a = y^{-1}a$.

Proposition 1.7.15. *The following two properties of a constellation are equivalent:*
1. *The automorphism group acts transitively on the darts.*
2. *The automorphism group is isomorphic to the cartographic group.*

If a constellation possesses either (and then both) of these properties, then its darts are in bijection with the elements of the cartographic group, and the group acts on itself by multiplication on the right, while the automorphism group acts by multiplication on the left, as in Lemma 1.7.14.

In the situation of the above lemma and proposition the roles of the cartographic group and of the automorphism group are symmetric: we might just as well declare the action on the left to be cartographic, and that on the right to be the action of the automorphisms.

Proof of Proposition 1.7.15. The cartographic group G is a transitive permutation group of degree n, therefore $|G| \geq n$. On the other hand, an automorphism is entirely determined by fixing a dart and its image under this automorphism. Therefore, $|A| \leq n$, where A denotes the automorphism group. Thus, if $A \cong G$ then $|A| = |G| = n$ and A acts transitively.

Now suppose that A acts transitively. Take an arbitrary dart and mark it by the identity of the automorphism group. The inequality $|A| \leq n$ implies that for any other dart an element $y \in A$ which sends the identity element to this dart is unique. We mark this dart by y^{-1}. Now the action of A on darts becomes equivalent to its action on itself by left multiplications. Therefore, as we have seen in Lemma 1.7.14, the cartographic group G is isomorphic to A and acts on itself by right multiplications.

Definition 1.7.16 (Regular constellation). A constellation having the properties of Proposition 1.7.15 is called *regular*.

The quotient of a regular constellation by its automorphism group consists of a single dart.

The above construction permits us to affirm the following:

Proposition 1.7.17. *Every finite group is the automorphism group of a regular constellation.*

In [64] it is proved that every finite group is the automorphism group of a *map*. Our construction implies this fact only for the groups which can be generated by two generators, one of which is an involution without fixed points. Note however that the latter property is true for all simple groups: see in this respect [209].

We finish this section with the following statement, quite surprising from the geometric point of view in spite of an entirely trivial proof.

Proposition 1.7.18. *A constellation with the cartographic group G can be covered by a regular constellation for which the group G serves as the automorphism group.*

Proof. Let C be a constellation with the cartographic group G. The regular constellation is constructed by the action of the group G on itself by right multiplications. Now take as imprimitivity blocks the subgroups of G which are stabilizers of the darts of C, and apply the construction of Ritt's theorem.

1.8 Review of Riemann Surfaces

This section is a kind of an interlude between the main part of Chapter 1, in which only combinatorial and topological structures played a role, and the rest of the book (mainly Chapters 2, 4, and 5). By necessity, it is very fragmentary. It is mainly a recollection of notions and facts which play an important role in the subsequent exposition. We also make a particular stress on some parts of the theory that might otherwise get lost in the abundance of the information given in textbooks on Riemann surfaces and algebraic curves. We supply our text with appropriate references whenever possible. The general references are [104], [240], [163], [124], [123] (starting from more elementary ones).

Definition 1.8.1 (Riemann surface). A *Riemann surface* is a complex analytic manifold of complex dimension one. Two Riemann surfaces are *isomorphic* if there exists a biholomorphic bijection between them (which is also called a complex isomorphism).

Remark 1.8.2. In order to avoid confusion, we must attract the reader's attention to the following phenomenon. In algebraic geometry, the same object can carry the names of both surfaces and curves. For example, $\mathbb{C}$ is termed both "the complex line" and "the complex plane", and a Riemann surface is also called a complex curve. The Riemann complex sphere $\overline{\mathbb{C}} = \mathbb{C} \cup \{\infty\}$ is the same thing as the complex projective line $\mathbb{C}P^1$. We hope that this language jumble will cause no misunderstanding.

Convention 1.8.3. We will work only with *compact* and *connected* Riemann surfaces, and will call them just Riemann surfaces, without additional adjectives.

Being a compact one-dimensional complex manifold, a Riemann surface is also a compact two-dimensional real manifold, and, what is more, it is orientable. (In fact, it is even already oriented: a complex coordinate gives us the positive orientation, which is "from 1 to i".)

However, it is extremely important to understand that *a topological isomorphism (i.e., a homeomorphism) and a complex isomorphism (i.e., a biholomorphic bijection) are very different things*. Two Riemann surfaces may well be homeomorphic (or, equivalently, have the same genus), but be different as complex manifolds. For genus $g = 0$ there exists only one Riemann surface, namely, the Riemann sphere $\overline{\mathbb{C}} = \mathbb{C} \cup \{\infty\}$. But already for genus $g = 1$ there exist infinitely many non-equivalent Riemann surfaces, which are called elliptic curves. They are all homeomorphic to each other, each of them is a torus, but they are not equivalent as Riemann surfaces. They have different *complex structures*. In order to "introduce a complex structure", we must explain how we choose a particular complex coordinate in a neighborhood of each point. In order to compare two complex structures ... well, it is a much more complicated matter, on which we will elaborate gradually.

Thus, the problem of a constructive representation of a Riemann surface becomes very important. What data must we supply in order to be able to say: this is *the* Riemann surface such and such? There are two convenient ways to specify a Riemann surface: to define it by a system of polynomial equations in a complex projective space, or to present it as a ramified covering of the complex projective line. In order to discuss these two possibilities in more detail, we shall need the notion of meromorphic function.

Definition 1.8.4 (Meromorphic function). Let X be a Riemann surface. A *meromorphic function* on X is a holomorphic mapping $f : X \to \overline{\mathbb{C}}$. A *zero* of f is a point $x \in X$ such that $f(x) = 0$; a *pole* of f is a point $x \in X$ such that $f(x) = \infty$.

The following fact gives an adequate intuitive image of what a meromorphic function is:

Fact 1.8.5. If $X = \overline{\mathbb{C}}$, then any meromorphic function is a rational function (with obvious zeros and poles). A meromorphic function on $\overline{\mathbb{C}}$ with a single pole at infinity is a polynomial.

The fundamental statement below plays a crucial role in working with Riemann surfaces.

Fact 1.8.6. There exist many nonconstant meromorphic functions on each Riemann surface.

This is an easy consequence of the famous Riemann–Roch theorem. It may be used, for example, to prove the following statement.

Proposition 1.8.7. *Each Riemann surface can be realized as an algebraic curve in a complex projective space; that is, it can be given by a system of polynomial equations.*

In fact, if we consider two linearly independent meromorphic functions f_1, f_2 on X, these functions are always related by a polynomial relation $P(f_1, f_2) = 0$. Hence, any pair of linearly independent functions determines a mapping (f_1, f_2) from the Riemann surface to a plane algebraic curve. This simple remark, however, is not sufficient for the proof of the proposition: the mapping may turn out not to be an isomorphism, the algebraic curve may prove to be singular, and so on. But it gives at least a correct direction. A rigorous proof requires some more sophisticated techniques, and we do not present it here; see, for example, [163].

Note that the fact itself is very surprising: a priori there is no reason for an analytic manifold to be algebraic. Compactness plays an important role here, but also some other phenomena that we don't describe. The fact that the notion of Riemann surface is the synonym of that of complex algebraic curve is one of the miracles of this theory. For us the following aspect of this statement is very important: it provides us with one of the possible constructive ways of representing a Riemann surface, namely, by a system of algebraic equations. However, a Riemann surface admits many such representations, and it is not easy to identify two of them representing the same surface.

Definition 1.8.8. If it is possible to realize a Riemann surface X by a system of equations with coefficients in a subfield $K \subseteq \mathbb{C}$, then we say that X is *defined over* K.

Meromorphic functions on a complex algebraic curve embedded in a projective space have a particularly simple form: they are restrictions to the curve of rational functions P/Q, where P and Q are homogeneous polynomials of the same degree in projective coordinates, and where Q does not vanish identically on the curve. Two such rational functions determine the same meromorphic function if and only if their difference belongs to the *ideal of the curve*. The latter means that in the ring of homogeneous polynomials we take the ideal generated by those polynomials which belong to the system of equations defining the curve. If both the curve X and the polynomials P, Q are defined over a subfield $K \subset \mathbb{C}$, then we say that the corresponding meromorphic function is defined over K.

One may easily guess that all the techniques of working with ideals in polynomial rings (such as Gröbner bases, etc.) may help in a concrete work with these objects. In certain books meromorphic functions are called rational functions even in the case of Riemann surfaces of genus $g > 0$. Once again, the fact that the "meromorphic" and "rational" are synonyms is highly nontrivial: in complex analysis they study meromorphic functions on noncompact curves, such as the complex line $\mathbb{C}$, and these functions are not necessarily rational.

Now let us discuss another approach to the constructive representation of Riemann surfaces. The following statement is obvious.

Proposition 1.8.9. *A nonconstant meromorphic function $f : X \to \mathbb{C}P^1$, considered as a mapping of the underlying topological space, is a ramified covering of the sphere $S^2 \cong \mathbb{C}P^1$.*

Hence, all notions related to ramified coverings (the degree of the covering, the ramification points, and so on) make sense for meromorphic functions as well. The only difference with the situation described in Sec. 1.2 is that the new construction involves more subtle structures: X is not just a topological surface but a Riemann surface; f is not just a continuous function but an analytic function; Y is not just a topological sphere S^2 but the complex projective line $\mathbb{C}P^1$.

For meromorphic functions, it is convenient to express topological notions in analytic terms. To define a *critical point* of a function $f : X \to \mathbb{C}P^1$ let us choose local coordinates around $x \in X$ and $y = f(x) \in \mathbb{C}P^1$ in such a way that $x \neq \infty$ and $y \neq \infty$. Then x is critical if and only if $f'(x) = 0$. In fact, these local coordinates s and t in vicinities of the points x and y may be chosen such that we obtain $x = 0$, $y = 0$, and f acquires the form $f(s) = s^d$ in these coordinates. The value d is called the *degree*, or *multiplicity*, or *order* of the critical point. As usual, the values of f at critical points are the *critical values*. The *ramification locus* of the covering f coincides with the set of its critical values. Below, we also use the term a *complex ramified covering* for a meromorphic function.

Notation 1.8.10. Taking into account that among the three objects X, f, and Y the third element Y is (almost) always the same, namely, $\overline{\mathbb{C}}$, we will often denote a ramified covering of the complex sphere by (X, f).

Definition 1.8.11. Two complex ramified coverings (X_1, f_1) and (X_2, f_2) of the complex sphere will be called *isomorphic* if there exists a biholomorphic isomorphism $u : X_1 \to X_2$ such that the following diagram is commutative:

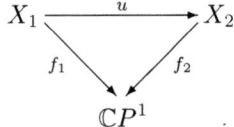

It is clear that isomorphic complex ramified coverings have the same set R of critical values. The above definition repeats word for word that of the isomorphism of topological coverings, see Definition 1.2.18, with the only difference that now the mapping $u : X_1 \to X_2$ must be not a homeomorphism but a complex isomorphism. Usually it will always be clear from the context which of the two notions of isomorphism is meant.

Remark 1.8.12. Be careful: isomorphic complex ramified coverings produce isomorphic Riemann surfaces (by definition), but the converse is certainly false. The same Riemann surface may be obtained by many different and pairwise non-isomorphic coverings. It suffices to take the same X and to consider different meromorphic functions f defined on X.

Example 1.8.13. Consider the case $X = \overline{\mathbb{C}}$, when f is an odinary rational function. Suppose $f(x) = P(x)/Q(x)$, where P and Q are coprime polynomials. Then $\deg f = \max\{\deg P, \deg Q\}$. The fact that $y \neq \infty$ is a critical value of f means that the equation $f(x) = y$ has fewer than n solutions (including ∞ as a possible solution); or, equivalently, it has multiple solutions. If $x \neq \infty$ is one of the multiple solutions, then $f'(x) = 0$. Thus a critical value of f is a value of f at a zero of its derivative. Such a point x is called a *critical point* of f. The *multiplicity*, or the *degree* of a critical point x coincides with its multiplicity as a zero of $f(x) - y$; namely, it is equal to $d \geq 2$ if

$$f'(x) = 0, \ f''(x) = 0, \ \ldots, \ f^{(d-1)}(x) = 0, \ \text{and} \ f^{(d)}(x) \neq 0.$$

Among the points $f^{-1}(y)$ some are critical, others are not; we count the multiplicity of the latter as 1.

74   1 Constellations, Coverings, and Maps

The case of infinity should be considered separately. The multiple roots of $Q(x)$ are the critical points sent by f to the critical value $y = \infty$, with the corresponding multiplicities. If $\deg P > \deg Q$, then f sends ∞ to ∞ with multiplicity $\deg P - \deg Q$. If $\deg P < \deg Q$, then f sends ∞ to 0 with multiplicity $\deg Q - \deg P$. We leave it to the reader to consider the case $\deg P = \deg Q$.

The only isomorphisms of the Riemann sphere $\overline{\mathbb{C}}$ are linear fractional mappings $x \mapsto (ax + b)/(cx + d)$. Therefore two rational functions f_1 and f_2 give isomorphic coverings if and only if

$$f_1(x) = f_2\left(\frac{ax+b}{cx+d}\right)$$

for some a, b, c, d such that $ad - bc \neq 0$.

Fact 1.8.6 and Proposition 1.8.9 show that every Riemann surface may be represented by a ramified covering of the complex sphere. The following theorem, which affirms the converse statement, is one of the most fundamental results of the whole theory.

Theorem 1.8.14 (Riemann's existence theorem). *Suppose a base star is fixed in $\mathbb{C}P^1$, and the sequence of its terminal vertices is $R = [y_1, \ldots, y_k]$. Then for any constellation $[g_1, \ldots, g_k]$, $g_i \in S_n$, there exists a compact Riemann surface X and a meromorphic function $f : X \to \mathbb{C}P^1$ such that $y_1, \ldots, y_k$ are the critical values of f, and $g_1, \ldots, g_k$ are the corresponding monodromy permutations. The ramified covering $f : X \to \mathbb{C}P^1$ is independent of the choice of the base star in a given homotopy type and is unique up to isomorphism.*

Proof. We have seen already in Proposition 1.2.15 that a constellation uniquely determines a topological ramified covering $f : X \to S^2$. The only additional feature we require is a complex structure on X such that f is a meromorphic function. Puncturing $\mathbb{C}P^1$ at the critical values y_i we make f into an unramified covering $f : X \setminus f^{-1}(R) \longrightarrow \mathbb{C}P^1 \setminus R$. The complex structure on $f : X \setminus f^{-1}(R)$ is reconstructed by lifting the complex structure from $\mathbb{C}P^1 \setminus R$. Now, if $x_0 \in f^{-1}(y_i)$ is a preimage of a critical value, then we can introduce a complex coordinate z in a neighborhood of x_0 by setting $z = (f(x) - f(x_0))^{1/d}$, where d is the degree of f at x_0. Any point in this neighborhood of x_0 admits a neighborhood so small that the restriction of f to this neighborhood is a homeomorphism. On this small neighborhood the two complex coordinates obviously agree.

This way of reconstructing a complex structure on X is, in fact, unique, and the uniqueness of the complex ramified covering follows.

A remarkable feature of this theorem is the fact that there are no restrictions whatsoever on the critical values $y_1, \ldots, y_k$: they may be chosen arbitrarily. Later we will see that for the critical points on X this is not the case: very often they must satisfy certain constraints, which sometimes may be very restrictive. But there are no constraints at all on critical values. A profound discussion of numerous aspects of this theorem may be found, for example, in [288].

For us the most important aspect of this theorem is the fact that it provides us with a second way of a constructive representation of Riemann surfaces. In order to represent such a surface we must supply a data of two kinds:

- a constellation $C = [g_1, \ldots, g_k]$, $g_i \in S_n$, and
- a sequence of k complex numbers $R = [y_1, \ldots, y_k] \in \mathbb{C}P^1$ and a base star on this sequence defined up to homotopy.

Let us call this data (C, R) the *Riemann data*. This data proves to be much more manageable than the previous one, a system of equations. While for a system of equations there are all too many traps consisting in various kinds of singularities and degeneracy, for the Riemann data we have nothing to worry about: it works equally well for *any* constellation, and for *any* sequence $y_1, \ldots, y_k$.

Note, however, that the object represented by the Riemann data is not only the Riemann surface X, but the Riemann surface X *together with a meromorphic function* $f : X \to \mathbb{C}P^1$. The set of Riemann data is in a natural one-to-one correspondence with the set of pairs (X, f) endowed with a base star.

The constellation may be considered as the combinatorial part of the Riemann data. This is the part of the information that may change only discretely, while the parameters y_i may change continuously. Such continuous parameters are often called continuous *moduli*.

Remark 1.8.15. The Riemann existence theorem remains valid, both in the existence and the uniqueness parts, for coverings of an arbitrary Riemann surface Y (of an arbitrary genus), arbitrary points $y_1, \ldots, y_k \in Y$ ($k \geq 0$), and for an arbitrary generalized constellation as its combinatorial data. However, if we want to use this fact in order to represent a surface X together with a covering $f : X \to Y$, we must first represent Y itself and the points y_i on it. Other facts concerning coverings, such as, for example, Ritt's theorem, also have their natural counterparts for Riemann surfaces: see, for example, Theorem 5.4.15.

Sometimes we will also be interested in an equivalence relation on meromorphic functions which is weaker than isomorphism. This equivalence relation permits automorphisms in the image as well.

Definition 1.8.16 (Complex equivalence). Two ramified coverings (X_1, f_1) and (X_2, f_2) are called *complex equivalent* if there exist two complex isomorphisms $u : X_1 \to X_2$ and $v : \mathbb{C}P^1 \to \mathbb{C}P^1$ such that the following diagram is commutative:

76   1 Constellations, Coverings, and Maps

$$X_1 \xrightarrow{u} X_2$$
$$\downarrow f_1 \qquad \downarrow f_2$$
$$\mathbb{C}P^1 \xrightarrow{v} \mathbb{C}P^1$$

If the ramified coverings f_1, f_2 are endowed with base stars, then we require that the isomorphism v take the first star to the second one (up to homotopy).

* * *

The discussion that follows (as well as the long quote after it) is much more a prelude to Chapter 2 than a postlude to Chapter 1.

Complex equivalence preserves many important properties of the coverings. First of all, complex equivalent Riemann surfaces X_1 and X_2 are isomorphic (by definition). Therefore, if we are interested only in a representation of the corresponding Riemann surface (or of its complex structure), we may choose an arbitrary covering among equivalent ones, although there still remains a lot of ways to realize the same surface. If we are also interested in the coverings themselves, we may observe that complex equivalence preserves the constellation C, and only changes the sequence $R = [y_1, \ldots, y_k]$, replacing it by a new sequence $R' = [y'_1, \ldots, y'_k]$.

The automorphisms v of the complex sphere $\overline{\mathbb{C}}$, which are linear fractional mappings, permit us to take three arbitrary elements of $\overline{\mathbb{C}}$ to any three arbitrarily chosen positions; after that the mapping becomes completely determined. Let us make the following operation: take the last three critical values y_{k-2}, y_{k-1} and y_k, and, making a linear fractional mapping, put them to the following positions fixed once and for all: 0, 1, and ∞. The other critical values are thus transformed into $y'_1, \ldots, y'_{k-3}$. This operation gives us a *canonical representative* of the equivalence class of coverings. Thus, considered up to equivalence, the Riemann data contains $k - 3$ continuous parameters.

At this point it becomes clear that the case $k = 3$ is of special interest. No continuous parameters remain any more, and all the information about the Riemann surface and its complex structure is encoded in purely combinatorial data. The corresponding phenomenon is called *rigidity*. Having at hand a number of continuous parameters we can change a covering "slightly". But it is impossible to change "slightly" a triple of permutations.

This "rigid" situation will be studied in detail in Chapter 2. It turns out that the corresponding Riemann surfaces are all defined over the field $\overline{\mathbb{Q}}$ of algebraic numbers, and therefore the absolute Galois group $\text{Aut}(\overline{\mathbb{Q}} | \mathbb{Q})$ (that is, the automorphism group of the field $\overline{\mathbb{Q}}$) acts on them, and thus on 3-constellations as well. Incidentally, as we have already seen, a 3-constellation also describes a classical object of combinatorial theory, namely, the hypermap. The action of the absolute Galois group on hypermaps, and also on maps, came as a complete surprise to combinatorialists, but also to the experts in Galois theory. The mysterious nature of the group and the simplicity of the objects on which it acts, gave rise to the following term which may look

1.8 Review of Riemann Surfaces

a bit strange: *theory of dessins d'enfants*. The quotation from Grothendieck cited below illustrates very vividly his emotional reaction to the above facts. The reader may consider it as an epigraph to Chapter 2.

> Now taking the Riemann sphere, or the projective complex line, as reference sphere, rigidified by the three points 0, 1 and ∞ < ... >, and recalling that every finite ramified covering of a complex algebraic curve itself inherits the structure of a complex algebraic curve, we arrive at this fact, which eight years later still appears to me as extraordinary: *every "finite" oriented map is canonically realized on a complex algebraic curve*! Even better, as the complex projective line is defined over the absolute base field $\mathbb{Q}$, as are the admitted points of ramification, the algebraic curves we obtain are defined not only over $\mathbb{C}$, but over the algebraic closure $\overline{\mathbb{Q}}$ of $\mathbb{Q}$ in $\mathbb{C}$. As for the map we started with, it can be found on the algebraic curve, as the inverse image of the real segment $[0, 1]$ < ... >.
>
> This discovery, which is technically so simple, made a very strong impression on me, and it represents a decisive turning point in the course of my reflections, a shift in particular of my centre of interest in mathematics, which suddenly found itself strongly focused. I do not believe that a mathematical fact has ever struck me quite so strongly as this one, nor had a comparable psychological impact.[1] This is surely because of the very familiar, non-technical nature of the objects considered, of which any child's drawing scrawled on a bit of paper < ... > gives a perfectly explicit example. To such a dessin, we find associated subtle arithmetic invariants, which are completely turned topsy-turvy as soon as we add one more stroke.
>
> ---
> [1] With the exception of another "fact", at the time when, around the age of twelve, I was interned in the concentration camp of Rieucros (near Mende). It was there that I learnt, from another prisoner, Maria, who gave me free private lessons, the definition of the circle. It impressed me by its simplicity and its evidence, whereas the property of "perfect rotundity" of the circle previously had appeared to me as a reality mysterious beyond words. < ... >

(A. Grothendieck [129], pages 252–253 and 280)

2
Dessins d'Enfants

2.1 Introduction: The Belyi Theorem

A substantial part of the theory of dessins d'enfants could have been developed in the 19th century. However, the fundamental theorem that lies at the basis of the theory was established by G. V. Belyi[1] [21] only in 1979. Below we formulate the theorem, but postpone its proof until the very end of the chapter (Sec. 2.6). This order of exposition permits the reader to begin by acquiring some experience, by seeing examples and following the accompanying discussion.

Theorem 2.1.1 (Belyi theorem). *A Riemann surface X admits a model over the field $\overline{\mathbb{Q}}$ of algebraic numbers if and only if there exists a covering $f : X \to \overline{\mathbb{C}}$ unramified outside $\{0, 1, \infty\}$. In such a case, the meromorphic function f can also be chosen in such a way that it will be defined over $\overline{\mathbb{Q}}$.*

Remark 2.1.2. The formulation of the theorem is fraught with various traps. The expressions "is defined over", or "there exists a model", mean that *there exists a solution* over the corresponding field, and not that all the solutions possess this property. It is very easy to "spoil" a good function, such as, for example, $\overline{\mathbb{C}} \to \overline{\mathbb{C}} : x \mapsto x^2$ (two critical values: 0 and ∞), by replacing it with $x \mapsto \pi x^2$ (the same critical values). Similarly, the function

$$f(x) = \frac{1}{a^4}(x-a)^2(x+a)^2$$

has the same critical values 0, 1, and ∞ for any a, be it rational, algebraic or transcendental. The important thing is the possibility to chose a algebraic (in this particular case it may be chosen even rational).

In what follows we will always try to find the "best" possible field, or the minimal one. Such a field will be called the *field of moduli*. However, the field

---

[1] The majority of authors use this simple spelling, though a more adequate one, corresponding to the standard of transliteration adopted by the AMS, is Belyĭ.

which is the best from the point of view of Galois theory does not always permit to construct a model over it. This point will be discussed later.

If the Riemann surface X is the Riemann sphere, we will be interested only in the moduli field of the function.

Definition 2.1.3 (Belyi function; Belyi pair). A meromorphic function $f : X \to \overline{\mathbb{C}}$ unramified outside $\{0, 1, \infty\}$ is called a *Belyi function*. A pair (X, f), where f is a Belyi function on the Riemann surface X, is called a *Belyi pair*.

Now, take the segment $[0, 1] \subset \overline{\mathbb{C}}$, color the point 0 in black ($\bullet$), color the point 1 in white ($\circ$), so that the segment itself looks like $\bullet$———$\circ$, and take the preimage $H = f^{-1}([0, 1]) \subset X$. According to what we have seen in Sec. 1.8, H is a hypermap drawn on the Riemann surface X. The black (resp. white) vertices of H are the preimages of 0 (resp. of 1), their valencies being equal to the multiplicities of the corresponding critical points. Certain black (or white) vertices may have valency one; then they are still preimages of 0 (resp. of 1), but they are not critical points. Also, each face of the hypermap H contains exactly one pole, that is, a preimage of ∞. The multiplicity of the pole is equal to the valency of the corresponding face. Sometimes we will explicitly mark the poles by an asterisk ($*$), and we will call them the *centers of faces*. Outside of the set of black and white vertices and the centers of the faces there are no other critical points of the function f on X.

A hypermap which is considered not just as a topological or combinatorial object but as a representation of a particular Belyi pair (which is, by the way, unique up to isomorphism) is often called *dessin d'enfant*. A very attractive feature of the theory of dessins d'enfants is the abundance of beautiful examples. This is the main reason why below we dedicate several sections to examples. The reader may partake of them in their totality, or choose only a sample according to his or her own taste.

2.2 Plane Trees and Shabat Polynomials

2.2.1 General Theory Applied to Trees

We start with the most simple case, when the Riemann surface X is of genus zero, that is, it is the complex sphere $\overline{\mathbb{C}}$, and the rational function in question is a polynomial. A polynomial of degree n always has a unique pole of multiplicity n at ∞; therefore, the corresponding hypermap has a single face; that is, it is a bicolored plane tree (see Sec. 1.5.2). In such circumstances we may just forget about ∞ and work on the usual complex plane $\mathbb{C}$, looking for polynomials with not three but *two finite critical values*.

2.2 Plane Trees and Shabat Polynomials

Convention 2.2.1. While speaking of polynomials, we will consider only finite critical values and put aside the point at infinity. (Thus, we will say about a polynomial which is a Belyi function that it has only two critical values.) We will not insist on the critical values always being 0 and 1 but admit as critical values arbitrary complex numbers y_1 and y_2.

2.2.1.1 Critical Points and Critical Values of Complex Polynomials

Consider a polynomial P of degree n with complex coefficients. It maps the complex plane of the variable x onto another one, that of the variable y. Fix a point $y_0 \in \mathbb{C}$ and consider its preimage $P^{-1}(y_0) = \{x \mid P(x) = y_0\}$.

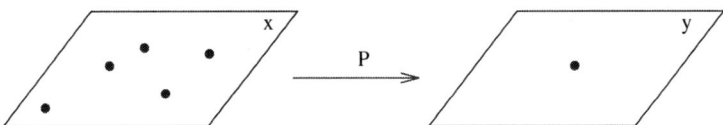

Fig. 2.1. Preimage of a point

Usually, this set consists of n distinct points, i.e., solutions of the degree n equation $P(x) = y_0$ (see Fig. 2.1). But for some specific values of y this equation may have multiple roots. A multiple root of the equation $P(x) = y$ is also a root of the equation $P'(x) = 0$.

Definition 2.2.2 (Critical points and critical values). A point $x \in \mathbb{C}$ at which $P'(x) = 0$ is called a *critical point* of the polynomial P. The value $y = P(x)$ of the polynomial at its critical point x is called a *critical value*. We say that a critical point x has *multiplicity*, or *order*, or *degree* $k \geq 2$ if at this point $P'(x) = 0$, $P''(x) = 0$, ..., $P^{(k-1)}(x) = 0$, $P^{(k)}(x) \neq 0$.

Recall that we usually assign to each critical value y a partition of n whose parts are the multiplicities of the roots of the equation $P(x) = y$, and that the sequence of these partitions is called a *passport*.

Since the derivative P' is a polynomial of degree $n - 1$, the polynomial P "usually" (that is, in the generic case) has $n - 1$ critical points of order 2 and $n - 1$ distinct critical values. But in some degenerate cases one may observe critical points of higher order, and/or the values of the polynomial at different critical points may coincide.

What we are interested in is the most degenerate case, when the set of critical values of P is as small as possible.

Example 2.2.3. The polynomial $P(x) = x^n$ has only one critical point of order n (namely, $x = 0$), and only one critical value $y = 0$.

Up to obvious changes of variables, this polynomial is the only one that has exactly one critical value. The next case, and a much more interesting one, is that of two critical values.

Example 2.2.4 (Chebyshev polynomials). It is well known that $\cos n\varphi$ can be expressed as a polynomial of degree n in $\cos\varphi$. For example,

$$\cos 2\varphi = 2\cos^2\varphi - 1, \quad \cos 3\varphi = 4\cos^3\varphi - 3\cos\varphi, \quad \ldots,$$

and in general

$$\cos n\varphi = T_n(\cos\varphi),$$

where $T_n(x)$ is the n-th *Chebyshev polynomial*,

$$T_0(x) = 1, \quad T_1(x) = x, \quad T_2(x) = 2x^2 - 1, \quad T_3(x) = 4x^3 - 3x, \quad \ldots,$$

and

$$T_n(x) = 2xT_{n-1}(x) - T_{n-2}(x) \quad \text{for} \quad n \geq 2.$$

The graph of the polynomial $T_n(x)$ on the segment $[-1, 1]$ resembles that of $\cos n\varphi$ on the segment $[-\pi, 0]$: all its maxima are equal to 1, and all its minima are equal to -1. Thus, $T_n(x)$ has $n-1$ critical points of order 2 (namely, $x = \cos\frac{k\pi}{n}$), but only two critical values: $y_{1,2} = \pm 1$.

The goal of this section is to describe further examples of such polynomials.

Definition 2.2.5 (Shabat polynomial). A polynomial with at most two critical values is called a *Shabat polynomial*, or a *generalized Chebyshev polynomial*.

In other words, for any such polynomial there exist two distinct complex numbers y_1, y_2 such that

$$P'(x) = 0 \quad \Rightarrow \quad P(x) \in \{y_1, y_2\}.$$

(For $P(x) = x^n$ one of the values (say, y_1) should be taken equal to $y_1 = 0$, while the other one may be chosen arbitrarily; by convention, we take $y_2 = 1$.)

Remark 2.2.6. Any Shabat polynomial P may be easily normalized to have the critical values 0 and 1: one should take $p(x) = (P(x) - y_1)/(y_2 - y_1)$. The additional liberty given by an arbitrary choice of critical values is often convenient in computations.

2.2.1.2 Preimage of a Segment

Take two arbitrary complex numbers y_1, y_2 on the y-plane, and join them by a straight line segment. In what follows we will denote this segment by $[y_1, y_2]$, even in the case when one or both of its endpoints do not belong to $\mathbb{R}$, or when they are both real but $y_2 < y_1$. Suppose there are no critical values on this

2.2 Plane Trees and Shabat Polynomials   83

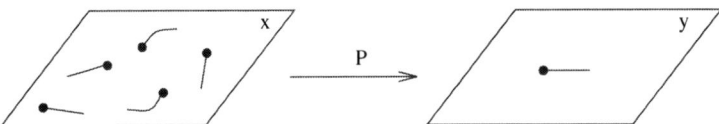

Fig. 2.2. Preimage of a segment

segment; then its preimage $P^{-1}([y_1, y_2])$ is the disjoint union of n curvilinear segments on the x-plane (see Fig. 2.2). We color the ends of $[y_1, y_2]$ in black and white in order to distinguish between their preimages on the x-plane.

Now suppose there are still no critical values *inside* the segment $[y_1, y_2]$, but one or both of its ends become critical. As a result some "curvilinear segments" are glued one to another on the x-plane (see Fig. 2.3).

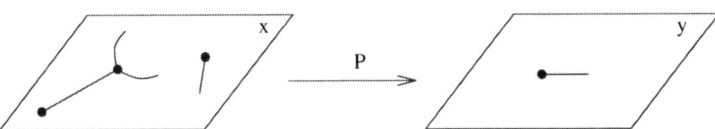

Fig. 2.3. The ends of some segments become critical points, and the segments themselves are glued together

And what happens if the polynomial $P(x)$ in question is a Shabat polynomial, and y_1, y_2 are its (only) critical values? The answer is that *all the segments* are glued together, and we will see a *bicolored plane tree* on the x-plane: for the definition of this combinatorial object see Sec. 1.5.2.

Example 2.2.7. For $P(x) = x^n$ the inverse image of the segment $[y_1, y_2] = [0, 1]$ is a "star-tree". For $P(x) = T_n(x)$, a Chebyshev polynomial, the inverse image of the segment $[y_1, y_2] = [-1, 1]$ is a "chain-tree" (see Fig. 2.4).

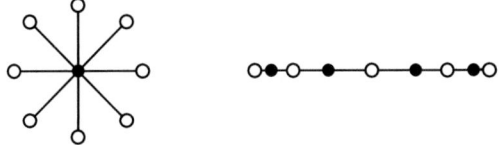

Fig. 2.4. The trees corresponding to x^n and to $T_n(x)$

Many other examples of Shabat polynomials and of the corresponding trees will be given below.

2.2.1.3 Main Theorem

In Sec. 1.8 we have given a definition of the complex equivalence of two complex ramified coverings, or, otherwise, of two meromorphic functions, see Definition 1.8.16. When both the source and the target Riemann surface is the sphere, its automorphisms u and v might be arbitrary linear fractional transformations. If, however, we would like to deal not with arbitrary rational functions but only with polynomials, we must preserve the point at ∞. Thus the only admissible automorphisms become affine: $u : x \mapsto ax + b$ and $v : y \mapsto Ay + B$. These considerations lead to the following definition.

Definition 2.2.8 (Equivalent Shabat polynomials). Let P and Q be two Shabat polynomials, and let y_1, y_2 and z_1, z_2 be their critical values. We call the two *pairs* $(P, [y_1, y_2])$ and $(Q, [z_1, z_2])$ *equivalent* if there exist constants $A, B, a, b \in \mathbb{C}$, $A, a \neq 0$, such that

$$Q(x) = AP(ax + b) + B, \quad \text{and} \quad z_1 = Ay_1 + B, \quad z_2 = Ay_2 + B.$$

By abuse of language, instead of saying that the pairs $(P, [y_1, y_2])$ and $(Q, [z_1, z_2])$ are equivalent, we shall say that the polynomials P and Q themselves are equivalent.

Theorem 2.2.9. *There is a bijection between the set of combinatorial bicolored plane trees and the set of equivalence classes of Shabat polynomials.*

Proof. In fact, we have nothing to prove: this theorem is a particular case of Riemann's existence theorem, see Theorem 1.8.14. Let us follow one by one the main stages of the correspondence "tree $\mapsto$ polynomial". First, a bicolored plane tree being given, we encode it by a triple of permutations, which is a 3-constellation. Next, this 3-constellation, together with an arbitrarily chosen pair of complex numbers y_1 and y_2, determines a ramified covering. The constellation being planar, this is a covering $f : X \to \overline{\mathbb{C}}$ in which $X = \overline{\mathbb{C}}$. In such a case the meromorphic function f is in fact rational. Finally, a rational function with a single pole at infinity is a polynomial. The uniqueness of the polynomial (up to affine change of the variable x) also follows from Riemann's existence theorem.

As we shall see below, in practice, when one wants to find a polynomial corresponding to a tree, one must solve a system of algebraic equations with the coefficients of the Shabat polynomials as the unknowns. The system may be very complicated, and a priori it is not clear at all why it should have a solution. But a solution does exist, thanks to Riemann's existence theorem. This is one more manifestation of a rather enigmatic phenomenon, when some very general and abstract considerations like that of looking for a complex structure corresponding to a ramified covering of the Riemann sphere, lead to very concrete consequences which are anything but evident.

Let us answer here one frequently asked question. It is clear that after gluing a number of segments at their ends we always obtain *a graph*. The

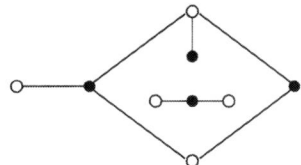

Fig. 2.5. A graph with a circuit

question is: why is this graph always connected? Although this is an immediate consequence of the general theory developed in Chapter 1, the fans of complex analysis will probably find the following argument amusing.

First of all, note that the number of solutions of two equations $P(x) = y_1$ and $P(x) = y_2$ counted with multiplicities is $2n$ (because $\deg P = n$), and counted with *multiplicities diminished by one* is $n-1$ (because $\deg P' = n-1$). Therefore, being counted without multiplicities, the number of solutions is $2n - (n-1) = n+1$. Thus, we have a graph with n edges and $n+1$ vertices. Such a graph is either connected, and then it is a tree; or it is disconnected, and then it possesses a circuit (see Fig. 2.5).

Using, if necessary, an affine transformation of the y-plane, assume the points y_1 and y_2 to be real. Then our polynomial takes only real values on the boundary of the domain bounded by our circuit. This means that its imaginary part $\Im P(x)$, which is a harmonic function on $\mathbb{R}^2$, is identically equal to zero on the boundary. Therefore, it is identically equal to zero inside, which is impossible.

The theorem is proved.

2.2.1.4 Remarks on Computation of Shabat Polynomials

In subsequent sections we will see many concrete examples of Shabat polynomials, and sometimes we will compute them explicitly. Here we give only some very general remarks concerning the computation procedure.

Let a passport

$$[\alpha, \beta, n] = [(\alpha_1, \alpha_2, \ldots, \alpha_p); (\beta_1, \beta_2, \ldots, \beta_q), n]$$

of a tree be given; recall that $p + q = n + 1$. Let us also fix $y_1 = 0, y_2 = 1$. Denote by $a_1, a_2, \ldots, a_p$ and $b_1, b_2, \ldots, b_q$ the coordinates of the black and white vertices respectively. Then we have

$$P(x) = C(x - a_1)^{\alpha_1}(x - a_2)^{\alpha_2} \ldots (x - a_p)^{\alpha_p},$$

$$P(x) - 1 = C(x - b_1)^{\beta_1}(x - b_2)^{\beta_2} \ldots (x - b_q)^{\beta_q}.$$

The equalities between the coefficients provide us with n algebraic equations in $n+2$ unknowns $C, a_1, \ldots, a_p, b_1, \ldots, b_q$. Two additional "degrees of freedom" correspond to the possibility of making an affine transformation of the x-plane. We may use it in the way we find convenient. We may, for example, fix the positions of any two vertices, or make some other choice that places the tree in an unambiguous position on the complex plane. We may use as well some "general" normalization, such as, for example,

$$C = 1, \quad \sum_{i=1}^{p} \alpha_i a_i = 0.$$

In addition to the above set of equations, we must impose the condition that all the coordinates a_i, b_j are distinct.

These considerations show that we have as many equations as we need. But the method itself, taken literally, is far too complicated, and the main reason for this is that the vertex coordinates a_i, b_j belong to a much bigger field than the coefficients of P. The following approach is much better. Let us represent the partition α in the form

$$1^{d_1} 2^{d_2} \ldots n^{d_n};$$

here d_i is the number of occurrences of i in the partition. The polynomial P may be presented in the *square-free form*:

$$P = p_1 p_2^2 \ldots p_n^n.$$

The polynomial p_i is of degree d_i, has distinct roots, and is equal to the product

$$p_i(x) = \prod_{\text{black vertices of degree } i} (x - a_k)$$

(of course similar operations should be made with the white vertices and with the polynomial $P - 1$). The interest of this representation is explained by the following lemma.

Lemma 2.2.10. *Let $P \in K[x]$ be a polynomial with the coefficients in a subfield $K \subseteq \mathbb{C}$. Then $p_i \in K[x]$ for all i.*

Proof. Computing the derivative and the greatest common divisor don't take outside $K[x]$. The greatest common divisor of P and P' is

$$P_1 = C_1 p_2 p_3^2 \ldots p_n^{n-1};$$

the greatest common divisor of P_1 and P_1' is

$$P_2 = C_2 p_3 p_4^2 \ldots p_n^{n-2},$$

and so on. This sequence of polynomials permits to extract the factors p_i. For example, P/P_1 is proportional to $p_1 p_2 \ldots p_n$, while P_1/P_2 is proportional to $p_2 \ldots p_n$, so their ratio gives us p_1.

2.2 Plane Trees and Shabat Polynomials

We conclude that in order to compute a Shabat polynomial P it is advisable to represent both P and $P - 1$ in the square-free form, to develop, and write the algebraic equations representing the equalities among their coefficients.

Remark 2.2.11. There is one more advantage of the above method, of a rather practical nature: the systems of symbolic computations, like Maple for example, can't cope with multiple roots. Make the following experiment: take $\sqrt{2}$; compute it with a reasonable accuracy (say, with 20 digits); take the polynomial $P(x) = (x - \sqrt{2})^{15}$; develop it, and solve the equation $P(x) = 0$. The results will be amusing: your system will find you 15 roots, many of them very far away from $\sqrt{2}$.

An advanced method, well-known in the computational algebraic number theory [57] and used in computation of Shabat polynomials in [68], [122], [212], is based on the LLL-algorithm (called so after its authors A. K. Lenstra, H. W. Lenstra, and L. Lovász). The idea is as follows. A system of algebraic equations is first solved numerically, with a very great accuracy (up to several thousand digits after the decimal point). And then, if, say, a is one of the unknowns, the LLL-algorithms permits to find a linear dependence (over $\mathbb{Z}$) between 1, a, a^2, a^3 etc., that is, to find the number field to which a belongs. In practice this method turns out to be much more efficient than, for example, the Gröbner bases approach.

Returning back to the vertex coordinates, we may note the following. All the coefficients of the polynomials P and $P - 1$, except the free term, are pairwise equal. Therefore, all the elementary symmetric functions of their roots, up to degree $n - 1$, are equal. Hence, all the power sums of the roots, up to degree $n-1$, are also equal. This simple observation gives us the following beautiful system of equations:

$$\alpha_1 a_1 + \alpha_2 a_2 + \ldots + \alpha_p a_p = \beta_1 b_1 + \beta_2 b_2 + \ldots + \beta_q b_q;$$
$$\alpha_1 a_1^2 + \alpha_2 a_2^2 + \ldots + \alpha_p a_p^2 = \beta_1 b_1^2 + \beta_2 b_2^2 + \ldots + \beta_q b_q^2;$$
$$\ldots \quad \ldots$$
$$\alpha_1 a_1^{n-1} + \alpha_2 a_2^{n-1} + \ldots + \alpha_p a_p^{n-1} = \beta_1 b_1^{n-1} + \beta_2 b_2^{n-1} + \ldots + \beta_q b_q^{n-1}.$$

This system was discovered by Couveignes [68].

We would finally like to remark that, while the usual approach of enumerative combinatorics considers the objects like trees as a flock of more or less indistinguishable entities, computing Shabat polynomials demands to look at them in a more individual way, to make a "personal acquaintance" of sorts.

2.2.1.5 Geometry of Plane Trees

Maybe the most surprising consequence of the main theorem is the following fact: *every plane tree has a unique and canonical geometric form.* Geometrically, any tree can be drawn on the plane in infinitely many different ways

88   2 Dessins d'Enfants

(see, e.g., Fig. 2.6). But one of its geometric forms is distinguished. Indeed, the affine transformation $x \mapsto ax + b$ may change the size of a tree and its position on the x-plane, but it does not change its geometric form.

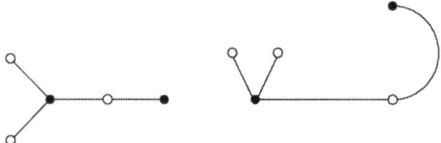

Fig. 2.6. Two "geometric forms" of the same combinatorial tree

In the catalog [26] the computer-made "true" geometric forms are given for all plane trees with ≤ 8 edges. On page 89 we give an excerpt from this catalog.

2.2.2 Simple Examples

2.2.2.1 A Very Simple Example

The first tree which is not a star and not a chain is the tree with the passport $[31, 21^2, 4]$ (see Fig. 2.8).

Let us take $y_1 = 0$, so that the coordinates of the black vertices are the roots of the polynomial we are looking for, the vertex of valency 3 being the root of multiplicity 3. The liberty of an affine transformation permits us to place this vertex at the point $x = 0$, and the other black vertex, at the point $x = 1$. Then we may take $P(x) = x^3(x-1)$, and we are done.

Now, if we also want to compute the coordinates of the white vertices, then let us compute the derivative $P'(x) = x^2(4x - 3)$. The derivative has a double root at $x = 0$ (which is only natural), and one more root $x = 3/4$; this is exactly the position of the white vertex of valency 2.

Now we may compute the second critical value of P: namely,

$$y_2 = P\left(\frac{3}{4}\right) = -\frac{27}{256}.$$

Finally, to get the coordinates of the other two white vertices, we solve the equation

$$x^3(x-1) = -\frac{27}{256}.$$

It has a double root $x = 3/4$, and two simple roots, $(-1 \pm \sqrt{-2})/4$.

Looking at the coordinate $3/4$ of the white vertex and at the negative value of y_2, we may conclude that it would have been more clever to take, from the very beginning, the position of the black vertex of valency 1 at $x = 4$,

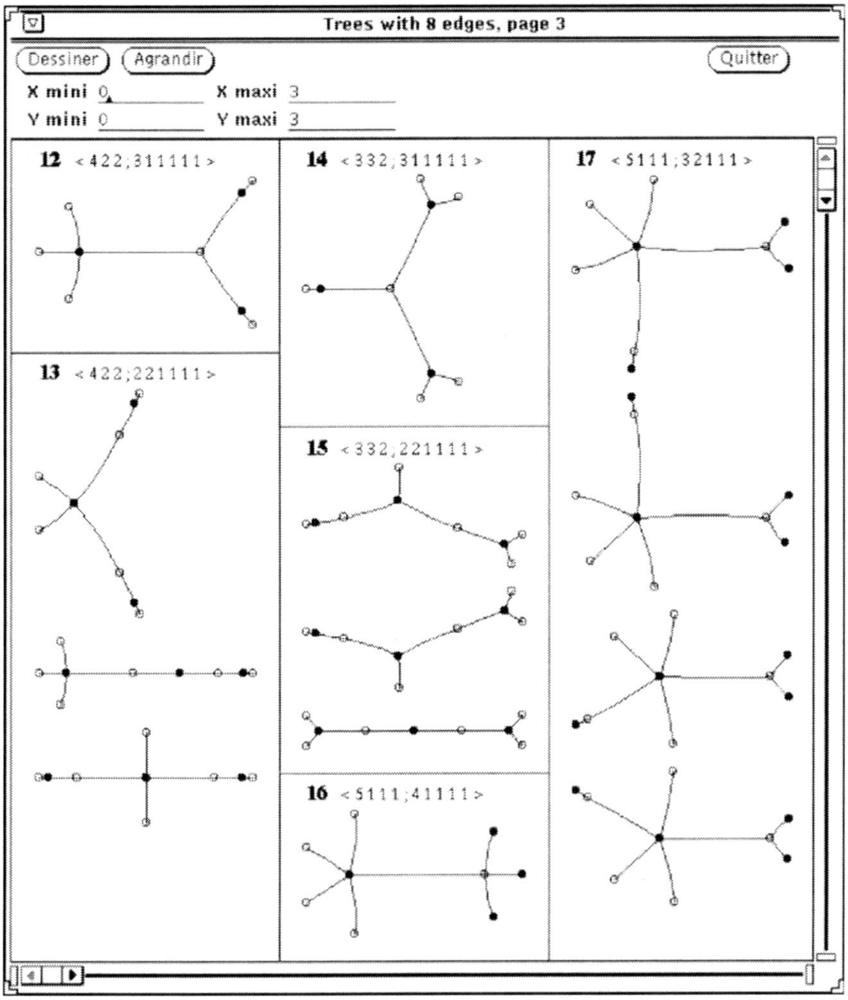

2.2 Plane Trees and Shabat Polynomials   89

Fig. 2.7. Excerpt from the catalog [26]

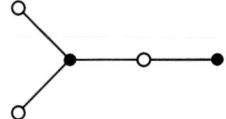

Fig. 2.8. A simple tree

and to change the sign: $P(x) = x^3(4-x)$. Then the white vertex of valency 2 finds itself at $x = 3$, the second critical value is $y_2 = P(3) = 27$, and the coordinates of the other two white vertices become equal to $-1 \pm \sqrt{-2}$.

2.2.2.2 First Example of Conjugate Trees

Consider the tree with the passport $[32^2, 2^2 1^3, 7]$ shown in Fig. 2.9. As usual, take $y_1 = 0$. Let us place the black vertex of valency 3 at the point $x = 0$. Place the two black vertices of valency 2 in such a way that the midpoint of the segment joining them is $x = 1$; or, in other words, the sum of the corresponding complex numbers is equal to 2.

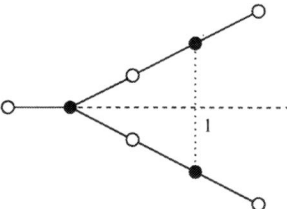

Fig. 2.9. A tree with the passport $[32^2, 2^2 1^3, 7]$

Then these two points are the roots of a quadratic polynomial $x^2 - 2x + a$, with a yet unknown value of the parameter a. Thus, the corresponding Shabat polynomial may be taken equal to

$$P(x) = x^3(x^2 - 2x + a)^2.$$

What are the additional conditions that enable us to find a? To answer the question, let us compute the derivative:

$$P'(x) = x^2(x^2 - 2x + a)(7x^2 - 10x + 3a).$$

The derivative has (as it should have) a double root at $x = 0$, two simple roots, those of $x^2 - 2x + a$, and two other simple roots, namely the roots of the polynomial

$$Q(x) = 7x^2 - 10x + 3a.$$

Obviously, these latter roots are the coordinates of the two *white* vertices of valency 2. Thus, we have two conditions to ensure:

- the roots of Q must be distinct, hence

$$\text{discriminant}(Q) = 21a - 25 \neq 0;$$

- the values of P at the roots of Q must be equal to each other: this common value is the second critical value y_2.

2.2 Plane Trees and Shabat Polynomials

Our first intention is to compute the roots of Q and to substitute them into P. But this is far too cumbersome. The following way is better: let us divide P by Q,
$$P = S \cdot Q + R,$$
where R is the remainder, and, hence, R is linear (since Q is quadratic): $R(x) = Ax + B$, with coefficients A and B depending on the parameter a. Whenever $Q = 0$, the values of P coincide with those of R, hence the values of R must be equal to each other at the roots of Q. This means that we have $A = 0$, and the remainder R must be a constant.

The results of the computation are as follows:
$$A = -\frac{16}{7^6}(21a - 25)(49a^2 - 476a + 400),$$
$$B = -\frac{196}{7^6}a(28a - 25)(7a - 10).$$

Therefore the equality $A = 0$, together with the condition $21a - 25 \neq 0$, leads to the equation
$$D(a) = 49a^2 - 476a + 400 = 0,$$
whence
$$a = \frac{1}{7}(34 \pm 6\sqrt{21}),$$
while the second critical value is equal to $y_2 = B = B(a)$.

A priori it is not obvious which one of the two values of a we must take. It turns out that, instead of trying to make numerical estimations, it is much more interesting to take both values. Then, besides the tree already shown in Fig. 2.9, we also obtain the one shown in Fig. 2.10.

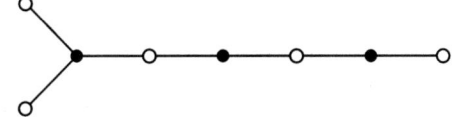

Fig. 2.10. The tree conjugate to that of Fig. 2.9

What do these two trees have in common? The answer is, they have a common passport, namely, $[3^2 2, 2^2 1^3, 7]$. If we look more closely at the process of computation, we will see that the set of valencies is the only information about the trees we actually used.

Basing on the above example, we would like to introduce some notions.

- The coefficients of the Shabat polynomial (and the critical value y_2) belong to the number field $\mathbb{Q}(\sqrt{21})$. We express this fact by saying that the trees in question are *defined over* the field $\mathbb{Q}(\sqrt{21})$.

- The polynomial $D(a) = 49a^2 - 476a + 400$ whose roots generate this field, is called a *defining polynomial*.
- The trees themselves are called *conjugate* (to allude to the fact that the two values of the parameter a are algebraically conjugate).
- The set consisting of these two trees is a *Galois orbit*, i.e., an orbit of the action of the universal Galois group $\Gamma = \mathrm{Gal}(\overline{\mathbb{Q}}|\mathbb{Q})$ on trees.

A more precise meaning of each term will be explained later. Let us only add a few words about the notion of Galois orbit. When the Galois group Γ sends $\sqrt{21}$ to $-\sqrt{21}$, one of the two Shabat polynomials is transformed into the other, and thus the corresponding tree is also transformed into the other tree.

Remark 2.2.12. Let us return to the case $21a - 25 = 0$. Why does the factor $21a - 25$ enter the coefficient A? Obviously, $a = 25/21$ corresponds to a double root of Q, which *does not contradict* the condition that "the values of P at the roots of Q are equal". What it does contradict is the set of white valencies of the tree. Taking $a = 25/21$, we obtain a tree in which, instead of two white vertices of valency 2, we get one white vertex of valency 3, thus forming a tree with the set of white valencies 31^4 instead of $2^2 1^3$. Such a solution is called *parasitic*. The corresponding tree is shown in Fig. 2.11.

Fig. 2.11. A parasitic solution of the system defining the above orbit

2.2.2.3 A Cubic Orbit

In the same vein, we may consider a *cubic orbit*, i.e., a set of three conjugate trees whose Shabat polynomials are defined over a cubic field. Let us take the set of trees of the type $[321, 2^2 1^2, 6]$ (see Fig. 2.12).

Place the black vertex of valency 3 at $x = 0$, the black vertex of valency 2 at $x = 1$, and denote the position of the black vertex of valency 1 by a. Then we get a Shabat polynomial of the form

$$P(x) = x^3(x-1)^2(x-a),$$

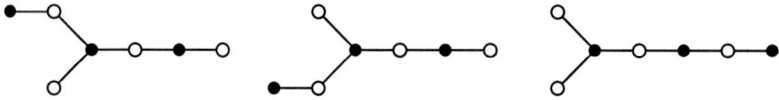

Fig. 2.12. A cubic orbit

whose derivative is equal to

$$P'(x) = x^2(x-1)Q(x), \quad \text{where} \quad Q(x) = 6x^2 - (5a+4)x + 3a.$$

The two white vertices of valency 2 are the roots of Q. They must be different, hence the discriminant of Q must be non-zero:

$$25a^2 - 32a + 16 \neq 0.$$

The remainder after division of $P(x)$ by $Q(x)$ is $Ax + B$, where

$$A = -\frac{1}{6^5}(25a^2 - 32a + 16)(25a^3 - 12a^2 - 24a - 16),$$

$$B = \frac{1}{2^5 3^4} a(5a - 8)(25a^3 - 6a^2 + 8).$$

Hence the condition $A = 0$ (together with the condition on the discriminant of Q) gives us the cubic equation

$$D(a) = 25a^3 - 12a^2 - 24a - 16 = 0.$$

Notation 2.2.13. Let ζ be an algebraic number. Then $\mathbb{Q}(\zeta)$ denotes the minimal extension of $\mathbb{Q}$ containing ζ, and $\mathbb{Q}\langle\zeta\rangle$ denotes the minimal Galois extension of $\mathbb{Q}$ containing ζ (see Sec. 2.4).

The roots of the above cubic equation generate a cubic number field which is in fact equal to $\mathbb{Q}\langle\sqrt[3]{2}\rangle$. It is well known that its Galois group is S_3. But it is nice to note that in this particular example the Galois group may be "seen" geometrically. Indeed, the complex conjugation exchanges the first and the second tree, and leaves the third one fixed. Therefore the Galois group contains a simple transposition and cannot be cyclic. (However, we need to know that the defining polynomial is irreducible over $\mathbb{Q}$.)

As in the previous example, here also we obtain parasitic solutions. The roots of the equation $25a^2 - 32a + 16 = 0$, which are equal to $a = \frac{4}{25}(4 \pm 3\sqrt{-1})$, lead to the identification of the roots of Q; geometrically this means that the two white vertices of valency 2 are glued into one (white) vertex of valency 3. Thus we obtain the type $[321, 31^3, 6]$. The corresponding "quadratic orbit" is shown in Fig. 2.13.

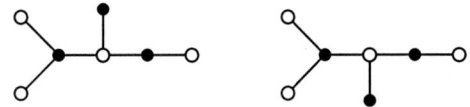

Fig. 2.13. A parasitic solution for the cubic orbit

2.2.3 Further Discussion

2.2.3.1 Several Questions

By now, the reader has certainly a few questions to ask.

Question 1. Is it always the case that the set of *all* trees of the same type (i.e., with the same passport) forms an orbit of the Galois group action? Or, in other words, are the coefficients of the corresponding Shabat polynomials always algebraically conjugate?

The answer is no. Or, better to say, very often but not always. Here is the simplest counter-example.

Example 2.2.14. Take the passport $[41^2, 2^2 1^2, 6]$. There are two trees of that type (see Fig. 2.14).

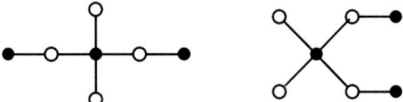

Fig. 2.14. One family but two orbits

On the basis of our previous experience we might suppose that they would form a quadratic orbit, the coefficients of the corresponding Shabat polynomials being conjugate quadratic irrationalities. But this is not true. Both polynomials have rational coefficients: for the first tree it is

$$P(z) = x^4(x^2 - 1),$$

and for the second one,

$$P(z) = x^4(x^2 - 2x + \frac{25}{9}).$$

Thus, instead of having one (quadratic) orbit of the Galois group action, in this example we have two distinct orbits.

We will use the following terminology: the set of all trees with the same passport is called a *family*. Every family is either a Galois orbit, or a disjoint union of several different orbits.

Question 2. Is the combinatorial symmetry always reflected in the geometric symmetry?

The answer is yes. A hint is to be found in the form of the Shabat polynomial for the first tree of the above example: it is a polynomial in x^2. Let a tree having the symmetry of order k be given. Place its center at $x = 0$ and apply the polynomial $y = x^k$; as a result you get a tree "k times smaller". Now find a Shabat polynomial $z = p(y)$ corresponding to this smaller tree. The polynomial $P(x) = p(x^k)$ thus obtained is a Shabat polynomial for the "bigger" symmetric tree.

By this procedure we *can* realize a combinatorially symmetric tree in a geometrically symmetric way. Now we may cite the main theorem (see Theorem 2.2.9): for every tree the corresponding polynomial is unique (up to an affine transformation of x).

Question 3. Is symmetry the only possible reason for the splitting of a family of trees into several orbits?

Not at all! The main interest of the theory of dessins d'enfants is the search for combinatorial invariants of the Galois action. Later we will see many of them. For the moment, just one more example.

Example 2.2.15. Take the passport $[421, 2^2 1^3, 7]$. The corresponding family contains 4 trees, see Fig. 2.15.

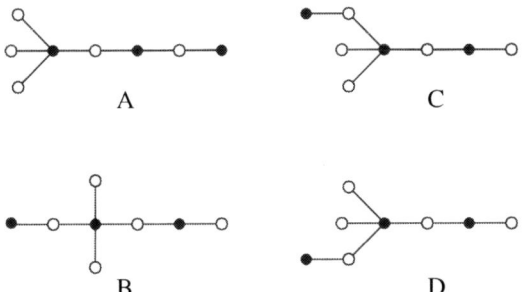

Fig. 2.15. Two quadratic orbits

This is not an orbit of degree 4. Computations show that this family splits into two quadratic orbits. One of them consists of the trees (A) and (B), with the corresponding field $\mathbb{Q}(\sqrt{21})$, and the second one, of the trees (C) and (D), with the field $\mathbb{Q}(\sqrt{-7})$. This time the splitting is explained by the fact that for the trees (A) and (B) the cartographic group is A_7, while for the trees (C) and (D) it is $\mathrm{PSL}_3(2)$.

Question 4. Is it true that for a family having N trees we are always able to write a system of algebraic equations (for coefficients of Shabat polynomials) of degree N?

Yes, it is true, but the proof is not simple; it is given in Sec. 2.4. This property is valid for trees and Shabat polynomials, but is not in general valid for maps and Belyi functions.

Question 5. Suppose that there are 6 trees in a family, we did obtain an equation of degree 6, and found 6 solutions. Why do all the solutions correspond to different trees? Maybe two of them correspond to the same tree, and one of the trees does not have a representation at all?

We cite once more Theorem 2.2.9: the above situation would lead to a contradiction. According to this theorem, for any tree there exists at least one solution of our equation.

The Riemann existence theorem is incredibly powerful: the methods of proof are not algebraic in nature, but they have very strong algebraic consequences.

Question 6. Suppose there is a quadratic orbit, and we have found Shabat polynomials with their coefficients belonging to a quadratic field. Is it true that other quadratic fields would also do?

No! See Example 2.2.16 below, and the theory developed in Sec. 2.4.

2.2.3.2 Mistakes Promote Understanding

Sometimes making a mistake and then trying to understand what the matter is, considerably helps to penetrate into the subtleties of the subject. Here we describe three mistakes (we don't pretend they were the only ones) we have made in the process of understanding the algebraic structures related to dessins d'enfants.

Example 2.2.16 (Mistake 1). Consider the trees of type $[421, 31^4, 7]$ (see Fig. 2.16).

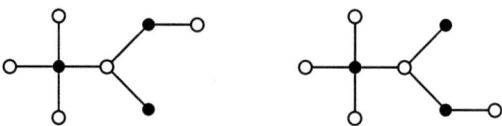

Fig. 2.16. A quadratic orbit; is it true that any quadratic field will do?

Positioning the black vertex of valency 4 at $x = 0$ and the white vertex of valency 3 at $x = 1$, we get a Shabat polynomial P whose derivative is equal to $P'(x) = x^3(x-1)^2(x-a)$. Computing P as the integral, and adding the condition $P(a) = 0$, we get $a = (14 \pm \sqrt{-14})/10$. It is quite natural to have the complex conjugation, since the trees are mirror symmetric to each other. But the square root $\sqrt{-14}$ seems to be a kind of exaggeration, and we started to look for a possible simplification of the field, hoping to reduce it to $\mathbb{Q}(\sqrt{-1})$.

Alas, all our efforts were in vain. A dozen different approaches to writing a system of algebraic equations corresponding to this orbit all led to the same field $\mathbb{Q}(\sqrt{-14})$. This is what made us grasp the notion of the *field of moduli* (see Definition 2.4.3). It is an *objective characteristic* of the orbit by the nature of things, and cannot be subjected to an arbitrary change.

Example 2.2.17 (Mistake 2). Consider the passport $[321^2, 321^2, 7]$. We drew the six corresponding plane trees. Three of them were symmetric with respect to an edge midpoint (half-symmetric), the other three were not. It was rather simple to deal with the half-symmetric ones: they turned out to be defined over the field $\mathbb{Q}\langle\sqrt[3]{28}\rangle$. But the three asymmetric trees gave us enormous trouble. All our efforts to find a cubic field proved to be futile: we kept getting a field of degree 6. Finally we gave up and drew a computer-made picture. And we saw immediately the source of our mistake: yes, indeed, there are three *asymmetric plane* trees; but the number of the *bicolored asymmetric plane* trees is six. Thus we learned that the proper combinatorial structure related to number fields is that of a bicolored plane tree, and not that of a plane tree as such. In Fig. 2.17 we show two (distinct!) elements of this orbit.

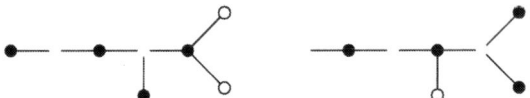

Fig. 2.17. One plane tree but two bicolored plane trees

Obviously, such a situation may occur only when the black and white partitions α and β coincide. By the way, the enumerative formula (1.5) gives in this case $9 = 6 + 3$ trees in total, and not $6 = 3 + 3$. Equations are more intelligent than we are.

Example 2.2.18 (Mistake 3). While preparing the catalog [26], the computer produced a picture of an orbit containing two trees of type $[32^2, 2^2 1^3, 7]$. The picture looked awkward: one of the trees was very narrow and horizontal, while the other was also very narrow but vertical, and the page between the two was almost empty. We decided to change the position of the second tree, making it horizontal. The result should have been easy to predict: the first one immediately became vertical. Thus we learned that an orbit is an "indivisible" entity; we can only work with the whole orbit, but not with its parts.

2.2.3.3 Unambiguous Positioning

In all previous examples we took the liberty of placing some vertices at certain points of complex plane. This procedure is very simple when applied to the so-called "bachelors": a vertex is a *bachelor* if it is the only vertex of a given color and valency. For vertices that are not unique, i.e., when there are several of them of the same color and valency, we may encounter certain difficulties.

Consider the family of type $[3^22, 2^21^4, 8]$. There are three trees of that type (see Fig. 2.18), one of them is centrally symmetric.

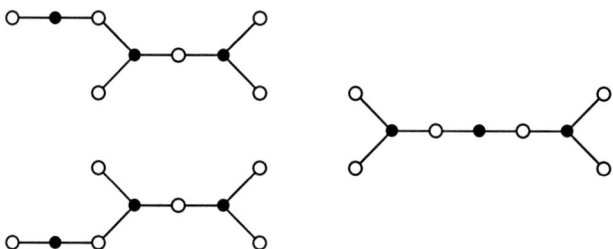

Fig. 2.18. How to position these trees?

There is no problem with the symmetric one: we place its center at $x = 0$, and the two other black vertices, at the points $x = \pm 1$, thus getting the Shabat polynomial $P(x) = x^2(x^2 - 1)^3$.

Now let us turn to the other two trees. First trial: put the two black vertices of valency 3 at the points $x = \pm 1$, and the other black vertex, at the point $x = a$. We obtain a polynomial of the form $P(x) = (x^2-1)^3(x-a)^2$. Proceeding in the same manner as in all the previous examples we get, quite unexpectedly, a field of degree 4, with the defining polynomial $D(a) = a^4 - 8a^2 + 32$. Producing a picture with the help of a computer, we see each of the trees twice: once with the "long branches" on the left, the second time on the right.

Let us make another trial; in order to "break the symmetry" put one of the black vertices at $x = 0$, and the other one at $x = 1$. The result is even worse. We still have 4 different values of a, though this time the symmetry is less obvious: they are symmetric with respect to the axis $\Re x = 1/2$ instead of the axis $\Re x = 0$.

Several other possibilities lead to similar results. We may put, for example, the black vertex of valency 2 at $x = 0$, and one of the black vertices of valency 3, at $x = 1$, and we still get a field of degree 4 instead of 2. The point is that the black vertex of valency 2 is a bachelor, and we may put it anywhere we like, for example, at $x = 0$. But the two black vertices of valency 3 form a "couple"; there is no unambiguous way of positioning only one of them, because we cannot algebraically distinguish between them. It is easy to point at one of these vertices on the picture, but it is impossible to do that algebraically. Our only possibility is to "place" the quadratic polynomial that has these two vertices as its roots. And, indeed, taking this polynomial equal to $p(x) = x^2 - 2x + a$, that is, the Shabat polynomial of the form $P(x) = x^2(x^2 - 2x + a)^3$, we get the quadratic field $\mathbb{Q}(\sqrt{-1})$ we were looking for, and $a = (7 \pm \sqrt{-1})/8$. If we now ask a computer to make a picture, both trees will appear in it only once.

Note that for maps the situation described above is sometimes inevitable. The "best" Belyi function we are able to find may produce the same map more than once; see Sec. 2.4.

2.2.3.4 Inverse Enumeration Problem

For a given passport, the number of trees in its family gives the upper bound for the degree of the corresponding field (or fields, if there are several orbits). For example, if the tree is unique, it is defined over $\mathbb{Q}$. This remark leads to the following problem: *For a given k, enumerate all passports $[\alpha, \beta, n]$ such that there exist exactly k trees with this passport.* We call this problem "the inverse enumeration problem" referring to bicolored plane trees. It goes without saying that the similar question may be asked about plane maps, plane hypermaps, maps of genus 1, etc., and, in general, about every type of constellations.

For trees and for $k = 1$ this problem was solved by N. Adrianov in 1989 (the proof remains unpublished, while the answer was first published in [258]).

Proposition 2.2.19. *Here is the complete list of passports for which the corresponding tree is unique:*

1. $[n, 1^n, n]$ (the "star-trees");
2. $[r1^{t-1}, t1^{r-1}, n]$ (here $n = r + t - 1$);
3. $[2^m, 2^{m-1}1^2, 2m]$ or $[2^m 1, 2^m 1, 2m + 1]$ (the "chain-trees");
4. $[r1^p, s^q t, n]$ (here $n = r + p = qs + t$);
5. $[rs1^p, 2^q, n]$ (here $n = r + s + p = 2q$);
6. $[r^2 1^p, 3^q, n]$ (here $n = 2r + p = 3q$);
7. $[3^3 1^5, 2^7, 14]$.

The trees themselves are shown in Fig. 2.19.

We leave to the reader the pleasure to compute the corresponding Shabat polynomials; all of them are defined over $\mathbb{Q}$. We see that the six first classes are, in fact, infinite series of passports and trees, while the seventh one is a single tree which does not enter into any series; one may call it a *sporadic* tree. We may also note that the second class is in fact a particular case of the fourth one, with $s = 1$.

The above pattern is generalized in [7].

Definition 2.2.20 (Diameter). The *diameter* of a tree is the number of edges in the longest path in this tree.

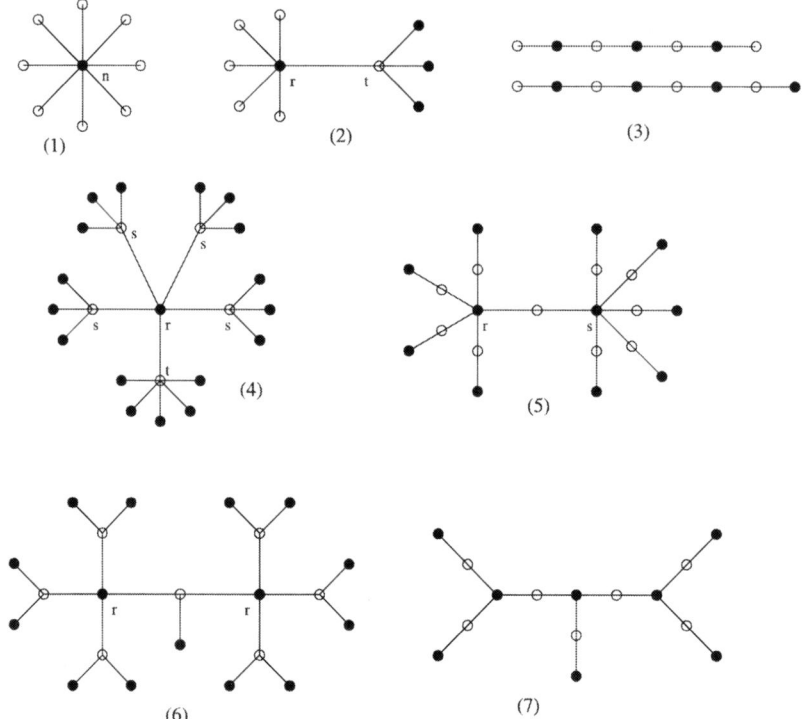

Fig. 2.19. The trees uniquely determined by their passport

For the above classification, the trees of the first series are of diameter 2, of the second one, of diameter 3, of the third one, of diameters $2m$ and $2m+1$ respectively, for the fourth one, of diameter 4, for the fifth and sixth ones, of diameter 6, and the sporadic tree of the seventh type is of diameter 8.

Proposition 2.2.21 ([7]). *For any $k \geq 2$ the families of trees containing exactly k elements each are classified as follows: there are several infinite series of trees of diameter 4, several infinite series of diameter 6, and a finite set of "sporadic" families, with number of edges $n \leq 16k+1$.*

Example 2.2.22 (D. Péré, unpublished). The following is the complete list of the passports for which there are exactly 2 trees. It consists of: three series of diameter 4:

1. $[31^{n-3}, rst, n]$ (here $n = r + s + t$);
2. $[41^{n-4}, r^2s^2, n]$ (here $n = 2r + 2s$; one of the trees is symmetric; therefore, both are defined over $\mathbb{Q}$);
3. $[51^{n-5}, r^3s^2, n]$ (here $n = 3r + 2s$);

three series of diameter 6:

1. $[3^p, rs1^t, n]$ (here $n = 3p = r + s + t$ and $p + t + 2 = n + 1$);
2. $[4^p, r^2 1^s, n]$ (here $n = 4p = 2r + s$ and $p + s + 2 = n + 1$; one of the trees is symmetric; therefore, both are defined over $\mathbb{Q}$);
3. $[5^p, r^2 1^s, n]$ (here $n = 5p = 2r + s$ and $p + s + 2 = n + 1$);

and six sporadic pairs:

1. $[3^2 1, 2^2 1^3, 7]$ (diameter 5);
2. $[3 2^2, 2^2 1^3, 7]$ (diameter 6);
3. $[3^2 1^3, 2^4 1, 9]$ (diameter 6);
4. $[3^2 2 1^3, 2^5, 10]$ (diameter 8);
5. $[4^3 1^8, 2^{10}, 20]$ (diameter 8; one of the trees is symmetric; therefore, both are defined over $\mathbb{Q}$);
6. $[5^3 1^{11}, 2^{13}, 26]$ (diameter 8).

It seems that similar "inverse enumerative" results concerning maps are not known.

To conclude, let us mention that for $n = 24$, there are 23411 passports and 26873059986 trees. Thus, for a "randomly chosen" passport of degree 24 there are more than million trees. Dealing with systems of algebraic equations of a degree more than million is beyond our capabilities.

2.2.4 More Advanced Examples

In this section we consider only trees of diameter 4. Each tree of diameter 4 has a black center of valency, say, p, from which p branches go outside to white vertices of certain valencies; and these white vertices are connected only with black leaves and the center. Thus, the black partition is of the form $\alpha = p 1^{n-p}$, while the white one is of the form $\beta = m_1^{k_1} \ldots m_r^{k_r}$. Here

$$k_1 m_1 + \ldots + k_r m_r = n, \quad k_1 + \ldots + k_r = p.$$

In the particular examples below we will often use other letters for white valencies than $m_1, \ldots, m_r$. For example, if there are only three branches, we may denote the corresponding white valencies as m, n, and k (therefore n does not necessarily mean the total number of edges).

2.2.4.1 Factorization of Discriminants

The following statement was experimentally observed in [258] and later proved by Birch in [30], [31]. Let the parameters $k_1, \ldots, k_r$ be fixed, while $m_1, \ldots, m_r$ remain variable but distinct: this is a way to describe not a particular passport but an infinite series of passports. We look for a defining polynomial for these trees (i.e., for a minimal polynomial whose splitting field contains the coefficients of the Shabat polynomials).

Proposition 2.2.23. *The discriminant Δ of the defining polynomial is itself a polynomial in $m_1, \ldots, m_r$. The defining polynomial may be found in such a way that Δ splits into linear factors of the form*

$$a_1 m_1 + \ldots + a_r m_r, \quad 0 \leq a_i \leq k_i, \quad i = 1, \ldots, r$$

(note that each factor is smaller than or equal to the total number of edges).

Note that the discriminants of different defining polynomials obtained, for example, using different placements may differ by a quadratic factor. In the examples we usually give the results of our concrete computations.

Example 2.2.24. Consider the series of trees with 5 branches and with white partition (m, n, k, k, k) (m, n, and k are supposed to be distinct). There are 4 trees with these data. The defining polynomial we have computed is of degree 4, and its discriminant is equal to

$$\Delta = 2^{10} 3^3 m^3 n^3 k^6 (m+k)^2 (n+k)^2 (m+2k)(n+2k)(m+n+k)(m+n+2k)^2 (m+n+3k)^3.$$

This nice property permits one to obtain many interesting consequences.

Example 2.2.25. Consider the trees with 3 branches issuing from the center, and with white vertices of pairwise distinct valencies m, n, k. There are, obviously, exactly two plane trees of that type: they correspond to the two cyclic orders of the letters m, n, k, namely,

$$(m, n, k) \quad \text{and} \quad (m, k, n).$$

We may guess that we deal with a quadratic Galois orbit. Even more than that: the trees are mirror symmetric to each other, so we may guess that we will get an *imaginary* quadratic field, i.e., it will be generated by the square root of a negative number. Note that these trees cannot be defined over $\mathbb{Q}$: the complex conjugation sends a tree defined over $\mathbb{Q}$ to itself.

It would be too annoying to give all details of computations every time. But some hints will be helpful to show that the job is not difficult. Put the center at $x = 0$, and the vertices of valencies m, n, k, at the points a, b, c respectively. Take $y_2 = 0$; thus the Shabat polynomial has the following form:

$$P(x) = (x-a)^m (x-b)^n (x-c)^k.$$

Its derivative is

$$P'(x) = (x-a)^{m-1} (x-b)^{n-1} (x-c)^{k-1} Q(x),$$

where Q is a quadratic polynomial. The point $x = 0$ being a critical point of order 3, the polynomial Q must be proportional to x^2: its coefficient in front of x and its free term must be equal to 0. In this way we obtain two equations on a, b, c. Add one more equation according to your taste: for example, $a = 1$, or $a + b + c = 1$ if you don't want to break the symmetry. (In fact the symmetry will be broken anyway because you will reduce the system to a quadratic equation by eliminating two variables out of three.)

Finally we get a quadratic defining polynomial with the discriminant

$$\Delta = -mnk(m+n+k),$$

and the corresponding field is thus $\mathbb{Q}(\sqrt{\Delta})$.

What else can we derive from this answer? Adrianov proposed to take a square-free d, and to set

$$m = 1, \quad n = 16d, \quad k = 64d^2.$$

Then m, n, k are indeed all distinct, and we have

$$\Delta = -mnk(m+n+k) = -1024d^3(1 + 16d + 64d^2) = -1024d^3(1+8d)^2,$$

so that

$$\sqrt{\Delta} = 32d(1+8d)\sqrt{-d}.$$

This observation leads to the following proposition.

Proposition 2.2.26. *By choosing appropriate values of m, n, k we may obtain a quadratic orbit of trees of diameter 4 with the white partition $\beta = (m, n, k)$ defined over an arbitrary imaginary quadratic field.*

Example 2.2.27. Let us take the family of the two trees with 5 branches and with white valencies two times m and three times n, the numbers m and n being distinct. The two trees of this family correspond to the two cyclic orders of the white valencies:

$$(m, m, n, n, n) \quad \text{and} \quad (m, n, m, n, n).$$

Note that this time the trees are not symmetric to each other with respect to any axis (while each of them is mirror symmetric to itself). Therefore we may guess that the corresponding field will be real quadratic.

Making the same type of calculations as in the previous example, we get a quadratic equation with the discriminant

$$\Delta = 3(m+2n)(2m+3n).$$

Let d be an arbitrary positive integer. Then we may take, say,

$$m = 6ds^2 - 3t^2, \quad n = 2t^2 - 3ds^2, \quad (2.1)$$

where t and s are arbitrary positive integers satisfying the inequalities

$$\sqrt{\frac{3}{2}d} < \frac{t}{s} < \sqrt{2d}, \quad \frac{t}{s} \neq \sqrt{\frac{9}{5}d} \quad (2.2)$$

(the first one means m and n are positive, the second one, $m \neq n$). Then we obtain

$$m + 2n = t^2, \quad 2m + 3n = 3ds^2,$$

and $\sqrt{\Delta}$ reduces to $\sqrt{d}$.

Proposition 2.2.28. 1. *By choosing appropriate values of m and n we may obtain a quadratic orbit of trees of diameter 4 with the white partition $\beta = (m, m, n, n, n)$ defined over an arbitrary real quadratic field.*

2. If the number d in (2.1) and (2.2) is a perfect square, then the corresponding family splits into two orbits, both defined over $\mathbb{Q}$.

The trees corresponding to d being a square don't have any specific combinatorial properties. It would be more suitable to say that this time the reasons for splitting into two orbits are *Diophantine*.

Example 2.2.29. Let us now consider a cubic family, that with 7 branches and with 2 white vertices of valency m and 5 white vertices of valency n, $m \neq n$. All the three trees are mirror symmetric to themselves: we guess that we will get a totally real cubic field.

As usual, we take $P(x) = p_1(x)^m p_2(x)^n$, where $\deg p_1 = 2$ and $\deg p_2 = 5$, and compute the derivative $P'(x) = p_1(x)^{m-1} p_2(x)^{n-1} Q(x)$, where $\deg Q = 6$. We set $Q(x) \sim x^6$ and eliminate from the system of equations thus obtained all unknowns but one, finally obtaining the following cubic equation for the last variable:

$$D(a) = 15n^3 a^3 - 45n^2(m+3n)a^2 + 15n(m+3n)(m+4n)a$$
$$- (m+3n)(m+4n)(m+5n) = 0. \qquad (2.3)$$

The discriminant of the polynomial D is

$$\Delta = 8100 n^6 (m+3n)^2 (m+4n)(2m+3n)(2m+5n)^2.$$

A cubic field is called *cyclic* if its Galois group is cyclic, i.e., C_3. Recall that a cubic field is cyclic if and only if the discriminant of a cubic polynomial whose roots generate the field is a perfect square. Noting that $8100 = 90^2$, we see that the only possibly non-square factor of Δ is $(m+4n)(2m+3n)$. To make it a square we may take, for example,

$$m = 4t^2 - 3s^2, \quad n = 2s^2 - t^2$$

for some integers $t, s > 0$ satisfying $\sqrt{3/4} < t/s < \sqrt{2}$, $t/s \neq 1$. Taking, for example, $t = 4$, $s = 3$ we get $m = 37$ and $n = 2$, the trees thus having $2m + 5n = 84$ edges.

In this way we obtain an infinite series of cyclic cubic fields. To be sure that we deal indeed with a cubic orbit, we must also verify that the cubic polynomial in a is irreducible. Substituting the above values $m = m(t, s)$ and $n = n(t, s)$ into its coefficients, we obtain a polynomial in three variables a, t, s, which is irreducible. Hence, because of the Hilbert irreducibility theorem, there exist infinitely many values of t and s such that the corresponding polynomial is irreducible as a polynomial in a.

Example 2.2.30. Consider one more cubic family, that with 4 branches and with white vertices of valencies m, m, n, k. Among the three trees of this family, two are mirror symmetric to each other, while the third one is mirror symmetric to itself. Therefore without any computations we may assert that the cubic field cannot be cyclic. And indeed, the discriminant

$$\Delta = -432 m^3 n^2 k^2 (m+n)(m+k)(m+n+k)^3 (2m+n)^2 (2m+n+k)^2$$

this time is negative and therefore cannot be a square.

A cubic field is called *purely cubic* if it has the form $\mathbb{Q}\langle \sqrt[3]{c} \rangle$ (recall Notation 2.2.13) with rational c. It is well known that a cubic field is purely cubic if and only if the discriminant of the corresponding cubic polynomial has the form

$$\Delta = -3 \times (\text{perfect square}).$$

Taking into account the fact that $432 = 3 \times 12^2$, the only thing we must ensure is that the product

$$m(m+n)(m+k)(m+n+k)$$

be a square. A small example is given by $m = 1$, $n = 2$, $k = 5$.

A complete solution of the Diophantine equation

$$m(m+n)(m+k)(m+n+k) = r^2$$

is a bit long. However, we cannot resist the pleasure of giving a *partial* solution. It is sufficient (though not necessary) to set

$$m = x^2, \quad m+n = y^2, \quad m+k = z^2, \quad m+n+k = t^2.$$

These equations are not independent: we have the relation

$$x^2 + t^2 = y^2 + z^2 = 2m + n + k.$$

Incidentally, $2m + n + k$ is the number of edges of the tree. Thus, we come to the following sufficient condition: *the number of edges has two different representations as the sum of two squares.*

The problem of representation of a number as the sum of two squares is one of the most classical problems of number theory. The following statement was first announced by Fermat (1640) and proved 100 years later by Euler (see [294]): *any prime number of the form $4N+1$ has a unique representation as the sum of two squares.* The statement to follow was already known to Diophantus: *the product of two numbers representable as sums of two squares has a non-unique representation as the sum of two squares.* Let us take, for example, the number of edges equal to $5 \times 13 = 65$. Then we have $65 = 1 + 64 = 16 + 49$, so we may take $m = 1$, $n = 15$, $k = 48$. In this way we can construct an infinite series of examples of purely cubic fields.

2.2.4.2 Leila's Flowers

The following series of examples has an interesting history and was a source of several publications. We give here only a brief account of the events.

Example 2.2.31. In [248] Leila Schneps considered the family of trees of type $[51^{15}, 65432, 20]$. There are 24 trees in the family; each of them is uniquely determined by a cyclic permutation of its five branches around the center. If we place the center at the point $x = 0$, and place, for example, the white vertex of valency 6 at the point $x = 1$, then the 24 trees correspond to the 24 permutations of the branches of valencies 5, 4, 3, 2.

Instead of having one orbit of degree 24, we obtain in this example two separate orbits of degree 12 each. It turns out that one of them corresponds to the set of even permutations, i.e., to the elements of the group A_4, while the set of odd permutations forms the other orbit.

Three years later Kochetkov [176] found many similar examples, and then he remarked that in all of them the product

$$d = klmnp(k + l + m + n + p)$$

was a perfect square, where k, l, m, n, p were the valencies of the white vertices, which were supposed to be pairwise distinct. (In the initial example we have $d = 6 \cdot 5 \cdot 4 \cdot 3 \cdot 2 \cdot (6 + 5 + 4 + 3 + 2) = 120^2$.)

Kochetkov proved that in all these cases there were indeed at least two orbits of size 12 and 12. What remained to be proved was the fact that the orbits corresponded to the even and odd permutations of branches. This fact was established by Zapponi [303] using Strebel differentials. Recently Kochetkov found an explicit factorization of the defining polynomial into two factors of degree 12 over the field $\mathbb{Q}(\sqrt{d})$ even in the case when the product d was not a perfect square.

2.2.4.3 Using Elliptic Curves

In order to understand the example of this section the reader must have some preliminary knowledge about elliptic curves.

Let us return to Example 2.2.29 and to the cubic polynomial D of Eq. (2.3). We would like to discuss *whether it is possible for this polynomial to have a rational root?* If yes, then this family will split into two orbits, one rational and one quadratic (and possibly even in three rational ones, though we have never observed such a possibility).

First of all we may note that the polynomial D is homogeneous in m and n. Hence we may divide it by n^3 and introduce a new variable $b = m/n$, thus having

$$f(a,b) = 15a^3 - 45a^2(b+3) + 15a(b+3)(b+4) \\ - (b+3)(b+4)(b+5). \quad (2.4)$$

The most important observation is that we have obtained a plane cubic. Our problem may be reformulated as follows: *Find all rational points (a,b) on the cubic $f(a,b) = 0$ satisfying two additional conditions: $b \neq 1$ and $b > 0$.*

The first condition is very easy to satisfy: the substitution of $b = 1$ reduces the equation to $a^3 - 12a^2 + 20a - 8 = 0$, which does not have any rational roots. The second condition is rather nasty and causes some trouble.

Let us mention two basic facts: (a) any plane cubic reduces to an elliptic curve; (b) the problem of finding rational points on elliptic curves is classical (see, for example, [259]), but it is also notoriously difficult, and many questions remain open. An algorithm of reducing a cubic to one of the standard forms is described in [57], Sec. 7.4.2. To start with, it needs at least one rational point on the cubic. We may use one of the following:

$$(a,b) = (0,-3),\ (0,-4),\ (0,-5),\ (1,0);$$

in our computations we have used the point $(1,0)$.

After a series of rather tedious transformations we finally obtain the following elliptic curve equivalent to (2.4):

$$y^2 = x^3 - 2475x - 5850. \tag{2.5}$$

An inquisitive reader may be interested in the following additional information. The polynomial $x^3 - 2475x - 5850$ does not have rational roots, and it has three real roots. Its discriminant is equal to

$$59719680000 = 2^{17} \cdot 3^6 \cdot 5^4.$$

Finally, the J-invariant of the curve is equal to

$$J = \frac{898425}{512} = \frac{3^3 \cdot 5^2 \cdot 11^3}{2^9}.$$

Now begins the hunt for rational points. We will not describe all its stages but just communicate the results.

The torsion group of the curve is isomorphic to C_3, and the two torsion points of order three are

$$Q = (75, 480) \quad \text{and} \quad -Q = (75, -480).$$

The point
$$P = (-21, 192),$$

which belongs to the curve, is not of finite order since, for example, the point $3P$ has fractional coordinates. Thus, the rank of the curve is at least one, and the curve contains infinitely many rational points.

To each point (x,y) on the curve (2.5) there corresponds a point (a,b) on the initial curve (2.4). The parameter a does not interest us very much. On the other hand, $b = m/n$ represents the valencies of the white vertices, and therefore its value does have an interest. This parameter may be found by the following formula:

$$b = 30 \cdot \frac{111x^2 - 6090x - 29385 - 3xy + y}{x^3 - 1305x^2 + 63675x + 299925}.$$

An elementary analysis of this expression produces the domains of its positivity on the real plane (x, y). We may observe that on the real part of the curve there are three segments of positivity of b, one on the compact oval and two on the infinite oval. The rational point P belongs to the compact oval, and this fact ensures that the rational points are dense on both ovals (since they form a group). Therefore, we may be sure that *there exist infinitely many rational and positive values of b*.

We have looked through all the points $pP + qQ$ with $-25 \le p \le 25$ and $q = 0, 1, 2$, i.e., through over 150 rational points on the curve. Among them, there were 11 points that led to positive values of b. The smallest example is given by the point

$$-4P + Q = \left(\frac{8155}{121}, \frac{486280}{1331}\right).$$

The corresponding value of b is $33/124$. Therefore, the vertex degrees of the corresponding family of trees are $m = 33$ and $n = 124$, while the total number of edges is

$$2m + 5n = 686 = 2 \cdot 7^3.$$

The biggest example found corresponds to the point $22P$. The values of m and n do not fit into a line even with the smallest font; however, the number of edges may be written in the following way:

$$2m + 5n = 2^2 \cdot 5 \cdot 2584416196841212744132447585056811154400766793^3$$
$$\approx 3.4523 \times 10^{134}$$

(the long integer in the last formula is a prime).

Remark 2.2.32. In all the 11 examples mentioned above the number of edges is "almost" a perfect cube. Why? – The following explanation was suggested by Yu. Kochetkov. Substituting $a = (c + 15(m + 3n))/15n$ in Eq. (2.3) we obtain the equation

$$c^3 - 225(m + 3n)(2m + 5n)c - 450(m + 3n)(2m + 5n)(4m + 11n) = 0.$$

The numbers m and n are coprime; therefore, $m + 3n$ and $2m + 5n$ are also coprime, and the same is true for $2m + 5n$ and $4m + 11n$, except a possible common factor 2. If a solution c is rational, it is in fact an integer, and it is divisible by any prime factor $p \ne 2, 3, 5$ of $2m + 5n$. Hence, if c contains a factor p^k and $2m + 5n$ contains p^l, then c^3 contains p^{3k}, while the second term of the equation contains p^{k+l}. From this one can infer that $l = 3k$. The similar arguments show that $m + 3n$ is also "almost" a perfect cube (the only non-cubic factors can be 2, 3, or 5).

Other details concerning the example discussed in this section may be found in [308].

It is incredible how many interesting phenomena one may observe while working with such a seemingly trivial structure as the bicolored plane tree.

2.3 Belyi Functions and Belyi Pairs

By now the reader has seen so many trees that he is certainly eager to see at least a few maps. We start this section with a very simple example, then add several more complicated ones, some of them corresponding to dessins of a higher genus.

Example 2.3.1 (Simple map). Consider a very simple map which is not a tree: see Fig. 2.20.

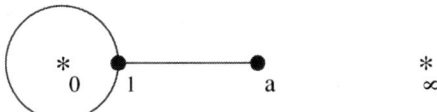

Fig. 2.20. A simple map

First of all we must place it on the complex plane. The placement is also shown in the figure: the center of the outer face is placed at $x = \infty$, the center of the face of degree one, at $x = 0$, and the vertex of degree 3, at $x = 1$. When three elements of a dessin become fixed, the entire position becomes also fixed. We denote the position of the vertex of degree 1 by a; this is an unknown parameter.

The Belyi function for this map is a rational function of degree 4 (since the map has 4 darts); the vertices of the map are its roots (with the multiplicities taken into account); the face centers are its poles. Thus the function has the form

$$f(x) = K \frac{(x-1)^3(x-a)}{x}.$$

(The fact that the numerator is of degree 4 and the denominator is of degree 1 ensures that infinity is a pole of order 3.)

We need some additional information in order to find K and a. This information is provided by the fact that the function $f - 1$ has two double roots: they are the edge midpoints, or, if you prefer, they are the two white vertices which we did not draw but which do exist inside the edges. (We should warn the reader that the term "midpoint" is misleading: for example, the point

which happens to be inside the segment $[1, a]$ is not its geometric middle.) We conclude that
$$f(x) - 1 = K\frac{(x^2 + bx + c)^2}{x}.$$

Equating the expressions
$$\frac{K(x-1)^3(x-a)}{x} - 1 = \frac{K(x-1)^3(x-a) - x}{x} = \frac{K(x^2 + bx + c)^2}{x}$$

we obtain four equations for the four unknowns K, a, b, c.

In fact the situation is not as dramatic as it seems to be. Computing the derivative
$$f'(x) = K\frac{(x-1)^2(3x^2 - 2ax - a)}{x^2}$$

we may note that the polynomial $3x^2 - 2ax - a$ must be proportional to $x^2 + bx + c$, and we immediately find that $b = -2a/3$ and $c = -a/3$. Finally, the calculations give us $a = 9$ and $K = -1/64$, so

$$f(x) = -\frac{(x-1)^3(x-9)}{64x}, \qquad f(x) - 1 = -\frac{(x^2 - 6x - 3)^2}{64x}. \qquad (2.6)$$

The midpoints of edges are the roots of $x^2 - 6x - 3$, that is, $3 \pm 2\sqrt{3}$.

Remark 2.3.2. Let us explain once more what we call *midpoints*. If a dessin corresponds to a hypermap, we draw its white vertices explicitly. If it is a map, we often omit the white vertices (all having degree 2) which implicitly exist inside every edge. Such an "invisible" white vertex is usually called the *midpoint* of the corresponding edge. Note that a midpoint is not necessarily the geometric middle of the edge, even if the edge in question is a straight line segment; it is by definition a preimage of 1. In the above example, $3 + 2\sqrt{3}$ is not the geometric middle of the segment $[1, 9]$.

A Belyi function is called a *pure Belyi function* if it corresponds to a map, that is, if all preimages of 1 are of multiplicity 2.

Example 2.3.3 (A dessin on an elliptic curve). Let us turn immediately to a dessin of higher genus: concretely, of genus one. Consider the elliptic curve E defined by the equation
$$y^2 = x(x-1)(x-9),$$

and the meromorphic function on E which is the projection on the first coordinate:
$$p : E \to \overline{\mathbb{C}} \ : \ (x, y) \mapsto x.$$

The projection p is a covering of degree 2. It is not a Belyi function since it has four critical values: 0, 1, 9, and ∞ (see Remark 2.3.4 below). But we already have a function which sends all four of them to 0, 1, and ∞: it is the function f of the previous example, see Eq. (2.6). Therefore, taking (x,y) to x, and then x to

$$z = -\frac{(x-1)^3(x-9)}{64x}$$

we obtain a Belyi function on E. Now it becomes clear why we have taken 9, and not any other value of x, as the fourth critical value. The dessin on the torus corresponding to this Belyi function is shown in Fig. 2.21. We may note that the vertices of degree 3 and 1 of the plane dessin became vertices of degree 6 and 2 respectively, which is the result of the ramification of order 2, and the faces of degree 3 and 1 are also transformed into faces of degree 6 and 2.

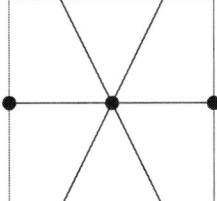

Fig. 2.21. A dessin on an elliptic curve

A more interesting result will be obtained if we apply the same trick to the curve E_1 defined by the equation

$$y^2 = x(x-1)(x-(3+2\sqrt{3})).$$

Now the projection $p : (x,y) \mapsto x$ has the critical values 0, 1, $3+2\sqrt{3}$, and ∞, and the same function f as before sends them to 0, 1, and ∞. But this time the curve E_1 has a conjugate, namely, the curve E_2 defined by the equation

$$y^2 = x(x-1)(x-(3-2\sqrt{3})).$$

The two corresponding dessins are shown in Fig. 2.22. For these dessins we show the white vertices explicitly because one of them becomes of degree 4. This degree is explained by the ramification of order 2 over an edge midpoint. Thus, the two dessins on the torus are this time not maps but hypermaps. It goes without saying that they have the same passport: $[61^2, 42^2, 62]$.

It is important to understand one thing. Not only a Belyi pair (a curve with a Belyi function on it) produces a dessin, but the correspondence also works in the opposite direction: a dessin fixes a curve. It would be impossible to obtain the same dessin on any other curve. (Once more: suppose an elliptic curve is given. Topologically it is a torus, and we may draw on this torus whatever we like; for example, we may draw the hypermaps of Fig. 2.22. But

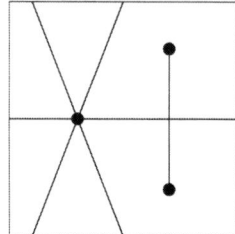 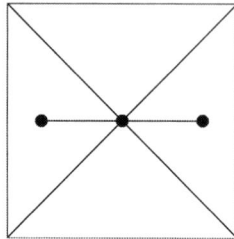

Fig. 2.22. Two conjugate dessins of genus 1; they are defined over the field $\mathbb{Q}(\sqrt{3})$

if we want to obtain this hypermap as the preimage of the segment $[0,1]$ via a meromorphic function with the critical values 0, 1, and ∞, it will not work unless the curve is E_1 for the left hypermap and E_2 for the right one.) A combinatorial object (a hypermap) is one of the ways to represent a complex structure.

Returning now to the construction of our Belyi pairs we may note that the pattern used here is of rather general nature. In the proof of Belyi's theorem, see Sec. 2.6, we take an arbitrary curve, make a projection, and then try, by successive compositions, to send all critical values of this projection to 0, 1, and ∞. On the other hand, in constructing examples we proceed in a similar but a somewhat different manner. This time we try from the very beginning to take such a curve that the critical values of its projection could be very easily taken to 0, 1, and ∞. Concretely this means that they should be vertices, or edge midpoints, or face centers of an already known plane dessin.

Remark 2.3.4. For a reader not very familiar with algebraic curves we add some remarks concerning the previous example. Consider the curve E defined by the equation

$$y^2 = x(x-1)(x-\lambda) = x^3 - (\lambda+1)x^2 + \lambda x, \tag{2.7}$$

with $\lambda \neq 0, 1$. For every fixed finite $x \in \mathbb{C}$, $x \neq 0, 1, \lambda$, there exist two solutions of the equation for y, and thus two points (x, y) on the curve. Therefore the projection to the x-axis is a covering of degree 2. For $x = 0, 1, \lambda$ there is only one solution for y (of multiplicity 2); thus these three points are critical values of the projection. But what happens at infinity?

In fact, Eq. (2.7) is merely a simplified way of writing the homogeneous equation

$$y^2 z = x^3 - (\lambda+1)x^2 z + \lambda x z^2 \tag{2.8}$$

for the homogeneous coordinates $(x:y:z)$ in the complex projective plane $\mathbb{C}P^2$. The "projection" is the function

$$p: \mathbb{C}P^2 \to \mathbb{C}P^1 \;:\; (x:y:z) \mapsto (x:z)$$

which is defined everywhere except the point $(0\!:\!1\!:\!0)$ (since there is no point $(0\!:\!0)$ in $\mathbb{C}P^1$). This exceptional point $(0\!:\!1\!:\!0)$ obviously belongs to E, see Eq. (2.8).

Now, it is easy to verify two things. First, if a point $(x\!:\!y\!:\!z)$ on E tends to $(0\!:\!1\!:\!0)$, then $(x\!:\!z)$ tends to $(1\!:\!0)$. Indeed, since $y \neq 0$ near $(0\!:\!1\!:\!0)$, we may normalize by $y = 1$; then Eq. (2.8) and the equations $x = o(1)$, $z = o(1)$ show that $z \sim x^3$ and hence $z/x \to 0$. Therefore, it is natural to extend the projection p on E by continuity, setting

$$p : (0\!:\!1\!:\!0) \mapsto (1\!:\!0),$$

and the point $(1\!:\!0)$ represents $\infty \in \overline{\mathbb{C}} = \mathbb{C}P^1$. And, second, when $x = 1$ is fixed and z is small, we have $y^2 \approx 1/z$; hence, when a small z turns around 0, the value of y changes its sign. Therefore, the point $\infty = (1\!:\!0) \in \mathbb{C}P^1$ is a critical value, and its preimage $(0\!:\!1\!:\!0) \in E$ is a critical point of multiplicity 2.

Example 2.3.5 (Icosahedron). To give a more advanced example, here is the Belyi function for the icosahedron map:

$$f(x) = 1728 \, \frac{\left(x^{10} - 11\,x^5 - 1\right)^5 x^5}{\left(x^{20} + 228\,x^{15} + 494\,x^{10} - 228\,x^5 + 1\right)^3}$$

(note that one of the 12 vertices of degree 5 is placed at infinity). The only thing that needs to be verified is the following factorization of $f - 1$:

$$f(x) - 1 = -\frac{\left(x^{10} + 1\right)^2 \left(x^{20} - 522\,x^{15} - 10006\,x^{10} + 522\,x^5 + 1\right)^2}{\left(x^{20} + 228\,x^{15} + 494\,x^{10} - 228\,x^5 + 1\right)^3}.$$

Curiously enough, this function, as well as Belyi functions for the other four Platonic maps, were found by Felix Klein in 1875 [170]. In his famous "icosahedron book" [170], Klein also explains how this function is related to the modular function invariant under the action of the icosahedron group. This subject certainly deserves a longer discussion, but we cannot dwell on it here.

Example 2.3.6 (Pseudorhombicuboctahedron). An Archimedean solid is a polytope whose faces are regular polygons, and whose vertices are "indistinguishable": the latter means that the same types of polygons in the same cyclic order (or its inverse) are adjacent to every vertex. The history of Archimedean solids is rather complicated. Traditionally, their discovery is attributed to Archimedes (Heron and Pappus refer to his manuscript which perished in a fire of the library at Alexandria). They were rediscovered (and baptized by their now traditional and rather fancy names) by Johannes Kepler [167]; see also [73]. Before Kepler, partial lists were found by Piero della Francesca [234], Albrecht Dürer, and W. Jamnitzer [73]. Pictures of some of the Archimedean solids made by Leonardo da Vinci (they served as illustrations for a book by Luca Pacioli) may be found in many books on the history

114   2 Dessins d'Enfants

of mathematics. Their list, besides prisms and antiprisms and the five Platonic solids, consists of 13 polytopes.

The 14th figure was overlooked for more than 2000 years and found only in 1957 [15]: see Fig. 2.23. Its vertex type is the same as for the rhombicuboctahedron, so it was called pseudorhombicuboctahedron. This polytope is not always accepted to the Archimedean club, because its isometry group is isomorphic to C_4 and does not act transitively on the vertices. It is exactly for the same reason that the computation of its Belyi function becomes very difficult. It was computed in [207] and [122]. The corresponding dessin may be seen in Fig. 2.24.

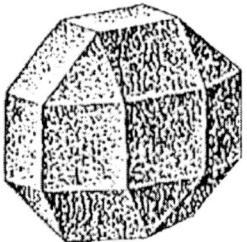

Fig. 2.23. The pseudorhombicuboctahedron

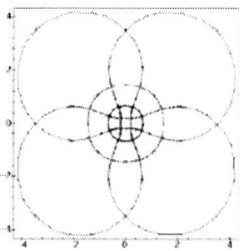

Fig. 2.24. The dessin of the pseudorhombicuboctahedron

It turns out that the dessin is defined over the field $\mathbb{Q}\langle\sqrt[4]{12}\rangle$, and must therefore have three more conjugates. We show two of them in Fig. 2.25; the third one is mirror symmetric to the map on the right.

All the three maps have the vertex partition 4^{24} and the face partition $4^{18}3^8$, the same as for the above polytope. According to the well-known theorem of Steinitz [266], every 3-connected planar map without loops and multiple edges and without vertices of degree 1 and 2 can be realized as a convex polytope. However, the faces of such polytope for the above three maps cannot be regular polygons, because there are some vertices surrounded by four

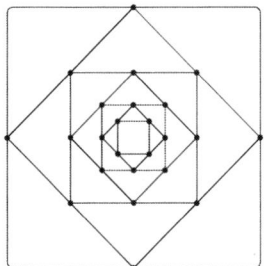 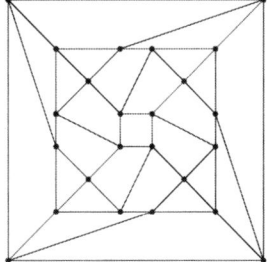

Fig. 2.25. Maps conjugate to the pseudorhombicuboctahedron

quadrilaterals. We don't know about any other similar "distant relatives" for other Archimedean solids.

We interrupt here our flow of examples in order to develop a bit of theory. More examples of Belyi functions and Belyi pairs will be given in the subsequent sections.

2.4 Galois Action and Its Combinatorial Invariants

It would be a suicidal act on our part to start here a careful exposition of Galois theory: we would never arrived at the next chapter. But it is also out of the question to leave this most important aspect of the theory of dessins d'enfants without any description at all. Therefore, according to the general guidelines of this book, we will try our best to *explain* the corresponding notions and phenomena.

2.4.1 Preliminaries

2.4.1.1 The Universal Galois Group

The *universal Galois group*, or the *absolute Galois group* is the group of automorphisms of the field $\overline{\mathbb{Q}}$ of algebraic numbers. It is denoted by $\Gamma = \text{Gal}(\overline{\mathbb{Q}} \mid \mathbb{Q})$. Obviously, this group leaves fixed the field $\mathbb{Q}$. The group Γ is one of the most important (and most intractable) objects in mathematics. Actually, this single group embodies the whole of the classical Galois theory over $\mathbb{Q}$.

Let $k \subset \overline{\mathbb{Q}}$ be a finite extension of $\mathbb{Q}$. (Finite extensions of $\mathbb{Q}$ are also called *number fields*.)

Fact 2.4.1. *Every automorphism of the field k may be extended to an automorphism of the field $\overline{\mathbb{Q}}$.*

Two algebraic numbers $a, b \in \overline{\mathbb{Q}}$ are called *conjugate* if there exists an automorphism $\alpha : \overline{\mathbb{Q}} \to \overline{\mathbb{Q}}$ which sends a to b.

For any finite extension k of $\mathbb{Q}$ there exists a subgroup Γ_k of finite index in Γ consisting of all the automorphisms of $\overline{\mathbb{Q}}$ that leave the elements of k fixed. In the opposite direction, to any subgroup of finite index of Γ we may associate the subfield of $\overline{\mathbb{Q}}$ consisting of all the algebraic numbers that are fixed under the action of this subgroup.

Fact 2.4.2 (The main theorem of Galois theory). Subgroups of Γ of finite index are in one-to-one correspondence with finite extensions of $\mathbb{Q}$ inside $\overline{\mathbb{Q}}$.

If H is a normal subgroup of Γ, the corresponding field K is called a *normal*, or *Galois* extension of $\mathbb{Q}$. We may also use the following construction: take a field k, take the corresponding subgroup $\Gamma_k < \Gamma$, and then take the maximal normal subgroup $H \triangleleft \Gamma$ contained in Γ_k. Then the number field K corresponding to H is the minimal normal extension of k. Though k has many normal extensions, most often we will work with the minimal one. The finite quotient group $G = \Gamma/H$ is called the *Galois group* of the field k.

Let D be a polynomial irreducible over $\mathbb{Q}$. Adding some of its roots to $\mathbb{Q}$ (and making all possible arithmetic operations) gives a finite extension k of $\mathbb{Q}$. Adding to $\mathbb{Q}$ *all its roots* gives a normal extension K. (For example, adding only the real root of $D(a) = a^3 - 2$ we get an extension which is not normal, but adding all the three roots of D gives a normal one.) In such a case the field K is called the *splitting field* of D, and the polynomial D is called a *defining polynomial* of K. (The defining polynomial of a field is in no way unique.) The Galois group G of K is also called the Galois group of the polynomial D. For a D of degree n it is a transitive permutation group of degree n.

Is it true that every finite group is a Galois group of some finite extension of $\mathbb{Q}$? This question is the famous "inverse problem of Galois theory". There are numerous works concerning this subject (the paper of Belyi [21] among them), but the complete answer is still unknown.

2.4.1.2 How $\mathrm{Gal}(\overline{\mathbb{Q}} \,|\, \mathbb{Q})$ Acts on Dessins d'Enfants

Consider first the planar case. Let M be a dessin, and let f be a corresponding Belyi function. Then f is a rational function, and according to Belyi's theorem f may be chosen in such a way that all its coefficients are algebraic numbers. Let K be the normal extension of $\mathbb{Q}$ generated by the coefficients. Now, let us act on all these numbers simultaneously by an automorphism of K; note that, according to Fact 2.4.1, this is the same thing as to act by an automorphism α of $\overline{\mathbb{Q}}$, that is, by an element $\alpha \in \Gamma$. We will see below that the result f^α of such an action is once more a Belyi function. The dessin M^α which corresponds to f^α is, by definition, the result of the action of α on M.

For a dessin of genus $g \geq 1$ the procedure is the same, though now we must act on a Belyi pair. Let M be a dessin, and let (X, f) be a corresponding Belyi pair. According to Proposition 1.8.7 the curve X may be realized as an algebraic curve in $\mathbb{C}P^k$, that is, as a solution of a system of homogeneous algebraic equations in homogeneous coordinates $(x_0 : x_1 : \ldots : x_k)$. According to Belyi's theorem the coefficients of these equations may be chosen as algebraic numbers. The function f is a rational function in the variables $x_0, \ldots, x_k$ whose coefficients, once more according to Belyi's theorem, may be chosen as algebraic numbers. Now we act on all the above algebraic numbers simultaneously by an automorphism $\alpha \in \Gamma$, and we get a new Belyi pair (X^α, f^α) which produces a new dessin M^α. We explain why this is indeed an action in the next section.

The most important observation is the fact that *all orbits of the action of Γ on dessins are finite.* Indeed, the passport is an invariant (see below), and the number of hypermaps with a given passport is finite.

Let M be a dessin. Consider its stabilizer $\Gamma_M \leq \Gamma$. Due to the fact that the orbit of M under the action of Γ is finite, the subgroup Γ_M is of finite index in Γ. Let $H \leq \Gamma_M$ be the maximal normal subgroup of Γ contained in Γ_M, and let K be the Galois extension of $\mathbb{Q}$ which corresponds to the subgroup $H \trianglelefteq \Gamma$. Note that the group H is the stabilizer of all the elements of the orbit.

Definition 2.4.3 (Field of moduli). The field k which corresponds to the group Γ_M, where Γ_M is the stabilizer of a dessin M, is the *field of moduli of the dessin M*. The field K which corresponds to the maximal normal subgroup $H \trianglelefteq \Gamma$ contained in Γ_M, is called the *field of moduli of the orbit* of M. (See also Notation 2.2.13.)

We see that according to this definition, together with the main theorem of Galois theory (Fact 2.4.2) the field of moduli of an orbit cannot be chosen arbitrarily: it is an "objective" characteristic of the orbit; cf. Example 2.2.16.

Remark 2.4.4. The terminology of this notion did not settle at once. In many publications what we call here a field of moduli is called a "field of definition". We don't use the latter term, but we use the term "is defined over L" as a synonym of the term "is realized over L" (which means that all the coefficients of equations defining X, and all the coefficients of f belong to L). In practice, a dessin may be realized over a field bigger than its field of moduli.

Remark 2.4.5. For a Galois orbit having N elements, the subgroup Γ_M is of index N, and therefore the field of moduli of the orbit is generated by the roots of a polynomial of degree N. In particular, if an orbit contains a single element, its field of moduli is $\mathbb{Q}$ itself.

Now we must explain why the passport and many other combinatorial characteristics of dessins are invariants of the Galois action.

2.4.2 Galois Invariants

2.4.2.1 A Descriptive Language

Let us introduce a descriptive formal language $\mathcal{L}$ with the following ingredients:

- variables for complex numbers;
- the four arithmetic operations (and parentheses);
- logical operations $\vee$, &, and $\neg$ (therefore, also the implication $\Rightarrow$), and quantifiers $\exists$ and $\forall$;[1]
- a unique predicate, that of equality $=$ (but by means of the negation we may also use $\neq$);
- two constants 0 and 1 (though by use of the arithmetic operations we can obtain any rational number).

What we are interested in is the expressive power of this language. For example, we are able to say $x^2 = 17$, thus saying that x is one of the two square roots of 17. But is it possible to specify which one, $\sqrt{17}$ or $-\sqrt{17}$? The answer is no. Indeed, it is trivial to verify that the mapping $a + b\sqrt{17} \mapsto a - b\sqrt{17}$, $a, b \in \mathbb{Q}$, is an automorphism of the field $\mathbb{Q}(\sqrt{17})$. According to Fact 2.4.1 this automorphism can be extended to an automorphism of $\overline{\mathbb{Q}}$. And this means that any algebraic relation *with rational coefficients* we might write about $\sqrt{17}$ is also valid for $-\sqrt{17}$. Being conjugate in $\overline{\mathbb{Q}}$ means being *indistinguishable* by any expression in the language $\mathcal{L}$. This principle is valid not only for individual numbers but also for more complicated objects.

Principle 2.4.6. Two objects constructed of algebraic numbers are not conjugate if and only if there exists an expression in the language $\mathcal{L}$ which permits to distinguish between them, or, in other words, which is true for one object and false for the other.

Example 2.4.7. The property "to be positive" is not expressible, otherwise we would be able to distinguish between $\sqrt{17}$ and $-\sqrt{17}$. The property "to be real" is not expressible; indeed, the three roots of $x^3 - 2 = 0$ are conjugate, but only one of them is real. The property "to be less than 1 in absolute value" is not expressible, and so on.

Let $P \in \mathbb{Q}[x]$ be a polynomial. It is easy to express the fact that $x_1, \ldots, x_n$ are the (distinct) roots of P, though the expression is rather nasty:

---

[1] The formal theory we are presenting admits the *elimination of quantifiers*: any proposition with quantifiers is equivalent to some proposition without quantifiers. For example, "$\exists x \, (P(x) = 0) \, \& \, (Q(x) = 0)$" is equivalent to "Resultant$(P, Q) = 0$". Elimination of quantifiers is a very efficient tool for certain algebraic problems.

$$(P(x_1) = 0) \ \& \ \ldots \ \& \ (P(x_n) = 0) \ \&$$
$$(x_1 \neq x_2) \ \& \ \ldots \ \& \ (x_{n-1} \neq x_n) \ \&$$
$$\forall x \ (x \neq x_1) \ \& \ \ldots \ \& \ (x \neq x_n) \Rightarrow P(x) \neq 0.$$

Another question: if we renumber the roots in a different manner, i.e., permute their indices, will we be able to distinguish between these two numberings? Let us consider an example. The *discriminant* of a polynomial P is defined as
$$\text{Discriminant}(P) = \prod_{i<j}(x_i - x_j)^2,$$
where $x_1, \ldots, x_n$ are the roots of P. It is a symmetric function of $x_1, \ldots, x_n$ and therefore it can be expressed via the coefficients of P. Let us take two examples:
$$P(x) = x^3 + 2 \quad \Longrightarrow \quad \text{Discriminant}(P) = -108,$$
while
$$Q(x) = x^3 + x^2 - 2x - 1 \quad \Longrightarrow \quad \text{Discriminant}(Q) = 49.$$

In the second case the discriminant of Q is a square in $\mathbb{Q}$, and we can therefore compute the product $\prod(x_i - x_j)$, which will be equal to 7 or to -7, depending on the numbering of the roots. Now, all even permutations of the roots do not change the sign of the last product, and we cannot distinguish among the corresponding numberings, while odd permutations change the sign, and we can distinguish between the sequence of roots and the same sequence oddly permuted. As to the polynomial $P(x) = x^3 + 2$, the fact that its discriminant is equal to -108 does not provide us with the same possibility. In fact, the Galois group of P is S_3, while the Galois group of Q is $A_3 = C_3$. Of course, we did not prove this fact, but we explained in what respect P and Q were different. In general, the Galois group of an irreducible polynomial consists of such permutations of its roots that the corresponding sequences of roots are indistinguishable.

We come closer to the objects of our study. How to express in $\mathcal{L}$ the fact that P is a Shabat polynomial? – This is easy, we just repeat the definition:
$$\exists y_1, y_2 \ \forall x \ (P'(x) = 0) \Rightarrow (P(x) = y_1 \vee P(x) = y_2).$$

While for Belyi functions the critical values are fixed at 0, 1 and ∞, for Shabat polynomials we did not impose any constraints on the critical values y_1 and y_2. Now we would like to say that the critical value y_1 corresponds to the partition $\alpha = (\alpha_1, \ldots, \alpha_p)$ of the passport (and not to the β). To do that, we must say (in $\mathcal{L}$) that "there exist distinct $x_1, \ldots, x_p$ such that $P(x_1) = y_1$, $P'(x_1) = 0, \ldots, P^{(\alpha_1 - 1)}(x_1) = 0, P^{(\alpha_1)}(x_1) \neq 0, \ldots$ and similar statements about $x_2, \ldots, x_p$". This statement permits to distinguish y_1 corresponding to α from y_2 corresponding to β. Note however that this method does not work when $\alpha = \beta$. This reminds us of the situation we have already met in Example 2.2.17.

A more important question remains for the time being without answer. The group Γ acts on the set of Shabat polynomials by replacing their coefficients with conjugate algebraic numbers. Why is this action an action on trees?

Let us formulate the question more accurately.

Let P_1 and Q_1 be two Shabat polynomials corresponding to the same tree T_1. Let $\varphi \in \Gamma$ be an automorphism of $\overline{\mathbb{Q}}$, and let φ take P_1 to P_2 and Q_1 to Q_2. Why do the polynomials P_2 and Q_2 correspond to the same tree T_2? – The fact that P_1 and Q_1 correspond to the same tree (or, in other words, Q_1 is equivalent to P_1) means that

$$Q_1(x) = AP_1(ax + b) + B, \quad A \neq 0, a \neq 0.$$

Therefore

$$Q_2(x) = \varphi(A)P_2(\varphi(a)x + \varphi(b)) + \varphi(B),$$

which gives a polynomial equivalent to P_2.

For plane Belyi functions the manipulations are similar, with the only difference that we must use linear fractional transformations instead of affine ones, and apply the usual changes of variables while dealing with infinity. Concerning Belyi pairs, in order to express such simple statements as "x is a critical point of f of a given multiplicity", and other expressions of this sort, we need some elementary techniques of manipulation with algebraic curves, which we have not introduced. But we hope that by now the reader is ready to take our word for it.

2.4.2.2 Combinatorial Invariants of Galois Action

We will gradually formulate various combinatorial properties of dessins d'enfants which are Galois invariant. We do not formulate the corresponding assertions as theorems or propositions because the proofs, or "explanations", will have different degrees of accuracy.

Symmetric trees. The property of a tree to be centrally symmetric, with the symmetry of order k, is Galois invariant. Indeed: place the center at $x = 0$; then its Shabat polynomial is a polynomial in x^k; cf. Question 2 in Sec. 2.2.3.

"Half-symmetric" trees. The property of a tree to be symmetric with respect to an edge midpoint is Galois invariant. Indeed, place this midpoint at $x = 0$; if P is the corresponding Shabat polynomial, then P' is an even polynomial.

Self-dual maps. The property of a map to be self-dual is Galois invariant. In general, if f is a Belyi function of a dessin, the Belyi function corresponding to the dual dessin is $1/f$. Indeed, $1/y$ sends ∞ to 0 and 0 to ∞, thus exchanging the roles of vertices and face centers. To be self-dual does not mean that $f = 1/f$: the map and its dual do not coincide. They are isomorphic, though differently placed on the plane. The word "isomorphic" means that there exists a linear fractional transformation $x \mapsto (ax+b)/(cx+d)$ such that

$$\frac{1}{f(x)} = f\left(\frac{ax+b}{cx+d}\right).$$

The latter phrase is a sentence of the language $\mathcal{L}$.

Compositions. The property of a function $F : X \to Z$ to be decomposable as

$$F : X \xrightarrow{f} Y \xrightarrow{g} Z$$

can be expressed in $\mathcal{L}$. If we want F to be a Belyi function, then only g must also be a Belyi function, while f may have an arbitrary number of critical values. But these critical values of f must be placed among the g-preimages of 0, 1, and ∞, that is, among the vertices, midpoints and face centers of the dessin corresponding to g. We may add to the above description any expressible properties of f and g, and also the properties like "a critical value of f corresponding to the partition λ of the f-passport is placed at a vertex of degree k of the g-dessin".

All these descriptions are direct in the planar case, and involve some machinery of algebraic curves when X, or both X and Y are curves of higher genera.

Automorphism group. Symmetry is a particular case of composition. We have already seen that for trees. In general this fact is based on the possibility to construct a quotient dessin; see the details in Sec. 1.7.3.

Cartographic group. It is in the same Sec. 1.7.3 that we find the origins of the fact that the cartographic group is an invariant. Indeed, for a dessin M with the cartographic group G there exist a dessin M^* with the *automorphism* group G and a ramified covering $f : M^* \to M$. The Galois group acts on all the three elements M^*, f, and M simultaneously, transforming them in N^*, g, and N such that g is a covering $g : N^* \to N$. Now, N^* has the same automorphism group G, since the automorphism group is an invariant. Therefore, the cartographic group of N is G. Thus once more composition plays a crucial role here.

Now let us consider several examples.

Example 2.4.8. The following beautiful example is borrowed from the paper by Filimonenkov and Shabat [101]. Consider the family of maps with the passport $[5^2 1^2, 2^6, 5^2 1^2]$; see Fig. 2.26. All of them are self-dual; in addition, the maps (a), (b), and (c) are symmetric, with the automorphism group C_2,

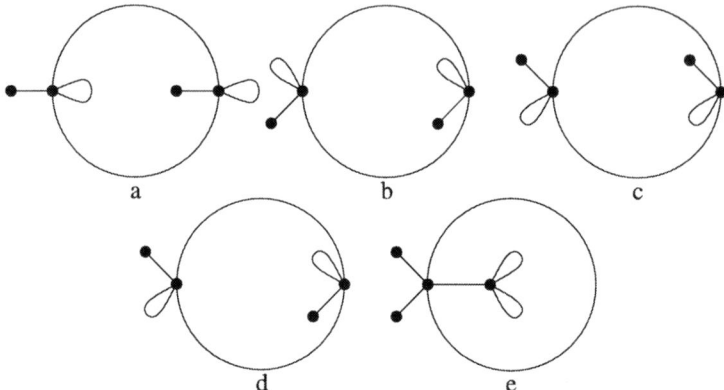

Fig. 2.26. The family of maps with the passport $[5^2 1^2, 2^6, 5^2 1^2]$

while the maps (d) and (e) are asymmetric. Nevertheless the set consisting of the maps (a), (b), (c) splits into two orbits: one consisting of the map (a) (with the field of moduli $\mathbb{Q}$), and the other one consisting of the two maps (b) and (c) (with the field of moduli $\mathbb{Q}(\sqrt{-15})$). What is the reason?

Let us make the following operation with each of the three maps: draw in the same picture the map itself and its dual. Because of the autoduality the automorphism group of the resulting map must contain the automorphism group of the initial map as a subgroup of index 2. This is what occurs. But for the map (a) this bigger group is the cyclic group C_4, while for the maps (b) and (c) it is the Klein group $C_2 \times C_2$. We hope that the reader has recognized in the map (a) the map of Fig. 1.33.

By the way, if f is a Belyi function corresponding to a map, then the Belyi function correponding to the "double map" (the initial map together with its dual) is $F = 4f/(f+1)^2$. Indeed, the Belyi function $g(y) = 4y/(y+1)^2$ sends 0 to 0, 1 to 1, and ∞ to 0, thus creating new vertices at the initial face centers. This remark permits us to express the properties of the "double" map using the language $\mathcal{L}$ as usual, though applying all expressions to the function F.

Example 2.4.9 (Non-definability over the field of moduli). For the two remaining maps (d) and (e) of the previous example there exist only two possibilities: either they form two different orbits, and then the field of moduli for both is $\mathbb{Q}$; or they form one orbit, and then the field of moduli of this orbit is quadratic (in reality it is the second possibility which turns out to be true). We may say even more: in the latter case the field must be *real quadratic*, since the map (e) is invariant under complex conjugation (indeed, this field is $\mathbb{Q}(\sqrt{5})$). But the map (d) cannot be placed on the complex plane in such a way as to be mirror symmetric to itself: the complex conjugation transforms it into another, though isomorphic map. Therefore it is impossible to find a Belyi function for the map (d) which is defined over $\mathbb{Q}$ or over a real quadratic field; in other words, this map does not have a Belyi function with the coefficients in the field of moduli.

The first example of this kind appeared in [68]. Note a characteristic property of this dessin: the non-existence of a "bachelor".

Example 2.4.10. The invariants we discussed above may sometimes be incredibly strong. Let us return to Example 1.5.18, that is, to the family of trees with the passport $[2^8 1^7, 4^4 2^2 1^3, 23]$. We know that this family contains 60060 trees. Therefore we may expect that after writing the corresponding system (and eliminating the parasitic solutions) we will obtain something of degree 60060. There is little hope to carry out this computation to the end, isn't there? But now, we may predict that this "something", which we are probably even unable to write down explicitly, splits into at least two factors, one of them of degree 4 (corresponding to the 4 trees with the cartographic group M_{23}), the other one of degree 60056.

But this is not the end yet. Looking into the character table of M_{23} (see [62]) we find that all irrationalities present there are expressed in terms of $\sqrt{-7}$, $\sqrt{-11}$, $\sqrt{-15}$, and $\sqrt{-23}$. For the conjugacy classes of the cyclic permutations of length 23 (present in our passport) there is only one of them, namely, $\sqrt{-23}$; two other conjugacy classes of our passport are rational. Now, Theorem 8.2.1 on page 84 of [253] implies that the field of moduli of our 4 trees must contain $\sqrt{-23}$. And indeed, Matiyasevich and his student Vsemirnov [212], after a breath-taking computation using the LLL algorithm, found that the field of moduli for this orbit is $\mathbb{Q}\left\langle \sqrt{-23/2 - (5/2)\sqrt{-23}} \right\rangle$.

Exercise 2.4.11. Looking into the character tables of the Mathieu groups M_{11} and M_{12} we see that all irrationalities present there are expressed in terms of $\sqrt{-11}$.

1. There are two maps with the passport $[3^3 1^3, 2^6, (11, 1)]$. Verify that their cartographic group is M_{12}. Compute their Belyi functions and show that they are defined over $\mathbb{Q}(\sqrt{-11})$.

2. Do the same thing for the two maps with the passport $[4^2 1^4, 2^6, (11, 1)]$: the cartographic group is once more M_{12}, and the field of moduli, $\mathbb{Q}(\sqrt{-11})$.

3. There are 10 trees with the passport $[4^2 1^3, 2^4 1^3, 11]$. Two of them have cartographic group M_{11}. Find them, and show that they are defined over $\mathbb{Q}(\sqrt{-11})$ (the latter computation is very difficult).

2.4.3 Two Theorems on Trees

Theorem 2.4.12. *For any bicolored plane tree there exists an associated Shabat polynomial whose coefficients belong to the field of moduli of the tree.*

This theorem was first published by Couveignes in 1994 [68]. Another proof existed earlier in the unpublished notes (in Russian) of a Moscow seminar

[256]. We remind to the reader that the analogous statement for maps is not valid; see Example 2.4.9. Our proof follows [68].

Proof. Let $p_k(x)$ and $q_k(x)$ be the monic polynomials whose roots are the black and white vertices of valency k, respectively. The coefficients of these polynomials will be our unknowns. These coefficients can be expressed as elementary symmetric functions of the coordinates of the vertices of the same color and valency. For brevity we will call these symmetric functions *vertex combinations*. We recall once more that any vertex combination includes only vertices of the same color and valency.

We look for a Shabat polynomial of the form

$$P(x) = C \prod p_k(x)^k$$

such that

$$P(x) - 1 = C \prod q_k(x)^k.$$

The latter equalities give us the number of equations fewer by two than we need. We must add two additional equations in such a way as to obtain the simplest possible field.

Let σ be an arbitrary *sum* of vertices of the same color and valency, i.e., an arbitrary vertex combination *of degree* 1. The first equation we add is

$$\sigma = 0.$$

Now the position of the tree is determined up to a transformation $x \mapsto Ax$ with arbitrary $A \in \mathbb{C}$.

Let Σ be a vertex combination of the smallest degree m which is not equal to 0. In fact, m is the order of symmetry of the tree. The second equation we add is

$$\Sigma = 1.$$

Now the position of the tree is determined up to $x \mapsto Ax$ with A satisfying $A^m = 1$.

Let $\overline{P}$ be a *conjugate polynomial*, i.e., another solution of the same system of equations, and *suppose that $\overline{P}$ gives the same tree as P itself*. This means that

$$\overline{P}(x) = P(Ax).$$

Now the fact that the tree is symmetric of order m implies that $P(x)$ is in fact a polynomial in x^m; hence $P(Ax) = P(x)$. In other words, if two conjugate polynomials give the same tree, then they are equal; two distinct conjugate polynomials give distinct trees. The theorem is proved. □

On the practical level, if one has obtained a field and would like to know if it is the field of moduli or some bigger field, the best way to proceed is to draw a computer-made picture. If every tree of the orbit is repeated only once, it is the field of moduli; if more than once, it is a bigger field.

The following generalization of the above theorem was proved for the planar case in [68], and for the general case in [300].

Definition 2.4.13 (Bachelor). For a hypermap, a *bachelor* is a black vertex, or a white vertex, or a face center which is unique for its type and degree.

Theorem 2.4.14. *If a hypermap has a bachelor, then there exists an associated Belyi function defined over the field of moduli of this hypermap.*

Obviously, for a tree its unique face is a bachelor.

Theorem 2.4.15 (H. W. Lenstra, Schneps [248]). *The action of the universal Galois group Γ on the set of trees is faithful.*

This means that for any non-identity element of the group Γ there is a bicolored plane tree which is transformed into another tree by the action of this element. The class of fields of moduli of the trees is not a special class of fields: all number fields belong to this class.

We give a sketch of the proof, following [248].

Lemma 2.4.16. *Let f be a polynomial of degree n and let $d \mid n$. Suppose there exists a polynomial h such that $h(0) = 0$, h is monic, $\deg h = d$ and for some polynomial g, $f = g(h)$. Then h is unique.*

Indeed, $f = a_m h^m + a_{m-1} h^{m-1} + \ldots + a_0$, and the terms of degrees $n, n-1, \ldots, n-d+1$ all come from the leading term $a_m h^m$, which provides us with enough information to find all the coefficients of h.

Lemma 2.4.17. *Let g, h, $\overline{g}$, and $\overline{h}$ be polynomials such that $g(h) = \overline{g}(\overline{h})$ and $\deg h = \deg \overline{h}$. Then there exist constants c and d such that $\overline{h} = ch + d$.*

It suffices to turn h and $\overline{h}$ into monic polynomials with the free terms equal to 0, and then apply Lemma 2.4.16, since

$$g(h) = g_1(h/c_1 - d_1) = \overline{g}(\overline{h}) = g_2(\overline{h}/c_2 - d_2).$$

Lemma 2.4.18. *For any polynomial p with algebraic coefficients there exists a polynomial f with rational coefficients such that $f(p)$ is a Shabat polynomial.*

This lemma constitutes one of the main parts of the proof of Belyi's theorem; it is a particular case of Theorem 2.6.1 proved in Sec. 2.6.

Proof of Theorem 2.4.15. Let k be an arbitrary number field, and let α be a primitive element of k, i.e., an algebraic number that generates the field. Let us take a polynomial $p_\alpha(x)$ such that

$$p'_\alpha(x) = x^3(x-1)^2(x-\alpha),$$

and find a Shabat polynomial $P_\alpha(x) = f(p_\alpha(x))$, $f \in \mathbb{Q}[x]$ (which exists according to Lemma 2.4.18). Denote by T_α the corresponding tree. Let β be a number conjugate to α. In the same way as before, we construct the polynomials p_β, $P_\beta = f(p_\beta)$ and the tree T_β.

We must show that $\alpha \neq \beta$ implies that the trees T_α and T_β are distinct, or, equivalently, if T_α and T_β are isomorphic then $\alpha = \beta$. Indeed, let T_α and T_β be isomorphic. Then

$$P_\beta(x) = P_\alpha(ax+b) \quad \text{for some} \quad a, b.$$

Now, applying Lemma 2.4.17 to $g = \bar{g} = f$, $h = p_\alpha(ax+b)$, and $\bar{h} = p_\beta(x)$, we see that there exist constants c and d such that

$$p_\alpha(ax+b) = cp_\beta(x) + d.$$

The right-hand side has three critical points, namely, 0 (of order 4), 1 (of order 3), and β (of order 2). The left-hand side has three critical points x_0, x_1, and x_α, of orders 4, 3, 2 respectively, such that

$$ax_0 + b = 0, \quad ax_1 + b = 1, \quad ax_\alpha + b = \alpha.$$

Since critical points must match (respecting their orders), we have

$$x_0 = 0, \quad x_1 = 1, \quad x_\alpha = \beta,$$

which leads to $a = 1$, $b = 0$, and $\beta = \alpha$.

The theorem is proved.

2.5 Several Facets of Belyi Functions

The main interest of the theory of dessins d'enfants is the Galois action on maps and hypermaps. But there also exist other aspects of the theory that make Belyi functions an extremely interesting object of study independently of any Galois theory. The goal of this section is to present some of these aspects.

2.5.1 A Bound of Davenport–Stothers–Zannier

This is one of the most spectacular applications of Belyi functions. Let P and Q be two complex polynomials. We are interested in the following question that was posed in 1965 in [32]: *what is the smallest possible degree of the polynomial $P^3 - Q^2$ (for the given degrees of P and Q)?*

Obviously, we may suppose that both P and Q are monic, and $\deg P = 2m$, $\deg Q = 3m$: then $\deg P^3 = \deg Q^2 = 6m$, and the leading terms cancel. In [32] it was conjectured that if $P^3 \neq Q^2$, then $\deg(P^3 - Q^2) \geq m+1$, and this inequality is sharp, that is, the equality is attained for infinitely many values of m. One part of the conjecture, the inequality itself, was proved the same year by Davenport [75]. The sharpness part turned out to be significantly more difficult: it was proved only 16 years later by Stothers [268], and then reproved independently and generalized by Zannier [302]. It turns out that the equality is attained for every m. The proof is given below.

Let us reformulate the problem in terms close to ours. Let us set

$$P^3(x) - Q^2(x) = R(x),$$

and

$$f(x) = \frac{P^3(x)}{R(x)}, \quad \text{so that} \quad f(x) - 1 = \frac{Q^2(x)}{R(x)}.$$

These expressions resemble those we have seen many times. Therefore let us ask ourselves: *what if f is a Belyi function?* The translation of the conditions imposed on the polynomials into the combinatorial language then runs as follows:

- the numerator of f has the form P^3, $\deg P = 2m$; if P does not have multiple roots, then the corresponding dessin has $2m$ vertices, all of degree 3;
- the numerator of $f - 1$ has the form Q^2, $\deg Q = 3m$; under the same condition of not having multiple roots, this means that the dessin in question has $3m$ edges;
- now the Euler formula implies that the number of faces of the map is $m + 2$; one of the faces, the "outer" one, has its center at infinity; the other centers are the roots of R; in order to have $\deg R = m + 1$ we must ensure that all the faces except the outer one are of degree 1.

This simple translation permits us to reformulate our problem in purely combinatorial terms: *Does there exist, for every m, a plane map with $3m$ edges, with $2m$ vertices of degree 3, and with all the faces except the outer one having degree 1?*

The long-standing open problem suddenly becomes trivial: the two stages of constructing a map in question are shown in Fig. 2.27. We leave it to the

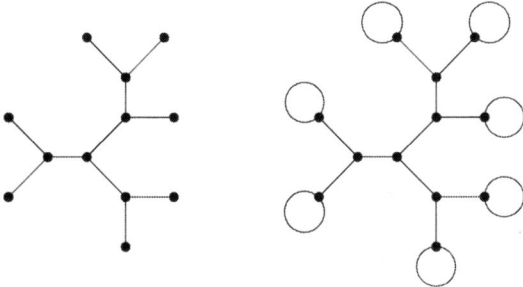

Fig. 2.27. First stage: draw a tree with all the internal vertices of degree 3; second stage: attach a loop to each leaf

128   2 Dessins d'Enfants

reader to verify the numerical characteristics (number of edges, number of vertices, ...). The statement is proved.

Of course, the main interest would be to find polynomials defined over $\mathbb{Q}$. Basing our judgement on the experience of working with Belyi functions we may say that such a situation is practically improbable (except for very small m). Nevertheless sometimes we may observe some interesting phenomena. For example, let us take $m = 5$ (see Fig. 2.28). There exist 4 corresponding maps. The map (a) is centrally symmetric while the three other maps are not; thus (a) forms a separate Galois orbit. The maps (b) and (c) are both symmetric with respect to a midpoint while the two other maps are not; thus, they form one more (quadratic) Galois orbit. The map (d) remains alone in its orbit and therefore its field of moduli is $\mathbb{Q}$. The "bachelor criterion" may be applied here: the outer face may serve as a bachelor. Therefore, in this case the polynomials P, Q, R are defined over the field of moduli and thus have rational coefficients. We did not compute these polynomials and presume that the task is not easy.

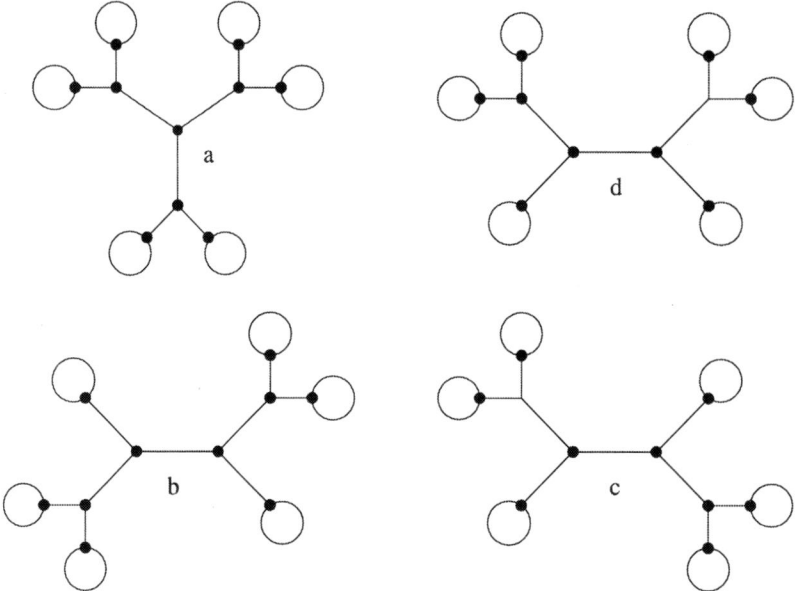

Fig. 2.28. For $m = 5$ there are three Galois orbits: (a), $\{(b), (c)\}$, and (d)

The question itself may be generalized. Let $\alpha, \beta \vdash n$ be two partitions of n, $\alpha = (\alpha_1, \ldots, \alpha_p)$, $\beta = (\beta_1, \ldots, \beta_q)$. Consider two polynomials of degree n:

$$P(x) = \prod_{i=1}^{p}(x - a_i)^{\alpha_i}, \quad Q(x) = \prod_{j=1}^{q}(x - b_j)^{\beta_j}. \tag{2.9}$$

The question is, what is the smallest possible degree of the polynomial $R = P - Q$? (The special case considered before corresponded to $n = 6m$, $\alpha = 3^{2m}$, $\beta = 2^{3m}$.) The complete answer was given by Zannier [302]. The following lower bound was already found in [268].

Proposition 2.5.1 (Lower bound). *We have*
$$\deg R \geq (n+1) - (p+q). \tag{2.10}$$

Remark 2.5.2. We see from this proposition that if we allow some of the roots a_i of P, or some of the roots b_j of Q, to merge (with the corresponding α_i or β_j being added), then the parameter p (resp. q) becomes smaller, and the bound (2.10) itself gets worse. In the same way, if P and Q are not coprime, then their common factor is also present in $R = P - Q$, and we are unable to control its degree. Therefore we suppose that all a_i and b_j are pairwise distinct.

Remark 2.5.3. Note that if $p + q \geq n + 1$, inequality (2.10) does not contain any valuable information: it states that $\deg R \geq 0$.

Proposition 2.5.4. *If $p + q \geq n + 1$, then the bound $\deg R \geq 0$ is attained. That is, for any n, and for any $\alpha, \beta \vdash n$ such that $p + q \geq n + 1$, there exist polynomials P and Q satisfying (2.9) such that their difference $R = P - Q = $ Const.*

Proof. If $p + q = n + 1$, the passport $[\alpha, \beta, n]$ is a passport of a tree. Let P be the corresponding Shabat polynomial with one of the critical values $y_1 = 0$; then we may take $Q = P - y_2$. Now, if $p + q > n + 1$, we can construct a passport of the form $[\alpha, \beta, \gamma, n]$. For it to be a valuable passport of a cactus, the partition γ must have $(2n+1) - (p+q)$ parts; such a passport does exist since the latter number is smaller than n. If S is a polynomial corresponding to such a cactus, and y_1, y_2, y_3 are its critical values corresponding to α, β, γ respectively, we may take $P = S - y_1$ and $Q = S - y_2$.

Now, let us have a look at the case when the numbers α_i and β_j, $i = 1, \ldots, p$, $j = 1, \ldots, q$ have a non-trivial greatest common divisor $d > 1$. This means that $P = f^d$ and $Q = g^d$. Then $R = f^d - g^d$ factors into d factors $f - \zeta g$, where ζ runs over the d-th roots of unity. If one factor, which we may assume without loss of generality to be $f - g$, has degree $< n/d$, then all the remaining factors have degree exactly n/d. So the only thing we may hope for is to diminish the degree of the factor $f - g$, by carefully choosing f and g.

Suppose that the numbers α_i and β_j, $i = 1, \ldots, p$, $j = 1, \ldots, q$ are coprime.

Theorem 2.5.5 (Main result). *Let the greatest common divisor of the numbers α_i, β_j, $i = 1, \ldots, p$, $j = 1, \ldots, q$ be 1, and $p + q \leq n + 1$. Then the bound (2.10) is attained. That is, there exist polynomials P and Q satisfying (2.9) such that*
$$\deg R = (n+1) - (p+q),$$
where $R = P - Q$.

Return once more to the case when α_i, β_j are *not* coprime, that is, $P = f^d$ and $Q = g^d$. We may now apply the above results to the polynomials f and g. For them, the degree is $\deg f = \deg g = n/d$, while the parameters p and q (the numbers of distinct roots) remain unchanged.

If $p + q \leq (n/d) + 1$, then, according to Theorem 2.5.5,

$$\min \deg(f - g) = \left(\frac{n}{d} + 1\right) - (p + q).$$

For the degree of $R = P - Q$ this gives

$$\min \deg R = \left(\frac{n}{d} + 1\right) - (p + q) + \frac{(d-1)n}{d} = (n + 1) - (p + q),$$

that is, in this case the bound attained coincides once more with (2.10).

On the contrary, if $p + q > (n/d) + 1$, then, according to Proposition 2.5.4, $\min \deg(f - g) = 0$; thus, for the degree of R we find

$$\min \deg R = \frac{(d-1)n}{d}.$$

Note that in this case the number $(d-1)n/d$ is strictly greater than the number $(n+1) - (p+q)$, their difference being equal to $(p+q) - ((n/d)+1)$. This is the only case when the bound (2.10) is not attained.

We summarize these simple observations in the following

Corollary 2.5.6. *Let the numbers α_i, β_j, $i = 1, \ldots, p$, $j = 1, \ldots, q$ have a non-trivial greatest common divisor $d > 1$. Then*

- *if $p + q \leq \frac{n}{d} + 1$, then the bound (2.10) is attained;*
- *if $p + q > \frac{n}{d} + 1$, then we have the inequality*

$$\deg R \geq \frac{(d-1)n}{d}, \tag{2.11}$$

and this bound is attained.

Example 2.5.7. Let $n = 6$, $\alpha = 4^2$, $\beta = 2^3$. We have $p = 2$, $q = 3$, and $d = 2$. The bound (2.10) would give us $\deg R \geq (6 + 1) - (2 + 3) = 2$, while the bound (2.11) gives $\deg R \geq 6/2 = 3$, and it is this latter bound that is attained.

Proof of Proposition 2.5.1. Suppose that

$$R(x) = \prod_{k=1}^{r} (x - c_k)^{\gamma_k}, \quad m = \deg R = \sum_{k=1}^{r} \gamma_k$$

(thus $m \geq r$). Consider the rational function

$$f = \frac{P}{R}, \quad \text{while} \quad f - 1 = \frac{Q}{R},$$

and let $y_1, \ldots, y_s$ be all critical values of f different from 0, 1, and ∞. Let $\lambda_1, \ldots, \lambda_s$ be the partitions of n describing the multiplicities of the preimages of $y_1, \ldots, y_s$. Then according to the Riemann–Hurwitz formula (1.2) (for $g = 0$) we obtain

$$r = (n+1) - (p+q) + \sum_{l=1}^{s} v(\lambda_l).$$

Taking into account that $m \geq r$, and that all the numbers $v(\lambda_l)$ are strictly positive, we obtain the desired inequality.

The last expression indicates the only way in which we may hope to obtain the equality $m = (n+1) - (p+q)$. Two conditions must be satisfied:

- $s = 0$; equivalently, there are no critical values of f besides 0, 1, and ∞, and f is a Belyi function;
- all $\gamma_l = 1$; equivalently, all the faces of the corresponding hypermap, except the outer one, are of degree 1.

We may forget about polynomials. In order to prove Theorem 2.5.5, it remains to prove a purely combinatorial statement: for any two partitions $\alpha, \beta \vdash n$ satisfying the assumptions of Theorem 2.5.5, there exists a hypermap for which α and β are the sets of degrees of black and white vertices, and whose all faces except the outer one are of degree 1. This statement is not very easy to prove. But at least it has the merit of being completely elementary, and may be discussed even with high school students. We leave to the reader the pleasure of searching for a solution.

Remark 2.5.8. The context of the Davenport–Stothers–Zannier bound is one more occasion where the inverse enumeration problem would be of great interest. We would like the polynomials P, Q, R to have rational coefficients. This will be achieved if the hypermap constructed above is unique (although this requirement is not necessary). Therefore, we may raise the following questions. Let p, q, r be positive integers, and $p + q + r = n + 1$. Then for any two partitions $\alpha, \beta \vdash n$ having, respectively, p and q parts, do the following:

- enumerate the hypermaps with the passport $[\alpha, \beta, (n-r)1^r]$;
- classify the passports of the type $[\alpha, \beta, (n-r)1^r]$ for which there exists a unique hypermap;
- compute the corresponding Belyi functions which are automatically defined over $\mathbb{Q}$.

A particular example of this type of Belyi functions is presented in the next section.

2.5.2 Jacobi Polynomials

The series of maps considered in this section was studied in [206].

In the previous section we saw the special interest of maps and hypermaps all of whose faces except the outer one are of degree 1, and which are de-

2 Dessins d'Enfants

fined over $\mathbb{Q}$. Here we consider a specific class of such maps, and we will see what kind of phenomena can be encountered. We call the maps in question *double flowers*. They consist of the segment $[-1, 1]$ subdivided into k parts, like in the chain tree, of l loops ("petals") attached to the point 1, and of m petals attached to the point -1: see Fig. 2.29. The total number of edges is $k + l + m = n$. The uniqueness of the map with the corresponding passport is "geometrically evident", and therefore its Belyi function must be defined over $\mathbb{Q}$.

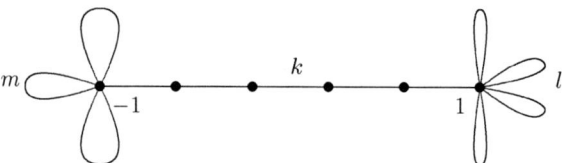

Fig. 2.29. A double flower

The initial motivation for studying these maps was not related to the Davenport–Stothers–Zannier bound. Rather, it was an attempt to answer another frequently asked question: "OK, we see that Chebyshev polynomials play a certain role in this theory. But are there any other orthogonal polynomials that also make their appearance?"

At the beginning we used to answer in the negative. Indeed, all roots of orthogonal polynomials are distinct and real; and it is difficult to imagine a tree, besides the chain tree, whose vertices would all lie on the real axis. Then, after a more careful reflexion, an idea came to our mind: why trees? Perhaps some more general maps would do? This is how the double flowers appeared. The results turned out to be gratifying: the interior vertices and the edges midpoints of the double flowers are zeros of a well-known class of orthogonal polynomials, the Jacobi polynomials.

Let us remind our reader what these are. For $a, b \in \mathbb{R}$, $a, b > -1$, the Jacobi polynomials $J_n^{(a,b)}$ are orthogonal polynomials with respect to the measure on the segment $[-1, 1]$ given by the density $(1-x)^a(1+x)^b$. The particular case $a = b = -1/2$ corresponds to *Chebyshev polynomials of the first kind*; that of $a = b = 1/2$, to *Chebyshev polynomials of the second kind*; the case $a = b = 0$ gives *Legendre polynomials*; finally, the case $a = b$ gives a one-parameter family of polynomials which are called *Gegenbauer polynomials*. All the roots of Jacobi polynomials lie inside the open segment $(-1, 1)$.

The above property of double flowers was first proved in [206]. Here we give another proof, proposed by Don Zagier. The text below (up to Exercise 2.5.11) is his.

The Jacobi polynomials, in their standard normalization, can be given in several different ways – by various explicit formulas such as

$$J_n^{(a,b)}(x) = \sum_{k=0}^{n} \binom{n+a+b+k}{k}\binom{n+a}{n-k}\left(\frac{x-1}{2}\right)^k \quad (2.12)$$

or

$$J_n^{(a,b)}(x) = 2^{-n} \sum_{k=0}^{n} \binom{n+a}{k}\binom{n+b}{n-k}(x+1)^k(x-1)^{n-k}, \quad (2.13)$$

by the generating function

$$\sum_{n=0}^{\infty} J_n^{(a,b)}(x) z^n = \frac{1}{S}\left(\frac{2}{1-x+S}\right)^a \left(\frac{2}{1+x+S}\right)^b \quad (S = \sqrt{1-2xz+z^2}),$$

or as the unique polynomial solution of the differential equation

$$L_n^{(a,b)}\left(J_n^{(a,b)}(x)\right) = 0 \quad (2.14)$$

with $J_n^{(a,b)}(1) = \binom{n+a}{n}$, where $L_n^{(a,b)}$ is the second order differential operator

$$L_n^{(a,b)} = (1-x^2)\frac{d^2}{dx^2} + (b - a - (a+b+2)x)\frac{d}{dx} + n(n+a+b+1).$$

(For these and other properties, see [274], the "bible" of orthogonal polynomials, or any of various standard reference works such as [2], Chapter 22.) From each of these definitions it follows that $J_n^{(a,b)}(x)$, which was initially defined only for a and b real and greater than -1, is in fact a polynomial in a and b and hence makes sense for any real or complex values of a and b. We can now formulate:

Proposition 2.5.9. *The vertices of valency 2 of the double flower are the roots of $J_{k-1}^{(l+\frac{1}{2}, m+\frac{1}{2})}(x)$, and the edges midpoints are the roots of $J_{k+l+m}^{(-l-\frac{1}{2}, -m-\frac{1}{2})}(x)$.*

Proof. The Belyi function associated to the double flower corresponds to an identity of the form $(1-x)^{2l+1}(1+x)^{2m+1}P(x)^2 - Q(x)^2 = R(x)$ where P, Q, and R are polynomials of degree $k-1$, $k+l+m$, and $l+m$, respectively, and these polynomials are determined uniquely (up to scalar multiples) by this identity. We have to show that P and Q are multiples of the Jacobi polynomials $J_{k-1}^{(l+\frac{1}{2}, m+\frac{1}{2})}$ and $J_{k+l+m}^{(-l-\frac{1}{2}, -m-\frac{1}{2})}$, respectively.

From the property (2.14) we find that both the functions

$$y_1 = J_{k+l+m}^{(-l-\frac{1}{2}, -m-\frac{1}{2})}(x) \quad \text{and}$$

$$y_2 = \left(\frac{x-1}{2}\right)^{l+\frac{1}{2}}\left(\frac{x+1}{2}\right)^{m+\frac{1}{2}} J_{k-1}^{(l+\frac{1}{2}, m+\frac{1}{2})}(x)$$

($x > 1$) are solutions of the differential equation $L_{k+l+m}^{(-l-\frac{1}{2},-m-\frac{1}{2})}(y) = 0$. But one checks easily that if $y(x)$ is any function with an asymptotic expansion $y = C_0 x^d + C_1 x^{d-1} + \ldots$ at infinity, then

$$L_{k+l+m}^{(-l-\frac{1}{2},-m-\frac{1}{2})}(y) = (k+l+m-d)(d+k)C_0 x^d + O(x^{d-1}) \quad \text{as} \quad x \to \infty,$$

so if $L_{k+l+m}^{(-l-\frac{1}{2},-m-\frac{1}{2})} y = 0$ we must have either $d = k+l+m$ or $d = -k$. For $y = y_1$, of course, we have $d = k+l+m$, but Eq. (2.12) or (2.13) implies that $y_2 - y_1 = o(x^{k+l+m})$ as $x \to \infty$ and hence that necessarily $y_2 - y_1 \sim C x^{-k}$ for some $C \neq 0$. (The value of C, unimportant here, is $(-1)^l \frac{(2k+2l-1)!!(2k+2m-1)!!}{2^{k+l+m}(2k+l+m)!}$.) It follows that $y_2^2 - y_1^2 = O(x^{l+m})$ and since y_1^2 and y_2^2 are polynomials we have

$$\left(\frac{x-1}{2}\right)^{2l+1} \left(\frac{x+1}{2}\right)^{2m+1} J_{k-1}^{(l+\frac{1}{2},m+\frac{1}{2})}(x)^2 - J_{k+l+m}^{(-l-\frac{1}{2},-m-\frac{1}{2})}(x)^2$$

$$= (-1)^l \frac{(2k+2l-1)!! \, (2k+2m-1)!!}{2^{2k+2l+2m-1}(k-1)! \, (k+l+m)!} R_{k,l,m}(x) \quad (2.15)$$

for some polynomial $R_{k,l,m}(x)$ of degree $l+m$. This proves the proposition.

Remark 2.5.10 (For fanatics only). The numerical factor on the right-hand side of (2.15) was included because the polynomial $R_{k,l,m}(x)$ then has a nicer form: for l and m fixed it is not only a polynomial in x, but also an odd rational function of $K = 2k+l+m$ which vanishes at infinity and has only simple poles at integral arguments, e.g.,

$$R_{k,1,2}(x) = \frac{1}{K} x^3 + \frac{K^2+3}{K(K^2-1)} x^2 - \frac{K^2+3}{K(K^2-1)} x - \frac{K^4-2K^2+9}{K(K^2-1)(K^2-9)}$$

$$= \frac{(x-1)^3}{K} + \frac{(2x-1)^2}{2}\left(\frac{1}{K-1} + \frac{1}{K+1}\right) - \frac{1}{2}\left(\frac{1}{K-3} + \frac{1}{K+3}\right)$$

with $K = 2k+3$. More precisely,

$$R_{k,l,m}(x) = \sum_{i=0}^{l+m} p_i(2k+l+m,l,m) \, x^{l+m-i}$$

where each $p_i(K,l,m)$ is a linear combination of $(K+j)^{-1}$ ($|j| \leq i$) with coefficients which are polynomials in l and m, the first few terms of the expansion being

$$R_{k,l,m}(x) = \frac{1}{K} x^{l+m} + \left(\frac{D}{K} + \frac{D(S+1)}{K(K^2-1)}\right) x^{l+m-1}$$

$$+ \left(\frac{D^2-S}{2K} + \frac{D^2(2S+1) - S(S+2)}{2K(K^2-1)} + \frac{3(D^2-1)S(S+2)}{2K(K-1)(K^2-4)}\right) x^{l+m-2} + \ldots,$$

where $K = 2k+l+m$, $S = l+m$, $D = m-l$.

Exercise 2.5.11. When the total number of petals in a double flower is equal to zero, $l = m = 0$, the above proposition gives us $a = b = 1/2$, that is, Chebyshev polynomials of the second kind, while it is known that the Shabat polynomials for the chain trees are Chebyshev polynomials of the first kind. Is there a contradiction?

Jacobi polynomials are very well studied. For example, their roots possess an electrostatic interpretation, their discriminants are explicitly written, and so on (see, e.g., [274]). This information may give some ideas about Belyi functions. It would also be nice to generalize the results of this section to other classes of orthogonal polynomials and special functions.

2.5.3 Fermat Curve

The following beautiful example is borrowed from [159]. Let us consider the Fermat curve X defined by the equation $x^n + y^n = 1$. Let $p : X \to \overline{\mathbb{C}} : (x, y) \mapsto x$ be the projection onto the x coordinate. The critical values of p are easy to determine: the equation $y^n = 1 - x^n$, considered as an equation in y, the "parameter" x being fixed, has n distinct solutions for all x such that $1 - x^n \neq 0$. Therefore, the critical values of p are the n-th roots of unity.

Remark 2.5.12. Let us explain more explicitly what takes place "above infinity". In fact, our curve must be considered as the projective curve

$$x^n + y^n = z^n$$

where $(x:y:z)$ are homogeneous coordinates in $\mathbb{C}P^2$. The term "homogeneous coordinates" means that x, y, z may not be all equal to zero simultaneously, and the triples $(x:y:z)$ and $(\alpha x:\alpha y:\alpha z)$, $\alpha \neq 0$, represent the same point in $\mathbb{C}P^2$. The projection p is the mapping $p : (x:y:z) \mapsto (x:z)$, where $(x:z)$ are the homogeneous coordinates on the projective line $\overline{\mathbb{C}} = \mathbb{C}P^1$. This mapping is not defined at the point $(0:y:0) \in \mathbb{C}P^2$, $y \neq 0$, because we must not obtain $(0:0)$ as an image. Fortunately, the point $(0:y:0)$ does not belong to the curve X. The point $\infty \in \overline{\mathbb{C}}$ is represented by a pair $(x:z)$, where $z = 0$, while $x \neq 0$. Thus, our equation is reduced to the form $x^n + y^n = 0$, and for $x \neq 0$ it does have n distinct solutions in y. Therefore, infinity is not a critical value of the projection p.

To specify an appropriate Belyi function is now easy. Let us take the mapping $f : x \mapsto x^n$; it sends all the n-th roots of unity (that is, all the critical values of p) to 1, and it creates two new critical values: 0 and ∞. Therefore, the function $F : X \to \overline{\mathbb{C}} : (x, y) \mapsto x^n$ is a Belyi function defined on the Fermat curve.

In this example it is more instructive to consider the preimage not of the segment $[0, 1]$, but of the whole real axis. More exactly, let us take the preimage of the elementary canonical triangulation of the sphere $\overline{\mathbb{C}}$, consisting of three vertices 0, 1, and ∞, of three edges $[0, 1]$, $[1, \infty]$, and $[\infty, 0]$, and of two triangles, namely, the upper and the lower half-planes.

1. The point 0 is a vertex of degree 2 of the elementary triangulation. We have $f^{-1}(0) = 0$, but $\deg f = n$, and therefore 0 becomes a vertex of degree $2n$. Being a generic point in the image of p, it gives rise to n distinct points $p^{-1}(0)$, all being vertices of degree $2n$. We color them in black, because all of them are the preimages (under F) of 0.
2. The point 1 is a vertex of degree 2 of the elementary triangulation. Taking its preimage under $f = x^n$, we get n distinct points (roots of unity), all of them being vertices of degree 2. Now, they are critical values of p; therefore their preimages are n vertices, all of degree $2n$ (because $\deg p = n$). We color these points in white, because they are preimages (under F) of 1.
3. The situation with ∞ resembles that of 0: we get n distinct vertices, each of degree $2n$; we mark them by an asterisk.
4. There are no critical values inside the edges and the faces of the elementary triangulation. Taking into account that $\deg F = n^2$, we obtain, as a preimage under F, $2n^2$ faces (all of them triangles) and $3n^2$ edges.

The genus of the curve may now be easily computed:

$$2 - 2g = 3n - 3n^2 + 2n^2 \implies g = \frac{(n-1)(n-2)}{2},$$

which is the genus of all non-singular plane algebraic curves of degree n.

Let us look more attentively at the resulting map. First of all, the graph itself is the *complete tripartite graph* $K_{n,n,n}$. Indeed, it has n vertices of type •, n vertices of type ∘ and n vertices of type ∗, and every vertex (i) is not connected to any other vertex of the same type (by construction), and (ii) is connected to all the vertices of the two other types (because its degree is $2n$, and there are no multiple edges). The second observation is that all the faces of our map are the "smallest possible" (indeed, they are triangles), therefore their number is the biggest possible, and therefore the genus of the surface is the smallest possible. Thus, we have obtained an embedding of the least possible genus of the complete tripartite graph $K_{n,n,n}$. The fact that the least possible genus of $K_{n,n,n}$ is equal to $(n-1)(n-2)/2$ was proved by purely combinatorial methods a long time ago: see [297], [242]. But what had probably never come to the mind of the authors of these two papers is the fact that the problem they considered was in some way related to the Fermat equation.

Let us repeat it once more. Among infinitely many non-isomorphic complex curves of genus $g = (n-1)(n-2)/2$, the least genus embedding of the graph $K_{n,n,n}$ (which is topologically unique) specifies a unique complex curve, namely, *the* Fermat curve. It would be justifiable to say that there exist two ways of representation of complex curves: by systems of algebraic equations, and by dessins (though the latter method works only for the curves defined over $\overline{\mathbb{Q}}$).

2.5.4 The *abc* Conjecture

The *abc* conjecture may well replace the Fermat theorem for the future generation of mathematicians. At least, it shares with this great theorem the simplicity of formulation; as to the difficulty of the proof, the future will show.

There exist several versions of the problem; here we choose one of them (see [226]). Let a, b, c be three positive coprime integers such that $a + b = c$. Note that $c = \max(a, b, c)$. Let

$$a = p_1^{\alpha_1} p_2^{\alpha_2} \ldots p_k^{\alpha_k}, \quad b = q_1^{\beta_1} q_2^{\beta_2} \ldots q_m^{\beta_m}, \quad c = r_1^{\gamma_1} r_2^{\gamma_2} \ldots r_n^{\gamma_n}$$

be their factorizations into prime factors. Denote by R the product of all the prime factors of a, b, c *taken only once*, that is,

$$R = p_1 p_2 \ldots p_k q_1 q_2 \ldots q_m r_1 r_2 \ldots r_n.$$

The conjecture affirms that R cannot be too small compared to c. More precisely: let $c = R^\alpha$, or, equivalently, $\alpha = \log c / \log R$.

Conjecture 2.5.13 (abc). *For any $\alpha_0 > 1$ there exist at most a finite number of triples (a, b, c) such that the corresponding value of the parameter α satisfies the inequality $\alpha > \alpha_0$.*

Note that if we replace $\alpha_0 > 1$ with 1, the statement turns wrong:

Example 2.5.14. Take the equality $1 + (2^{6n} - 1) = 2^{6n}$. The contribution of $a = 1$ and of $c = 2^{6n}$ to R is only 2, while $b = 2^{6n} - 1$ is divisible by $2^6 - 1 = 63$ and hence divisible by 9. Therefore the contribution of b is not bigger than $b/3$, and we conclude that $R < c$, so $\alpha > 1$.

The next example shows what kind of consequences may be obtained from the conjecture.

Example 2.5.15. Consider the equality $x^n + y^n = z^n$. Obviously,

$$R \leq xyz < z^3;$$

therefore

$$\alpha = \frac{\log c}{\log R} > \frac{\log z^n}{\log z^3} = \frac{n}{3}.$$

Thus, the *abc* conjecture implies that all Fermat equations with $n \geq 4$ taken together may have at most a finite number of coprime solutions.

Keeping in mind the conjecture, it is interesting to look for triples a, b, c with the parameter α as big as possible. The world record up to now belongs to Eric Reyssat: $2 + 3^{10} \cdot 109 = 23^5$, with $\alpha = 1.6299\ldots$. This triple was found by a brute force search. A number of papers were dedicated to a more systematic search: see, for example, [224], [225], [46] (a complete list of known instances with $\alpha > 1.4$), [292] (a complete list of instances with the prime factors not greater than 13), etc. A rather long list of instances with $\alpha > 1.2$ was collected by Noam Elkies several years ago.

It is clear that in a "good" example the numbers a, b, c must be well factorizable in prime factors. Several authors proposed to use, in a search for good triples, Belyi functions defined over $\mathbb{Q}$. The polynomials constituting a Belyi function are already well factorized. Substituting rational numbers into them, we may, with a little bit of luck, find interesting examples. This line of research was conducted by N. Magot [206] and L. Habsieger (private communication). The method turned out to be rather prolific. Let us illustrate it by taking one particular (and very simple) Belyi function.

Example 2.5.16. We take the dessin of Example 2.3.1, though for technical reasons we have changed its position: we put the vertex of degree 1 at infinity, the vertex of degree 3, at 0, and the center of the face of degree 1, at 1. The corresponding Belyi function is

$$f(x) = \frac{64x^3}{(x+9)^3(x+1)}, \quad \text{and} \quad f(x) - 1 = -\frac{(x^2 - 18x - 27)^2}{(x+9)^3(x+1)}.$$

The above two equalities, with $x = a/b$, may be rewritten as

$$64a^3 b + (a^2 - 18ab - 27b^2)^2 = (a + 9b)^3(a + b).$$

Substituting $(a, b) = (-32, 23)$ we get, after rearranging the terms,

$$11^2 + 3^2 \cdot 5^6 \cdot 7^3 = 2^{21} \cdot 23,$$

which gives the value of α equal to $1.62599\ldots$. This is not the record, but this is the second best value known up to now. However, this triple was found before the search undertaken by N. Magot.

In general, the above Belyi function has produced 43 triples with $\alpha > 1.2$; among them 11 triples were "new" (with respect to the list of Elkies as it was in 1996). The best among the "new" triples was

$$2^{15} \cdot 5 \cdot 67 + 11^2 \cdot 23^8 = 7^3 \cdot 3023^3, \quad \alpha = 1.35670\ldots.$$

Several hundred "new" triples with $\alpha > 1.2$ and not present in the Elkies list were found by L. Habsieger.

A much more profound application is given in [93]. The title indicates that the author deduces from the *abc* conjecture (formulated for number fields) the former Mordell conjecture, now the Faltings theorem. This theorem, which is one of the most important results of the last decades, affirms that an algebraic curve of genus $g \geq 2$ defined over a number field may have at most a finite number of points over this field. The proof of Elkies is based in a very important way on Belyi's theorem.

2.5.5 Julia Sets

One of the nice properties of Belyi functions is the possibility to compose them.

Proposition 2.5.17. *Let Riemann surfaces X, Y, and Z be all equal to $\overline{\mathbb{C}}$. If $f : X \to Y$ and $g : Y \to Z$ are Belyi functions, and $g(\{0, 1, \infty\}) \subseteq \{0, 1, \infty\}$, then the composition $h : X \to Z$, $h = g \circ f$ is a Belyi function.*

The proof is obvious: the critical values of h can be obtained either by taking critical values of g, or by applying g to the critical values of f. Note that any one of the 27 possible mappings $g : \{0, 1, \infty\} \to \{0, 1, \infty\}$ will do.

The condition that the points $0, 1, \infty \in Y$ lie among the points of the g-preimage of $0, 1, \infty \in Z$, has an obvious geometrical meaning: the points $0, 1, \infty \in Y$ are "occupied" by some geometric elements (black or white vertices or face centers) of the hypermap H_g corresponding to g and drawn on Y. Note that it is always possible to choose any three elements and put them at 0, 1, and ∞ using a linear fractional transformation.

A natural idea would be to consider complex dynamical systems involving Belyi functions, that is, consider the iterations

$$f^n = f \circ f \circ \ldots \circ f \quad (n \text{ times}).$$

Following [43], introduce the notion of dynamical Belyi function.

Definition 2.5.18 (Dynamical Belyi function). A Belyi function $f : \overline{\mathbb{C}} \to \overline{\mathbb{C}}$ is called *dynamical* if $f(\{0, 1, \infty\}) \subseteq \{0, 1, \infty\}$.

A complex dynamical system obtained from the iterations of a dynamical Belyi function does not leave the framework of Belyi functions.

Let us recall some basic definitions concerning complex dynamical systems. For an open domain $U \subseteq \overline{\mathbb{C}}$, a family $\mathcal{F}$ of holomorphic functions $f : U \to \overline{\mathbb{C}}$ is called a *normal family* on U if for any sequence $f_n \in \mathcal{F}$, $n \geq 1$ and for any compact subset $K \subset U$ there exists a subsequence converging uniformly on K. For a rational function f, the set $F_f \subseteq \overline{\mathbb{C}}$ consisting of all points $x \in \overline{\mathbb{C}}$ having a neighborhood in which the sequence f^n, $n \geq 1$ of iterations of f forms a normal family, is called the *Fatou set* of f. The *Julia set* of f is the complement of its Fatou set: $J_f = \overline{\mathbb{C}} \setminus F_f$. Informally speaking, the Fatou set is the set of "regular behavior" of the iterations of f, while the Julia set is the set of their "irregular behavior".

Remark 2.5.19. Don't confuse the above sets with the *Mandelbrot set*, which is the set of values of a parameter for iterations of quadratic polynomials.

All of us have seen pictures of beautiful fractals. Such pictures usually represent either the Mandelbrot set or the Julia sets of various rational functions. However, there also exists a situation which is probably not so beautiful

140   2 Dessins d'Enfants

visually, but still very interesting theoretically. We mean the situation of "complete chaos", when $J_f = \overline{\mathbb{C}}$. The first example of that kind was constructed by Lattès as early as in 1918 [194]. In 1985 Sullivan [272] gave a very general sufficient condition which guarantees that $J_f = \overline{\mathbb{C}}$.

Call a point x *periodic* if there exists a positive integer m such that $f^m(x) = x$. Call a point x *eventually periodic* if there exists a p such that the point $f^p(x)$ is periodic. Finally, call a point *strictly preperiodic* if it is eventually periodic but not periodic.

Theorem 2.5.20 ([272]). *If all critical points of f are strictly preperiodic, then $J_f = \overline{\mathbb{C}}$.*

Now let us try to apply this theorem to Belyi functions.

At the first stage f sends all its critical points to 0, 1, or ∞. Thus, already all the critical points different from these three cannot be periodic (the fact that they are eventually periodic will be explained in a minute).

All subsequent iterations move the points inside the set $\{0, 1, \infty\}$. It is therefore impossible to do without periodic points: some of them will necessarily be periodic. Is everything lost? Not at all: let us make them non-critical! These points are occupied by some geometric elements of the hypermap H_f corresponding to f: by its vertices or face centers. If these elements are of valency 1, then the condition is satisfied, and we are done.

Example 2.5.21. The following example appeared in the original paper by Sullivan:
$$f(x) = \left(\frac{x-2}{x}\right)^2.$$
Incidentally it is the Belyi function corresponding to the hypermap which has a black vertex of degree 2 at $x = 2$, two white vertices of degree 1, at $x = 1$ and at $x = \infty$, and a face of degree 2 with the center at $x = 0$. The only critical points of f are 0 and 2, and we have
$$2 \mapsto 0 \mapsto \infty \mapsto 1 \mapsto 1,$$
and 1 is not a critical point.

Example 2.5.22. The Belyi function of Eq. (2.6) from Example 2.3.1 does not work. Indeed, its critical points are $3 \pm 2\sqrt{3}$, 1, and ∞, and we have
$$3 \pm 2\sqrt{3} \mapsto 1 \mapsto 0 \mapsto \infty \mapsto \infty;$$
unfortunately, ∞ is itself a critical point. But we may change the position of the dessin. Let us take
$$f(x) = -\frac{(x-1)^3(9x-1)}{64x^3}.$$

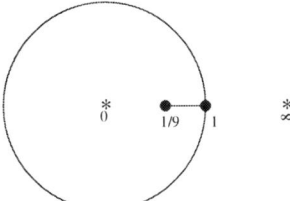

Fig. 2.30. The dessin for the Belyi function $f(x) = -\frac{(x-1)^3(9x-1)}{64x^3}$

This Belyi function corresponds to the dessin shown in Fig. 2.30. The only critical points that are not marked in the figure are the edge midpoints: their coordinates are $-1 \pm (2/3)\sqrt{3}$. Thus we have

$$-1 \pm (2/3)\sqrt{3} \mapsto 1 \mapsto 0 \mapsto \infty \mapsto \infty.$$

The only periodic point is once more infinity; but this time it is not a critical point, because it is the center of a face of degree 1. Therefore, $J_f = \overline{\mathbb{C}}$.

We must say that the methods proposed, for example, in [142], permit to construct many more examples than the one proposed here. But a very attractive side of our method is the possibility to work directly with pictures, without any computations. For that, we need a map or a hypermap having an element (say, a vertex) of degree 1.

Example 2.5.23. Let us consider the "dessin d'enfant" from Fig. 2.31.

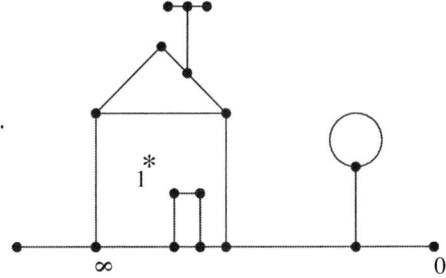

Fig. 2.31. A dessin d'enfant

We have marked three points and placed them at 0, 1, and ∞ (using if necessary a linear fractional transformation). The center of one of the faces is placed at 1. The fact that it is a face center means that $f : 1 \mapsto \infty$. One of the vertices is placed at ∞; the fact that it is a vertex means that $f : \infty \mapsto 0$. Finally, one of the vertices is placed at 0; the fact that it is a vertex means that $f : 0 \mapsto 0$. We see that all the critical points "finish" at 0. But 0 is not a critical point, because the vertex which is placed there is of degree 1. We may conclude that the Belyi function corresponding to this dessin satisfies the conditions of Sullivan's theorem. Of course, we are completely unable to compute the function in question.

It would be extremely interesting to study the Galois action on complex dynamical systems. This is true not only for dynamical Belyi functions but also for more general *critically finite* rational functions which do not possess the so-called "Thurston's obstructions", see [86], since they are also rigid and thus defined over $\overline{\mathbb{Q}}$. This line of research is only in the bud; see, however, [235].

2.5.6 Pell Equation for Polynomials

In the 6th century B.C. Pythagoras proved that the equation $x^2 - 2y^2 = 0$ did not have non-trivial integral solutions. One may say to oneself: "OK, if this one does not work, let us try $x^2 - 2y^2 = 1$". By some historical error this equation, as well as the more general one, $x^2 - dy^2 = 1$, with a square-free d, was called the Pell equation. In fact, this equation was already studied by ancient Greeks, Indians, and Arabs, and then by Fermat, Euler, Lagrange and others. The Pell equation played a very important role in the development of number theory, especially in introducing the notion of the norm of an algebraic number. For any square-free d there exist infinitely many solutions, and all of them can be obtained as numerators and denominators for incomplete quotients of the development of $\sqrt{d}$ into a continued fraction.

Exercise 2.5.24. Let (x_1, y_1) and (x_2, y_2) be two solutions of the equation $x^2 - dy^2 = 1$. Show that the product

$$(x_1 + y_1\sqrt{d})(x_2 + y_2\sqrt{d}) = a + b\sqrt{d}$$

gives one more solution (a, b) of the equation.

Many other interesting properties and examples concerning this equation may be found in textbooks on number theory.

In 1826 Abel [1] posed a similar question for polynomials: for a given square-free polynomial $D(x)$, find polynomials $P = P(x)$ and $Q = Q(x)$ satisfying

$$P^2 - D(x)Q^2 = 1.$$

We call this equation the *Pell–Abel equation*.

Example 2.5.25. Take $D(x) = x^2 - 1$. Let T_n and U_n be the Chebyshev polynomials of the first and the second kind respectively. (Recall that by definition of U_n we have $\sin(n+1)\varphi = U_n(\cos\varphi)\sin\varphi$.) Taking $x = \cos\varphi$ we get

$$T_n^2(\cos\varphi) - (\cos^2\varphi - 1)U_{n-1}^2(\cos\varphi) =$$

$$\cos^2 n\varphi + \sin^2\varphi \frac{\sin^2 n\varphi}{\sin^2\varphi} = \cos^2 n\varphi + \sin^2 n\varphi = 1.$$

Thus, the pairs (T_n, U_{n-1}) form an infinite series of solutions of the Pell–Abel equation with $D(x) = x^2 - 1$.

One of the first remarks of Abel was the fact that, in contrast to the case of integers, the equation for polynomials does not always have a solution. He proved the following theorem [1]:

Theorem 2.5.26 (Abel). *The following three statements concerning any square free polynomial $D(x)$ are equivalent:*

1. The Pell–Abel equation has a solution (in this case it has infinitely many solutions).

2. The square root $\sqrt{D(x)}$ can be represented as an eventually periodic continued fraction of the form

$$\sqrt{D(x)} = d_0(x) - \cfrac{1}{d_1(x) - \cfrac{1}{d_2(x) - \cfrac{1}{d_3(x) - \cdots}}}$$

with $d_k \in \mathbb{C}[x]$.

3. There exists a polynomial p, $\deg p \leq \deg D - 2$, such that the integral

$$\int \frac{p(x)}{\sqrt{D(x)}} dx$$

can be computed in terms of elementary functions (more exactly, in radicals and logarithms).

The last statement is the most interesting one. Usually such integrals which are *elliptic* if $\deg D = 3$ or 4, and *hyperelliptic* if $\deg D \geq 5$, cannot be expressed in elementary functions. The condition $\deg p \leq \deg D - 2$ is imposed in order to exclude the trivial case $p(x) = D'(x)$. Integrals satisfying the conditions of the above theorem are called *quasi-elliptic* (for $\deg D = 3$ or 4) or *quasi-hyperelliptic*. The search of quasi-elliptic integrals was in fact the main motivation of Abel's work. Important results in this direction also belong to Chebyshev [275]. In our days this subject continues to attract attention of specialists in systems of symbolic calculations; see, for example, [76].

Returning to our main subject (dessins d'enfants), we note that to any plane tree there corresponds a polynomial $D(x)$ for which a solution (P, Q) does exist, namely,

$$D(x) = \prod(x - c_k),$$

where the product is taken over *all vertices of odd valencies* (regardless of their colors), and c_k are the coordinates of these vertices.

Indeed, let us take a bicolored plane tree and normalize the corresponding Shabat polynomial P in such a way that it has ± 1 as the critical values. Then

$$P(x) + 1 = C \prod_{i=1}^{p} (x - a_i)^{\alpha_i}$$

and

$$P(x) - 1 = C \prod_{j=1}^{q} (x - b_j)^{\beta_j};$$

therefore,

$$P^2 - 1 = C^2 \prod_{i=1}^{p} (x - a_i)^{\alpha_i} \prod_{j=1}^{q} (x - b_j)^{\beta_j} = D(x) Q^2,$$

where $D(x)$ just collects all the factors $(x - a_i)$ or $(x - b_j)$ of odd degrees (each of the factors $(x - a_i)$ or $(x - b_j)$ being taken only once). If we now want to obtain infinitely many solutions for the same polynomial D, we may subdivide all the edges of our tree into m parts (or, equivalently, compose P and Q with Chebyshev polynomials): this operation does not change the geometric form of the tree, and the vertices of odd valencies will remain on their places.

Our first example, that of $D(x) = x^2 - 1$, is obviously related to the chain-tree. The next interesting class of trees to be considered is that of trees having a "center" of valency 3 (or 4), out of which grow 3 (respectively 4) simple chains of an arbitrary length. These trees lead to the quasi-elliptic case. For example, the tree shown in Fig. 2.32 gives $D(x) = x(x-4)(x^2 + 2x + 3)$.

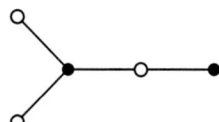

Fig. 2.32. A tree with three branches

Profound results in this direction were obtained by Pakovitch [229]. He exploits the relations of the Pell–Abel equation to torsion points on elliptic curves. Once again we assume the reader has at least a rudimentary knowledge of the theory of elliptic curves.

Consider the elliptic curve

$$y^2 = D(x) = x^4 - 6Ax^2 + 4Bx + C,$$

and the isomorphic curve E in the Weierstrass form:

$$w^2 = 4v^3 - g_2 v - g_3,$$

where

$$g_2 = 3A^2 + C, \quad g_3 = -AC + A^3 - B^2.$$

Then the Pell–Abel equation has a solution (P, Q) with an indecomposable polynomial P of $\deg P = n$ if and only if the point $(v, w) = (A, B)$ lies on the curve E and is a torsion point of order n on this curve.

Example 2.5.27. Of course the examples given here require some computations. But the starting point is always a tree: see Fig. 2.33.

(a) The tree (a) gives the elliptic curve $w^2 = 4v^3 + 540v + 10665$, and the point $(A, B) = (21, -243)$ of order 5 on it.
(b) The tree (b) gives the curve $w^2 = 4v^3 - 261228v + 3497176$, with the point $(A, B) = (-37, 3600)$ of order 6 on it.
(c) The tree (c) has also 6 edges, like the tree (b); but it is decomposable (that is, the corresponding Shabat polynomial is decomposable), this is why our procedure gives only a point of order 2.

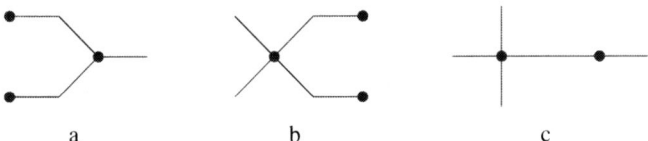

Fig. 2.33. Trees from Example 2.5.27

But the examples are not the most interesting part of the story. Below we give a theorem which, to our knowledge, has no analogues. It is often possible to say that in such and such a case there exists at least this number of orbits (when there are certain invariants or other reasons for splitting); but it is much more difficult to say that the number of orbits is *at most* such: the equations we must solve may have, for example, rational solutions because of merely numeric coincidences. What follows is, however, an exception to the rule. It is well known (Mazur's theorem) that the order of torsion points over $\mathbb{Q}$ cannot be greater than 12 (and also cannot be equal to 11): see, for example, [259]. Take a tree with three branches, with n edges, and with the coprime lengths of the branches (this condition ensures that the tree is indecomposable). Such a tree inevitably produces a polynomial $D(x)$, $\deg D = 4$, and a solution (P, Q) of the Pell–Abel equation, where P is the corresponding Shabat polynomial, $\deg P = n$. In addition, we get a torsion point (A, B) of order n on the

146   2 Dessins d'Enfants

corresponding elliptic curve. Now, if the tree is defined over $\mathbb{Q}$, then the torsion point is also defined over $\mathbb{Q}$ (by construction). For $n = 11$ and $n > 12$ this is impossible. For $n \leq 12$, $n \neq 11$ we may compute the Shabat polynomials case by case. The result is summarized in the following theorem.

Theorem 2.5.28 ([229]). *There exist only four trees with three branches and with the coprime lengths of the branches which are defined over $\mathbb{Q}$. These trees have the following branch lengths:*

$$1,1,1; \quad 2,1,1; \quad 2,2,1; \quad 3,1,1.$$

For more details, and also for the hyperelliptic case, see Pakovitch [229].

We hope that Sec. 2.5 has given enough evidence to demonstrate the interest and beauty of dessins d'enfants.

2.6 Proof of the Belyi Theorem

The Belyi theorem, see Theorem 2.1.1, has two parts, which are traditionally called the "difficult part" and the "obvious part". Paradoxically enough, it is the "difficult part" which is quite simple and elementary, while the "obvious part" is not simple at all. The "difficult part" (namely, the "only if" part) was proved by Belyi in [21]. It consists of several ingenious tricks, but they are very elementary in nature and accessible to an undergraduate student. Now, the "if" part is indeed obvious for a specialist in algebraic geometry and Galois theory with many years of experience. For any other mathematician it remains rather mysterious and enigmatic, and its explicit proof makes use of some subtle constructions and complicated theories.

We start the proof with the "difficult" "only if" part.

2.6.1 The "Only If" Part of the Belyi Theorem

Theorem 2.6.1. *If a Riemann surface X is defined over the field $\overline{\mathbb{Q}}$ of algebraic numbers, then there exists a meromorphic function $f : X \to \overline{\mathbb{C}}$ unramified outside $\{0, 1, \infty\}$.*

The proof of the theorem consists of three steps.

Step 1. Take an *arbitrary* meromorphic function $h : X \to \overline{\mathbb{C}}$ defined over $\overline{\mathbb{Q}}$. For example, if X is represented as an algebraic curve, we may take as h the projection onto one of the coordinates.

Among the critical values of h certain are rational, others irrational (though algebraic). We forget temporarily the rational critical values and will deal only with the (algebraic) irrational ones. As to these algebraic points, we add to them all their conjugates. Let S_0 denote the set of all irrational critical values of h together with all their conjugates, and let $|S_0| = N$.

Step 2. The polynomial P_0 which annihilates the set S_0 is defined over $\mathbb{Q}$, and $\deg P_0 = N$. This polynomial has at most $N - 1$ critical values (the values of the polynomial itself at the roots of its derivative); denote this set by S_1. An important observation is that the set S_1 already contains all the conjugates of its elements, so its annihilating polynomial P_1 is of degree at most $N - 1$ and is defined over $\mathbb{Q}$.

Exercise 2.6.2. Write explicitly the polynomial $P_1(t)$ annihilating the set S_1 of the critical values of the polynomial $P_0(x)$ as the determinant of the Sylvester matrix of polynomials $P_0(x) - t$ and $P_0'(x)$ considered as polynomials in x, and t being a parameter. Verify that this determinant is indeed a polynomial of degree $N - 1$ in t. See also Lemma 5.1.6.

The set S_2 of the critical values of P_1 has at most $N-2$ elements. Applying the same trick to this set, then to S_3, etc., we finally obtain a composition of polynomials
$$P_{N-1} \circ \ldots \circ P_1 \circ P_0, \quad \deg P_m \leq N - m,$$
which sends all the critical values of the initial function h to rational numbers. Note that the same is also true of the *rational* critical values of h (that were "temporarily forgotten"), because all P_m have rational coefficients.

Step 3. What remains to do now is to send all the rational critical values obtained during the steps 1 and 2 to 0, 1 or ∞. First of all, by applying an affine transformation we push them all (except infinity) inside the segment $[0, 1]$. Now consider the mapping
$$p_{m,n}(x) = \frac{(m+n)^{m+n}}{m^m n^n} x^m (1-x)^n,$$
which is, by the way, a Shabat polynomial. We have $p_{m,n}(0) = 0$, $p_{m,n}(1) = 0$, $p_{m,n}(\infty) = \infty$ and $p_{m,n}([0,1]) = [0,1]$. Finally, the mapping $p_{m,n}$ sends the rational number $m/(m+n)$ to 1, and all the other rational numbers that remain to be treated, to some other rational numbers. Obviously, this operation decreases the number of rational numbers under consideration. Specifically, if one supposes that the critical values are $x_1, \ldots, x_k$ with $x_1 = 0$, $x_2 = 1$, $x_3 = \infty$, and $0 < x_i < 1$ for other i, then setting $x_k = m/(m+n)$ and applying $p_{m,n}$ decreases k by 1. Applying successively these mappings as many times as we need, we get a Belyi function. The theorem is proved.

2.6.2 Comments to the Proof of the "Only If" Part

The following proposition is, in a way, a comment to the first step of the above proof. Its slight modification permits us to prove a much stronger result [4].

148   2 Dessins d'Enfants

Proposition 2.6.3. *For any Riemann surface defined over $\overline{\mathbb{Q}}$ there exists a Belyi function with a single pole (which corresponds to a map or a hypermap with a single face).*

Proof. It is sufficient to take a meromorphic function h having a single pole, which exists thanks to the Riemann–Roch theorem. All the subsequent mappings are polynomial and thus send infinity to infinity. (We need here the Riemann–Roch theorem *over number fields*: see, for example, Chapter 14 of [191] or Chapter 19 of [289].)

The trick used in the second step can also be generalized to other situations when one of the fields is an algebraic closure of the other. Let one of the fields be $\mathbb{R}$, and the other one be $\mathbb{C}$. If we have a finite set of complex numbers, add to each number its complex conjugate. The set S_0 thus obtained is annihilated by a polynomial P_0 with real coefficients. After that, the set S_1 of critical values of the polynomial P_0 is also annihilated by a polynomial with real coefficients.

The reader will certainly find amusing the following "comment" to the third step of the proof.

Example 2.6.4. Let us take the set $\{0, 1/5, 2/5, 3/5, 4/5, 1\}$ and try to move all its elements to 0 or 1 using the technique of Step 3. The choice of the first polynomial to apply is not very important, since all of them will be of degree 5. Take, for example, $1/5$, and apply $p_{1,4}$. The result is as follows:

$$0 \mapsto 0, \quad \frac{1}{5} \mapsto 1, \quad \frac{2}{5} \mapsto \frac{81}{128}, \quad \frac{3}{5} \mapsto \frac{3}{16}, \quad \frac{4}{5} \mapsto \frac{1}{64}, \quad 1 \mapsto 0.$$

For the next one, we can apply $p_{m,n}$ for $m/(m+n) = 81/128$, $3/16$, or $1/64$. The simplest choice is $p_{3,13}$ because it has the smallest degree. Applying it, we get

$$0 \mapsto 0, \quad 1 \mapsto 0, \quad \frac{3}{16} \mapsto 1, \quad \frac{1}{64} \mapsto \alpha, \quad \frac{81}{128} \mapsto \beta,$$

where

$$\alpha = \frac{3^{23} \times 7^{13}}{2^{32} \times 13^{13}} \quad \text{and} \quad \beta = \frac{3^9 \times 47^{13}}{2^{48} \times 13^{13}}.$$

This time we must take the polynomial corresponding to α; it is of degree $2^{32} \times 13^{13}$. It will send α to 1, and it will send β to a number ω which the reader may write explicitly if he wants. After that it will remain to apply one more polynomial whose degree is equal to the denominator of ω, in order to send ω to 1. The degree of the resulting composition is approximately $10^{5 \times 10^{25}}$.

We see that the method of Step 3 is not very practical if we want to find a Belyi function algorithmically. It was Belyi himself [22] who found a remedy to this difficulty. We present his method in a slightly simplified form. It will be convenient to suppose that the numbers we deal with are no longer rational but integral. For $c_1, \ldots, c_n \in \mathbb{Z}$ we will construct a rational function of the form

$$g(x) = \prod_{k=1}^{n}(x - c_k)^{r_k}, \quad r_k \in \mathbb{Z}. \tag{2.16}$$

Note that the value of this function at infinity is equal to 0, 1, or ∞, depending on the sign of the sum $r = \sum_{k=1}^{n} r_k$. The proposed method will always give $r = 0$, and therefore the value of g at infinity will always be 1.

The only critical points of the function g are the roots and the poles of its logarithmic derivative

$$\frac{g'}{g} = \sum_{k=1}^{n} \frac{r_k}{x - c_k}.$$

While the poles of the logarithmic derivative are immediately visible, the same is not true of its roots. We will try to find such r_i that g'/g will be equal to

$$\frac{\text{Const}}{\prod_{k=1}^{n}(x - c_k)},$$

and thus will have a unique zero of order n at ∞. But, by comparing the residues, one may easily see that

$$\frac{1}{\prod_{k=1}^{n}(x - c_k)} = \sum_{k=1}^{n} \frac{C_k^{-1}}{x - c_k} \quad \text{where} \quad C_k = \prod_{i \neq k}(c_i - c_k). \tag{2.17}$$

Therefore, making the r_k integers with no common factors *proportional to* C_k^{-1} we get what we want and thus obtain a Belyi function of the form (2.16) and of a reasonable degree. (Note that if $r = \sum r_k \neq 0$, then the asymptotic behaviour at infinity of the sum in (2.17) would be $O(1/x)$, while in fact it is $O(1/x^n)$.)

Example 2.6.5. Returning to the previous example, instead of the initial set $\{0, 1/5, 2/5, 3/5, 4/5, 1\}$ we may take $c_1, \ldots, c_6 = 0, 1, 2, 3, 4, 5$. Then for c_1 and c_6 we get

$$C_1, C_6 = \pm\, 1 \cdot 2 \cdot 3 \cdot 4 \cdot 5 = \pm\, 120,$$

for c_2 and c_5 we get

$$C_2, C_5 = \pm\, 1 \cdot 1 \cdot 2 \cdot 3 \cdot 4 = \pm\, 24,$$

and for c_3 and c_4 we get

$$C_3, C_4 = \pm\, 2 \cdot 1 \cdot 1 \cdot 2 \cdot 3 = \pm\, 12.$$

These results give us the following values for the numbers r_k:

$$r_1 = -1, \quad r_2 = 5, \quad r_3 = -10, \quad r_4 = 10, \quad r_5 = -5, \quad r_6 = 1.$$

Instead of the polynomial of a monstrous degree obtained before, this time we get a rational function of degree 16:

$$g(x) = \frac{(x-5)(x+3)^5(x-1)^{10}}{(x+5)(x-3)^5(x+1)^{10}}.$$

(Of course, not every example will give us such a small degree. But anyway, degrees produced by this method are far smaller than those given by the previous approach.)

2.6.3 The "If", or the "Obvious" Part of the Belyi Theorem

We follow here the explanations given in [300] and [177]. The reader may also find there additional details, comments and corollaries. In particular, Wolfart [300] explains why the usual reference to the paper [293] by Weil is not entirely justified.

2.6.3.1 The Group Aut($\mathbb{C}$)

We start the proof by considering a rather strange object: the automorphism group Aut($\mathbb{C}$) of the field $\mathbb{C}$, or, otherwise, the Galois group Gal($\mathbb{C}\,|\,\mathbb{Q}$). We underline the fact that only the field structure of $\mathbb{C}$ is taken into account; its topological properties are put aside.

Every automorphism of the field $\mathbb{C}$ preserves 0 and 1, and therefore it also preserves the subfield $\mathbb{Q}$. Now suppose we have added to $\mathbb{Q}$ a non-rational complex number x. If x is transcendental, then we may perform with it (and with rational numbers) all the four arithmetic operations, thus obtaining expressions having the form of rational functions in x, and no two such expressions will be equal. The result of such an extension is therefore isomorphic to the field $\mathbb{Q}(x)$ of rational functions in one variable. Adding one more transcendental number y which is algebraically independent of x we obtain the field isomorphic to the field $\mathbb{Q}(x,y)$ of rational functions in two variables. The exchange of x and y is one of the automorphisms of this extension. It is not in any way unique: we may, for example, replace y with $x+y$.

In the same way we may add to $\mathbb{Q}$ uncountably many variables representing algebraically independent transcendental numbers. In fact, this operation does not exhaust the field $\mathbb{C}$. Indeed, having added x we cannot represent $\sqrt{x}$; and if we add also $y = \sqrt{x}$ we cannot pretend any longer that our variables are algebraically independent. Never mind: what we are interested in is the possibility to construct *at least* an extension of $\mathbb{Q}$ by many independent variables. In the same way we may affirm that *at least* all the bijections between these variables can be extended to automorphisms of such an extension of $\mathbb{Q}$.

Exactly the same may be said if we add transcendental numbers not to $\mathbb{Q}$ but to $\overline{\mathbb{Q}}$. We will also consider the Galois group Gal($\mathbb{C}\,|\,\overline{\mathbb{Q}}$), i.e., the group of automorphisms of $\mathbb{C}$ preserving $\overline{\mathbb{Q}}$.

2.6 Belyi's Theorem 151

For algebraic numbers being added to $\mathbb{Q}$ the situation is different: they can be replaced only with their algebraic conjugates. The group $\text{Aut}(\overline{\mathbb{Q}})$ is a quotient of the group $\text{Aut}(\mathbb{C})$; the factorization consists in forgetting the action of an automorphism of $\mathbb{C}$ on transcendental numbers. Summing up the above discussion we may say that *a complex number is algebraic if and only if its orbit under the action of the group* $\text{Aut}(\mathbb{C})$ *is finite*.

We would like to extend the above principle to a larger context. In spite of the complicated nature of the group $\text{Aut}(\mathbb{C}) = \text{Gal}(\mathbb{C}\,|\,\mathbb{Q})$, the general principles of Galois theory remain valid for it too. For example, there still exists the *Galois correspondence* between subgroups of $\text{Gal}(\mathbb{C}\,|\,\mathbb{Q})$ and extensions of $\mathbb{Q}$: a subgroup fixes a subfield of $\mathbb{C}$, and the stabilizer of a subfield is a subgroup of $\text{Gal}(\mathbb{C}\,|\,\mathbb{Q})$. In fact, for this correspondence to be bijective one needs to introduce a special class of subgroups of $\text{Gal}(\mathbb{C}\,|\,\mathbb{Q})$ which are called *closed subgroups*; the same thing may also be said about Galois theory in $\overline{\mathbb{Q}}$. But don't worry: we only need to know that all the subgroups of finite index are closed, and they correspond to finite extensions of $\mathbb{Q}$. Therefore, if the group $\text{Aut}(\mathbb{C})$ acts on the set of certain "objects", and if the orbit of an "object" is finite, then the stabilizers of the elements of the orbit are subgroups of finite index, and we may assign to the orbit a finite extension of $\mathbb{Q}$.

Let us now turn to Belyi pairs. Consider a pair (X, f) where X is an algebraic curve and f is a Belyi function defined on X. The curve X can be represented by a system of polynomial equations in a projective space $\mathbb{C}P^d$; the function f is a rational function of the coordinates in the same projective space. Let us list all the coefficients of the equations defining X, and all the coefficients of f, in a set denoted by $\mathcal{P}(X, f)$ ($\mathcal{P}$ stands for "parameters"). The action of an automorphism $\alpha \in \text{Aut}(\mathbb{C})$ on (X, f) consists in acting by this automorphism simultaneously on all the elements of $\mathcal{P}(X, f)$, thus creating a new Belyi pair (Y, g) defined by new parameters. The fact that (Y, g) is indeed a Belyi pair is explained by the property of $\text{Aut}(\mathbb{C})$ to preserve all the algebraic relations between numbers. The property to have all the critical values in $\{0, 1, \infty\}$ can be written explicitly in algebraic terms. The same is true for the possibility to express the fact that various critical points have given multiplicities. Therefore, we may conclude that the passport of the Belyi pair is also an invariant of the action.

We may affirm even more: *the group* $\text{Aut}(\mathbb{C})$ *acts not only on individual Belyi pairs but also on their equivalence classes*. Indeed, if (X, f) and (X', f') are two isomorphic Belyi pairs, and $u : X \to X'$ is an isomorphism between them (so that $f' \circ u = f$), then the group $\text{Aut}(\mathbb{C})$ acts on the set of parameters $\mathcal{P}(X, f, X', f', u)$ preserving all algebraic relations between them, and thus creates a couple of Belyi pairs (Y, g) and (Y', g') together with an isomorphism $v : Y \to Y'$ between them.

But we know that for every equivalence class of Belyi pairs the corresponding orbit is finite, since an element of the orbit is described by a triple of permutations with a given passport. Thus we have proved the following

Proposition 2.6.6. *The field of moduli of any given dessin is a number field.*

We have made an important step, but our work is not finished. It remains to prove that for any equivalence class of Belyi pairs there exists a representative (X, f) such that the corresponding parameter set $\mathcal{P}(X, f)$ contains only algebraic numbers. In such a situation the number field K obtained as the extension of $\mathbb{Q}$ by the set $\mathcal{P}(X, f)$ is called a *field of definition* of (X, f). It can also be said that (X, f) *is defined over* K, or *is realized over* K.

2.6.3.2 Field of Definition

Proposition 2.6.7. *Every dessin can be realized over a number field.*

Let us fix a dessin M, and let (X, f) be a corresponding Belyi pair. If the parameter set $\mathcal{P}(X, f)$ contains only algebraic numbers, we have nothing to prove. Suppose that it contains several transcendental numbers; let us collect them in the set denoted by $\mathcal{T}(X, f)$. Now let us act on (X, f) by an element of the group $\text{Gal}(\mathbb{C} \,|\, \overline{\mathbb{Q}})$ preserving the algebraic numbers. We obtain another Belyi pair (X', f'), with the corresponding set $\mathcal{T}(X', f')$ of transcendental parameters. First of all note that (X', f') corresponds to the same dessin M, since the field of moduli of M is a subfield of $\overline{\mathbb{Q}}$, and this latter field remains fixed. Therefore, there exists an isomorphism $u : X \to X'$ such that $f' \circ u = f$.

The great liberty of action by automorphisms of $\mathbb{C}$ permits us to choose the parameters in $\mathcal{T}(X', f')$ which have no algebraic relations whatsoever with the elements of $\mathcal{T}(X, f)$. However, certain relations (over $\overline{\mathbb{Q}}$) inside the set $\mathcal{T}(X', f')$ may exist. According to Hilbert's basis theorem the set of such independent relations is finite; therefore we can write them all in an explicit way.

As for the isomorphism u, we cannot claim that its coefficients can be made independent of $\mathcal{T}(X, f)$. Obviously, the construction of u involves the parameters of both $\mathcal{T}(X, f)$ and $\mathcal{T}(X', f')$. However, what we can claim is the fact that the number of such isomorphisms is finite: it does not exceed the number of automorphisms of the dessin M. Thus, the elements of $\mathcal{T}(u)$ belong to a *finite extension* L of the field K generated by $\overline{\mathbb{Q}} \cup \mathcal{T}(X, f, X', f')$. The same thing may be said of the inverse isomorphism u^{-1}.

The field L being a finite extension of K, there exists a primitive element ξ generating L which is a root of a polynomial $D \in K[x]$. All the coefficients of u and u^{-1} may be expressed in terms of ξ. The verification of the fact that $u \circ u^{-1} = u^{-1} \circ u = \text{id}$ makes use only of the properties of the polynomial D and algebraic relations inside $\mathcal{T}(X, f, X', f')$.

Now we make a decisive step: we replace the *numbers* belonging to the set $\mathcal{T}(X', f')$ by *variables*. These variables represent a point of an algebraic variety $\mathcal{V}$ definied over $\overline{\mathbb{Q}}$. (If there were no relations between the elements of $\mathcal{T}(X', f')$, then this variety is simply an affine space.) We claim that *almost every point of $\mathcal{V}$ corresponds to a Belyi pair for the dessin M*. Indeed, for every such point we have all the ingredients we need: a curve, a Belyi function, and an isomorphism with the inital Belyi pair. We only must exclude certain algebraic subvarieties of $\mathcal{V}$. They correspond to two possibilities: (1) the polynomial D degenerates to a constant, so there is no solutions for ξ; (2) in the process of verification of the equality $u \circ u^{-1} = u^{-1} \circ u = \mathrm{id}$ a division by zero takes place.

In any case the remaining part of $\mathcal{V}$ is not empty, since at least one solution does exist (this fact is ensured by Riemann's existence theorem). Let us recall once more that $\mathcal{V}$ is an algebraic variety *definied over* $\overline{\mathbb{Q}}$. Such a variety, with some excluded subvarieties, necessarily contains an algebraic point. (A more accurate statement should be that the remaining set is open in Zariski topology; we will not dwell on it here.)

The Belyi theorem is proved.

Remark 2.6.8. The principle applied here is of general nature: if something is true for "randomly chosen transcendental parameters", then it is true for almost all parameters. Also, if an object is described by certain parameters some of which are transcendental, we must not consider this object individually. We must include it into a family of similar objects, replacing the transcendental parameters by variables. Then the description of the whole family will not contain transcendental entities. See one of the manifestations of this principle in Sec. 5.5, where *families* of coverings with four critical values will be described by dessins d'enfants.

Remark 2.6.9. Everywhere in the above proof we might replace the field $\overline{\mathbb{Q}}$ with the field of moduli k of the dessin M. Then in the last phrase we could say: a variety defined over k necessarily contains a point defined over a finite extension of k. More precise information about the degree of such an extension may be found in [300] and [177].

$$* \quad * \quad *$$

The appropriate context for the theory of dessins d'enfants is the inverse Galois theory and the phenomenon of rigidity. The reader may find a recent exposition of this theory in [208] and [288]. In our presentation we tried to emphasize the combinatorial side of the subject. Among the many omissions of our exposition the most notable are: the triangular groups aspect, and many beautiful but difficult computations of dessins of higher genera. But this chapter is already long enough as it is, so we stop here.

3
Introduction to the Matrix Integrals Method

This chapter is devoted to a description of the intriguing connection between map enumeration and matrix integrals. This connection was first established in [143] for the purposes of matrix models of quantum gravity. It was later reinvented by mathematicians [138] in the computation of the Euler characteristic of moduli spaces of complex curves. Our presentation follows [25] and we show, along the lines of [138], how it leads to the enumeration of one-face maps. We also relate, following [87], [299], the universal one-matrix model to the Korteweg–de Vries (KdV) hierarchy of partial differential equations. The results of this chapter will be used in the next one in the description of the Harer–Zagier computation of the Euler characteristic of moduli spaces of curves, as well as in the study of Witten's conjecture, which is now Kontsevich's theorem.

3.1 Model Problem: One-Face Maps

Consider a square. Gluing its edges pairwise we obtain a two-dimensional (oriented) surface. Recall that in order to obtain an oriented surface we glue edges with opposite directions (with respect to the orientation of the square). All three ways to glue the edges are presented in Fig. 3.1.

Here we join the edges to be glued together by an arc connecting them. Note that *we distinguish between a gluing and another one that can be obtained from the first one by a rotation of the square.*

The first and the third ways of gluing result in the sphere, see Fig. 3.1. The second way gives the torus.

Similarly, there exist fifteen ways of gluing the hexagon, see Fig. 3.2. The first five result in the sphere, while the last ten ways give the torus. This statement can be easily verified in the following way. Consider the boundary of the original polygon. After gluing, this boundary becomes an embedded graph on the resulting surface. We know the number of edges in this graph: for a polygon with 6 sides this number equals 3 since two sides of the polygon

156   3 Matrix Integrals

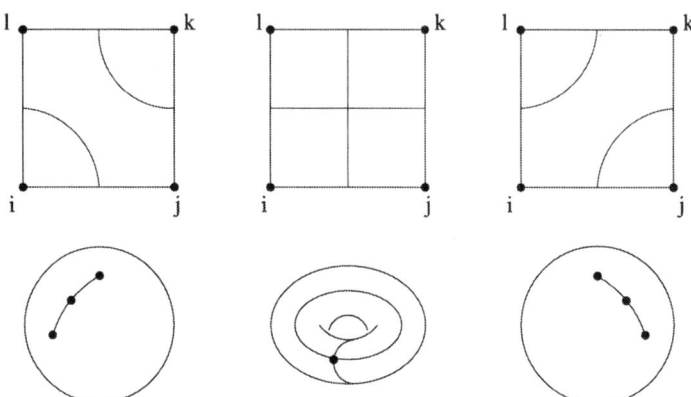

Fig. 3.1. Three possible gluings of the square

are glued together, producing an edge of the graph. The number of faces equals one. Therefore, the genus of the surface is uniquely determined by the number of vertices of the embedded graph.

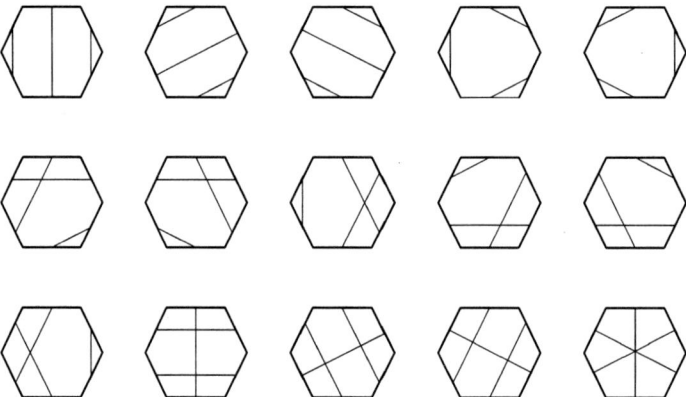

Fig. 3.2. All possible ways to glue the edges of the hexagon

This number can be easily computed looking at the picture of the gluing, see Fig. 3.3. Gluing the sides iq and lk we must impose the condition $i = k, q = l$. Similarly, the gluing of the sides ij and pq gives $j = p, i = q$, and the sides jk, lp produce the equalities $j = p, k = l$. Finally, we conclude that $i = k = l = q$ and $j = p$, and the resulting embedded graph has, therefore, two vertices. Hence, its Euler characteristic is $2 - 2g = V - E + F = 2 - 3 + 1 = 0$ and the surface is the torus.

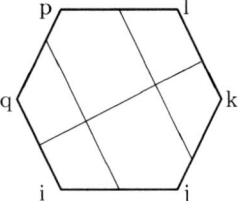

Fig. 3.3. A labelling of glued vertices

In Table 3.1 one can see the number of ways to obtain a surface of genus g from a regular $2n$-gon (we denote this number by $\varepsilon_g(n)$) for small values of n. The table is borrowed from [138].

The sum of all numbers in a row is $(2n-1)!! = 1 \cdot 3 \cdots (2n-1)$ since there exist exactly $(2n-1)!!$ ways to split the $2n$ edges of the polygon into disjoint pairs. Indeed, there are $2n-1$ ways to find a pair for a fixed edge of the polygon. For any other edge there are $2n-3$ ways to find a pairing edge that is not used yet, and so on.

Table 3.1. The number $\varepsilon_g(n)$ of genus g gluings of a $2n$-gon ([138])

$n\backslash g$	0	1	2	3	4	5
1	1	0	0	0	0	0
2	2	1	0	0	0	0
3	5	10	0	0	0	0
4	14	70	21	0	0	0
5	42	420	483	0	0	0
6	132	2310	6468	1485	0	0
7	429	12012	66066	56628	0	0
8	1430	60060	570570	1169740	225225	0
9	4862	291720	4390386	17454580	12317877	0
10	16796	1385670	31039008	211083730	351683046	59520825

Exercise 3.1.1. Prove the equivalence of the following statements:
- the resulting surface of a gluing is the sphere;
- the resulting graph is a tree;
- chords connecting pairs of gluing sides do not intersect.

The problem we consider now is

3 Matrix Integrals

Problem 3.1.2. Enumerate the ways to obtain a surface of genus g by gluing the $2n$-gon.

Note that in the dual setting, gluing a surface from a polygon with $2n$ sides is replaced with gluing a surface from a star with $2n$ darts. In this case the number of vertices is always 1, the number of edges equals n, and the number of faces depends on a gluing, thus determining the genus.

A reader familiar with enumerative combinatorics may note that the first column of Table 3.1 consists of the famous Catalan numbers

$$\mathrm{Cat}_n = \frac{1}{n+1}\binom{2n}{n}.$$

The zeroth Catalan number is usually set to $\mathrm{Cat}_0 = 1$. See Sec. 3.8 for a more detailed description of the Catalan numbers.

Exercise 3.1.3. Prove that the number of ways to glue the sphere from a polygon with $2n$ sides equals Cat_n. [**Hint:** Make use of Exercise 3.1.1.]

We will give the answer to Problem 3.1.2 in the following form. Introduce the sequence of polynomials

$$T_1(N) = N^2,$$
$$T_2(N) = 2N^3 + N,$$
$$T_3(N) = 5N^4 + 10N^2,$$
$$T_4(N) = 14N^5 + 70N^3 + 21N,$$
$$T_5(N) = 42N^6 + 420N^4 + 483N^2,$$

and so on. The polynomials T_n are the generating polynomials for the rows of the table:

$$T_n(N) = \sum_\sigma N^{V(\sigma)} = \sum_{g=0}^\infty \varepsilon_g(n) N^{n+1-2g},$$

where the first sum is taken over all possible ways σ of gluing and $V(\sigma)$ is the number of vertices in the resulting map. Generating functions are discussed in more detail in Sec. 3.8.

Exercise 3.1.4. Prove that the polynomial T_n is odd for n even, and is even for n odd.

Consider the (exponential) generating function for the sequence of polynomials $T_n(N)$:

$$T(N, s) = 1 + 2Ns + 2s \sum_{n=1}^\infty \frac{T_n(N)}{(2n-1)!!} s^n$$

$$= 1 + 2Ns + 2N^2 s^2 + \frac{2}{3}(2N^3 + N)s^3 + \frac{2}{15}(5N^4 + 10N^2)s^4 + \ldots$$

(3.1)

3.1 One-Face Maps

Theorem 3.1.5 ([138]). *The generating function $T(N,s)$ is equal to*

$$T(N,s) = \left(\frac{1+s}{1-s}\right)^N.$$

The theorem will be proved in Sec. 3.2, and now we are only going to verify that the first few coefficients of the expansion do coincide with those given by Eq. (3.1). We have

$$\left(\frac{1+s}{1-s}\right)^N = (1+s)^N(1-s)^{-N}$$

$$= \left(1 + Ns + \frac{N(N-1)}{2!}s^2 + \frac{N(N-1)(N-2)}{3!}s^3 + \ldots\right)$$

$$\cdot \left(1 + Ns + \frac{N(N+1)}{2!}s^2 + \frac{N(N+1)(N+2)}{3!}s^3 + \ldots\right)$$

$$= 1 + 2Ns + 2N^2s^2 + \frac{2}{3}(2N^3 + N)s^3 + \ldots \quad .$$

Theorem 3.1.5 implies a convenient recursion formula for computing the values of $\varepsilon_g(n)$.

Corollary 3.1.6. *The numbers $\varepsilon_g(n)$ satisfy the recurrence relation*

$$(n+2)\varepsilon_g(n+1) = (4n+2)\varepsilon_g(n) + (4n^3 - n)\varepsilon_{g-1}(n-1), \qquad (3.2)$$

and

$$\varepsilon_g(0) = \begin{cases} 1 & \text{if } g = 0, \\ 0 & \text{otherwise.} \end{cases}$$

The recursion follows from the obvious formula

$$\left(\frac{1+s}{1-s}\right)^N = (1+s)(1+s+s^2+\ldots)\left(\frac{1+s}{1-s}\right)^{N-1}.$$

Remark 3.1.7 (D. Zagier). No combinatorial interpretation of the recursion Eq. (3.2) is known. Such an interpretation would lead to a new, and a simple proof of Theorem 3.1.5. In fact, one can go further. Introduce new coefficients $C_g(n)$ by the formula $2^g \varepsilon_g(n) = \text{Cat}_n C_g(n)$. Then Eq. (3.2) becomes

$$C_g(n+1) = C_g(n) + \binom{n+1}{2}C_{g-1}(n-1).$$

This implies by induction that $C_g(n)$ is a positive integer and, also, for fixed g, a polynomial of degree $3g$ in n with roots at $n = -1, 0, 1, \ldots, 2g-1$, for $g > 0$. For example, $C_2(n) = (n+1)n(n-1)(n-2)(n-3)(5n-2)/360$. Here are further interesting problems concerning the coefficients $C_g(n)$.

1. Do the positive integers $C_g(n)$ have some combinatorial interpretation making the recursion for them evident?
2. Is there some combinatorial reason why $2^g \varepsilon_g(n)$ is divisible by Cat_n?
3. (Assuming one has answered 1.) Why the quotient is precisely $C_g(n)$?
4. Give a closed formula for the polynomials $C_g(\cdot)$.

Corollary 3.1.8. *The value $\varepsilon_g(n)$ is $\frac{(2n)!}{(n+1)!(n-2g)!}$ times the coefficient of s^{2g} in the generating function*

$$\left(\frac{s/2}{\tanh s/2}\right)^{n+1}.$$

There exist different ways to prove Theorem 3.1.5, see, e.g., [149], [301], [192], and also Sec. A.2.3. Below in Sec. 3.2 we present the main steps of the original proof from [138]. This proof is based on an expression of the polynomials T_n through matrix Gaussian integrals.

3.2 Gaussian Integrals

Important integrals coming from various applications often have a nontrivial combinatorial meaning. In this section we study the combinatorial interpretation of integrals over the Gaussian measure. Even simple integrals of this kind on the real line have a combinatorial interpretation, and integration over the Gaussian measure on the space of Hermitian matrices happens to be closely related to enumeration of embedded graphs with respect to their genera. Although we concentrate mainly on the space of Hermitian matrices, which corresponds to the orientable case, the reader must bear in mind that there exist other important applications of Gaussian integrals in map enumeration. For example, integration over the space of real symmetric matrices enumerates graphs embedded in both orientable and non-orientable surfaces, and we elaborate this line in exercises. The standard reference on probability theory, including Gaussian measures, is [99].

3.2.1 The Gaussian Measure on the Line

On the real line $\mathbb{R}$, the standard Gaussian measure is the measure μ with the density

$$d\mu(x) = \frac{1}{\sqrt{2\pi}} e^{-\frac{x^2}{2}} dx, \quad x \in \mathbb{R}.$$

You may see the graph of the density function in Fig. 3.4.

This measure has the following properties:

1. It is *normalized*, that is, the integral of the density over the whole line is equal to 1:

3.2 Gaussian Integrals

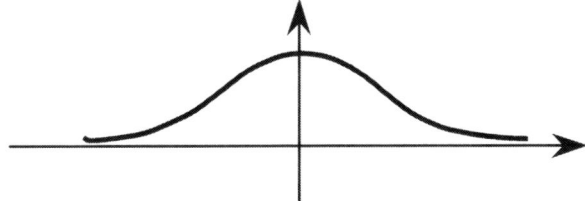

Fig. 3.4. The standard Gaussian density on the line

$$\int_{\mathbb{R}} d\mu(x) = \frac{1}{\sqrt{2\pi}} \int_{-\infty}^{\infty} e^{-\frac{x^2}{2}} dx = 1.$$

2. The *mean* is equal to 0:

$$\int_{\mathbb{R}} x\, d\mu(x) = 0.$$

3. The *variance* is equal to 1:

$$\int_{\mathbb{R}} x^2 d\mu(x) = 1.$$

4. The *characteristic function* (or Fourier transform) is

$$\varphi(t) = \int_{\mathbb{R}} e^{itx} d\mu(x) = e^{-\frac{t^2}{2}}.$$

5. The integral

$$\int_{\mathbb{R}} p(x) e^{-\frac{x^2}{2}} dx$$

converges for any polynomial p.

6. For any positive real number b, we have

$$\int_{\mathbb{R}} e^{-\frac{bx^2}{2}} dx = \frac{1}{\sqrt{b}} \int_{\mathbb{R}} e^{-\frac{x^2}{2}} dx.$$

Exercise 3.2.1. Prove these properties.

As an example, we present a proof of the first property.

$$\left(\int_{-\infty}^{\infty} e^{-\frac{1}{2}x^2} dx \right)^2 = \int_{-\infty}^{\infty} e^{-\frac{1}{2}x^2} dx \int_{-\infty}^{\infty} e^{-\frac{1}{2}y^2} dy$$

$$= \int_{-\infty}^{\infty}\int_{-\infty}^{\infty} e^{-\frac{1}{2}(x^2+y^2)} dx dy = \int_{0}^{\infty}\int_{0}^{2\pi} r e^{-\frac{1}{2}r^2} d\varphi dr$$

$$= 2\pi \int_{0}^{\infty} r e^{-\frac{1}{2}r^2} dr = 2\pi \int_{0}^{\infty} e^{-\frac{1}{2}r^2} d\left(\frac{1}{2}r^2\right) = 2\pi.$$

Notation 3.2.2. We will regularly use the following notation borrowed from physics: for any measure μ on X, and for any function $f : X \to \mathbb{R}$ or $f : X \to \mathbb{C}$ we will denote by $\langle f \rangle$ the *mean*, or the *average* value, of f with respect to the measure μ:

$$\langle f \rangle = \int_X f(x) d\mu(x).$$

The measure μ and the space X, not explicitly mentioned in this notation, will usually be clear from the context. For example, the above formulas may be rewritten as

$$\langle 1 \rangle = 1, \quad \langle x \rangle = 0, \quad \langle x^2 \rangle = 1, \quad \varphi(t) = \langle e^{itx} \rangle = e^{-\frac{t^2}{2}}.$$

Exercise 3.2.3. Prove that

$$\int_{-\infty}^{\infty} x^{2n} d\mu(x) = (2n-1)!!. \tag{3.3}$$

[**Hint:** Taking the integral by parts twice, prove the recurrence formula

$$\int_{-\infty}^{\infty} x^{2n} d\mu(x) = (2n-1) \int_{-\infty}^{\infty} x^{2n-2} d\mu(x).]$$

Note that $(2n-1)!!$ is precisely the number of ways to glue an oriented surface from the regular $2n$-gon. As we will see below, this coincidence is not accidental.

3.2.2 Gaussian Measures in $\mathbb{R}^k$

These measures will be one of the main objects of our study. Consider a point $x = (x_1, x_2, \ldots, x_k) \in \mathbb{R}^k$. By (x, y) we denote the ordinary scalar product in $\mathbb{R}^k$, that is, $(x, y) = x_1 y_1 + \ldots + x_k y_k$.

Here we work only with non-degenerate Gaussian measures; an approach to degenerate ones can be found in Sec. 3.8. For a positively defined symmetric $k \times k$-matrix B the measure is defined by the *density*

$$d\mu(x) = \text{Const} \times \exp\left\{-\frac{1}{2}(Bx, x)\right\} dv(x), \tag{3.4}$$

where $dv(x) = dx_1 \ldots dx_k$ is the standard volume form. For this measure to be normalized, i.e., the measure of the whole space $\mathbb{R}^k$ to be equal to 1, we must take the constant factor in the last formula equal to

$$\text{Const} = (2\pi)^{-k/2} (\det B)^{1/2}. \tag{3.5}$$

Indeed, using an orientation preserving orthogonal transformation $x = Oy$ in $\mathbb{R}^k$ we obtain

$$\int_{\mathbb{R}^k} e^{-(Bx,x)/2} dv(x) = \int_{\mathbb{R}^k} e^{-(BOy,Oy)/2} dv(Oy)$$
$$= \int_{\mathbb{R}^k} e^{-(O^{-1}BOy,y)/2} dv(y),$$

since an orthogonal transformation preserves both the scalar product and the volume form. Choosing O that diagonalizes the matrix B, we obtain a diagonal matrix $B_1 = O^{-1}BO$ with positive elements $b_1, \ldots, b_k$, the eigenvalues of B, on the diagonal. Now the integration can be done separately in each of the variables $y_1, \ldots, y_k$, and the result will be

$$\int_{\mathbb{R}^k} e^{-(Bx,x)/2} dv(x) = \prod_{i=1}^{k} \left(\frac{2\pi}{b_i}\right)^{1/2} = (2\pi)^{k/2} \left(\prod_{i=1}^{k} b_i\right)^{-1/2}$$
$$= (2\pi)^{k/2} (\det B)^{-1/2},$$

i.e., precisely what we expected.

In probability theory, the inverse matrix $C = (c_{ij}) = B^{-1}$ is called the *covariance matrix*, and we have $\langle x_i \rangle = 0$, $\langle x_i x_j \rangle = c_{ij}$. These properties follow from the obvious fact that they are satisfied in the case B and C are diagonal because of the properties of the Gaussian measure on the line. A Gaussian measure in $\mathbb{R}^k$ is called *standard* if both B and C are the identity matrices.

3.2.3 Integrals of Polynomials and the Wick Formula

A great part of the machinery of quantum physics consists of methods of computing integrals with respect to a Gaussian measure (not necessarily in the finite-dimensional case); see, for example, [260]. The goal of this section is to develop a technique for integrating polynomials.

Knowing that $\langle x_i \rangle = 0$, $i = 1, \ldots, n$, and $\langle x_i x_j \rangle = c_{ij}$, $i, j = 1, \ldots, n$, one can easily compute the integral of any polynomial in $x_1, \ldots, x_n$ of degree 2. What about higher degrees?

Lemma 3.2.4. *If $f(x)$ is a monomial of odd degree, then $\langle f \rangle = 0$.*

Indeed, a monomial of odd degree contains a variable in an odd degree. Changing the sign of this variable we conclude that the integral is equal to itself taken with the opposite sign, and hence it is zero.

The following theorem reduces the integration of any polynomial to that of polynomials of degree 2.

164   3 Matrix Integrals

Theorem 3.2.5 (Wick formula). *Let $f_1, f_2, \ldots, f_{2n}$ be a set of* (not necessarily distinct) *linear functions of $x_1, \ldots, x_k$. Then*

$$\langle f_1 f_2 \ldots f_{2n} \rangle = \sum \langle f_{p_1} f_{q_1} \rangle \langle f_{p_2} f_{q_2} \rangle \cdots \langle f_{p_n} f_{q_n} \rangle,$$

where the sum is taken over all permutations $p_1 q_1 p_2 \ldots q_n$ of the set of indices $1, 2, \ldots, 2n$ such that $p_1 < p_2 < \ldots < p_n$, $p_1 < q_1, \ldots, p_n < q_n$.

The number of summands on the right-hand side is equal to $(2n-1)!!$. A partition of the set $1, 2, \ldots, 2n$ into couples (p_i, q_i) satisfying the conditions of the Wick formula is called a *Wick coupling*.

Example 3.2.6. As an application of the Wick formula let us compute the one-dimensional integral

$$\frac{1}{\sqrt{2\pi}} \int_{-\infty}^{\infty} x^4 e^{-\frac{x^2}{2}} dx = \langle x^4 \rangle.$$

We have $x^4 = f_1 f_2 f_3 f_4$, where $f_1 = f_2 = f_3 = f_4 = x$. Hence,

$$\langle f_1 f_2 f_3 f_4 \rangle = \langle f_1 f_2 \rangle \langle f_3 f_4 \rangle + \langle f_1 f_3 \rangle \langle f_2 f_4 \rangle + \langle f_1 f_4 \rangle \langle f_2 f_3 \rangle.$$

Now each $\langle f_i f_j \rangle = \langle x^2 \rangle = 1$; therefore the result is $1^2 + 1^2 + 1^2 = 3$, which is consistent with our previous calculations.

In the same way we may find $\langle x^{2n} \rangle = (2n-1)!!$ (cf. Exercise 3.2.3).

Proof of the Wick formula. We outline here a proof of the Wick formula. Another proof given in Sec. 3.8 is based on a formula for the logarithm of a power series, and has various combinatorial applications.

After an orthogonal change of variables both the matrix B and the covariance matrix $C = B^{-1}$ can be made diagonal. Note now that both sides in the Wick formula are linear in each variable f_i. It is therefore sufficient to prove the formula for products of coordinate functions, i.e., for monomials $x_1^{n_1} \ldots x_k^{n_k}$. For a diagonal matrix C we have $\langle x_i^2 \rangle = c_{ii}$, while $\langle x_i x_j \rangle = 0$ for $i \neq j$. Now, the integral on the left-hand side of the Wick formula splits into the product of one-dimensional integrals separately in each variable, and the formula becomes obvious:

$$\langle x_1^{2n_1} \ldots x_k^{2n_k} \rangle = \langle x_1^{2n_1} \rangle \ldots \langle x_k^{2n_k} \rangle = \left(\sum \langle x_1^2 \rangle^{n_1} \right) \ldots \left(\sum \langle x_k^2 \rangle^{n_k} \right),$$

where the ith sum is taken over all $(2n_i - 1)!!$ possible couplings of $2n_i$ copies of the monomial x_i. All the other couplings are equal to zero.

3.2.4 A Gaussian Measure on the Space of Hermitian Matrices

Let $H = (h_{ij})$ denote an Hermitian $N \times N$ matrix, that is, a matrix whose entries are complex numbers: $h_{ij} \in \mathbb{C}$, and $h_{ij} = \overline{h}_{ji}$, where the bar denotes complex conjugation. We denote by $\mathcal{H}_N$ the space of all such matrices. This space is an N^2-dimensional real vector subspace of the real $2N^2$-dimensional space of complex $N \times N$ matrices.

Every Hermitian matrix can be described by N^2 real parameters:

$$x_{ii} = h_{ii} \in \mathbb{R}, \quad i = 1, \ldots, N,$$

for diagonal entries, and

$$x_{ij} = \Re(h_{ij}), \quad y_{ij} = \Im(h_{ij}), \quad 1 \leq i < j \leq N,$$

for over-diagonal entries (the under-diagonal entries being obtained from the over-diagonal ones by complex conjugation). Thus, the space $\mathcal{H}_N$ is isomorphic to the vector space $\mathbb{R}^{N^2}$. We introduce the ordinary volume form in $\mathcal{H}_N$:

$$dv(H) = \prod_{i=1}^{N} dx_{ii} \prod_{i<j} dx_{ij} dy_{ij}.$$

In principle, the space $\mathcal{H}_N \cong \mathbb{R}^{N^2}$ does not differ from any other vector space of the same dimension. But it is convenient to express some characteristics of the Gaussian measure on it in terms of matrix operations. In order to introduce a Gaussian density in the form (3.4), we need a nondegenerate quadratic form on $\mathcal{H}_N$. Let us take the following one:

$$\mathrm{tr}(H^2).$$

This quadratic form is simply the restriction of the ordinary scalar product on the space $\mathbb{R}^{2N^2}$ of complex $N \times N$ matrices to the subspace of Hermitian matrices. For two complex matrices X and Y this scalar product is $(X, Y) = \mathrm{tr}(X\overline{Y}^t)$.

Let us see what this form looks like in the coordinates x_{ij}, y_{ij}. The term $(\cdot)_{ik}$ of H^2 is equal to $\sum_{j=1}^{N} h_{ij} h_{jk}$; hence the diagonal term $(\cdot)_{ii}$ is equal to $\sum_{j=1}^{N} h_{ij} h_{ji}$, and the trace of H^2 is

$$\mathrm{tr}(H^2) = \sum_{i,j=1}^{N} h_{ij} h_{ji} = \sum_{i,j=1}^{N} h_{ij} \overline{h}_{ij}$$

$$= \sum_{i=1}^{N} x_{ii}^2 + \sum_{i \neq j} (x_{ij}^2 + y_{ij}^2) = \sum_{i=1}^{N} x_{ii}^2 + 2 \sum_{i<j} (x_{ij}^2 + y_{ij}^2).$$

Let us specify the matrix B of this quadratic form (see Eq. (3.4)). It must be of size $N^2 \times N^2$. But the above expression shows that this matrix is very simple; indeed, it is diagonal! We have N diagonal terms equal to 1 (they correspond to the coordinates x_{ii}), and $N^2 - N$ terms equal to 2 (they correspond to the coordinates x_{ij}, y_{ij} for $i < j$):

$$B = \begin{pmatrix} 1 & & & & & \\ & \ddots & & & & \\ & & 1 & & & \\ & & & 2 & & \\ & & & & 2 & \\ & & & & & \ddots \\ & & & & & & 2 \end{pmatrix} \quad (3.6)$$

Note that there is nothing unexpected in the appearance of "twos" in this matrix. Indeed, the subspace of the off-diagonal Hermitian matrices in the space of the off-diagonal complex matrices resembles the diagonal $x = y$ in the two-dimensional real plane $\mathbb{R}^2$. Restricting the ordinary scalar product in $\mathbb{R}^2$ to the diagonal and taking x as the coordinate on the diagonal we obtain $(e,e) = 2$ for the unit diagonal vector e, see Fig. 3.5.

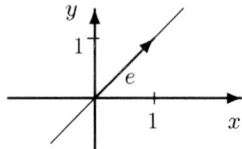

Fig. 3.5. The length of the unit diagonal vector

For the matrix B from Eq. (3.6) it is very easy to find all the necessary ingredients for subsequent work. For example,

$$\det B = 2^{N^2 - N},$$

so the density (3.4), (3.5) takes the form

$$d\mu(H) = (2\pi)^{-N^2/2} 2^{(N^2-N)/2} \exp\left\{-\frac{1}{2}\text{tr}(H^2)\right\} dv(H). \quad (3.7)$$

Exercise 3.2.7. Compute the covariance matrix $C = B^{-1}$. Infer from it that $\langle x_{ii}^2 \rangle = 1$, and $\langle x_{ij}^2 \rangle = \langle y_{ij}^2 \rangle = 1/2$ for $i < j$.

We will also need the following preparatory lemma:

Lemma 3.2.8. *Taking for linear functions of the coordinates x_{ij}, y_{ij} the matrix elements h_{ij}, we have*

$$\langle h_{ij} h_{ji} \rangle = 1,$$

while all the other second moments are equal to zero:

$$\langle h_{ij} h_{kl} \rangle = 0 \quad \text{whenever} \quad (i,j) \neq (l,k).$$

Proof. If $i = j$, then
$$\langle h_{ii}^2 \rangle = \langle x_{ii}^2 \rangle = 1.$$

If $i \neq j$, then
$$\langle h_{ij} h_{ji} \rangle = \langle h_{ij} \overline{h}_{ij} \rangle = \langle x_{ij}^2 + y_{ij}^2 \rangle = \frac{1}{2} + \frac{1}{2} = 1.$$

All the other second moments of the type $\langle h_{ij} h_{kl} \rangle$, $(i,j) \neq (l,k)$ involve only off-diagonal terms of the covariance matrix C, which are zero.

Exercise 3.2.9. Here we start a series of exercises which allow the reader to develop on his/her own the theory of integration over the space of symmetric matrices, which will be parallel to that of Hermitian matrices.

Let $\mathcal{S}_N$ denote the space of real symmetric $N \times N$ matrices, i.e., for a matrix $S = (s_{ij}) \in \mathcal{S}_N$, we have $s_{ij} = s_{ji}$ for all $i, j = 1, \ldots, N$.

1. Verify that $\mathcal{S}_N$ is a vector subspace in the space of all real $N \times N$ matrices. Find the dimension of this subspace.

2. Verify that the trace form $\text{tr}(S^2)$ determines a non-degenerate quadratic form on the space of symmetric matrices coinciding with the restriction of the ordinary scalar product $(X, Y) = \text{tr}(XY^t)$ in the space of all real matrices.

3. Find the matrix of the trace form $\text{tr}(S^2)$ and the scalar products $\langle s_{ij}^2 \rangle = \langle s_{ij} s_{ji} \rangle$.

4. Prove that $\langle s_{ij} s_{kl} \rangle \neq 0$ if and only if either $i = k$, $j = l$ or $i = l$, $j = k$.

In what follows we denote by
$$d\mu(S) = c_N \exp\{-\frac{1}{2} \text{tr}(S^2)\} dv(S)$$

the Gaussian measure on the space of symmetric matrices. Here
$$dv(S) = \prod_i ds_{ii} \prod_{i<j} ds_{ij}$$

and c_N is the normalizing constant chosen in such a way that
$$\int_{\mathcal{S}_N} d\mu(S) = 1.$$

5. Find the value of the constant c_N.

3.2.5 Matrix Integrals and Polygon Gluings

Let us compute the integral
$$\int_{\mathcal{H}_N} \text{tr}(H^4) d\mu(H).$$

168   3 Matrix Integrals

The integrand here is the sum of monomials

$$\operatorname{tr}(H^4) = \sum_{i,j,k,l=1}^{N} h_{ij}h_{jk}h_{kl}h_{li}.$$

The entries h_{pq} are linear functions in x_{pq}, y_{pq} and we can apply Wick's formula:

$$\begin{aligned}\langle h_{ij}h_{jk}h_{kl}h_{li}\rangle &= \langle h_{ij}h_{jk}\rangle\langle h_{kl}h_{li}\rangle \\ &+ \langle h_{ij}h_{kl}\rangle\langle h_{jk}h_{li}\rangle \\ &+ \langle h_{ij}h_{li}\rangle\langle h_{jk}h_{kl}\rangle.\end{aligned} \quad (3.8)$$

Consider the first term on the right-hand side of (3.8). The first factor $\langle h_{ij}h_{jk}\rangle$ differs from zero iff $i = k$. In this case it equals 1 (see Lemma 3.2.8). The same statement is valid for the second factor $\langle h_{kl}h_{li}\rangle$. Since indices i, j, k, l vary from 1 to N, the contribution of the first term summed over all indices equals N^3: the values of j, l and $i = k$ are independent.

The second term differs from zero if and only if $i = j = k = l$. Therefore, the contribution of this term is N. The third term is similar to the first one and its contribution is therefore N^3.

We may note that there is a one-to-one correspondence between the three summands on the right-hand side of Eq. (3.8) and the three ways of gluing the edges of the square shown in Fig. 3.1, and what is more, the contribution of a summand to the integral is equal to $N^{V(\sigma)}$, where $V(\sigma)$ is the number of vertices of the embedded graph in the corresponding gluing. As a result we obtain the equality

$$\int_{\mathcal{H}_N} \operatorname{tr}(H^4) d\mu(H) = 2N^3 + N = T_2(N).$$

Similar considerations for the $2m$-gon lead to the following statement.

Proposition 3.2.10.

$$\int_{\mathcal{H}_N} \operatorname{tr}(H^{2m}) d\mu(H) = T_m(N) \quad \text{for} \quad m = 1, 2, \ldots. \quad (3.9)$$

Note that substituting $N = 1$ in this formula we compute the value of the integral

$$\int_{\mathbb{R}} x^{2m} d\mu(x) = (2m-1)!!$$

from Sec. 3.1.

Thus we conclude that *integrating over the space $\mathcal{H}_N$ of Hermitian matrices allows one to separate the ways of gluing the regular $2m$-gon according to the genus of the resulting surface.*

The number of ways to obtain a surface of a given genus is a coefficient of the resulting polynomial in N. Since $T_m(N)$ is either an odd, or an even polynomial in N of degree $m+1$, in order to compute it, it suffices to know the values of the integral on the left-hand side of (3.9) at $\lfloor (m+1)/2 \rfloor$ distinct values of N.

In order to make Proposition 3.2.10 clearer, let us consider one more example.

Example 3.2.11. Let $m = 4$, so we deal with the sum of the N^8 products of the form
$$h_{i_1 i_2} h_{i_2 i_3} h_{i_3 i_4} h_{i_4 i_5} h_{i_5 i_6} h_{i_6 i_7} h_{i_7 i_8} h_{i_8 i_1}.$$
Let us choose an arbitrary Wick coupling (out of $7!! = 105$ ones). For example, we may couple $h_{i_1 i_2}$ with $h_{i_4 i_5}$; $h_{i_2 i_3}$ with $h_{i_5 i_6}$; $h_{i_3 i_4}$ with $h_{i_8 i_1}$; and $h_{i_6 i_7}$ with $h_{i_7 i_8}$; in other words, let us consider the product

$$\langle h_{i_1 i_2} h_{i_4 i_5} \rangle \langle h_{i_2 i_3} h_{i_5 i_6} \rangle \langle h_{i_3 i_4} h_{i_8 i_1} \rangle \langle h_{i_6 i_7} h_{i_7 i_8} \rangle. \qquad (3.10)$$

Recalling Lemma 3.2.8, we see that each factor of this product has quite good chances to be equal to 0 (and then the whole product will become 0). In order for the product to be non-zero, we need *all* of its factors to be equal to 1 (then the product itself is also equal to 1). And for that we must impose rather restrictive conditions on the indices (see Lemma 3.2.8 once more):

$$\begin{aligned}
\langle h_{i_1 i_2} h_{i_4 i_5} \rangle = 1 &\iff i_1 = i_5, \; i_2 = i_4, \\
\langle h_{i_2 i_3} h_{i_5 i_6} \rangle = 1 &\iff i_2 = i_6, \; i_3 = i_5, \\
\langle h_{i_3 i_4} h_{i_8 i_1} \rangle = 1 &\iff i_3 = i_1, \; i_4 = i_8, \\
\langle h_{i_6 i_7} h_{i_7 i_8} \rangle = 1 &\iff i_6 = i_8, \; i_7 = i_7.
\end{aligned}$$

Finally we have the following "chains of equalities":

$$\begin{aligned}
i_1 &= i_5 = i_3 = i_1, \\
i_2 &= i_4 = i_8 = i_6 = i_2, \\
i_7 &= i_7,
\end{aligned}$$

which gives us N^3 possible combinations of indices (the indices i_1, i_2, and i_7 may be chosen arbitrarily, while the values of all the other indices are determined by this choice). This fact is usually expressed by saying that the *contribution* of the coupling (3.10) is equal to N^3.

Consider now a polygon with 8 sides, and glue its sides in pairs as is shown in Fig. 3.6.

The fact that the side labelled $i_1 i_2$ is glued to the one labelled $i_4 i_5$ means that the polygon vertex i_1 is identified with i_5, as well as i_2 is identified with i_4. We express this by writing

$$i_1 = i_5, \quad i_2 = i_4.$$

170   3 Matrix Integrals

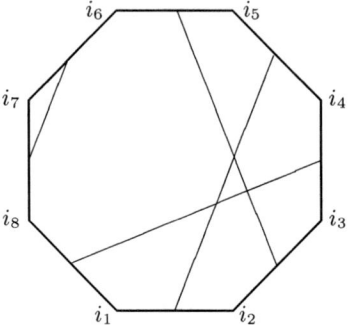

Fig. 3.6. A polygon whose sides are glued together in pairs

In the same vein, the side $i_2 i_3$ is glued to $i_5 i_6$, hence

$$i_2 = i_6, \quad i_3 = i_5,$$

and the two remaining identifications give us

$$i_3 = i_1, \quad i_4 = i_8, \quad i_6 = i_8, \quad i_7 = i_7.$$

Finally we see that the 8 original vertices of the polygon are identified in the following way (thus producing 3 vertices of the corresponding map):

$$\begin{aligned} i_1 &= i_5 = i_3 = i_1, \\ i_2 &= i_4 = i_8 = i_6 = i_2, \\ i_7 &= i_7. \end{aligned}$$

Hence, the contribution of the coupling (3.10) is $N^{V(\sigma)}$ for the gluing σ from Fig. 3.6.

Exercise 3.2.12 (Continuation of Exercise 3.2.9). 1. Compute the integrals

$$\int_{S_N} \operatorname{tr}(S^2) d\mu(S), \quad \int_{S_N} \operatorname{tr}(S^4) d\mu(S).$$

2. Draw all (both orientable and non-orientable) surfaces glued from the 2-gon and from the square.

3. Find the number of all (orientable and non-orientable) gluings of the $2n$-gon.

4. Prove that

$$\sum_\sigma N^{V(\sigma)} = 2^n \int_{S_N} \operatorname{tr}(S^{2n}) d\mu(S),$$

where the sum on the left-hand side is taken over all possible (orientable and non-orientable) gluings of the $2n$-star.

3.2.6 Computing Gaussian Integrals. Unitary Invariance

The main observation allowing one to compute the integral (3.9) is that the integrand $\operatorname{tr}(H^{2m})$ is invariant under the action of the unitary group U_N. In practice, functions coming from physical applications usually satisfy this assumption. The integral can then be simplified, which allows us (in some situations) to find its asymptotics or to compute it explicitly.

First let us define the action of the unitary group on the space of Hermitian matrices.

The *unitary group* U_N consists of *unitary matrices*, i.e., of complex $N \times N$ matrices U such that $U^t \overline{U} = I$. Here U^t is the transposed matrix, $\overline{U}$ is the complex conjugate matrix to U, and I is the identity $N \times N$ matrix.

Fact 3.2.13. The set U_N is a compact Lie group with respect to the matrix multiplication, of real dimension N^2.

The group U_N acts on the space $\mathcal{H}_N$ of Hermitian matrices of order N by the rule $U : H \mapsto U^{-1}HU$. This action is called the *adjoint action*. The set of eigenvalues of a Hermitian matrix H is invariant with respect to this action.

Exercise 3.2.14. Verify that this mapping determines an action of U_N on $\mathcal{H}_N$.

Each Hermitian matrix H is unitary equivalent to a diagonal matrix Λ, where the diagonal entries of Λ are the eigenvalues of H. The presentation of H as a pair (Λ, U) such that $H = U^{-1}\Lambda U$ is often referred to as the use of "polar coordinates" in the space $\mathcal{H}$ of Hermitian matrices: diagonal Hermitian matrices correspond to the radial coordinate, while unitary matrices correspond to the angle coordinate. In fact, the word "coordinates" is an abuse of language, since such a representation of a matrix H is not unique. There exists the following ambiguity:

- the matrix Λ is generally determined up to $N!$ permutations of the eigenvalues (each permutation causes the corresponding permutation of the columns of the matrix U). The word "generally" means: except for the set of matrices with at least two coinciding eigenvalues;
- for matrices H with distinct eigenvalues, the matrix U is determined up to multiplication by a diagonal unitary matrix with numbers $e^{i\theta_k}$, $\theta_k \in \mathbb{R}$ for $k = 1, \ldots, N$, on the diagonal. (The set of all such diagonal matrices is the N-dimensional torus subgroup $\mathrm{T}_N \subset \mathrm{U}_N$.)

As we will see below, the "Jacobian" of this change of variables also vanishes when some of eigenvalues of H coincide. Note, however, that the subset of Hermitian matrices with coinciding eigenvalues has measure zero and is therefore negligible from the integration point of view.

172   3 Matrix Integrals

For any matrix H without coinciding eigenvalues its preimage via the mapping $\mathbb{R}^N \times U_N \to \mathcal{H}_N$ given by the formula

$$(\Lambda, U) \mapsto H = U^{-1}\Lambda U$$

consists of $N!$ copies of the N-dimensional torus. Each of the copies has the form $T_N U_0$ for some unitary matrix $U_0 \in U_N$.

Let F be a unitary invariant function on $\mathcal{H}_N$, that is, $F(U^{-1}HU) = F(H)$ for any Hermitian matrix H and any unitary matrix U. The integration of F over $\mathcal{H}_N$ resembles the integration of "rotation-invariant" functions using polar coordinates: it is reduced to the integration of $F(\Lambda)$ over the space of the diagonal matrices Λ, which is nothing else but $\mathbb{R}^N$. Note also that if we denote by $\lambda_1, \ldots, \lambda_N$ the diagonal elements of Λ, the function $F(\Lambda)$ becomes a symmetric function of variables λ_i. In particular, if F is a polynomial, it can be expressed in terms of the power sums of λ_i, which are the traces $\operatorname{tr} \Lambda^k$. Thus, these traces, for $k = 1, \ldots, N$, generate the ring of U_N-invariant polynomials in H.

Proposition 3.2.15. *If F is a unitary invariant function on $\mathcal{H}_N$, then*

$$\int_{\mathcal{H}_N} F(H) d\mu(H) = c_N \int_{-\infty}^{\infty} \cdots \int_{-\infty}^{\infty} F(\Lambda) \prod_{1 \leq i < j \leq N} (\lambda_i - \lambda_j)^2 d\mu(\lambda_1) \ldots d\mu(\lambda_N)$$

(3.11)

for some constant c_N depending only on the dimension N of matrices. Here the matrix Λ is the diagonal $N \times N$-matrix with $\lambda_1, \ldots, \lambda_N$ on the diagonal, and

$$d\mu(\lambda) = \frac{1}{\sqrt{2\pi}} e^{-\lambda^2/2} d\lambda$$

is the standard Gaussian measure on the line.

In this form the proposition belongs to H. Weyl [295] (see also [296]); we are not entirely sure, however, that there were no predecessors. We present the computation of the constant c_N below in Sec. 3.5.

Let us remark that the λ-Jacobian

$$\prod_{1 \leq i < j \leq N} (\lambda_i - \lambda_j)^2 = \begin{vmatrix} 1 & 1 & \ldots & 1 \\ \lambda_1 & \lambda_2 & \ldots & \lambda_N \\ \lambda_1^2 & \lambda_2^2 & \ldots & \lambda_N^2 \\ \ldots & \ldots & \ddots & \ldots \\ \lambda_1^{N-1} & \lambda_2^{N-1} & \ldots & \lambda_N^{N-1} \end{vmatrix}^2$$

is the symmetric polynomial in λ of the smallest possible degree that vanishes whenever two eigenvalues coincide.

Proof of Proposition 3.2.15. We work in the real $2N^2$-dimensional space $\mathcal{Z}_N$ of $N \times N$ complex matrices $Z = (z_{ij})$, $z_{ij} = x_{ij} + iy_{ij}$. We introduce in this space the metric corresponding to the quadratic form

$$\operatorname{tr}(Z\overline{Z}^t) = \sum_{i,j=1}^{N} (x_{ij}^2 + y_{ij}^2)$$

and the corresponding volume form

$$\prod_{i,j=1}^{N} dx_{ij} dy_{ij}.$$

The restriction of the above quadratic form to the N^2-dimensional subspace $\mathcal{H}_N$ of Hermitian matrices coincides with our usual form

$$\operatorname{tr} H^2 = \sum_{i=1}^{N} x_{ii}^2 + 2 \sum_{i<j} (x_{ij}^2 + y_{ij}^2),$$

but if we want the volume form in $\mathcal{H}_N$ to be coherent with that in $\mathcal{Z}_N$, we must take not our usual form $dv(H)$ but the following one:

$$d\tilde{v}(H) = \prod_{i=1}^{N} dx_{ii} \prod_{i<j} (\sqrt{2} dx_{ij})(\sqrt{2} dy_{ij}) = \prod_{i=1}^{N} dx_{ii} \prod_{i<j} (2 dx_{ij} dy_{ij})$$

(cf. Fig. 3.5). The only change in the Gaussian integrals over $\mathcal{H}_N$ is that the factor $2^{(N^2-N)/2}$ disappears, being included into the new volume form.

Recall that we use matrix notation as a convenient tool for presenting a large number of independent coordinates. For a matrix $A = (a_{ij})$, denote by dA the matrix $dA = (da_{ij})$.

Exercise 3.2.16. 1. Verify the Leibniz identity for this differential: for any two $N \times N$ matrices A and B

$$d(AB) = dA \cdot B + A \cdot dB.$$

2. Verify that the matrix analog to the usual formula $d(1/x) = -(1/x^2)dx$ is the formula

$$d(A^{-1}) = -A^{-1} \cdot dA \cdot A^{-1}.$$

Returning to the proof, let us differentiate the equation $H = U^{-1} \Lambda U$. We obtain

$$\begin{aligned} dH &= d(U^{-1}) \cdot \Lambda U + U^{-1} \cdot d\Lambda \cdot U + U^{-1} \Lambda \cdot dU \\ &= U^{-1} \cdot U \cdot d(U^{-1}) \cdot \Lambda U + U^{-1} \cdot d\Lambda \cdot U + U^{-1} \Lambda \cdot dU \cdot U^{-1} \cdot U \\ &= U^{-1}(-\Omega \Lambda + d\Lambda + \Lambda \Omega)U \\ &= U^{-1} L U; \end{aligned}$$

here
$$\Omega = dU \cdot U^{-1}, \quad L = -\Omega\Lambda + d\Lambda + \Lambda\Omega,$$
and we use the relation $U \cdot d(U^{-1}) = -\Omega$ which follows from
$$0 = d(UU^{-1}) = d(U) \cdot U^{-1} + U \cdot d(U^{-1}).$$
Denote the entries of the matrix Ω by ω_{ij}, i.e., $\Omega = (\omega_{ij})$.

Exercise 3.2.17. Verify that the matrix Ω is skew symmetric: $\omega_{ij} = -\overline{\omega}_{ji}$. (It is also said that the tangent space to the Lie group $\mathbf{U}_N$ at the identity is the Lie algebra of skew symmetric matrices.)

The matrix $L = d\Lambda + \Lambda\Omega - \Omega\Lambda$ has diagonal entries $l_{ii} = d\lambda_i$ and off-diagonal entries $l_{ij} = (\lambda_i - \lambda_j)\omega_{ij}$. Therefore, denoting $\omega_{ij} = dx_{ij} + idy_{ij}$ we obtain

$$\operatorname{tr} L^2 = \sum_{i,j=1}^{N} l_{ij}l_{ji} = \sum_{i=1}^{N} d\lambda_{ii}^2 + \sum_{i\neq j}(\lambda_i - \lambda_j)(\lambda_j - \lambda_i)\omega_{ij}\omega_{ji}$$
$$= \sum_{i=1}^{N} d\lambda_{ii}^2 + 2\sum_{i<j}(\lambda_i - \lambda_j)^2(dx_{ij}^2 + dy_{ij}^2).$$

Finally, the corresponding volume form is

$$\prod_{1\leq i<j\leq N}(\lambda_i - \lambda_j)^2 \prod_{i=1}^{N} d\lambda_i \times \prod_{i<j}(2dx_{ij}dy_{ij}). \tag{3.12}$$

In order to finish the proof we must show that the adjoint action of a unitary matrix U on a matrix Z does not change the volume form for the matrix Z. Indeed, suppose the converse. Then

$$dv(U^{-1}ZU) = c_U dv(Z),$$

where c_U is a function depending on the coordinates U, Λ and taking real positive values. Applying this action k times we obtain

$$dv(U^{-k}ZU^k) = c_U^k dv(Z).$$

The compactness of the group $\mathbf{U}_N$ implies that U^k can be made arbitrarily close to the unit matrix by an appropriate choice of k. Therefore, the constant c_U^k takes values arbitrarily close to 1, and for a positive c_U this is possible only if $c_U = 1$.

Proposition 3.2.15 is proved.

Exercise 3.2.18. Here we propose one more way of proving that the action of U preserves the volume.

1. Let A, B, Z be three matrices; then the linear mapping $Z \mapsto AZB$ is represented (in an appropriate basis) by the matrix $A \otimes B$.
2. If A and B are real $m \times m$ and $n \times n$ matrices respectively, then

$$\det A \otimes B = (\det A)^n \cdot (\det B)^m.$$

3. If A is a complex $m \times m$ matrix, and $\mathbf{A}$ is the corresponding real $2m \times 2m$ matrix (every complex number $r \cdot e^{i\varphi}$ is replaced by a real 2×2 matrix of rotation through the angle φ, multiplied by r), then

$$\det \mathbf{A} = |\det A|^2.$$

Thus we may conclude that the determinant of the adjoint action is

$$|\det U \cdot \det U^{-1}|^{2N} = 1.$$

The integration of rotationally invariant functions in polar coordinates leads to a factor which involves the volume of the sphere: this factor represents the separate integration over the angular coordinates. In the same way, the integration of unitary invariant functions leads to a factor involving the volume of the unitary group. Indeed, Eq. (3.12) shows that the lifted volume form in the space of Hermitian matrices splits into a product of two volume forms: one in variables λ, and another one in coordinates along an orbit of the unitary group action.

Lemma 3.2.19. *The constant c_N in (3.11) is equal to*

$$c_N = \frac{\mathrm{vol}(\mathrm{U}_N)}{(2\pi)^{(N^2+N)/2} \cdot N!}.$$

We would rather expect $(2\pi)^N \cdot N!$ in the denominator of the above formula, which would reflect the ambiguity of the representation $H = U^{-1}\Lambda U$ discussed previously. But before looking closer at the coefficient we must explain what do we mean by the volume of the unitary group. This group is considered as a real N^2-dimensional surface in the space $\mathcal{Z}_N$. The metric in this space determines a volume form on U_N.

The idea now consists in integrating the function $F(H) = 1$. We even know the answer in advance: it is equal to 1. But we will use this fact later. Recall that the integration with $d\tilde{v}(H)$ makes the factor $2^{(N^2-N)/2}$ disappear. Thus we may write

$$\frac{1}{(2\pi)^{N^2/2}} \int_{\mathcal{H}_N} e^{-\frac{1}{2}\mathrm{tr}\, H^2} d\tilde{v}(H) =$$

$$\frac{1}{(2\pi)^{N^2/2}} \cdot \frac{\mathrm{vol}(\mathrm{U}_N)}{(2\pi)^N \cdot N!} \int_{\mathbb{R}^N} \prod_{i<j}(\lambda_i - \lambda_j)^2 e^{-\frac{1}{2}\sum_{k=1}^N \lambda_k^2} \prod_{k=1}^N d\lambda_k.$$

176   3 Matrix Integrals

It remains to divide the integral on the right-hand side of this equality by $(2\pi)^{N/2}$ (and to multiply the constant in front of it by the same factor) in order to transform the integral into

$$\int_{\mathbb{R}^N} \prod_{i<j}(\lambda_i - \lambda_j)^2 \prod_{k=1}^{N} d\mu(\lambda_k),$$

where μ is the standard Gaussian measure on the line. The lemma is proved.

Note that in order to compute the volume of U_N it remains to compute the latter integral. This will be done in Sec. 3.5.1.

Exercise 3.2.20 (Continuation of Exercise 3.2.12.). 1. Verify that SO_N, the group of orientation preserving orthogonal $N \times N$ matrices, acts on the space of $N \times N$ symmetric matrices by $S \mapsto O^{-1}SO$ for $O \in SO_N$, $S \in \mathcal{S}_N$.

2. Prove that the mapping $(\Lambda, O) \mapsto O^{-1}\Lambda O$, where $O \in SO_N$ and Λ is a diagonal $N \times N$ matrix, is a covering over the set of symmetric matrices with pairwise distinct eigenvalues. Find the Jacobian of this mapping for $N = 2$.

3. Prove that for a function F on the space of symmetric matrices, invariant with respect to the action of the orthogonal group,

$$\int_{\mathcal{S}_N} F(S)d\mu(S) = \tilde{c}_N \int_{\mathbb{R}^N} F(\Lambda) \prod_{i<j}(\lambda_i - \lambda_j) d\mu(\lambda_1)\ldots d\mu(\lambda_N),$$

where

$$d\mu(\lambda) = \frac{1}{\sqrt{2\pi}} e^{-\lambda^2/2} d\lambda$$

and $\tilde{c}_N$ is a constant depending only on N. Express $\tilde{c}_N$ in terms of the volume of SO_N.

3.2.7 Computation of the Integral for One Face Gluings

This section is devoted to the proof of Theorem 3.1.5.

The computation of the integral (3.9) splits into two stages. First of all we prove that for a fixed value of N the function

$$t(N, n) = \frac{T_n(N)}{(2n-1)!!}$$

is a polynomial in n of degree $N - 1$. In particular, as we have seen before, $t(1, n) \equiv 1$. Strangely enough, this proof is the only point where the matrix integral representation of the generating function is essential.

The second step is a purely combinatorial computation of the polynomial $t(N, n)$.

Lemma 3.2.21. *The function*

$$t(N,n) = \frac{T_n(N)}{(2n-1)!!}$$

is polynomial in n.

Proof. Equation (3.11) yields

$$\int_{\mathcal{H}_N} \operatorname{tr} H^{2n} d\mu(H) = c_N \int_{\mathbb{R}^N} \operatorname{tr} \Lambda^{2n} \prod_{i<j}(\lambda_i - \lambda_j)^2 \prod_{i=1}^N d\mu(\lambda_i).$$

The term Λ^{2n} in the integrand

$$\operatorname{tr} \Lambda^{2n} = \lambda_1^{2n} + \cdots + \lambda_N^{2n}$$

may be replaced with $N\lambda_1^{2n}$ since the values of the integral for all the summands are the same because of the symmetry. Expanding the square of the Vandermonde determinant as a polynomial in λ_1 and integrating over all the other variables $\lambda_2, \ldots, \lambda_N$, we obtain an integral over one variable λ_1. The integrand is a polynomial of degree $2n + 2N - 2$. The coefficients of the polynomial are constant for fixed N. The leading coefficient is Nc_N.

Integrating the monomial λ_1^{2n+2k} and dividing the result by $(2n-1)!!$ we obtain

$$\frac{(2n+2k-1)!!}{(2n-1)!!},$$

which is a polynomial of degree k in n. The lemma is proved.

Now suppose that the vertices of our polygon are colored in several colors. We say that a gluing *agrees with the coloring* if only vertices of the same color are glued together.

Lemma 3.2.22. *The number $T_n(N)$ is precisely the number of gluings of the $2n$-gon that agree with possible colorings of the vertices in (not more than) N colors.*

Indeed, let the boundary of the polygon after a gluing become an embedded graph with V vertices. Color each of the vertices in one of N colors. Any such coloring determines a coloring of the vertices of the polygon itself, and the gluing agrees with this coloring. And there are N^V ways to color the vertices of the graph in N colors. (Note that we do not require that neighboring vertices have different colors.)

Now introduce the function $\widetilde{T}_n(N)$, the number of gluings of the $2n$-gon that agree with colorings of its vertices in precisely N colors. Then

$$T_n(N) = \sum_{L=1}^N \binom{N}{L} \widetilde{T}_n(L).$$

Indeed, there are $\binom{N}{L}$ different ways to choose L colors from N given colors. Obviously, $\tilde{T}_0(N) = \tilde{T}_1(N) = \cdots = \tilde{T}_{N-2}(N) = 0$. Indeed, the resulting graph on the surface has at most $n + 1$ vertices (this happens only if it is a tree). Therefore, there is no coloring of the vertices of a $2n$-gon in more than $n + 1$ different colors that may agree with some gluing.

The function
$$\tilde{t}(N, n) = \frac{\tilde{T}_n(N)}{(2n-1)!!}$$
is a polynomial in n of degree $N - 1$. We know $N - 1$ roots $0, 1, 2, \ldots, N - 2$ of this polynomial. Therefore, there exists a constant A_N such that
$$\tilde{t}(n, N) = A_N n(n-1)(n-2)\ldots(n-N+2) = A_N \binom{n}{N-1}(N-1)!.$$

Substituting this formula in the expression for $T_n(N)$ yields
$$T_n(N) = (2n-1)!! \sum_{L=1}^{N} A_L \binom{n}{L-1}\binom{N}{L}(L-1)!.$$

In particular, the leading coefficient in N of the polynomial $T_n(N)$, i.e., the coefficient of N^{n+1}, equals
$$(2n-1)!! \cdot \frac{A_{n+1}}{(n+1)!} \cdot n!.$$

On the other hand, we know that this coefficient is the n-th Catalan number:
$$(2n-1)!! \cdot \frac{A_{n+1}}{(n+1)!} \cdot n! = \mathrm{Cat}_n = \frac{(2n)!}{n!(n+1)!}$$
(see Sec. 3.8). Then it follows that $A_{n+1} = 2^n/n!$. Thus,
$$T_n(N) = (2n-1)!! \sum_{L=1}^{N} 2^{L-1}\binom{N}{L}\binom{n}{L-1}.$$

Note that $\binom{n}{L-1}$ vanishes for $n < L - 1$, so the number of non-zero summands is $\min(N, n+1)$.

This is precisely the formula for the expansion of the function $\left(\frac{1+s}{1-s}\right)^N$ in powers of s. Indeed,
$$1 + 2Ns + 2s\sum_{n=1}^{\infty} \frac{T_n(N)}{(2n-1)!!} s^n = 1 + \sum_{L=1}^{N} 2^L \binom{N}{L} \sum_{n=L-1}^{\infty} \binom{n}{L-1} s^{n+1}$$
$$= \sum_{L=0}^{N} \binom{N}{L}\left(\frac{2s}{1-s}\right)^L = \left(\frac{1+s}{1-s}\right)^N,$$

and the Harer-Zagier theorem 3.1.5 is proved.

The history of this problem is very interesting. The paper [138] by Harer and Zagier was very influential and attracted much attention. First of all, as it soon became clear, the method of matrix integrals proposed in [138] was a reinvention of methods already used by physicists. Then, Jackson [149] found another proof based on the representation theory of the symmetric groups. This proof was rather complicated; a nice simplification and stronger results were presented by Zagier in [301]; see also Sec. A.2.3. A purely combinatorial proof was found in recent paper by Lass [192]. Much stronger results were proved by Goupil and Schaeffer in [119]: they enumerate one-face constellations when not only their genus but also their passport is given; see also Theorem A.2.9. For Harer and Zagier, the enumerative problem was only a step in the computation of the Euler characteristic of the moduli space of curves; the same result was achieved by Kontsevich ([178], Appendix D) by computing a one-dimensional Gaussian integral (see also Sec. 4.5). What then remains of the initial problem? For us the answer is, there remains the same quality as for all classical problems that have come to us since antiquity: its irresistible beauty. We are sure that this problem will continue to fascinate future researchers and will serve as a source of many new ideas.

3.3 Matrix Integrals for Multi-Faced Maps

In this section we discuss gluings of surfaces from arbitrary finite sets of polygons with arbitrary numbers of edges. However, it will be more convenient for us to work in dual terms, that is, we shall speak about gluing stars, not polygons. This language is more traditional for physicists.

3.3.1 Feynman Diagrams

Already in Example 3.2.11 it is not that easy to follow all indices in the integrand. Had the integrand been more complicated, both expressions on the left- and on the right-hand sides would have become a heap of sums, products, letters and indices. The Feynman diagram techniques has been invented specially for avoiding these cumbersome presentations. It is a method for expressing the integral by a compact geometric image.

We start with a simple example describing Feynman diagrams for the single 4-star. The three ways to glue the ends of the four darts into pairs are shown in Fig. 3.7. These three ways are in one-to one correspondence with the three terms on the right-hand side of the equation

$$\langle \operatorname{tr} H^4 \rangle = \sum_{i,j,k,l} \left(\langle h_{ij}h_{jk}\rangle\langle h_{kl}h_{li}\rangle + \langle h_{ij}h_{kl}\rangle\langle h_{jk}h_{li}\rangle + \langle h_{ij}h_{li}\rangle\langle h_{jk}h_{kl}\rangle \right).$$

The entries of the monomial $h_{ij}h_{jk}h_{kl}h_{li}$ may be considered as cyclically ordered and the right-hand side presents all possible couplings of these entries.

180   3 Matrix Integrals

In order to make the correspondence clearer 't Hooft [143] suggested doubling the darts of the star and marking each dart with the corresponding pair of indices. Note that each (broken) segment is then marked twice with the same index. A gluing of the darts then provides a gluing of pairs of segments, Fig. 3.7.

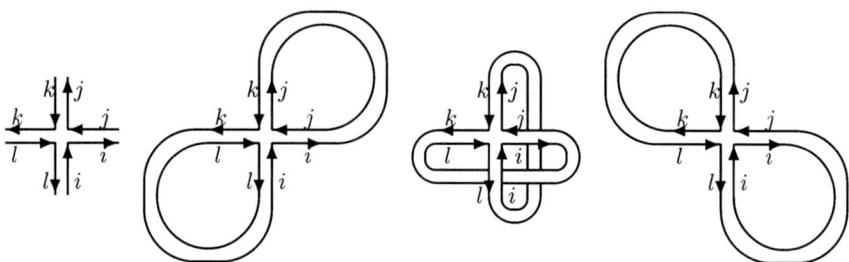

Fig. 3.7. Feynman diagrams for the 4-star

On the other hand, gluing darts of the 4-star is dual to gluing a surface from the square. The last equation may be then written down as

$$\int_{\mathcal{H}_N} \mathrm{tr}(H^4) d\mu(H) = \sum_\sigma N^{F(\sigma)},$$

where the sum is taken over all gluings σ of the 4-star and $F(\sigma)$ is the number of faces of the gluing σ.

3.3.2 The Matrix Integral for an Arbitrary Gluing

Suppose we investigate gluings of a set of stars. Then an appropriate matrix integral allows one to enumerate embedded graphs thus obtained according to their genera. It may happen, however, that the resulting graph is disconnected, so the notion of "genus" has to be made more precise.

Let us consider a set of α_1 copies of a 1-star, α_2 copies of a 2-star,..., α_k copies of a k-star ($\alpha_1, \ldots, \alpha_k$ are nonnegative integers). Consider a coupling σ of the ends of the darts. Such a coupling determines a set of embedded graphs $\Gamma_1, \ldots, \Gamma_m$: each graph in the set is a connected component of the glued graph. Define the number of faces $F(\sigma)$ as the sum of the numbers of faces for all connected components,

$$F(\sigma) = \sum_{i=1}^{m} F(\Gamma_i).$$

We are now in a position to state the enumerative formula.

3.3 Multi-Faced Maps

Proposition 3.3.1. *We have*

$$\sum_\sigma N^{F(\sigma)} = \int_{\mathcal{H}_N} (\operatorname{tr} H)^{\alpha_1} (\operatorname{tr} H^2)^{\alpha_2} \ldots (\operatorname{tr} H^k)^{\alpha_k} d\mu(H)$$

where the sum is taken over all possible couplings of the darts.

The proof of this proposition is practically the same as the proof of Proposition 3.2.10. We have

$$(\operatorname{tr} H)^{\alpha_1} = \sum_{i_1,\ldots,i_{\alpha_1}} h_{i_1 i_1} \ldots h_{i_{\alpha_1} i_{\alpha_1}},$$

$$(\operatorname{tr} H^2)^{\alpha_2} = \sum_{\substack{j_1,\ldots,j_{\alpha_2} \\ k_1,\ldots,k_{\alpha_2}}} h_{j_1 k_1} h_{k_1 j_1} \ldots h_{j_{\alpha_2} k_{\alpha_2}} h_{k_{\alpha_2} j_{\alpha_2}},$$

and so on. Hence, the integrand on the right-hand side may be presented as the sum of monomials of the following type:

$$h_{i_1 i_1} \ldots h_{i_{\alpha_1} i_{\alpha_1}} h_{j_1 k_1} h_{k_1 j_1} \ldots h_{j_{\alpha_2} k_{\alpha_2}} h_{k_{\alpha_2} j_{\alpha_2}} \ldots$$

(see Fig. 3.8).

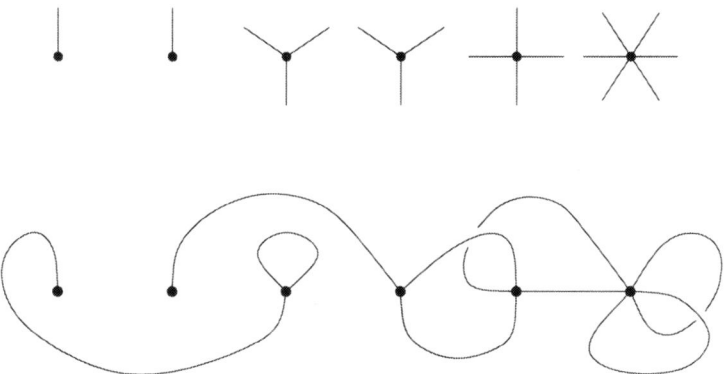

Fig. 3.8. A set of stars and a sample gluing

Further computations proceed as follows.
1. Label each dart of a star with a pair of indices.
2. Express the integral of each monomial through the Wick formula. Each monomial with non-zero contribution determines a Wick coupling σ of the darts.
3. Compute the number of possible values of indices producing a non-zero contribution to the value of the integral. This number equals precisely $N^{F(\sigma)}$.

4. Sum over all possible couplings.
Proposition 3.3.1 is proved.

Sometimes it is more convenient to express the left-hand side of the formula in Proposition 3.3.1 not as a sum over gluings, but as a sum over embedded graphs. In this case the contribution of each graph is inverse to the order of its automorphism group, and we can simply rewrite the statement in the following form.

Proposition 3.3.2. *We have*

$$\alpha_1!\ldots\alpha_k!1^{\alpha_1}\ldots k^{\alpha_k}\sum_\Gamma \frac{N^{|F(\Gamma)|}}{|\mathrm{Aut}(\Gamma)|} = \int_{\mathcal{H}_N}(\mathrm{tr}\,H)^{\alpha_1}(\mathrm{tr}\,H^2)^{\alpha_2}\ldots(\mathrm{tr}\,H^k)^{\alpha_k}d\mu(H),$$

where the sum is taken over all embedded graphs Γ with α_1 vertices of valency 1, α_2 vertices of valency 2, and so on, and $|\mathrm{Aut}(\Gamma)|$ is the number of automorphisms of the embedded graph.

Note that the factor $\alpha_1!\alpha_2!\ldots\alpha_k!1^{\alpha_1}2^{\alpha_2}\ldots k^{\alpha_k}$ is the number of automorphisms of the set of stars.

Example 3.3.3. There exist two different embedded graphs consisting of one vertex and two loops. The symmetry group of the spherical one is the cyclic group of order 2, while the toric graph has the cyclic group of order 4 as the symmetry group. Hence, in this case the formula from the proposition has the form

$$4\left(\frac{N^3}{2}+\frac{N}{4}\right) = \int_{\mathcal{H}_N}\mathrm{tr}\,H^4 d\mu(H)$$

in agreement with our previous calculations.

The trivial case $N=1$ of the last proposition is of special interest, and we formulate it as a separate statement.

Proposition 3.3.4. *We have*

$$\sum_\Gamma \frac{1}{|\mathrm{Aut}(\Gamma)|} = \frac{1}{\alpha_1!\alpha_2!\ldots\alpha_k!1^{\alpha_1}2^{\alpha_2}\ldots k^{\alpha_k}}\int_{-\infty}^{\infty}x^{1\cdot\alpha_1+2\cdot\alpha_2+\cdots+k\cdot\alpha_k}d\mu(x)$$

$$= \frac{1}{\alpha_1!\alpha_2!\ldots\alpha_k!1^{\alpha_1}2^{\alpha_2}\ldots k^{\alpha_k}}\int_{-\infty}^{\infty}x^{2E}d\mu(x)$$

$$= \frac{(2E-1)!!}{\alpha_1!\alpha_2!\ldots\alpha_k!1^{\alpha_1}2^{\alpha_2}\ldots k^{\alpha_k}},$$

where the sum is taken over all embedded graphs Γ with α_1 vertices of valency 1, α_2 vertices of valency 2, and so on, $|\mathrm{Aut}(\Gamma)|$ is the number of automorphisms of the embedded graph, and E is the (common) number of edges in Γ.

Note that if $1 \cdot \alpha_1 + 2 \cdot \alpha_2 + \cdots + k \cdot \alpha_k$ is odd, then the integral is zero, reflecting the fact that the odd number of darts cannot be split in pairs.

Exercise 3.3.5 (Continuation of Exercise 3.2.20). Prove that

$$\sum_\sigma N^{F(\sigma)} = 2^{\frac{1}{2}(1\cdot\alpha_1+2\cdot\alpha_2+\cdots+k\cdot\alpha_k)} \int_{\mathcal{S}_N} (\operatorname{tr} S)^{\alpha_1} (\operatorname{tr} S^2)^{\alpha_2} \ldots (\operatorname{tr} S^k)^{\alpha_k} d\mu(S),$$

where the sum is taken over all (both orientable and nonorientable) gluings of the darts.

3.3.3 Getting Rid of Disconnected Graphs

The formula in Proposition 3.3.1 enumerates both connected and disconnected graphs. This is not always convenient since we are, as a rule, interested in enumerating "elementary objects", connected graphs. Unfortunately there is no regular way to get rid of disconnected graphs in an isolated formula like this. However, disconnected graphs can be excluded if we consider the exponential generating function for a family of finite sets of glued stars, rather than for a single set of stars.

As an example consider the problem of gluing n copies of a 4-star for $n = 0, 1, 2, \ldots$, see Fig. 3.9.

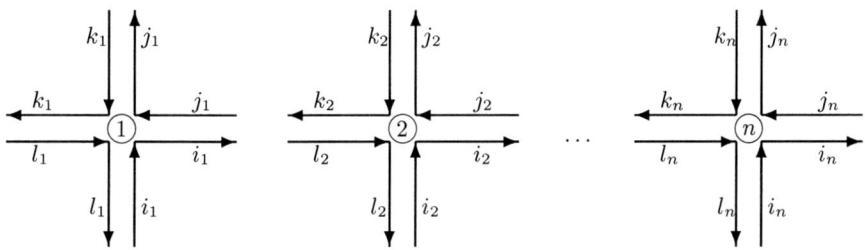

Fig. 3.9. An image for $(\operatorname{tr}(H^4))^n$

We know that the function

$$F_n = \int_{\mathcal{H}_N} (\operatorname{tr} H^4)^n d\mu(H)$$

can be represented as a polynomial in N:

$$F_n(N) = \sum_\sigma N^{F(\sigma)}.$$

As we have seen in Sec. 3.1, $F_1(N) = 2N^3 + N$.

Exercise 3.3.6. 1. Verify that $F_2(N) = 61N^2 + 40N^4 + 4N^6$.

2. Prove that the generating polynomial for connected gluings of two 4-stars is $F_2(N) - F_1^2(N) = 60N^2 + 36N^4$.

Now denote by $C_n(N)$, $n = 1, 2, \ldots$ the sum
$$C_n(N) = \sum_{\sigma_c} N^{F(\sigma_c)}$$
taken over all gluings σ_c that provide connected graphs. Consider two exponential generating functions
$$F(s, N) = \sum_{n=0}^{\infty} \frac{1}{n!} F_n(N) s^n,$$

$$C(s, N) = \sum_{n=1}^{\infty} \frac{1}{n!} C_n(N) s^n,$$
where s is a formal parameter. (Note that the second sum starts with the term $C_1(N)s$.)

Proposition 3.3.7. *The generating functions $F(s, N)$ and $C(s, N)$ are related by the equation*
$$F(s, N) = \exp(C(s, N)).$$

The equality in the proposition has in fact nothing special to do with gluings of 4-stars. The same statement is valid in a much more general situation and in Sec. 3.8 we will give two proofs of this more general statement.

For the problem of enumerating gluings of 4-stars we are interested in, the exponential generating function for all gluings is
$$\int_{\mathcal{H}_N} e^{-t \operatorname{tr} H^4} d\mu(H) = \int_{\mathcal{H}_N} \left(1 - \frac{t}{1!} \operatorname{tr} H^4 + \frac{t^2}{2!} (\operatorname{tr} H^4)^2 - \ldots\right) d\mu(H) \quad (3.13)$$

(the minus sign in the exponent is convenient for further usage), and the last proposition gives the following formula for the exponential generating function for connected gluings:
$$\log \int_{\mathcal{H}_N} e^{-t \operatorname{tr} H^4} d\mu(H). \quad (3.14)$$

The coefficient of t^n in the expansion of (3.13) is the polynomial $(-1)^n F_n(N)/n!$ in N, and the coefficient of t^n in the expansion of (3.14) is $(-1)^n C_n(N)/n!$. Note that the degree of F_n is $3n$, the maximal number of faces in a disconnected gluing, while the degree of C_n is $n + 2$. The coefficient of N^{n+2} in C_n is exactly the number of connected planar gluings of n 4-stars, the coefficient of N^n is the number of genus one gluings, and so on.

3.4 Enumeration of Colored Graphs

In this section we briefly mention some more interpretations of combinatorial problems in terms of matrix integrals. These interpretations are related to the so-called multi-matrix models, where more than one space of matrices are used.

3.4.1 Two-Matrix Integrals and the Ising Model

Let $\mathcal{H}_N$ be, as before, the space of $N \times N$ Hermitian matrices. In this section we consider Gaussian integration over the space $\mathcal{H}_N \times \mathcal{H}_N$ of pairs of such matrices. The matrices themselves will be denoted by H and G. First of all let us introduce a quadratic form on $\mathcal{H}_N \times \mathcal{H}_N$: it will be equal to

$$\operatorname{tr} H^2 + \operatorname{tr} G^2 - 2c \operatorname{tr} HG, \tag{3.15}$$

where $-1 < c < 1$.

Before discussing an integral we must better understand the corresponding Gaussian measure. We will do this step by step.

Step 1. Let us introduce the following real coordinates on $\mathcal{H}_N \times \mathcal{H}_N$: denote

$$h_{ii} = x_{ii},$$
$$h_{ij} = x_{ij} + iy_{ij}, \ h_{ji} = x_{ij} - iy_{ij} \ (\text{for } i < j),$$
$$g_{ii} = z_{ii},$$
$$g_{ij} = z_{ij} + it_{ij}, \ g_{ji} = z_{ij} - it_{ij} \ (\text{for } i < j).$$

Thus, we have $2N^2$ real coordinates: x_{ij} and z_{ij} for $i \leq j$, and y_{ij} and t_{ij} for $i < j$.

Step 2. Now let us compute the traces occurring in the quadratic form (3.15). We already know that

$$\operatorname{tr} H^2 = \sum_{i=1}^{N} x_{ii}^2 + 2 \sum_{i<j} (x_{ij}^2 + y_{ij}^2)$$

and

$$\operatorname{tr} G^2 = \sum_{i=1}^{N} z_{ii}^2 + 2 \sum_{i<j} (z_{ij}^2 + t_{ij}^2).$$

It is an easy exercise to compute also

$$\operatorname{tr} HG = \sum_{i=1}^{N} x_{ii} z_{ii} + 2 \sum_{i<j} (x_{ij} z_{ij} + y_{ij} t_{ij}).$$

186   3 Matrix Integrals

Step 3. The next operation consists in interpreting the monomials in the above traces (more exactly, the coefficients in front of these monomials) as entries of a matrix B of size $2N^2 \times 2N^2$. The rows and the columns of this matrix are marked by the coordinates $x_{ij}, y_{ij}, z_{ij}, t_{ij}$. The monomial, say, x_{ii}^2 in $\operatorname{tr} H^2$ means that we must put 1 on the intersection of the row corresponding to x_{ii} and of the column also corresponding to x_{ii}. We must put 2 on the intersection of the row and the column corresponding to x_{ij} for $i < j$. Consider the term $x_{ii} z_{ii}$ in $\operatorname{tr} HG$. First, we must not forget that this trace is multiplied by $-2c$ in (3.15). Second, we would like our matrix to be symmetric. Hence, we put the coefficient $-c$ at two places: on the intersection of the row x_{ii} and the column z_{ii}; and also on the intersection of the row z_{ii} and the coulmn x_{ii}. And so on.

We must take care of a convenient order of the coordinates along the matrix. A good order is to take first $x_{11}, z_{11}, x_{22}, z_{22}, \ldots, x_{NN}, z_{NN}$, and then $x_{12}, z_{12}, y_{12}, t_{12}, x_{13}, z_{13}, y_{13}, t_{13}, \ldots$. Then we find out that our matrix consists of blocks of size 2×2 along the diagonal: first there are N blocks of the type

$$\begin{pmatrix} 1 & -c \\ -c & 1 \end{pmatrix},$$

and all the other blocks are of the type

$$\begin{pmatrix} 2 & -2c \\ -2c & 2 \end{pmatrix}.$$

Step 4. In order to compute the covariance matrix $C = B^{-1}$ we need only to compute the inverse of a block:

$$\begin{pmatrix} 1 & -c \\ -c & 1 \end{pmatrix}^{-1} = \frac{1}{1-c^2} \begin{pmatrix} 1 & c \\ c & 1 \end{pmatrix},$$

and similarly for the other block:

$$\begin{pmatrix} 2 & -2c \\ -2c & 2 \end{pmatrix}^{-1} = \frac{1}{1-c^2} \begin{pmatrix} 1/2 & c/2 \\ c/2 & 1/2 \end{pmatrix}.$$

The matrix $C = B^{-1}$ consists of these inverse blocks arranged along the diagonal. The reader will easily find the determinant of B and the constant factor in front of the Gaussian measure

$$d\mu(H,G) = \operatorname{Const} \cdot e^{-\frac{1}{2}(\operatorname{tr} H^2 + \operatorname{tr} G^2 - 2c \operatorname{tr} HG)} dv(H) dv(G).$$

Step 5. The entries of the covariance matrix C also have the meaning of the mean values of the corresponding quadratic monomials. Let us list them explicitly:

$$\langle x_{ii}^2 \rangle = \langle z_{ii}^2 \rangle = \frac{1}{1-c^2};$$

$$\langle x_{ii}z_{ii} \rangle = \frac{c}{1-c^2};$$

$$\langle x_{ij}^2 \rangle = \langle y_{ij}^2 \rangle = \langle z_{ij}^2 \rangle = \langle t_{ij}^2 \rangle = \frac{1}{2}\frac{1}{1-c^2} \quad (i<j);$$

$$\langle x_{ij}z_{ij} \rangle = \langle y_{ij}t_{ij} \rangle = \frac{1}{2}\frac{c}{1-c^2} \quad (i<j).$$

All other covariances are equal to zero.

Step 6. Finally, we must find the mean values of quadratic monomials composed of the elements of the matrices H and G themselves. Combining the previous results, we find that for every combination of indices i and j, $i = 1, \ldots, N$, $j = 1, \ldots, N$, we have

$$\langle h_{ij}h_{ji} \rangle = \frac{1}{1-c^2}, \quad \langle g_{ij}g_{ji} \rangle = \frac{1}{1-c^2}, \quad \langle h_{ij}g_{ji} \rangle = \frac{c}{1-c^2}.$$

All other covariances between the matrix elements of H and G are equal to zero.

Our preparatory work is finished. Now we may consider the integration over the above Gaussian measure on $\mathcal{H}_N \times \mathcal{H}_N$, and interpret the integral in combinatorial terms. For example, what is the meaning of the integral

$$\int_{\mathcal{H}_N \times \mathcal{H}_N} e^{-t\,\text{tr}(H^4+G^4)} d\mu(H,G) \quad ?$$

The answer is, we must take, as before, four-valent maps, but this time to every vertex of the map we must assign one of the two "states". We mark these states by H and G; this marking simply means that the corresponding vertex represents either $\text{tr}\,H^4$ or $\text{tr}\,G^4$ respectively. Now, if two vertices are joined by an edge, and if they are in the same state, then the contribution of the edge to the whole sum (i.e., to the "partition function") is $1/(1-c^2)$, while if the vertices are in different states, the contribution of the edge is $c/(1-c^2)$. The summation must be carried over all maps and over all possible configurations of states of the vertices.

The model that we have obtained resembles very much the classical Ising model in statistical physics, though this time the underlying graph is not a regular lattice but an arbitrary map. Physicists call it the "Ising model on dynamical lattice". Many ordinary phenomena, such as phase transitions etc., take place also in this model.

This two-matrix model was first studied in [146]. Its interpretation as an Ising model was considered by many authors, see, e.g., [39], [40], [165].

3.4.2 The Gauss Problem

Consider an immersion $u : S^1 \to \mathbb{R}^2$ of the circle into the plane, or an immersion $u : S^1 \to S^2$ of the circle into the sphere, i.e., a mapping such that $\forall x \in S^1$ $u'(x) \neq 0$. Suppose also that the image $u(S^1)$ has only double points of self-intersection and that the intersection at each double point is transversal (i.e., the tangent vectors to the branches of the image are not collinear), see Fig. 3.10.

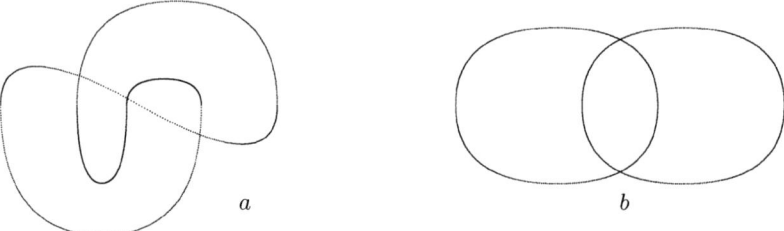

Fig. 3.10. (a) A unicursal curve, and (b) an Eulerian graph, which is not a unicursal curve

The image of such an immersion, considered up to isotopy of the target surface, is called a *unicursal curve*. Below we study only the case of nonoriented unicursal curves.

Let n be the number of double points in a unicursal curve. Table 3.2 borrowed from [12] contains the numbers of plane and spherical unicursal curves with given number n of self-intersection points. These numbers are obtained by direct drawing of the curves.

Table 3.2. The number of unicursal plane and spherical curves

n	0	1	2	3	4	5
$\mathbb{R}^2$	1	2	5	21	102	639
S^2	1	1	2	6	21	99

Considered up to isotopy of the plane, there are 2 curves with one double point, 5 curves with two double points, 21 curves with three double points, etc. Note that, considered as curves on the sphere, there is 1 curve with one double point, 2 curves with two double points and 6 curves with three double points. Spherical unicursal curves with up to two double points are drawn in Fig. 3.11.

The Gauss problem is: *How many unicursal curves (up to isotopy) are there with a given number of double points?*

This problem is still not resolved. We are going to write a Gaussian integral describing the number of isotopy classes of unicursal curves on the sphere.

3.4 Enumeration of Colored Graphs 189

Fig. 3.11. Spherical unicursal curves with up to two double points

Consider n copies of a 4-star on the plane. We know already how to select planar gluings of the darts, see Sec. 3.3. What we need is to select unicursal gluings. In order to do this let us color each pair of opposite darts in one of q colors. We consider further only *admissible* gluings, i.e., those that glue only darts of the same color.

Each gluing provides a curve with some number k of "components", that is, k superimposed unicursal curves. For example, the Eulerian graph from Fig. 3.10b has two "components". Unicursal curves are precisely those that consist of one "component" only. A curve with k "components" would have q^k admissible colorings.

In order to involve colorings in the integral, we introduce q copies of the space of Hermitian matrices, one copy for each color. Consider the integrand in the form

$$\sum_{i,j=1}^{q} \left(\mathrm{tr}(H_i H_j H_i H_j) \right)^n.$$

Integrating this expression over the product of spaces $\mathcal{H}_{N,i}$, $i = 1, \ldots, q$, we obtain a function which will be a polynomial both in N and in q. The coefficient at the monomial linear in q with the highest power of N, that is, the power $n+2$, is exactly the number of all spherical unicursal gluings. Note that a unicursal curve is automatically connected.

Some of these gluings provide the same unicursal spherical curve. Each unicursal curve is represented by precisely $n! \cdot 4^n$ gluings divided by the number of automorphisms of the curve. Indeed, in each gluing we are free to exchange the 4-stars and to rotate each star. Thus, the exponential generating function for connected unicursal curves is given by the following statement.

Theorem 3.4.1 ([186]). *The number of spherical unicursal curves with n double points, wheighted with the inverse order of the automorphism group, is equal to the coefficient of the monomial $(-1)^n q N^{n+2} s^n$ in the expansion of*

$$\int_{\mathcal{H}_N} \ldots \int_{\mathcal{H}_N} e^{-\frac{s}{4} \mathrm{tr} \sum_{i,j=1}^{q} H_i H_j H_i H_j} d\mu(H_1) \ldots d\mu(H_q).$$

3.4.3 Meanders

Let us fix a straight line on the plane and $2n$ points on it. Consider a simple connected not self–intersecting closed curve crossing the line at exactly those points. The equivalence class of such curves with respect to isotopies of the plane leaving the line fixed is called a *meander* of order n. All the meanders of order $1, 2, 3$ are shown in Fig. 3.12.

Fig. 3.12. Meanders with up to six intersection points

Enumeration of meanders is similar to that of unicursal curves from the previous section. The difference is that in this case we must introduce two sets of q colors instead of a single set. One set is ascribed to the horizontal line, the other one to the curve, the meander itself. As the result, the integration involves $2q$ copies of the Hermitian matrix space, and the exponent in the integrand is the sum of monomials of the form

$$\operatorname{tr} H_i G_j H_i G_j.$$

This form reflects the fact that the colors go around the vertex in the interchanging order. Two-colored 4-stars, with the vertical darts colored in a color belonging to one set, and the horizontal darts colored in a color belonging to the other set, have symmetry group of order 2, which leads to the factor $1/2$ in the exponent.

3.4 Enumeration of Colored Graphs

Theorem 3.4.2 ([188]). *The number of meanders of order n is equal to the coefficient of the monomial*

$$(-1)^n q^2 N^{2(n+1)} s^{2n}$$

in the expansion of

$$\int_{\mathcal{H}_N}\cdots\int_{\mathcal{H}_N}\int_{\mathcal{G}_N}\cdots\int_{\mathcal{G}_N} e^{-\frac{s}{2}\operatorname{tr}\sum_{i,j=1}^{q} H_i G_j H_i G_j}$$

$$\times d\mu(H_1)\ldots d\mu(H_q)d\mu(G_1)\ldots d\mu(G_q) \, .$$

As one of our colleagues has put it, we must take the leading term of the asymptotics when the dimension N tends to infinity, while the number q of matrices tends to zero.

Note that, similarly to the case of unicursal curves, a meander is automatically connected, and we do not need to take the logarithm. Unfortunately, the integral for the Gauss problem, as well as the integral for meanders, is not computed yet.

3.4.4 On Enumeration of Meanders

The enumeration of meanders was started in [189] in a rather direct and inefficient way. A new efficient algorithm using the transfer matrix was found in [154], from where we borrowed Table 3.3 of values of M_n up to $n = 24$.

n	M_n	n	M_n	n	M_n
1	1	9	933458	17	59923200729046
2	2	10	8152860	18	608188709574124
3	8	11	73424650	19	6234277838531806
4	42	12	678390116	20	64477712119584604
5	262	13	6405031050	21	672265814872772972
6	1828	14	61606881612	22	7060941974458061392
7	13820	15	602188541928	23	74661728661167809752
8	110954	16	5969806669034	24	794337831754564188184

Table 3.3. The number of plane meanders

The number of meanders M_n grows not faster than the square of the Catalan number Cat_n (cf. Exercise 3.8.2), that is, not faster than exponentially. Indeed, the horizontal line splits a meander of order n into an upper half-plane part and a lower half-plane part. Each of these parts is a system of n semicircles, and such systems are in obvious one-to-one correspondence with regular bracket structures of n pairs of brackets. As is well known, the latter are enumerated by the Catalan numbers. A subtler estimate from [189] reads

$$M_n \leq CR^n n^{-\frac{5}{2}}, \text{ where } R = \left(\frac{\pi}{4-\pi}\right)^2 \sim 13.394.$$

It was conjectured in [189] that the asymptotic behavior of meandric numbers is of the form

$$M_n \sim Cr^n n^{-\alpha}$$

for some numbers $r = 12.26\ldots$ and α. (The value $\alpha = 7/2$ conjectured there is certainly wrong.) A more precise conjecture from [83] reads:

$$r = 12.262874\ldots,$$

$$\alpha = \frac{29 + \sqrt{145}}{12} = 3.42013288\ldots.$$

These values fit much better the experimental data from Table 3.3 above, but they are not proved either; see the discussion at the end of Sec. 3.7.

The computation of the multi-matrix integrals, even in the planar approximation (that is, the computation of the main term of the $N \to \infty$ asymptotics) remains a very difficult task. The problem is solved for the two-matrix model. The same methods in principle permit solving the model with q matrices and with the Gaussian term of the form

$$\operatorname{tr}\left(\sum_{i=1}^{q} H_i^2 - 2 \sum_{i,j=1}^{q} c_{ij} H_i H_j\right),$$

when the graph with q vertices corresponding to non-zero coefficients c_{ij} is a tree. A trick due to Kazakov permits solving also the model when all c_{ij} are equal. Practically all other problems remain open.

Below we discuss the problems appearing in the computation of one-matrix integrals.

3.5 Computation of Matrix Integrals

3.5.1 Example: Computing the Volume of the Unitary Group

We have seen in Proposition 3.2.15 and Lemma 3.2.19 that

$$\frac{\operatorname{vol} U_N}{(2\pi)^{(N^2+N)/2} \cdot N!} \int_{\mathbb{R}^N} \prod_{i<j} (\lambda_i - \lambda_j)^2 \prod_{k=1}^{N} d\mu(\lambda_k) = 1,$$

where μ is the standard Gaussian measure on the line. Thus, in order to compute the volume of the unitary group we need to compute the integral

$$\int_{\mathbb{R}^N} \prod_{i<j}(\lambda_i - \lambda_j)^2 d\mu(\lambda_1)\ldots d\mu(\lambda_N). \tag{3.16}$$

This is the simplest integral of the type (3.11) we are interested in. Indeed, it corresponds to integrating over the space of Hermitian matrices of the function identically equal to one.

Hermite polynomials provide a very convenient tool to compute this integral. The n-th *Hermite polynomial* is the polynomial

$$H_n(x) = (-1)^n e^{\frac{x^2}{2}} \frac{d^n}{dx^n}\left(e^{-\frac{x^2}{2}}\right).$$

These are the first Hermite polynomials:

$$H_0(x) = 1$$
$$H_1(x) = x$$
$$H_2(x) = x^2 - 1$$
$$H_3(x) = x^3 - 3x$$
$$H_4(x) = x^4 - 6x^2 + 3$$

The usefulness of the Hermite polynomials is due to the properties they possess:

- H_n is a polynomial in x of degree n;
- the leading coefficient of H_n equals 1;
- the polynomials $H_0, H_1, \ldots, H_n$ form an orthogonal basis in the space of polynomials of degree $\leq n$ with respect to the standard Gaussian measure on the line: $(H_n, H_m) = 0$ whenever $n \neq m$; here (P, Q) denotes the scalar product for two polynomials P, Q:

$$(P, Q) = \int_{\mathbb{R}} PQ d\mu(\lambda);$$

- H_n is the unique sequence of polynomials satisfying the previous properties;
- $H_n = xH_{n-1} - H'_{n-1}$; $(H_n, H_n) = n(H_{n-1}, H_{n-1})$.

Exercise 3.5.1. Prove the above properties of the Hermite polynomials.

Elementary operations with rows allow one to replace the ith row in the Vandermonde determinant with an arbitrary monic (i.e., with leading coefficient 1) polynomial of degree i without changing the determinant. It is convenient to substitute Hermite polynomials in variables λ_i for the monomials in the Vandermonde determinant:

$$\prod_{i>j}(\lambda_i - \lambda_j) = \begin{vmatrix} 1 & 1 & \ldots & 1 \\ \lambda_1 & \lambda_2 & \ldots & \lambda_N \\ \lambda_1^2 & \lambda_2^2 & \ldots & \lambda_N^2 \\ \ldots & \ldots & \ddots & \ldots \\ \lambda_1^{N-1} & \lambda_2^{N-1} & \ldots & \lambda_N^{N-1} \end{vmatrix}$$

$$= \begin{vmatrix} H_0(\lambda_1) & H_0(\lambda_2) & \ldots & H_0(\lambda_N) \\ H_1(\lambda_1) & H_1(\lambda_2) & \ldots & H_1(\lambda_N) \\ H_2(\lambda_1) & H_2(\lambda_2) & \ldots & H_2(\lambda_N) \\ \ldots & \ldots & \ddots & \ldots \\ H_{N-1}(\lambda_1) & H_{N-1}(\lambda_2) & \ldots & H_{N-1}(\lambda_N) \end{vmatrix}.$$

(3.17)

The square of the last determinant is a polynomial in the variables $H_i(\lambda_j)$. Each term $H_i(\lambda_j)$ is of degree not more than 2 in each of the monomials of the square of the determinant's expansion. Such a monomial is a product of functions depending on one variable each, and the integral can be taken separately in each of the variables $\lambda_1, \ldots, \lambda_n$. Then the orthogonality property of the Hermite polynomials implies that only monomials of the form

$$H_0^2(\lambda_{\sigma(1)}) H_1^2(\lambda_{\sigma(2)}) \ldots H_{N-1}^2(\lambda_{\sigma(N)})$$

contribute to the integral. Here σ is a permutation on the set of indices $1, \ldots, N$. The number of permutations is $N!$ and the contribution of all these monomials is the same. Hence,

$$\int_{\mathbb{R}^N} \prod_{i<j}(\lambda_i - \lambda_j)^2 d\mu(\lambda_1) \ldots d\mu(\lambda_N) = N!(H_0, H_0)(H_1, H_1) \ldots (H_{N-1}, H_{N-1}).$$

It only remains to compute the scalar squares (H_i, H_i). The properties of the Hermite polynomials imply that

$$(H_n, H_n) = n!$$

and finally

$$\int_{\mathbb{R}^N} \prod_{i<j}(\lambda_i - \lambda_j)^2 d\mu(\lambda_1) \ldots d\mu(\lambda_N) = N!(N-1)! \ldots 2!1!. \quad (3.18)$$

Corollary 3.5.2. *The volume of the unitary group is equal to*

$$\operatorname{vol} U_N = \frac{(2\pi)^{(N^2+N)/2}}{\prod_{k=1}^{N-1} k!}.$$

3.5.2 Generalized Hermite Polynomials

The integral

$$\int_{\mathcal{H}_N} e^{-t\,\mathrm{tr}\,H^4} d\mu(H) = c_N \int_{\mathbb{R}^N} e^{-t(\lambda_1^4 + \cdots + \lambda_N^4)} \prod (\lambda_i - \lambda_j)^2 d\mu(\lambda_1) \ldots d\mu(\lambda_N),$$

where the normalizing constant c_N is given by the inverse to (3.18),

$$c_N = \frac{1}{\int_{\mathbb{R}^N} \prod_{i<j} (\lambda_i - \lambda_j)^2 d\mu(\lambda_1) \ldots d\mu(\lambda_N)}$$
$$= \frac{1}{N!(N-1)! \ldots 2!1!},$$

may be considered as a perturbation of the matrix integral of the constant function. In order to compute it, we introduce generalized Hermite polynomials.

Consider the measure

$$d\mu_V(\lambda) = e^{-tV(\lambda)} d\mu(\lambda)$$

on the line. For the time being we agree that

$$V = V(\lambda) = \lambda^4,$$

t being a parameter. For two functions P, Q, denote by $(P, Q)_V$ their scalar product with respect to this measure,

$$(P, Q)_V = \int PQ \, d\mu_V.$$

The *generalized Hermite polynomials* with respect to this measure are the polynomials (in λ) $P_{V,0}(\lambda), P_{V,1}(\lambda), \ldots$ such that

- the polynomial $P_{V,k}$ is a polynomial of degree k with the leading term λ^k;
- the polynomials $P_{V,k}$ are orthogonal with respect to the measure $d\mu_V$, $(P_{V,k}, P_{V,m})_V = 0$ for $m \neq k$.

In the case $V \equiv 0$ these polynomials are ordinary Hermite polynomials.

Exercise 3.5.3. Prove that for $V(\lambda) = \lambda^4$ the polynomials $P_{V,k}$ are even for even k, and are odd for odd k.

Notation 3.5.4. Practically all the objects studied below (polynomials, constants, ...) depend on V and on t, and in principle should carry an index (t, V). This would, however, make our notation too cumbersome. Therefore in the majority of cases we will omit the explicit mention of V, returning to it from time to time in order for it not to be entirely forgotten. We make a systematic exception for the measure μ_V, because μ denotes the usual Gaussian measure. Also, from time to time we will make the parameter t appear explicitly.

3 Matrix Integrals

Denote the scalar square of the polynomial P_k with respect to the measure μ_V by $h_k = h_k(t)$:
$$h_k = (P_k, P_k)_V.$$

The reason why the generalized Hermite polynomials are introduced is that, similarly to the unperturbed case, they allow one to simplify the calculation of the integral. Indeed, substituting linear combinations of the rows of the Vandermonde determinant instead of the original rows, we can replace the polynomials in the determinant with the generalized Hermite polynomials:

$$\Delta(\lambda) = \prod_{i>j}(\lambda_i - \lambda_j)$$

$$= \begin{vmatrix} 1 & 1 & \ldots & 1 \\ \lambda_1 & \lambda_2 & \ldots & \lambda_N \\ \vdots & \vdots & \ddots & \vdots \\ \lambda_1^{N-1} & \lambda_2^{N-1} & \ldots & \lambda_N^{N-1} \end{vmatrix}$$

$$= \begin{vmatrix} P_0(\lambda_1) & P_0(\lambda_2) & \ldots & P_0(\lambda_N) \\ P_1(\lambda_1) & P_1(\lambda_2) & \ldots & P_1(\lambda_N) \\ \vdots & \vdots & \ddots & \vdots \\ P_{N-1}(\lambda_1) & P_{N-1}(\lambda_2) & \ldots & P_{N-1}(\lambda_N) \end{vmatrix}.$$

The expansion of the determinant and separate integration over each of the N variables λ_k yield

$$\int_{\mathcal{H}_N} e^{-t\operatorname{tr} H^4} d\mu(H) = c_N N! h_{N-1} h_{N-2} \ldots h_0.$$

Note that this representation of the integral as a product is very convenient for our purposes since we will need taking its logarithm.

Our goal now is to calculate the values of the constants h_k. For small values of k these constants may be explicitly calculated as certain integrals but these formulas carry little information about their behavior as k increases. So we need a more efficient tool. Such a tool is a recurrence formula.

Let us consider the operator of multiplication by λ on the space of polynomials in λ and find the matrix of its presentation in the basis P_k. The following property is valid for every family of orthogonal polynomials.

Lemma 3.5.5. $\lambda P_k = P_{k+1} + R_k P_{k-1}$ for some constant $R_k = R_{V,k}(t)$.

Proof. Indeed, the polynomial $\lambda P_{V,k}$ is of degree $k+1$; so it is a linear combination of the polynomials $P_{V,0}, \ldots, P_{V,k+1}$. The coefficient of $P_{V,k+1}$ is 1 because the leading coefficients coincide. The coefficient of $P_{V,k}$ is zero because of the parity. The coefficient of $P_{V,m}$ for $m < k-1$ vanishes because of $(\lambda P_{V,k}, P_{V,m})_V = (P_{V,k}, \lambda P_{V,m})_V = 0$, since $\lambda P_{V,m}$ is a polynomial of degree $< k$, and the required assertion follows.

The constants h_k are related to the constants R_k by the formula

$$h_k = (P_k, \lambda P_{k-1})_V = (\lambda P_k, P_{k-1})_V$$
$$= (P_{k+1} + R_k P_{k-1}, P_{k-1})_V = R_k h_{k-1},$$

and the formula for the integral becomes

$$\int_{\mathcal{H}_N} e^{-t \operatorname{tr} H^4} d\mu(H) = c_N N! R_1^{N-1} R_2^{N-2} \ldots R_{N-1} h_0^N. \qquad (3.19)$$

For two polynomials P, Q, the integration by parts gives

$$(P, Q')_V = \frac{1}{\sqrt{2\pi}} \int PQ' e^{-\frac{1}{2}\lambda^2 - tV(\lambda)} d\lambda = -(P', Q)_V + (P, (\lambda + tV')Q)_V.$$

In particular,

$$k h_{k-1} = (P_{k-1}, P_k')_V$$
$$= -(P_{k-1}', P_k)_V + (P_{k-1}, (\lambda + 4t\lambda^3) P_k)_V$$
$$= (\lambda P_{k-1}, P_k)_V + 4t(\lambda P_{k-1}, \lambda^2 P_k)_V$$
$$= h_k + 4t(P_k + R_{k-1} P_{k-2}, \lambda^2 P_k)_V$$
$$= h_k + 4t[(\lambda P_k, \lambda P_k)_V + R_{k-1}(\lambda^2 P_{k-2}, P_k)_V]$$
$$= h_k + 4t(h_{k+1} + R_k^2 h_{k-1} + R_{k-1} h_k)$$
$$= h_{k-1}(R_k + 4t R_k(R_{k+1} + R_k + R_{k-1})),$$

and finally we obtain

$$k = R_k + 4t R_k (R_{k+1} + R_k + R_{k-1}). \qquad (3.20)$$

The last equation is very important. It is the *discrete Painlevé I equation*.

3.5.3 Planar Approximations

The coefficients of the expansion in t of

$$\log \int_{\mathcal{H}_N} e^{-t \operatorname{tr} H^4} d\mu(H)$$

are polynomials in N. We know that the leading coefficients describe planar connected gluings. These leading coefficients correspond to the power N^{n+2}. Indeed, a connected embedded graph with n vertices and $2n$ edges is planar if and only if the number of its faces equals $n + 2$.

Planar gluings of 4-stars are thus enumerated by the generating function

$$e_0(t) = \lim_{N \to \infty} \frac{1}{N^2} \log \int_{H_N} e^{-\frac{t}{N} \operatorname{tr} H^4} d\mu(H).$$

Substituting t/N for t in Eq. (3.20) and dividing both parts by N, we obtain

$$\frac{k}{N} = r_k(t)(1 + 4t(r_{k+1}(t) + r_k(t) + r_{k-1}(t))) \tag{3.21}$$

for the functions

$$r_k(t) = R_k(t)/N.$$

As N, k tend to infinity while $k/N \to x \in [0,1]$ with x fixed, the last equation yields

$$x = r(1 + 12tr), \tag{3.22}$$

where r is the limit function

$$r = r(x, t) = \lim_{N \to \infty} \frac{R_{xN}(t)}{N}.$$

Following physical literature, we assume here (and in similar cases below) that if N goes to infinity and k/N tends to x, the distribution given by the functions $R_k(t)/N$, $k = 0, 1, \ldots, N$, tends to a continuous function $r(x, t)$. We don't know any rigorous proof of this statement, though we were assured by some specialists that such a proof does exist. Anyway, to present a proof would lead us very far from our main preoccupation (embedded graphs); however, we regret being unable to provide an appropriate reference.

Taking $x = 1$ in Eq. (3.22) we obtain a function in t which turns out to be, up to a factor, the generating function for the Catalan numbers (cf. Exercise 3.8.2):

$$r(1, t) = \alpha(t) = \frac{-1 + \sqrt{1 + 48t}}{24t} = \sum_{n \geq 0} (-1)^n (12t)^n \operatorname{Cat}_n. \tag{3.23}$$

Let us now take the logarithm of both parts of Eq. (3.19). This leads to the equation

$$\frac{1}{N^2} \log \int_{\mathcal{H}_N} e^{-\frac{t}{N} \operatorname{tr} H^4} d\mu(H) = -\frac{1}{N} \sum_{k=1}^{N} \left(1 - \frac{k}{N}\right) \log \frac{R_k(t)}{R_k(0)} - \log \frac{h_0(t)}{h_0(0)}$$

and the limit as $N \to \infty$ (and t being fixed) is

$$e_0(t) = -\int_0^1 (1 - x) \log \left(\frac{r(x, t)}{x}\right) dx. \tag{3.24}$$

Now substitute in the latter integral the expression of x in terms of r taken from Eq. (3.22). This gives

$$e_0(t) = \int_0^\alpha (1 - r - 12tr)(1 + 24tr)\log(1 + 12tr)dr$$

$$= -\frac{1}{2}(-12t\alpha^3 + \alpha^2 - 2\alpha)(1 + 12t\alpha)\log(1 + 12t\alpha)$$

$$+ (18t^2\alpha^4 - 2t\alpha^3 + \frac{3}{2}\alpha)$$

$$= -\frac{1}{2}\log\alpha + \frac{1}{24}(1-\alpha)(\alpha - 9). \tag{3.25}$$

The substitution of the series for $\alpha(t)$ from (3.23) in the latter expression results in the series

$$e_0(t) = -\sum_{n=1}^\infty (-12)^n \frac{(2n-1)!\, t^n}{(n+2)!\, n!} = -\sum_{n=1}^\infty (-12)^n \frac{\mathrm{Cat}_n}{2n(n+2)} t^n.$$

Remark 3.5.6. Let us say more accurately what is counted by the number

$$(12)^n \frac{(2n-1)!}{(n+2)!}.$$

It is the number of plane maps with n vertices of degree 4, labelled by $1, 2, \ldots, n$; what is more, the darts at every vertex are labelled by 1, 2, 3, 4, and this labelling must agree with the cyclic order of the darts.

Similar, but considerably more complicated computations (see [25]) lead to the following results for higher genera:

Proposition 3.5.7. *We have*

$$e_1(t) = \frac{1}{12}\log(2 - \alpha(t)) = \frac{1}{24}\sum_{n>0}\left(\binom{2n}{n} - 4^n\right)\frac{(-12)^n}{n}t^n;$$

$$e_2(t) = \frac{1}{6!}\frac{(1-\alpha)^3}{(2-\alpha)^5}(82 + 21\alpha - 3\alpha^2)$$

$$= \sum_{n>0}\left(\frac{13(n-1)4^n}{2304} - \frac{28n^2 - 19n - 9}{4320}\binom{2n}{n}\right)(-12)^n t^n;$$

$$e_g(t) = \frac{(1-\alpha)^{2g-1}}{(2-\alpha)^{5(g-1)}}P_g(\alpha)$$

for some polynomial P_g, where α is given by (3.23).

3.6 Korteweg–de Vries (KdV) Hierarchy for the Universal One-Matrix Model

In this section we find ourselves on a rather swampy soil. Its material is borrowed from papers written by physicists, where neither the objects under consideration, nor the statements about these objects are presented in a mathematically clear way. The questions of existence of limit functions, of the types of these functions, and so on are even not discussed. Our attempts did not lead either to a justification of the results. However, we find it useful to include this material since it forms the base of the former Witten's conjecture (now Kontsevich's theorem) discussed in the next chapter and therefore its inclusion makes the text more self-contained. But the reader must be warned that the terms "theorem" and "proof" below do not have the usual mathematical meaning.

3.6.1 Singular Behavior of Generating Functions

Recall the expansion

$$\frac{1}{N^2} \log \int_{\mathcal{H}_N} e^{-\frac{t}{N} \operatorname{tr} H^4} d\mu(H) = e_0(t) + \frac{1}{N^2} e_1(t) + \frac{1}{N^4} e_2(t) + \ldots$$

from the previous section. Each of the functions $e_g(t)$, $g = 0, 1, 2, \ldots$ in this expansion is well defined since its coefficient in front of t^n is uniquely determined by the number of connected gluings of n 4-stars that give a genus g surface. These functions have a nice property: their unique finite singular point is the point $t_c = -1/48$, and they have either an algebraic, or a logarithmic singularity at this point.

The explicit presentation of the function $e_0(t)$ implies that in a vicinity of this point it is a sum of a regular function and a singular function with the leading term
$$\text{const} \cdot (t - t_c)^{5/2}.$$

Similarly, for higher genera the leading term of the singular part of the function $e_g(t)$ at t_c looks like
$$\text{const} \cdot (t - t_c)^{5(1-g)/2}.$$

(For the case $g = 1$, where the above exponent is 0, this leading term must be understood as $\log(t - t_c)$.)

For a fixed value of t, plane gluings dominate, and the functions $e_g(t)$ for $g \geq 1$ become "invisible". However, it is possible to bind the variables t and N by a relation, so that t varies as N grows, in such a way that the contributions of all genera become of the same order. What attracts the main interest of physicists is *the behavior of the functions e_g in the vicinity of the critical value t_c*. Our aim now is to construct a function describing this behavior and incorporating information about all genera. Let e_g denote the coefficient of the leading term in the expansion of the singular part of the function $e_g(\cdot)$ around the point $t = t_c$:

3.6 KdV Hierarchy

$$\text{sing}(e_0(t)) = e_0 \left(\frac{t-t_c}{t_c}\right)^{5/2} + \ldots,$$

$$\text{sing}(e_1(t)) = e_1 \log \frac{t-t_c}{t_c} + \ldots,$$

$$\text{sing}(e_2(t)) = e_2 \left(\frac{t-t_c}{t_c}\right)^{-5/2} + \ldots,$$

and so on; here dots denote terms of higher order in $t - t_c$. Now we introduce the function $E = E(y)$ by the following formal expansion at $y = 0$:

$$E(y) = e_0 y^{5/2} + e_1 \log y + e_2 y^{-5/2} + e_3 y^{-5} + \ldots,$$

where the coefficients e_i are the constants defined above.

The coefficients e_g can be found using the following important statement, which is more easily formulated in terms of the second derivative

$$u(y) = E''(y) = \frac{15}{4} e_0 y^{1/2} - e_1 y^{-2} + \frac{35}{4} e_2 y^{-9/2} + \ldots.$$

Theorem 3.6.1 ([45], [87], [125], [299]). *The second derivative $u(y)$ of the function $E(y)$ satisfies the Painlevé I equation*

$$y = u(y)^2 - \frac{1}{3} u''(y). \tag{3.26}$$

This equation can be understood as the following recurrence relation for the coefficients u_n in the expansion $u(y) = \sum_{n \geq 0} u_n y^{(1-5n)/2}$:

$$\frac{25n^2 - 1}{12} u_n = \sum_{i=0}^{n+1} u_i u_{n+1-i},$$

which immediately yields the following first few values of the coefficients:

$$u(y) = y^{1/2} - \frac{1}{24} y^{-2} - \frac{49}{1152} y^{-9/2} + \ldots.$$

Proof. We start with showing that the leading term of the perturbation of the function $r(x,t) = r_V(x,t)$ satisfies the Painlevé I equation, and then show that this leading term does coincide with the second derivative of E. Recall that $r(x,t)$ is the limit distribution of the coefficients of the multiplication by λ operator in the basis of generalized Hermite polynomials, see Eq. (3.21).

Equation (3.21) holds identically in k, t, N. In particular, it is valid for $k = N$, $t = t_c = -1/48$ (the *stationary* equation). Subtracting the stationary equation from the general one we obtain

$$\frac{k}{N} - 1 = r_k(t) - r_N(t_c) + 4(tr_k(t)r_{k+1}(t) -$$
$$t_c r_N(t_c)r_{N+1}(t_c) + tr_k^2(t) - t_c r_N^2(t_c) +$$
$$tr_k(t)r_{k-1}(t) - t_c r_N(t_c)r_{N-1}(t_c)). \tag{3.27}$$

Now let us restrict the last equation to the quasihomogeneous curve

$$\frac{k-N}{N} = \frac{t-t_c}{t_c} = N^{-4/5} y \tag{3.28}$$

parametrized by a parameter y close to $y = 0$. This means that we make the substitution

$$k = N(1 + N^{-4/5} y), \quad t = t_c(1 + N^{-4/5} y)$$

in Eq. (3.27). As a result, we obtain a more complicated equality but it depends now only on two parameters, y and N. Thanks to the choice of the exponents, the first nontrivial terms of the perturbations of all terms in the stationary equation becomes of the same order in N^{-1} as $N \to \infty$. The coefficients of these leading terms form the *double scaling limit*.

Denote by $u(y)$ the coefficient of the leading term of the difference $r_k(t) - r_N(t_c)$:

$$\frac{r_k(t) - r_N(t_c)}{r_N(t_c)} \sim N^{-2/5} u(y).$$

Then, equating to 0 the leading term of Eq. (3.27), that is, the coefficient of $N^{-4/5}$, we arrive at the Painlevé I equation (3.26).

On the other hand, our argument being similar to that of the derivation of Eq. (3.24), in the scaling limit we obtain

$$E(y) = \lim_{N \to \infty} N^{-2} \log \int_{\mathcal{H}_N} e^{-\frac{t}{N} \operatorname{tr} H^4} d\mu(H)$$
$$= \lim_{N \to \infty} \int_{N^{-4/5}}^{y} (\zeta - y) u(\zeta) d\zeta,$$

with the varying lower integration limit. Therefore, $u(y)$ is, indeed, the second derivative of $E(y)$.

3.6.2 The Operator of Multiplication by λ in the Double Scaling Limit

Let us return to the operator of multiplication by λ in the space of polynomials in one variable λ; we denote this operator by S. According to Lemma 3.5.5, in the basis of polynomials $P_k = P_{V,k}$ this operator has the form

$$SP_k = \lambda P_k = P_{k+1} + R_k P_{k-1}$$

for some functions R_k depending on t. We are interested in the limit form of this operator in the same double scaling limit as above as $N \to \infty$.

3.6 KdV Hierarchy

Let us modify slightly the basis in order to make the matrix form of the operator symmetric. Setting $\widetilde{P}_k = P_k/\sqrt{h_k}$, where $h_k = (P_k, P_k)_V$, we obtain, in the new basis $\widetilde{P}_k$:

$$S\widetilde{P}_k = \sqrt{R_{k+1}}\widetilde{P}_{k+1} + \sqrt{R_{k-1}}\widetilde{P}_{k-1}.$$

Using this formula we can act on polynomials represented by the basis expansion

$$a_0\widetilde{P}_0 + a_1\widetilde{P}_1 + \cdots + a_m\widetilde{P}_m,$$

by making shifts (which we denote $S_\pm$) and multiplications by $\sqrt{R_{k+1}}$ and $\sqrt{R_{k-1}}$ respectively.

Now consider the behavior of this operator under the double scaling limit. Its ingredients have the following expansion in N up to the first nontrivial orders:

$$R_{k\pm 1} \sim r(1, t_c)(1 - N^{-2/5}u(y))$$

$$S_\pm \sim 1 \pm N^{-1}\frac{d}{dx} + \frac{N^{-2}}{2}\frac{d^2}{dx^2},$$

and we have

$$S \sim \sqrt{r(1, t_c)(1 - N^{-\frac{2}{5}}u(y))}\left(2 + N^{-\frac{2}{5}}\frac{d^2}{dy^2}\right)$$

$$\sim 2\sqrt{r(1, t_c)} + N^{-\frac{2}{5}}\sqrt{r(1, t_c)}\left(u(y) - \frac{d^2}{dy^2}\right)$$

$$= \text{const} + N^{-\frac{2}{5}}S_2.$$

The first equality follows from the fact that the shift operator is the exponent of the derivative.

On the other hand, consider the operator D on the space of polynomials taking a polynomial to its derivative,

$$D : P(\lambda) \mapsto P'(\lambda).$$

Obviously, the commutator of the operators D and S is

$$[D, S] = 1.$$

As $N \to \infty$ in the double scaling limit, the operator D behaves as

$$D \sim \text{const} + N^{\frac{2}{5}}D_3.$$

It can be verified that D_3 in this expansion is a differential operator of order three starting with d^3/dy^3, and we have $[D_3, S_2] = 1$. Given a differential operator $S_2 = d^2/dy^2 - u(y)$ of order two, a third-order solution D_3 to the equation $[D_3, S_2] = 1$ exists if and only if the function u satisfies the equation

$$-\frac{1}{3}u''' + 2uu' = 1,$$

see below. And this is precisely the derivative of the Painlevé I equation (3.26), which, as we already know, holds for the function u.

3.6.3 The One-Matrix Model and the KdV Hierarchy

If the potential $V(\lambda)$ of the model is of a degree greater than 4 in λ (here we consider only even potentials), then, for some values of the parameters, the limit functions $e_g = e_g(t)$ can acquire more complicated singularities than that considered above. In the language of singularity theory, the case $V(\lambda) = \lambda^4$ corresponds to the so-called A_2-singularity, while a generic polynomial V of degree $2m$ in λ leads to an A_m-singularity with $m > 2$ (we do not describe in detail how this connection works; see [147], [80]). Fortunately, in these cases the main features of the pattern are preserved: there is a unique critical point t_c, common for all e_g, and the leading term of e_g at t_c is proportional to $(t - t_c)^{2+1/m}$. It is possible to elaborate the double scaling limit analysis in this more general case, and to arrive at a differential equation of order $2m - 1$ on the perturbation function $u(y)$.

In the language of the preceding section, the operator S of multiplication by λ, being expressed in the basis of generalized Hermite polynomials, still has the following behavior in the double scaling limit:

$$S \sim \text{const} + N^{-m/(2m+1)} S_2, \quad S_2 = \frac{d^2}{dy^2} - u_m(y),$$

while the derivative operator D behaves as

$$D \sim \text{const} + N^{m/(2m+1)} D_{2m-1},$$

where D_{2m-1} is a differential operator of order $2m - 1$ with the leading term d^{2m-1}/dy^{2m-1}. The solution $u_m(y)$ to the equation

$$[D_{2m-1}, S_2] = 1$$

is also the second derivative of the function $E_m(u)$ describing the double-scaling limit behavior of the functions $e_g = e_g(t)$.

Elaborating this procedure step by step for all kinds of singularities we arrive at the notion of the universal one-matrix model. Let us give its description. Consider the integral

$$\log \frac{1}{N} \int_{\mathcal{H}_N} e^{-\frac{1}{N} \operatorname{tr} H^4 - \operatorname{tr} V(H; \tau_1, \tau_2, \dots)} d\mu(H), \qquad (3.29)$$

where

$$V(\lambda; \tau_1, \tau_2, \tau_3, \dots) = \frac{\tau_1}{2} \lambda^2 + \frac{\tau_2}{4N} \lambda^4 + \frac{\tau_1}{6N^2} \lambda^6 + \dots$$

is the potential of the model. The *universal one-matrix model* is the power series in infinitely many variables $t_0 = y, t_1, t_2, \dots$ defined by

$$U(t_0, t_1, t_2, \dots) = \sum_{g=0}^{\infty} \sum_{\tau_i} \prod_{i=1}^{\infty} \frac{t_i^{\tau_i}}{\tau_i!} e_g(\tau_1, \tau_2, \dots).$$

Here the coefficients $e_g(\tau_1, \tau_2, \ldots)$ are defined in terms of the critical behavior of the integral in the same way as the coefficients $e_0, e_1, \ldots$ in Sec. 3.6.1, but for the potential depending on the parameters.

In a more general setting, consider a linear differential operator of order 2 on the line,
$$S_2 = \frac{\partial^2}{\partial y^2} + u(y).$$
Then for each k there exists a monic linear differential operator Q_k of order k on the line such that the operator $[Q_k, S_2]$ is a multiplication by a function (i.e., a differential operator of order 0). These operators Q_k can be found by means of the following procedure suggested in [110].

For an even $k = 2m$ one can set $Q_k = S_2^m$; obviously, $[Q_{2m}, S_2] = 0$. For an odd k the situation is more complicated. Namely, one can find a "pseudodifferential" operator $S_2^{1/2} = \partial + l_1 \partial^{-1} + l_2 \partial^{-2} + \ldots$, where $\partial = d/dy$. Here ∂^{-1} is the inverse operator to ∂. It commutes with ∂, and its commutator with the multiplication by a function $f(y)$ operator looks like
$$[\partial^{-1}, f(y)] = -f'\partial^{-1} + f''\partial^{-2} - \ldots.$$
The last rule allows one to represent each pseudodifferential operator A in the canonical form
$$A = a_k(y)\partial^k + \cdots + a_1(y)\partial + a_0(y) + a_{-1}(y)\partial^{-1} + a_{-2}(y)\partial^{-2} + \ldots,$$
where all the functional coefficients are placed to the left of powers of ∂. This presentation is unique. The *positive part* A_+ of the pseudodifferential operator A is the differential operator
$$A_+ = a_k(y)\partial^k + \cdots + a_1(y)\partial + a_0(y).$$

Now, one can simply set Q_k to be the positive part of $(S_2^{1/2})^k$, $Q_k = (S_2^{k/2})_+$. It happens that this choice of Q_k is unique up to adding linear combinations of the operators Q_l for $l < k$. In particular, $Q_1 = \partial$.

The operator D_{2m-1} simply coincides with Q_{2m-1} for a specific function $u_m(y)$, which is the solution to the ordinary differential equation determined by the operator identity
$$[D_{2m-1}, S_2] = 1.$$
The last system of equations may be considered as "initial conditions" for the system of partial differential equations
$$\frac{\partial S_2}{\partial t_m} = [Q_{2m+1}, S_2].$$
Here $S_2 = \partial^2/\partial y^2 + U$ where the function $U = U(y, t_0, t_1, \ldots)$ depends on y and on the infinite system of parameters t_i, and the left-hand side of this

equation is simply $\partial U/\partial t_m$. This system of equations is called the *Korteweg–de Vries*, or *KdV hierarchy*. These equations make sense since the right-hand sides also are just functions due to the definition of Q_k. Moreover, for arbitrary "initial value" these equations have a solution since they can be considered as an (infinite) set of *pairwise commuting* vector fields on the space of operators S_2 (= the space of functions U). Since $Q_1 = \partial$, a solution can be written in the form $U = U(y + t_0, t_1, t_2, \ldots)$.

The KdV hierarchy is equivalent to a system of equations

$$L_j U = 0, \qquad j = 1, 2 \ldots$$

on the unknown function U for some fixed differential operators L_j of order j.

We shall be interested in the solution corresponding to the "initial value" $U(y, 0, 0, \ldots) = 2y$ and the derivatives

$$\frac{\partial U}{\partial t_i}(y, 0, 0, \ldots) = u_k(y).$$

As we have seen, this function is the second derivative in y of an interesting function,

$$U(y + t_0, t_1, t_2, \ldots) = 2\frac{\partial^2}{\partial t_0^2} \log \tau(y + t_0, t_1, \ldots).$$

A function τ on the right-hand side of the last equation is called a *τ-function for the KdV hierarchy*. Witten's conjecture discussed in the next chapter identifies this function with another series in infinite number of variables constructed by means of the intersection theory on moduli spaces of curves.

3.6.4 Constructing Solutions to the KdV Hierarchy from the Sato Grassmanian

There is a standard way to construct solutions to the KdV hierarchy starting from a Sato subspace in the Sato Grassmanian. Kontsevich made use of it in his proof of Witten's conjecture. Below we follow the approach explained in [252].

Consider $\mathbb{C}((z))$, the infinite dimensional space of Laurent series in z, that is, series of the form

$$f(z) = \sum_{i=-\infty}^{\infty} a_i z^i.$$

A *Sato subspace* in $\mathbb{C}((z))$ is an infinite dimensional vector subspace possessing a basis $f_1, f_2, \ldots$ such that $f_j(z) = z^{-j}(1 + o(1))$ for all j starting from some number. All Sato subspaces constitute the *Sato Grassmanian*. Below we consider only Sato subspaces admitting a basis f_i such that *each* f_i starts with z^{-i}.

3.6 KdV Hierarchy

Let $T_1, T_2, \ldots$ be an infinite sequence of formal variables, and denote by $M(z; T_1, T_2, \ldots)$ the function

$$M(z; T_1, T_2, \ldots) = e^{T_1 z^{-1} + T_2 z^{-2} + \cdots}.$$

For a function f_j of the form

$$f_j = z^{-j} + \alpha_1 z^{-j+1} + \alpha_2 z^{-j+2} + \cdots$$

the function $M f_j \in \mathbb{C}((z))$ is well defined. For a Sato subspace $W \subset \mathcal{P}$, fix a basis $f_1, f_2, \ldots$ such that f_j starts with z^{-j}. Define the τ-*function* associated to the subspace W as the fraction

$$\tau_W(T_1, T_2, \ldots) = \frac{(\cdots \wedge M f_3 \wedge M f_2 \wedge M f_1) \wedge z^0 \wedge z^1 \wedge z^2 \wedge \cdots}{\cdots \wedge z^{-3} \wedge z^{-2} \wedge z^{-1} \wedge z^0 \wedge z^1 \wedge z^2 \wedge \cdots}.$$

Obviously, the τ-function depends indeed on the subspace W itself, and not on the specific choice of the basis $f_1, f_2, \ldots$ in it. The τ-function is non-zero. In practice its computation involves an evaluation of a determinant of finite size, as the following example shows.

Example 3.6.2. The first nontrivial example of a τ-function is given by the subspace in $\mathbb{C}((z))$ spanned by the vectors

$$f_1 = z^{-1} + a_1 z^0 + a_2 z^1, \quad f_2 = z^{-2} + \frac{a_2}{a_1} z, \quad f_j = z^{-j} \quad \text{for} \quad j \geq 3$$

(here a_1, a_2 are arbitrary non-zero complex numbers). Multiplying by the exponent M we obtain

$$M f_1 = \cdots + (T_1 + a_1(\tfrac{1}{2} T_1^2 + T_2) + a_2(\tfrac{1}{6} T_1^3 + T_1 T_2 + T_3)) z^{-2}$$
$$+ (1 + T_1 + a_2(\tfrac{1}{2} T_1^2 + T_2)) z^{-1} + \cdots$$
$$M f_2 = \cdots + (1 + \frac{a_2}{a_1} T_1) z^{-2} + \frac{a_2}{a_1} z^{-1} + \cdots,$$

where dots denote terms not containing z^{-2} and z^{-1}, and

$$M f_j = \cdots + z^{-j} \quad \text{for} \quad j \geq 3,$$

where dots denote terms of degree less than $-j$.

Hence, the τ-function is given by the determinant of the coefficients of z^{-2} and z^{-1} in the functions f_2 and f_1:

$$\tau_W(T_1, T_2, T_3, \ldots) = \begin{vmatrix} 1 + a_2 T_1/a_1 & a_2/a_1 \\ T_1 + a_1(\tfrac{1}{2} T_1^2 + T_2) & 1 + T_1 + a_2(\tfrac{1}{2} T_1^2 + T_2) \\ + a_2(\tfrac{1}{6} T_1^3 + T_1 T_2 + T_3) & \end{vmatrix}$$

$$= 1 + a_1 T_1 + a_2 T_1^2 + \frac{a_2^2}{a_1} \left(\frac{1}{3} T_1^3 - T_3 \right).$$

Proposition 3.6.3. *If a Sato subspace $W \subset \mathbb{C}((z))$ is invariant under the multiplication by z^{-2}, then*

- *the function τ_W does not depend on the variables T_i with even indices;*
- *the family of the second order differential operators*

$$S(y, T_1, T_3, T_5, \dots) = \frac{\partial^2}{\partial y^2} + 2\frac{\partial^2}{\partial T_1^2} \log \tau_W(y + T_1, T_3, T_5, \dots)$$

satisfies the KdV hierarchy

$$\frac{\partial S}{\partial T_{2k+1}} = [S_+^{(2k+1)/2}, S].$$

Example 3.6.4. In the above example, the vector space $W \subset \mathbb{C}((z))$ spanned by the functions $f_1, f_2, \dots$ is invariant under the multiplication by z^{-2}. This remark explains why the τ-function does not depend on $T_2, T_4, \dots$. It is easy to verify directly that the function

$$U(y; T_1, T_3, \dots) = 2\frac{\partial^2}{\partial T_1^2} \log \tau_W(y + T_1, T_3, T_5, \dots)$$

$$= 2\frac{\partial^2}{\partial T_1^2} \log\bigl(1 + a_1(y + T_1) + a_2(y + T_1)^2 +$$

$$\frac{a_2^2}{a_1}(\frac{1}{3}(y + T_1)^3 - T_3)\bigr)$$

satisfies the Korteweg–de Vries equation

$$\frac{\partial U}{\partial T_3} = \frac{3}{2}U\frac{\partial U}{\partial T_1} + \frac{1}{4}\frac{\partial^3 U}{\partial T_1^3}.$$

Now let us outline the proof of the proposition; we refer the reader to [252] for the details of the proof.

The first statement is obvious. Indeed, if the subspace W is invariant under the multiplication by z^{-2}, then it remains invariant under the multiplication by $e^{T_2 z^{-2} + T_4 z^{-4} + \cdots}$ as well, and only odd terms in the exponent contribute to the τ-function.

In order to prove the second statement, we introduce the so-called Baker function of the subspace W. By definition, the *Baker function* Ψ_W is the function of the form

$$\Psi_W(z; T_1, T_2, \dots) = M(z; T_1, T_2, \dots)\Phi_W(z; T_1, T_2, \dots),$$

where Φ_W is the unique function of the form

$$\Phi_W(z; T_1, T_2, \dots) = 1 + k_1(T_1, T_2, \dots)z + k_2(T_1, T_2, \dots)z^2 + \dots$$

such that the vector Ψ_W belongs to W for any set of fixed values of T_i. The function Φ_W is related to the τ-function by the formula

$$\Phi_W(z; T_1, T_2, \ldots) = \frac{\tau_W(T_1 - z, T_2 - \tfrac{1}{2}z^2, T_3 - \tfrac{1}{3}z^3, \ldots)}{\tau_W(T_1, T_2, T_3, \ldots)}. \qquad (3.30)$$

In particular, the first two coefficients in the expansion of Φ_W in powers of z are

$$k_1(T_1, T_2, \ldots) = -\frac{1}{\tau_W} \frac{\partial \tau_W}{\partial T_1} \qquad (3.31)$$

$$k_2(T_1, T_2, \ldots) = \frac{1}{2\tau_W} \left(\frac{\partial \tau_W}{\partial T_2} + \frac{\partial^2 \tau_W}{\partial T_1^2} \right). \qquad (3.32)$$

Since W is invariant under the multiplication by z^{-2}, the function Φ_W does not depend on the variables T_i with even indices.

To the Baker function one associates the pseudodifferential operator

$$K_W(T_1, T_2, \ldots) = 1 + k_1(T_1, T_2, \ldots)\partial^{-1} + k_2(T_1, T_2, \ldots)\partial^{-2} + \ldots,$$

whose coefficients coincide with that of Φ_W, as well as the family of genuine differential operators of order r

$$P_{Wr}(T_1, T_2, \ldots) = \partial^r + p_{r2}\partial^{r-2} + \cdots + p_{rr}, \qquad r = 1, 2, 3, \ldots.$$

(Similarly to the preceding section we denote by ∂ the partial derivative with respect to y, $\partial = \partial/\partial y$.) The operators P_{Wr} are uniquely determined by the condition

$$\frac{\partial \Psi_W(y + T_1, T_2, \ldots)}{\partial T_r} = P_{Wr} \Psi_W(y + T_1, T_2, \ldots). \qquad (3.33)$$

In fact, the operators P_{Wr} have the form

$$P_{Wr} = Q_{W+}^r,$$

where $Q_W = K_W \partial K_W^{-1}$ is a pseudodifferential operator of the form

$$Q_W = \partial + q_1(T_1, T_2, \ldots)\partial^{-1} + \ldots.$$

For the operator Q_W, Eq. (3.33) reads

$$\frac{\partial Q_W}{\partial T_r} = [Q_{W+}^r, Q_W]. \qquad (3.34)$$

The operator K_W, and therefore all operators P_{Wr} as well as Q_W do not depend on all variables T_i with even indices. This implies that Q_W^2 is, in fact, a genuine differential operator, thus coinciding with P_{W2}. Hence, the operator $Q_W^2(y + T_1, T_3, T_5, \ldots)$ is the second order differential operator S from the statement of the proposition.

The proposition is proved.

If there is an infinite number of the basis vectors f_i different from z^{-i}, then the computation of the τ-function involves computing the determinant of an infinite dimensional matrix, which is a cumbersome task. It can be simplified, however, if we choose an integer N and make the substitution

$$T_k = \frac{1}{k}(z_1^k + \cdots + z_N^k), \qquad k = 1, 2, \ldots.$$

After this substitution, the exponential function M becomes the product

$$M\left(\sum z_i, \frac{1}{2}\sum z_i^2, \frac{1}{3}\sum z_i^3, \ldots\right) = \frac{1}{1 - z_1/z} \cdots \frac{1}{1 - z_N/z},$$

and multiplication by M becomes a rather simple operator. It can be shown that it suffices to compute only the determinant of the $N \times N$-matrix of the first N coefficients of the first N functions:

$$\tau_W\left(\sum z_i, \frac{1}{2}\sum z_i^2, \frac{1}{3}\sum z_i^3, \ldots\right) = \frac{(\cdots \wedge z^{N+1} \wedge Mf_N \cdots \wedge Mf_1) \wedge z^0 \wedge \cdots}{\cdots \wedge z^{-3} \wedge z^{-2} \wedge z^{-1} \wedge z^0 \wedge z^1 \wedge z^2 \wedge \cdots}$$

(with the substitution made both in f_i and M). This means that the determinant will not change if we replace z^{N+1} with f_{N+1}, or z^{N+1} with f_{N+1} and z^{N+2} with f_{N+2}, and so on; hence this finite determinant coincides with the required infinite one. For example, if $N = 1$, then $\tau_W(z_1, z_1^2/2, \ldots)$ coincides with the coefficient of z^{-1} in Mf_0.

Knowing

$$\tau_W\left(\sum_{i=1}^N z_i, \frac{1}{2}\sum_{i=1}^N z_i^2, \frac{1}{3}\sum_{i=1}^N z_i^3, \ldots\right)$$

for a given N allows one to compute several first coefficients of the τ-function. Choosing an appropriately large N we can compute each coefficient by a finite procedure. In particular, if the substitution

$$T_k = \frac{1}{k}(z_1^k + \cdots + z_N^k), \qquad k = 1, 2, \ldots.$$

in a τ-function and in some power series $F(T_1, T_2, \ldots)$ gives the same result for all N, then the two functions coincide. This fact was used by Kontsevich in the proof of Witten's conjecture.

3.7 Physical Interpretation

We had a chance to give the explanations presented below at a number of seminars in the presence of physicists. Usually our statements did not provoke any serious protests. Hoping that the politeness of our colleagues-physicists was not the unique reason for that, we dare to conclude that our point of view is not too different from that of theoretical physics.

3.7.1 Mathematical Relations Between Physical Models

We mathematicians know that mathematical theories of very distant origins may be related to one another. However, we were surprised to discover that the same is true for physical theories. We don't see profound philosophical reasons for that: a physical model must in the first place reflect the natural phenomenon it has a mission to describe. Nevertheless, the relations between different physical models do exist. One of the examples we have already seen is the two-matrix model. In this model Feynman diagrams which usually serve as a purely technical instrument of evaluating Gaussian integrals suddenly become configurations of the Ising model on a "dynamical lattice". But the Ising model was initially invented in order to describe the ferromagnetism phenomenon and its disappearance under heating.

Another and much more famous example is the KdV equation and its higher analogues. Korteweg and his student de Vries invented their equation in order to describe waves in shallow water. Its relation to the Schrödinger equation discovered much later was already a big surprise. And now we come across the same equation in the theory of map enumeration. In Chapter 4 it will play an important role in the intersection theory on the moduli space of curves.

In the same vein, the matrix integrals approach to the map enumeration admits at least two physical interpretations: as a discretization of a path integration in string theory, and as a particular quantum field model.

3.7.2 Feynman Path Integrals and String Theory

In classical mechanics a point-like particle that moves from A to B "chooses" a trajectory $x(t)$ that minimizes the functional of action $S[x]$. In quantum mechanics (more exactly, in Feynman path integrals approach) a particle moves from A to B along all possible trajectories, the contribution of each trajectory into the "amplitude of probability" being

$$e^{-(i/\hbar)S[x]}.$$

In order to evaluate the amplitude (and other physical quantities) one must compute an integral over the space of all possible trajectories.

It turns out that the "space of all trajectories" is not well defined, and the measure on this space is complex-valued and not countably additive, so the result of the integration depends on the choice of the integration procedure. Never mind: the physical intuition often permits one to choose such a way of integration that the results not only become reasonable but even obtain an experimental confirmation.

One of the techniques of such computation consists in replacing trajectories by broken lines, thus reducing the problem to a finite-dimensional integration. If in the process of its evolution in time the particle breaks down in two, and/or

212   3 Matrix Integrals

two particles collide and form one particle, the space of trajectories becomes more complicated: we must integrate over all pairs of points M and N, and then over all trajectories from A to M, all pairs of trajectories from M to N, and all trajectories from N to B (see Fig. 3.13).

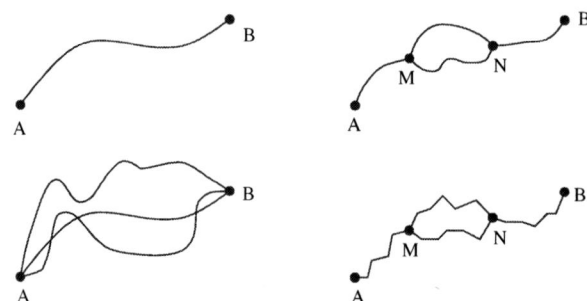

Fig. 3.13. Feynman paths and their discrete approximation

In the string theory approach a particle is no longer point-like, but is a small circle ("string"). As this circle evolves in time, a two-dimensional surface appears, which serves as an analog of a trajectory. The process of breaking down/colliding is described by the addition of a new handle to the surface (Fig. 3.14). We may suppose that the particle is born ex nihilo and disappears at the end of its life; or we may suppose that the time segment is finite. Then the corresponding surface becomes compact.

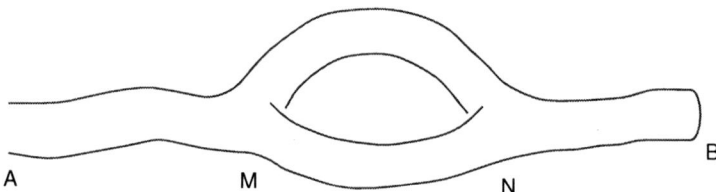

Fig. 3.14. Evolution of a string in time

The first impression is that the problem of integrating over the space of surfaces becomes hopeless. In fact, at least in certain respects it becomes simpler. One of its possible interpretations consists in considering the surfaces as complex curves. Then one may integrate over the moduli space of such curves, and this space is finite-dimensional. Another possibility consists in considering some kind of discrete approximations to the "paths". A very natural class of such approximations is represented by maps! This is how the maps enter the scene. In the cases when both approaches can be applied they give consistent results.

In fact, the models used by physicists are somewhat more sophisticated. For example, they use an integration over all metrics on a surface. But a map, e.g., a triangulation, represents a very natural way of discretizing a metric. A triangulation of a surface determines a piecewise flat metric on it: we simply suppose each triangle to be a regular plane triangle of a given area. Knowing the total area of the surface and the number of triangles one can easily calculate the area of a triangle, and therefore its side. Physicists usually suppose that for the number of triangles large enough, summation over the (finite) set of all triangulations is equivalent to integration over the space of all metrics. Here triangulations may well be replaced by maps or by some other types of cellular decompositions.

As we shall see in the next chapter, this approach, at least on the level of ideas, is consistent with the integration over the moduli space. One may introduce a cell decomposition of moduli spaces (in fact, of "decorated" moduli spaces: see the next chapter) of curves with marked points associated to embedded graphs structure. In these cell decompositions, triangulations naturally correspond to open cells, while other embedded graphs label cells of higher codimension. Thus, summing up over triangulations produces the most essential part of the integral.

An even more complicated class of models considers the surfaces not as abstract entities but as geometric objects embedded (or immersed) in a certain space. See in this respect [181], where embeddings of surfaces into a space of dimension -2 are considered! (Indeed, the level of abstraction in modern physics is often far beyond that of mathematics.)

3.7.3 Quantum Field Theory Models

There is a general recipe how to construct these models. First of all we must choose two spaces X and Y. The space X is our "model Universe". Fields defined on X take their values in Y. According to the nature of this space the field under consideration is scalar, or vector, or spinor, etc.

A field is not a function $f : X \to Y$, and is not even the space

$$\mathcal{F} = \{f \mid f : X \to Y\}$$

of such functions, but a measure on the latter space. Usually the density of the measure takes the form

$$\exp\{-(\text{quadratic functional of } f) - (\text{interaction term})\},$$

where the interaction term is of order greater than two. When there is no interaction, the measure becomes Gaussian (since it becomes exponent of a quadratic term), and is called the "free field". (Recall your school course of physics: there was kinetic energy which was quadratic, and potential energy which had an arbitrary form.) The interaction term quite often depends on some parameters. The goal is to integrate the measure over the space $\mathcal{F}$, thus finding what is called the *partition function*, and then to study its dependence on parameters.

The one-matrix model considered above has all the ingredients of this general scheme. The nature of the field is rather complicated, since the role of the space Y is played by $\mathcal{H}_N$, the space of Hermitian matrices. The free field and the corresponding Gaussian measure are described by the quadratic functional $\frac{1}{2}\mathrm{tr}(H^2)$. The interaction term is $t\,\mathrm{tr}(H^4)$, where t is a parameter. The most unusual part of the story is, however, the space X: *our model Universe consists of one point!*

The reader has probably come across such fancy things as a model of the Universe which takes the form of a 26-dimensional space, with certain number of usual "commutative" coordinates and other anti-commutative "super"-coordinates. In some models the Universe has a curvature, it may become a compact manifold, and so on. What we have seen in this chapter shows however that even a model in which the Universe consists of a single point may already be highly non-trivial. This observation permits better estimating the level of mathematical non-triviality of modern theoretical physics.

In the two-matrix model the Universe consists of 2 points, and in the many-matrix models it consists of several points (for example, $2q$ points in the meander problem).

3.7.4 Other Models

Certain combinatorial models expressed in terms of matrix integrals admit other physical interpretations. For example, the mere geometric nature of the meander problem suggests its interpretation as a model of polymer folding, and also as a specific model of the self-avoiding walk, both models being very important in statistical physics.

But the story does not stop here; it continues as follows. In [83] the authors explain that the meander problem may be interpreted as "a pair of two fully packed loop models". There is no proof (or probably there is one but we don't see it), but there are certainly very convincing physical arguments for such an interpretation. Now, a fully packed loop model has the "central charge" equal to $c = -2$, and "therefore" the central charge for the meander model is $c = -4$. Now, if we are ready to believe in the conformal invariance of the model, we may use the Knizhnik–Polyakov–Zamolodchikov (KPZ) equation which expresses the "string susceptibility" γ_{str} through the central charge:

$$\gamma_{\mathrm{str}}(c) = \frac{c - 1 - \sqrt{(25-c)(1-c)}}{12}.$$

Finally we may use the expression of the critical exponent α through the string susceptibility:
$$\alpha = 2 - \gamma_{\text{str}}.$$
For $c = -4$ all this gives
$$\alpha = \frac{29 + \sqrt{145}}{12} = 3.4201\ldots,$$
which is very close to the experimental value $\alpha = 3.4206\ldots$ (see our discussion in Sec. 3.4.4).

We hope that the reader will forgive us for not explaining the exact meaning of all physical terms we have used. All this remains very far from a rigorous mathematical proof. But without a physical intuition one would never guess which direction to follow.

The question of the mathematical rigor is often raised by our colleagues, and quite often we don't know what to answer. It is true that physicists usually don't provide a rigorous treatment of the subject. They don't pretend to do that, and are not supposed to. On the other hand, all the parts of the theory we were able to understand turned out to be rigorous. But what then is the difference between "understand" and "find a rigorous exposition"? Whatever are the answers to all these annoying questions, we feel it as our obligation to pursue the process of understanding of this remarkable theory.

3.8 Appendix

3.8.1 Generating Functions

Suppose we have a finite or infinite sequence of numbers enumerating some classes of objects. Then it is often convenient to treat this sequence as the sequence of coefficients of a polynomial or of a power series respectively. In this case the polynomial or the power series is called the *generating function* for the sequence. For example, the function
$$\left(\frac{1}{1-s}\right)^2 = 1 + 2s + 3s^2 + 4s^3 + \ldots$$
is the generating function for the sequence $1, 2, 3, 4, \ldots$, and the binomial formula is the generating polynomial for the binomial coefficients:
$$(1+s)^m = \binom{m}{0} + \binom{m}{1}s + \cdots + \binom{m}{m}s^m.$$
Newton's binomial formula simply generalizes the last one to the case of arbitrary (e.g., complex) degree α:
$$(1+s)^\alpha = 1 + \alpha s + \frac{\alpha(\alpha-1)}{2!} + \cdots + \frac{\alpha(\alpha-1)\ldots(\alpha-m+1)}{m!}s^m + \ldots.$$

We do not require that the power series corresponding to a sequence have a non-trivial convergence disk, that is, we work with *formal power series*. Natural operations with classes of objects often can be interpreted in terms of algebraic manipulations, like summing or differentiating, with formal power series. Moreover, the coefficients in formal power series can be taken from abstract rings, not necessarily commutative.

Below we give two important examples of generating functions; they are used in Secs. 3.1, 3.2.

Exercise 3.8.1. Prove that

$$\left(\frac{1}{1-s}\right)^m = \binom{m}{0} + \binom{m+1}{1}s + \binom{m+2}{2}s^2 + \cdots + \binom{m+k}{k}s^k + \cdots.$$

Exercise 3.8.2 (Catalan numbers). The sequence of *Catalan numbers* Cat_n is defined by the initial value $\mathrm{Cat}_0 = 1$ and the recurrence relation

$$\mathrm{Cat}_{n+1} = \mathrm{Cat}_0\,\mathrm{Cat}_n + \mathrm{Cat}_1\,\mathrm{Cat}_{n-1} + \cdots + \mathrm{Cat}_n\,\mathrm{Cat}_0.$$

The first Catalan numbers are 1, 1, 2, 5, 14, 42, 132, ...

1. Prove that
$$\mathrm{Cat}_n = \frac{1}{n+1}\binom{2n}{n} = \frac{(2n)!}{n!(n+1)!}.$$

2. Using the above recurrence relation deduce the generating function for the Catalan numbers:

$$\mathrm{Cat}_0 + \mathrm{Cat}_1\, s + \mathrm{Cat}_2\, s^2 + \cdots = \frac{1-\sqrt{1-4s}}{2s}.$$

3. Prove the following asymptotics for the nth Catalan number:

$$\mathrm{Cat}_n \sim \mathrm{const}\cdot 4^n \cdot n^{-3/2}.$$

Besides usual generating functions considered above, there are other types of generating functions. For example, there is a way to associate a class of generating functions to arbitrary fixed sequence $b_0, b_1, b_2, \ldots$: we assign to a sequence $a_0, a_1, \ldots$ the formal power series

$$a_0 b_0 + a_1 b_1 s + a_2 b_2 s^2 + \cdots.$$

The sequence $b_k = 1/k!$ is among the most widely used sequences b_i; the corresponding generating functions are called the *exponential generating functions*. For example,

$$e^{-x} = 1 - \frac{x}{1!} + \frac{x^2}{2!} - \cdots$$

is the exponential generating function for the sequence $1, -1, 1, -1, 1, \ldots$. The differentiating of an exponential generating function leads to the shift of the sequence of coefficients a_i. Exponential generating functions are extremely useful in the enumeration of labelled objects (see below).

For more about generating functions see [265], [113].

3.8.2 Connected and Disconnected Objects

The operation of taking the logarithm is very well known in combinatorics; see, for example, [136], [102], [24]. If one has an *exponential* generating function for a class of *labelled* objects, then its logarithm is the generating function for the *connected* objects of the same type.

Recall that the word "exponential" means "with $n!$ in the denominator". The word "labelled" means that the constituent parts of an object (like vertices, edges, darts, etc.) are marked in such a way that the automorphism group of every object in the class becomes trivial. Let us, however, add an important remark. One may consider objects (such as graphs) with additional structures (such as cyclic orders of edges around vertices) and with some properties imposed (such as prescribed vertex degrees). But one must be careful: a "property" has a legal right to be imposed when *it is valid for an object if and only if it is valid for all its connected components*, that is, if it is hereditary. For example, the property

Every vertex of a graph is colored in one of k colors

is acceptable; on the contrary, the property

Every vertex of a graph is colored in one of k colors, and *all k colors are used*

is not good, because it may be valid for a graph as a whole but not valid for some of its connected components.

Now let us be slightly more formal. Consider a graded set of *labelled objects* $\mathcal{O} = \mathcal{O}_1 \cup \mathcal{O}_2 \cup \mathcal{O}_3 \cup \ldots$ satisfying the following properties:

- each of the sets $\mathcal{O}_k$ is finite;
- an operation of union of the objects $\mathcal{O} \times \mathcal{O} \to \mathcal{O}$ is defined, which takes a pair (o_1, o_2) with $o_1 \in \mathcal{O}_k, o_2 \in \mathcal{O}_m$ into an object $o_1 + o_2 \in \mathcal{O}_{k+m}$;
- any object $o \in \mathcal{O}$ admits a unique decomposition into a union of connected objects $o = o_1 + o_2 + \cdots + o_k$. (A *connected object* is an object o that admits no decomposition $o = o_1 + o_2$.)

We don't give a formal definition of the notion of being labelled, limiting ourselves to the remarks given above. The intuitive meaning is clear, while the development of a coherent and rigorous theory would need a considerable effort (see, for example, [24]). Just note that the decomposition $o = o_1 + o_2 + \cdots + o_k$ of any object $o \in \mathcal{O}$ into connected components is, indeed, unique, as is said above, and not unique "up to the order", and this is so exactly because of the labelling.

Denote the set of connected labelled objects by $\mathcal{C} \subset \mathcal{O}$; we have $\mathcal{C} = \mathcal{C}_1 \cup \mathcal{C}_2 \cup \ldots$, with $\mathcal{C}_i = \mathcal{C} \cap \mathcal{O}_i$. Let $F : \mathcal{O} \to K$ be a multiplicative mapping of the set $\mathcal{O}$ to a ring K, i.e., $F(o_1 + o_2) = F(o_1)F(o_2)$. We can set, for example, $F(o) = t^n \in \mathbb{Z}[t]$ for $o \in \mathcal{O}_n$. Consider two exponential generating functions

$$F_{\mathcal{O}}(s) = 1 + \frac{1}{1!}s \sum_{o \in \mathcal{O}_1} F(o) + \frac{1}{2!}s^2 \sum_{o \in \mathcal{O}_2} F(o) + \ldots,$$

$$F_{\mathcal{C}}(s) = \frac{1}{1!}s \sum_{o \in \mathcal{C}_1} F(o) + \frac{1}{2!}s^2 \sum_{o \in \mathcal{C}_2} F(o) + \ldots$$

for the set of labelled objects and for the set of connected labelled objects respectively.

Proposition 3.8.3. *The above generating functions are related by the following formulas:*
$$F_{\mathcal{O}}(s) = \exp(F_{\mathcal{C}}(s)).$$
or, equivalently,
$$F_{\mathcal{C}}(s) = \log(F_{\mathcal{O}}(s)).$$

Idea of the proof. The set of indices $\{1, 2, \ldots, n\}$ is distributed among the connected components $o_1, \ldots, o_k$ in such a way that each component obtains the number of indices equal to its grading. Let us look at the coefficient of s^n in the exponent on the right-hand side of the first formula. A contribution to this coefficient is given by a set of connected labelled objects $o_1, \ldots, o_k$ such that $o = o_1 + \cdots + o_k \in \mathcal{O}_n$. Suppose $o_1 \in \mathcal{O}_{i_1}, \ldots, o_k \in \mathcal{O}_{i_k}$, where $i_1 + \cdots + i_k = n$. Then the contribution is equal to

$$\frac{1}{i_1!} \cdots \frac{1}{i_k!}.$$

On the other hand, the disconnected object o admits

$$\frac{n!}{i_1! \ldots i_k!}$$

labellings. Therefore its contribution to the coefficient of s^n in the generating function on the left-hand side is

$$\frac{1}{i_1! \ldots i_k!}.$$

A more rigorous proof would need more precise definitions.

In practice the "distribution of roles" between various labels and variables may turn out to be rather subtle. For example, while enumerating four-valent plane maps, we consider a map with n vertices as an object with grading n, its vertices being labelled by $1, 2, \ldots, n$. At the same time, the labelling of the darts by 1, 2, 3, 4 (see Remark 3.5.6) must be considered as a "property" imposed on the objects. Generating functions in Proposition 3.3.7 and in Eq. (3.13) are power series in two variables s and N, but the roles of these variables are different. While s is the formal parameter of the generating functions, their coefficients take values in the ring of polynomials in N: to each map we associate $N^{\#(\text{faces})}$. This function, defined on the set of "disconnected maps", is obviously multiplicative.

3.8.3 Logarithm of a Power Series and Wick's Formula

The Wick formula can be considered as a very particular case of a formula well known in probability theory, which expresses the moments for the probability distribution in terms of its semi-invariants. The latter one, in its turn, is a "probabilistic interpretation" of the formula of logarithm of a power series. For more details see, for example, [210], Chapter 2 "Semi-invariants and combinatorics".

We use the multi-index notation:

$$\alpha = (\alpha_1, \ldots, \alpha_n), \quad \alpha_i \in \mathbb{N},$$
$$|\alpha| = \alpha_1 + \ldots + \alpha_n,$$
$$\alpha! = \alpha_1! \times \ldots \times \alpha_n!,$$
$$t^\alpha = t_1^{\alpha_1} \ldots t_n^{\alpha_n}, \quad \text{where} \quad t = (t_1, \ldots, t_n),$$
$$\frac{\partial^{|\alpha|} f}{\partial t^\alpha} = \frac{\partial^{|\alpha|} f}{\partial t_1^{\alpha_1} \ldots \partial t_n^{\alpha_n}}.$$

Let

$$f(t) = 1 + \sum_{|\alpha|>0} \frac{m_\alpha}{\alpha!} t^\alpha$$

be the Taylor series for a function $f(t)$ satisfying $f(0) = 1$; here

$$m_\alpha = \left. \frac{\partial^{|\alpha|} f}{\partial t^\alpha} \right|_{t=0}.$$

Let the following be the Taylor series for $\log f(t)$:

$$\log f(t) = \sum_{|\alpha|>0} \frac{s_\alpha}{\alpha!} t^\alpha.$$

The formula below is interesting in itself. It expresses the coefficients m_α in terms of s_α.

Proposition 3.8.4. *The Taylor coefficients m_α of a function f are expressed in terms of the Taylor coefficients s_α of $\log f$ by the following formula:*

$$m_\alpha = \sum_{k=1}^{|\alpha|} \frac{1}{k!} \sum_{\substack{\beta^1 + \ldots + \beta^k = \alpha, \\ |\beta^i| > 0}} \frac{\alpha!}{\beta^1! \ldots \beta^k!} \prod_{i=1}^{k} s_{\beta^i}.$$

Here β^i are also multi-indices; the interior sum is taken over all ordered k-tuples of multi-indices (but the factor $1/k!$ "kills the order").

Remark 3.8.5. The above formula is valid for formal power series as well as for analytic functions.

Proposition 3.8.6. *Let μ be a probability distribution in $\mathbb{R}^n$ (an arbitrary one, not necessarily Gaussian), and let*

$$\varphi(t) = \langle e^{i(t,x)} \rangle = \int_{\mathbb{R}^n} e^{i(t,x)} d\mu(x) \tag{3.35}$$

be its characteristic function. If φ is analytic in a neighborhood of 0, then all the moments $m_\alpha = \langle x^\alpha \rangle$ exist, and

$$\varphi(t) = 1 + \sum_{|\alpha|>0} \frac{m_\alpha}{\alpha!} (it)^\alpha.$$

The coefficients s_α of the series

$$\log \varphi(t) = \sum_{|\alpha|>0} \frac{s_\alpha}{\alpha!} (it)^\alpha$$

are called *semi-invariants* of the distribution μ, so the formula of Proposition 3.8.4 expresses the moments of a probability distribution in terms of its semi-invariants.

Formulas and computations involving semi-invariants are often much simpler than those involving moments. A Gaussian distribution is a striking example of this phenomenon, since for the characteristic function (3.35), its logarithm is no longer an infinite series but a quadratic polynomial

$$\log \varphi(t) = -\frac{1}{2}(Ct, t), \tag{3.36}$$

so $s_\alpha \neq 0$ implies $|\alpha| = 2$, and for such an α we have $s_\alpha = m_\alpha = c_{ij}$ (here $\alpha_i = \alpha_j = 1$ if $i \neq j$, and $\alpha_i = 2$ if $i = j$, while all the other components of the multi-index α are equal to 0).

Remark 3.8.7. Formula (3.36) is often taken as the *definition* of the Gaussian measure. It remains meaningful even when the matrix C becomes degenerate and the definition given in Eq. (3.4) is no longer valid, since $B = C^{-1}$ does not exist. In this case the measure itself is also degenerate: it is concentrated on a subspace of a smaller dimension.

Lemma 3.8.8. *Let μ be a Gaussian measure in $\mathbb{R}^n$, and let $A : \mathbb{R}^n \to \mathbb{R}^k$ be a linear operator. Then A induces a measure μ_A in $\mathbb{R}^k$, which is also Gaussian.*

Proof. Let C be the covariance matrix of μ, and let $x, t \in \mathbb{R}^n$, $y, s \in \mathbb{R}^k$, and $y = Ax$. Then

$$\varphi_{\mu_A}(s) = \langle e^{i(s,y)} \rangle = \langle e^{i(s,Ax)} \rangle = \langle e^{i(A^*s,x)} \rangle = \varphi_\mu(A^*s),$$

where $A^* : \mathbb{R}^k \to \mathbb{R}^n$ is the operator adjoint to A. What remains is to substitute A^*s for t in $\varphi_\mu(t) = \exp\{-\frac{1}{2}(Ct,t)\}$, which gives the covariance matrix of μ_A equal to ACA^*.

Remark 3.8.9. In the above lemma, the dimension k may be less than n, equal to n, or greater than n, and the operator A may be of arbitrary rank.

Return now to the Wick formula. Let $y_1 = f_1(x)$, ..., $y_{2k} = f_{2k}(x)$. According to Lemma 3.8.8 the vector $(y_1, \ldots, y_{2k})$ has a Gaussian distribution, and our goal is to compute, for this distribution, the moment $\langle y_1 \ldots y_{2k} \rangle$ which corresponds to the multi-index $\alpha = (1, 1, \ldots, 1)$. Now we can apply the formula of Proposition 3.8.4 and split α into sums of multi-indices β^i, of which survive only those of the form $(0, \ldots, 0, 1, 0, \ldots, 0, 1, 0, \ldots, 0)$, having exactly two non-zero entries.

The Wick formula is proved.

We finish this chapter by yet one more facet of the logarithm. If you ask a physicist why the logarithm in Eq. (3.14) must be taken, he will most probably reply that the partition function does not have its proper physical meaning, but its logarithm has the meaning of energy. And the fact that the same operation leaves only connected Feynman diagrams is one of the many miracles we encounter in this theory almost everywhere.

$$* \quad * \quad *$$

It is possible that many enumerative results of this chapter could be obtained by purely combinatorial means, without having to resort to matrix integrals. However, without physical intuition one often does not know which direction to take. On the technical level, much remains to be done. First of all, the entire theory must be made mathematically rigorous (cf., e.g., [130]), or, at least, a careful distinction must be made between the results having rigorous proofs and those for which there are only plausible physical arguments. Then, it is necessary to develop more industrial and less handicraft methods of computing matrix integrals. This chapter is also a message to enumerative combinatorialists: among a large variety of objects one can enumerate, maps are of special interest because of their revealed relations to physics. If you hesitate what to enumerate, choose maps.

The history of mathematics shows that when its branch becomes related to physics, then in due time it acquires a fundamental importance for mathematics itself and starts to occupy a place close to its core. If you wish to be convinced that this remark is valid for embedded graphs, read the next chapter.

4

Geometry of Moduli Spaces of Complex Curves

In this chapter we present an overview of the connection between the geometry of moduli spaces of complex curves with marked points and the topology of embedded graphs. According to Harer [137], the idea of this connection belongs to Mumford. It proved to be extremely fruitful. The most celebrated results here are the calculation of the orbifold Euler characteristic of moduli spaces of smooth curves due to Harer and Zagier [138] (we present also Kontsevich's calculation based on similar ideas) and Kontsevich's proof of Witten's conjecture. We must note that a complete exposition of this proof containing all the details has not yet been published. Our text does not fulfill this mission either. In the next chapter we will show how the geometry of moduli spaces of curves is related to that of Hurwitz spaces, that is, the moduli spaces of meromorphic functions on complex curves. Since a meromorphic function is also associated to an embedded graph, we obtain another facet of the same connection.

4.1 Generalities on Nodal Curves and Orbifolds

This section contains additional information about complex curves we shall need in the present chapter and in the next one.

4.1.1 Differentials and Nodal Curves

Let X be a complex curve. A *meromorphic 1-form* (or, a *meromorphic differential*) over X is an object locally represented, in a coordinate z, as $\varphi(z)dz$, where φ is a meromorphic function. For example, for each meromorphic function $f : X \to \mathbb{C}P^1$ its differential df is a meromorphic 1-form. A coordinate change $z = \psi(z_1)$ leads to the following transformation of the coordinate presentation of the differential:

$$\varphi(z)dz = \varphi(\psi(z_1))\psi'(z_1)dz_1.$$

4 Geometry of Moduli Spaces

The *order of a pole* of a meromorphic differential is the order of the pole of the function $\varphi(z)$ in any local representation. For a meromorphic 1-form ω with a pole at a point $x_0 \in X$ its *residue* $\operatorname{Res}_{x_0} \omega$ is defined as the integral of ω over a small circle going around x_0 in the positive direction, divided by $2\pi i$. In other words, $\operatorname{Res}_{x_0} \omega$ coincides with the coefficient a_{-1} in any coordinate expansion

$$\omega = \left(\frac{a_{-k}}{z^k} + \cdots + \frac{a_{-1}}{z} + a_0 + a_1 z + \ldots \right) dz$$

of ω in a vicinity of x_0, and this value is independent of the choice of the coordinate z.

If a meromorphic 1-form admits a local representation without poles in each chart, then it is a *holomorphic 1-form*. Holomorphic differentials form a subspace in the infinite dimensional vector space of meromorphic differentials over $\mathbb{C}$. The space of holomorphic 1-forms is finite dimensional, and if X is a curve of genus g, then the complex dimension of the space of holomorphic 1-forms on X is g.

Besides smooth complex curves, we shall also need singular curves with simplest possible singularities, namely, nodal curves. While a smooth point of a curve admits a neighborhood biholomorphic to a disc, a *node*, or a *double point*, admits a neighborhood biholomorphic to a pair of disks with identified centers.

A *nodal curve* is a complex curve whose only singularities are nodes. The set of nodes of a curve is always finite. If we remove the node from a neighborhood which is biholomorphic to a pair of glued disks, then we obtain two disjoint punctured disks. They are called the *sheets* of the curve at the node.

It is convenient to think of nodal curves as of degenerations of smooth curves. For example, consider the family $t_1 t_2 = \varepsilon t_0^2$ on the projective plane (or the family $t_1 t_2 = \varepsilon$ in the affine chart $t_0 = 1$) depending on the complex parameter ε. Then for $\varepsilon \neq 0$ the member of the family is a smooth rational (i.e., of genus zero) curve. For $\varepsilon = 0$ the curve is no longer smooth: it acquires a node at the point $t_1 = t_2 = 0$; see the real part of these curves in Fig. 4.1. For a general nodal curve the construction is more complicated, and it requires

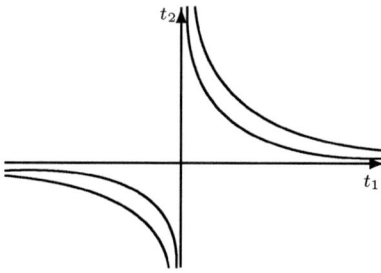

Fig. 4.1. The family of rational curves $t_1 t_2 = \varepsilon$ degenerating into the nodal curve $t_1 t_2 = 0$

studying families of curves in projective spaces of higher dimension. We do not describe it here.

Note that if we remove the node from the singular curve in the last example, then the curve splits into two disjoint rational curves given by the equations $t_1 = 0$ and $t_2 = 0$ respectively, each punctured at a single point. Attaching a point to each puncture, we obtain two curves that are called the *irreducible components* of the singular curve. However, a nodal curve does not necessarily have more than one irreducible component. For example, the singular curve $t_2^2 t_0 = t_1^2(t_1 + t_0)$ (or $t_2^2 = t_1^2(t_1 + 1)$ in the affine chart $t_0 = 1$) consists of a unique irreducible component. The real part of this curve is presented in Fig. 4.2.

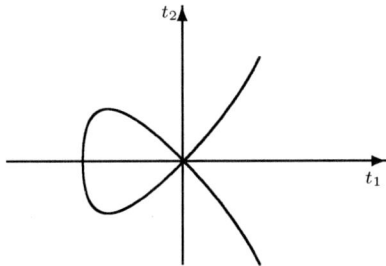

Fig. 4.2. The degenerate elliptic curve $t_2^2 = t_1^3 + t_1^2$

For a general nodal curve X, splitting a neighborhood of each double point into a pair of disjoint disks yields a smooth curve $\widetilde{X}$ consisting of a number of connected components. This curve, together with the mapping $\widetilde{X} \to X$ gluing in pairs the preimages of the double points, is called the *normalization* of the nodal curve. In the normalization, each node is replaced by a pair of points called the *normalized double point*. The connected components of the normalization $\widetilde{X}$ are in one-to-one correspondence with the irreducible components of the curve X. The two points in each pair can belong either to the same connected component of $\widetilde{X}$, or to two distinct connected components, and after gluing they form a double point in X. The normalization is a good way to say what a meromorphic function on a nodal curve is. A *meromorphic function on a nodal curve* X is simply a meromorphic function on the normalization $\widetilde{X}$ that takes coinciding values at each two points of $\widetilde{X}$ normalizing the same double point.

The singular curve in Fig. 4.2 can be considered as a rational curve with two points glued together; the rational curve in question is the normalization of the singular one. However, it is more convenient to treat it as an elliptic curve (that is, a curve of genus 1). It may be obtained as the result of degeneration of the family of elliptic curves $t_2^2 = t_1^2(t_1 + 1) + \varepsilon$.

Let us now turn to the following question: how to define the genus of a nodal curve? The easiest way to do this is to cut off a neighborhood of each node, that is, a pair of disks glued at a point, and to replace it with a cylinder preserving the orientation. We take the genus of the resulting smooth two-dimensional surface for the *genus of the nodal curve*. While defining the genus we are interested only in the topology of the resulting surface, not in the complex structure on it, and in the topological sense the gluing procedure is well-defined since the boundary of the pair of disks, as well as the boundary of the cylinder, consists of a pair of circles.

Another definition of the genus of a nodal curve, of which we shall also make an extensive use below, proceeds in terms of meromorphic differential 1-forms on its normalization. Namely, the genus g of a nodal curve X is the dimension of the space of meromorphic differentials on $\widetilde{X}$ possessing the following properties:

- the differentials are allowed to have poles only at the normalized double points;
- if there is a pole at a normalized double point, then the order of this pole is 1;
- the sum of the residues of the differentials at the two normalizations of the same double point is zero.

The last requirement has the following explanation. Consider the family of curves $t_1 t_2 = \varepsilon$ in the neighborhood of the origin. The restriction of the differential dt_1/t_1 to a nondegenerate curve of this family has no poles, while its restriction to the irreducible component $t_2 = 0$ of the degenerate curve corresponding to the value $\varepsilon = 0$ has a pole of order one at the origin, and its residue at the pole is $+1$. On the other hand, in the coordinate t_2, the same differential, when restricted to the curve $t_1 t_2 = \varepsilon$, has the coordinate presentation $-dt_2/t_2$. This shows that its restriction to the other irreducible component $t_1 = 0$ of the nodal curve also has a pole of order one, but with the residue -1. This local situation is, in a way, generic.

Exercise 4.1.1. Show that the two definitions of the genus of a nodal curve are equivalent. [**Hint:** Proceed by induction on the number of nodes.]

Exercise 4.1.2. 1. Verify that if a curve X is obtained as the union of two curves X_1 and X_2 of genera g_1 and g_2 respectively intersecting at a single point, then the genus of X is $g_1 + g_2$.

2. Show that the gluing of two smooth points of a connected nodal curve increases its genus by one.

4.1.2 Quadratic Differentials

A *meromorphic quadratic differential* on a complex curve X is an object with a local coordinate presentation $\rho = \varphi(z)(dz)^2$, where φ is a meromorphic function. A change of variable $z = \psi(z_1)$ leads to the following transformation of the coordinate presentation:

$$\varphi(z)(dz)^2 = \varphi(\psi(z_1))(\psi'(z_1))^2 (dz_1)^2.$$

4.1 Nodal Curves and Orbifolds

For each meromorphic differential $\omega = \varphi(z)dz$ its square $\omega^2 = (\varphi(z))^2(dz)^2$ is a quadratic differential. However, not each quadratic differential is the square of some 1-form. For instance, this is obviously not true for quadratic differentials with poles and/or zeroes of odd order.

Let $x_0 \in X$ be a pole of even order $2k$ of a quadratic differential ρ. Then it is possible to take the square root $\omega = \rho^{1/2}$ in a neighborhood of x_0, and it is a 1-form with a pole of order k,

$$\omega = \left(\frac{a_{-k}}{z^k} + \cdots + \frac{a_{-1}}{z} + a_0 + \ldots\right) dz.$$

This 1-form is well defined up to a sign. Being the residue of ω at x_0, the nonzero coefficient a_{-1} is also well defined up to a sign; hence its square is well-defined and does not depend on the coordinate presentation. We call this square the *quadratic residue* of a meromorphic quadratic differential at a pole of even order. For a pole of order two, the quadratic residue coincides with the coefficient of $(dz/z)^2$ in any coordinate presentation of the quadratic differential. At a pole of odd order (or if there is no pole at all), the quadratic residue is zero.

Similarly to meromorphic differentials, the notion of meromorphic quadratic differential extends to nodal curves as well. Namely, we allow a quadratic differential on a nodal curve to have poles of order not greater than two at the nodes, with coinciding quadratic residues at the two sheets of the curve.

Below, we shall be interested first of all in meromorphic quadratic differentials having poles of order at most two. The dimension of the space of meromorphic quadratic differentials on a smooth genus g curve having poles of order at most two at n marked points is $3g - 3 + 2n$. The space of meromorphic quadratic differentials on nodal curves, having poles of order at most two at the n marked and the double points, with coinciding quadratic residues at the two sheets at each double point, also has dimension $3g - 3 + 2n$.

4.1.3 Orbifolds

The moduli spaces of curves which we are going to study in this chapter are usually not smooth varieties, but *orbifolds*. Before giving a precise definition of an orbifold, in order to supply the reader with an impression of what an orbifold looks like, we present a simple example.

Example 4.1.3. Take the complex projective line $\mathbb{C}P^1$ with fixed coordinate z, and consider the mapping f of $\mathbb{C}P^1$ onto itself, $f : z \mapsto iz$. The iterations $id, f, f \circ f, f \circ f \circ f$ of this mapping define an action of the cyclic group $C_4 = \mathbb{Z}/4\mathbb{Z}$ on $\mathbb{C}P^1$. The quotient space of $\mathbb{C}P^1$ modulo this action is shown in Fig. 4.3.

Speaking informally, a d-dimensional complex *orbifold* is a topological space obtained as a union of quotients of open balls in $\mathbb{C}^d$ modulo discrete

228   4 Geometry of Moduli Spaces

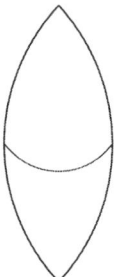

Fig. 4.3. The orbifold S^2/C_4

group actions so that these actions are consistent along intersections of the balls. A real orbifold has a similar meaning. The gluing data plays the role of an atlas in usual topology. In the above example each point of the orbifold, except two, admits a neighborhood, which is simply a disk in $\mathbb{C}^1$ endowed with the action of the trivial group. The action itself is, of course, trivial too. Each of the two exceptional points admits a neighborhood which is the quotient of a disk in $\mathbb{C}^1$ modulo the group C_4 action. The following definition (see, e.g. [214], Appendix A) generalizes this example.

Definition 4.1.4. A smooth d-dimensional complex *orbifold* O is a Hausdorff topological space M endowed with an atlas $\langle U_\alpha, V_\alpha, G_\alpha, \phi_\alpha \rangle$, where

- the set $\{U_\alpha\}$ is an open covering of M providing a base for the topology on M;
- the set $\{V_\alpha\}$ is a collection of open subsets in $\mathbb{C}^d$;
- each G_α is a finite group of diffeomorphisms of V_α; and
- $\phi_\alpha : V_\alpha \to U_\alpha$ is the continuous mapping whose fibers are the orbits of G_α.

The atlas must satisfy the following compatibility condition: whenever $U_\alpha \subset U_\beta$, there exists an injective homomorphism $h_{\alpha\beta} : G_\alpha \to G_\beta$ and a smooth embedding $\phi_{\alpha\beta} : V_\alpha \to V_\beta$ such that

- for all $g \in G_\alpha$ and $x \in V_\alpha$ we have $\phi_{\alpha\beta}(gx) = h_{\alpha\beta}(g)\phi_{\alpha\beta}(x)$;
- for all $x \in V_\alpha$ we have $\phi_\beta(\phi_{\alpha\beta})(x) = \phi_\alpha(x)$.

Two pairs (a topological space, an atlas) are equivalent if there is a homeomorphism of the underlying topological spaces respecting, in a natural sense, the group actions on the corresponding open domains in the complex spaces.

A group G of homeomorphisms of a Hausdorff topological space M is said to act *discretely* if for any two points $x_1, x_2 \in M$ (not necessarily distinct) there exist open neighborhoods $V_1, V_2 \subset M$ respectively such that the set

$$\{g \in G \mid g(V_1) \cap V_2 \neq \emptyset\}$$

is finite. In particular, the *stabilizer*

$$G_x = \{g \in G \mid g(x) = x\}$$

of each point x is finite. If a group G acts discretely on a smooth complex variety M of dimension d, then the quotient space M/G always carries a d-dimensional orbifold structure. Each point $[x] \in M/G$ admits a neighborhood of the form $\mathbb{C}^d/G_x$, where x is an arbitrary representative of the orbit $[x]$ in M. However, it is not true that each orbifold can be presented as the quotient space of a discrete group action on a smooth variety. We will not dwell on this here.

Let us consider one more example.

Example 4.1.5. Let M be the upper complex half-plane

$$\mathbb{C}^+ = \{z \in \mathbb{C} \mid \Im z > 0\}.$$

Consider the group G of fractional linear transformations of $\mathbb{C}^+$ of the form

$$z \mapsto \frac{az+b}{cz+d},$$

where a, b, c, d are integers such that $ad - bc = 1$ (it is called *the Klein modular group*). The quotient space $\mathbb{C}^+/G$ which is, by definition, the *modular curve* can be identified with the domain

$$\left\{ z \in \mathbb{C}^+ \,\bigg|\, -\frac{1}{2} \le \Re z \le \frac{1}{2},\ |z| \ge 1 \right\}$$

in the upper half-plane; the boundary points, that is, points lying on the circle $|z| = 1$ and on the two half-lines $\Re z = \pm 1/2$, are glued pairwise according to the equivalence relation $x + iy \sim -x + iy$ (see Fig. 4.4). Each internal point of the band admits a neighborhood isomorphic to the unit disk modulo the trivial group action. The same is true for all points of the boundary except two: the point i, and the point obtained by gluing two points $e^{2\pi i/3}$ and $e^{\pi i/3}$. The stabilizer subgroup of the point i is the cyclic group of order 2 and is generated by the transformation $z \mapsto -1/z$: indeed, one can easily verify that i is fixed by this transformation, and that the double application of $z \mapsto -1/z$ is the identity. Similarly, the stabilizer subgroup of the point $e^{\pi i/3}$ is the cyclic group of order 3 and is generated by the transformation $z \mapsto 1 - 1/z$. Note also what are the values of the angles of the quotient curve in neighborhoods of these points: at i this angle is π, while at $e^{\pi i/3}$ this angle is not $\pi/3$ as the figure might suggest but $2\pi/3$ (one must attach one more copy of the figure to its right side).

All topological notions, like (co)homology, Euler characteristic, vector bundles, and so on, have their orbifold counterparts that must agree with the groups actions. These objects behave naturally under orbifold coverings. If

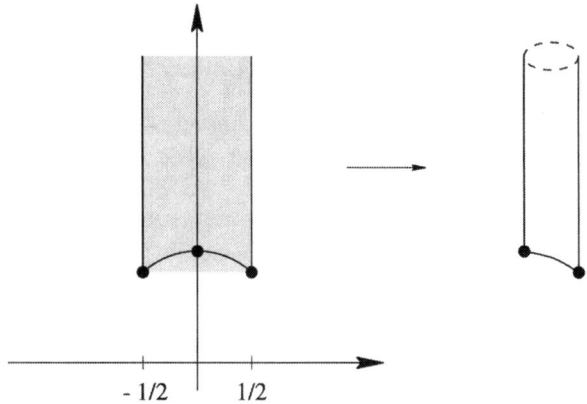

Fig. 4.4. The modular curve before and after the gluing

one wants to emphasize that an object refers to an orbifold, not to a variety, then the adjective *orbifold* is used: orbifold cohomology, orbifold Euler characteristic, and so on (the adjective *virtual* is also very common). In the orbifold case, characteristic cohomology classes of bundles are rational, not integer valued.

For example, the orbifold Euler characteristic is the orbifold analogue of the conventional Euler characteristic for smooth varieties. For a cell decomposition of an orbifold O into open cells of the form $\mathbb{R}^k/G$, where G is a finite group, the *orbifold Euler characteristic* can be defined as

$$\chi(O) = \sum \frac{(-1)^k}{|G|},$$

where the sum is taken over all open cells.

Example 4.1.6. The orbifold Euler characteristic of the orbifold in Fig. 4.3 is 1/2. Indeed, it is represented as the quotient space of the complex projective line $\mathbb{C}P^1$, whose Euler characteristic is 2, modulo a group of order 4 action. If the quotient mapping were an unramified covering, then the Euler characteristic of the covered space should be $\chi(\mathbb{C}P^1)/4$, and this is precisely what happens in the orbifold case. Another way to obtain the same answer is to represent this orbifold as the union of two cells: one, of dimension zero, consisting of an exceptional point, and the other, of dimension two, formed by the complement to this point. Each of them is the quotient space of a cell of the same dimension modulo a group of order 4 action. Hence, the orbifold Euler characteristic of each cell is 1/4, and we arrive at the same result. Note that although the group C_4 acts trivially on the 0-dimensional cell, the Euler characteristic is nevertheless divided by 4.

4.1 Nodal Curves and Orbifolds

Another way to define the orbifold Euler characteristic of an orbifold O is to split it into disjoint subvarieties O_G, $O = \bigsqcup O_G$, where $O_G \subset O$ is the subset of points admitting a neighborhood of the form $\mathbb{R}^k/G$, and the union is taken over all finite groups G. For such a splitting the Euler characteristic of O is

$$\chi(O) = \sum_G \frac{\chi(O_G)}{|G|},$$

where $\chi(O_G)$ is the conventional Euler characteristic of O_G. The same formula can be written in the form

$$\chi(O) = \sum_{k=1}^{\infty} \frac{\chi(O_{[k]})}{k},$$

where the subspace $O_{[k]}$ on the right consists of all points having a stabilizer of order k.

Example 4.1.7. The modular curve from Example 4.1.5 splits into three subvarieties: two singular points and the complement to them. The complement is homeomorphic to the sphere punctured at three points, which gives us the Euler characteristic -1. The total orbifold Euler characteristic of the modular curve is, therefore,

$$-1 + \frac{1}{2} + \frac{1}{3} = -\frac{1}{6}.$$

However, here we must attract the reader's attention to a rather subtle point. Consider two groups: the group $\mathrm{SL}_2(\mathbb{Z})$ of integer 2×2 matrices

$$\begin{pmatrix} a & b \\ c & d \end{pmatrix}$$

with determinant 1, and the group $\mathrm{PSL}_2(\mathbb{Z}) = \mathrm{SL}_2(\mathbb{Z})/\{\pm I\}$, where I is the unit matrix. The mapping

$$\begin{pmatrix} a & b \\ c & d \end{pmatrix} \mapsto \left(z \mapsto \frac{az+b}{cz+d} \right)$$

determines a representation of either of these groups in Klein's modular group. The representation of $\mathrm{PSL}_2(\mathbb{Z})$ is faithful, while that of $\mathrm{SL}_2(\mathbb{Z})$ has a kernel of order 2. Hence, if we consider this representation as the action of the group $\mathrm{SL}_2(\mathbb{Z})$, then the orders of all stabilizers in the calculation above must be multiplied by 2, and the orbifold Euler characteristic must be divided by 2, and it becomes $-1/12$. Note that although the subgroup $\{\pm I\} \subset \mathrm{SL}_2(\mathbb{Z})$ acts trivially on the upper half-plane, we obtain a different orbifold structure on the modular curve!

Exercise 4.1.8. Construct a cell decomposition of the modular curve and verify the calculation of its orbifold Euler characteristic in the example above.

4.2 Moduli Spaces of Complex Structures

Consider a genus g smooth complex curve X with n pairwise distinct points on it (n is allowed to be zero). We always suppose that the curve is compact, and the points usually have markings $\{x_1,\ldots,x_n\}$. Considered up to biholomorphic equivalence preserving the marked points, such a curve determines a point in the moduli space $\mathcal{M}_{g;n}$ of smooth marked complex curves. We denote this point by $(X; x_1,\ldots, x_n)$.

If our curve is rational ($g = 0$) with less than three marked points or elliptic ($g = 1$) without marked points, then it admits "infinitesimal isomorphisms" (see later). For all other values of g and n the curve is "stable": its automorphism group is finite (see Definition 4.3.1 below). We call a pair $(g; n)$ of non-negative integers *stable* if either $g = 0$ and $n \geq 3$, or $g = 1$ and $n \geq 1$, or $g \geq 2$. Working with the moduli spaces in the "unstable" cases $g = 0$, $n = 0, 1$, or 2 and $g = 1$, $n = 0$ requires more sophisticated techniques; below the unstable situation will rarely be mentioned.

Example 4.2.1. All rational (of genus $g = 0$) curves with $n = 3$ marked points are biholomorphically equivalent, therefore $\mathcal{M}_{0;3}$ consists of a single point. One can always choose a coordinate z in $\mathbb{C}P^1$ making the three marked points to have the coordinates $0, 1, \infty$. Any other triple of points acquires the same coordinate values under an appropriate linear fractional transformation.

Example 4.2.2. For two rational curves

$$(X'; x'_1, x'_2, x'_3, x'_4), \quad (X''; x''_1, x''_2, x''_3, x''_4)$$

with $n = 4$ marked points, there is a unique biholomorphic mapping $X' \to X''$ taking x'_1, x'_2, x'_3 to x''_1, x''_2, x''_3 respectively. Thus, the equivalence class of such a curve is uniquely determined by the position of the last marked point. This means that $\mathcal{M}_{0;4}$ can be naturally identified with the projective line $\mathbb{C}P^1$ punctured at three points since the marked points are not allowed to coincide. Note that all the choices of the punctures are equivalent. In other words, if we choose a coordinate z on $\mathbb{C}P^1$ in such a way that the first three marked points acquire coordinates 0, 1, and ∞ respectively, then there are no restrictions on the coordinate λ of the last marked point, except that it cannot take values $0, 1, \infty$. Two points of the moduli space $\mathcal{M}_{0;4}$ corresponding to two distinct values of λ are distinct as well. Hence, λ can be considered as a coordinate on the moduli space. The value of this coordinate at a curve $(\mathbb{C}P^1; x_1, x_2, x_3, x_4) \in \mathcal{M}_{0;4}$ endowed with an arbitrary coordinate z coincides with the cross-ratio

$$\lambda = \frac{(z(x_1) - z(x_4))(z(x_2) - z(x_3))}{(z(x_4) - z(x_3))(z(x_1) - z(x_2))};$$

it is independent of the choice of the coordinate z. If all the points x_1, x_2, x_3, x_4 are pairwise distinct, then λ cannot acquire the values $0, 1, \infty$.

4.2 Moduli Spaces of Complex Structures 233

Example 4.2.3. The case of elliptic (i.e., of genus $g = 1$) curves with $n = 1$ marked point is more complicated. Such a curve is a quotient space of a complex plane modulo a 2-lattice; the marked point coincides with the image of the lattice under the factorization mapping. A lattice is determined by a pair τ_1, τ_2 of noncollinear vectors in $\mathbb{C}$, $\tau_1/\tau_2 \notin \mathbb{R}$. We suppose that the numbering of the vectors determines the orientation of the frame (τ_1, τ_2) opposite to the orientation of $\mathbb{C}$. The group $\mathbb{C}^*$ of nonzero complex numbers acts on the set of lattices according to the rule $c : (\tau_1, \tau_2) \mapsto (c\tau_1, c\tau_2)$. The group $SL_2(\mathbb{Z})$ acts by

$$(\tau_1, \tau_2) \mapsto (a\tau_1 + b\tau_2, c\tau_1 + d\tau_2), \quad a, b, c, d \in \mathbb{Z}, \quad ad - bc = 1,$$

changing the base of the lattice but preserving the lattice.

Multiplying the pair (τ_1, τ_2) by $1/\tau_2$ we make it equal to $(\tau, 1)$, and the first base vector $\tau = \tau_1/\tau_2$ belongs to the upper half-plane. Two vectors τ and τ' determine the same elliptic curve if and only if they are taken one to another by the $SL_2(\mathbb{Z})$-action on the upper half-plane.

Hence, the moduli space $\mathcal{M}_{1;1}$ coincides with the modular curve from Example 4.1.7 considered as an orbifold, which is the *quotient space of the $SL_2(\mathbb{Z})$-action on the upper complex half-plane*. A generic point of $\mathcal{M}_{1;1}$ corresponds to a generic elliptic curve with a marked point. Such a curve admits an automorphism of order two. If the curve is presented as a plane cubic

$$y^2 = x^3 + ax + b,$$

then the automorphism looks like $y \mapsto -y$. The automorphism groups of the two exceptional curves corresponding to the points $\tau = e^{\pi i/3}$ and $\tau = e^{\pi i/2} = i$ are of orders 6 and 4 respectively.

Let us now turn to a formal definition of moduli spaces of curves. There are some properties one should expect from such a space. The most important of them is the following one. Consider a holomorphic "family" $p : P \to B$ of smooth genus g complex curves with n marked points. The word "family" means that

- P and B are orbifolds;
- the mapping p is holomorphic, and each fiber $p^{-1}(b) \subset P$, $b \in B$ is the quotient of a smooth genus g curve with n marked points modulo its automorphism group (in particular, the dimension of P is greater than that of B by one);
- n pairwise disjoint sections $\sigma_i : B \to P$, $p \circ \sigma_i = \mathrm{id}$ are given, which specify the marked points on each fiber: $x_i = p^{-1}(b) \cap \sigma_i(b)$.

An orbifold $\mathcal{M}_{g;n}$, whose points are in one-to-one correspondence with biholomorphic equivalence classes of genus g complex curves with n marked points, is called a *coarse* moduli space if for any such family the mapping $B \to \mathcal{M}_{g;n}$ taking a point $b \in B$ to the class of the fiber $(p^{-1}(b); \sigma_1(b), \ldots, \sigma_n(b))$ is holomorphic. The adjective "coarse" is used in contrast to a more subtle notion of "fine" moduli space, see the discussion in [139], Sec. 2.A. If a coarse moduli space exists, then it is unique up to biholomorphic equivalence. The notion of coarse moduli space extends to other classes of curves, say to stable curves, see below. The book [139] is a very good reference for the proof of the following statement, although only the case $n = 0$, $g > 1$ is discussed there in detail.

Theorem 4.2.4. *For each stable pair $(g;n)$ there exists the coarse moduli space $\mathcal{M}_{g;n}$ of smooth genus g complex curves with n marked points. This moduli space is an orbifold. It is a smooth variety if $n \geq n_0$ for some positive integer $n_0 = n_0(g)$. In particular, it is a smooth variety if $g = 0$. The dimension of $\mathcal{M}_{g;n}$ is $3g - 3 + n$.*

The dimension of $\mathcal{M}_{g;n}$ can be computed by means of the Riemann–Roch theorem, see [124].

4.3 The Deligne–Mumford Compactification

The moduli space $\mathcal{M}_{g;n}$ always comes together with the *universal curve*, i.e., with a smooth orbifold $\mathcal{C}_{g;n}$ of dimension $\dim \mathcal{M}_{g;n} + 1$ endowed with a projection $\mathcal{C}_{g;n} \to \mathcal{M}_{g;n}$. We attract the reader's attention to the fact that the word "curve" is somewhat misleading since the universal curve is, in fact, an orbifold of large dimension. The fiber of this projection over a point $(X; x_1, \ldots, x_n) \in \mathcal{M}_{g;n}$ is the quotient of the curve X modulo its automorphism group action. (This means that in the majority of cases the fiber coincides with the curve X itself.) The marked points $x_1, \ldots, x_n$ form n pairwise disjoint sections $\sigma_i : \mathcal{M}_{g;n} \to \mathcal{C}_{g;n}$. Each holomorphic family $p : P \to B$ of smooth curves leads not only to a holomorphic mapping $B \to \mathcal{M}_{g;n}$, but also to a holomorphic mapping $P \to \mathcal{C}_{g;n}$ of the total space of the family to the universal curve.

The examples above show that, generally speaking, the moduli space $\mathcal{M}_{g;n}$ is noncompact. It is natural to construct a compactification $\overline{\mathcal{M}}_{g;n}$ of the moduli space of curves together with a compactification $\overline{\mathcal{C}}_{g;n}$ of the universal curve so that the projection above would extend to a projection $\overline{\mathcal{C}}_{g;n} \to \overline{\mathcal{M}}_{g;n}$, and the sections above would extend to n pairwise disjoint sections $\sigma_i : \overline{\mathcal{M}}_{g;n} \to \overline{\mathcal{C}}_{g;n}$, $i = 1, \ldots, n$.

A noncompact topological space admits a lot of compactifications. The choice of a specific compactification depends on what we expect from it. In the case of the moduli spaces of curves, it is natural to require that the compactified space carries an orbifold structure, so that $\mathcal{M}_{g;n}$ is a suborbifold in it. Another important requirement is the "modularity" condition, that is, each point of the compactified space must correspond to some complex curve of genus g with n marked points (i.e., each point is not just an abstract point but the "module" of some curve). Of course, this requirement implies that singular curves should be allowed. These goals were achieved by Deligne and

Mumford [79] in their construction of the coarse moduli space of stable curves. Stable curves may be singular and reducible as well as smooth, but they are allowed to have only the simplest singularities, the nodes.

Definition 4.3.1. A (possibly singular) connected curve with marked points is called *stable* if

- its only singularities are simple nodes, that is, points of simple self-intersection; in other words, a stable curve is a nodal curve;
- the marked points are nonsingular;
- the curve does not admit infinitesimal automorphisms; this means that the group of (marked points preserving) automorphisms of the curve is finite.

One more way to express the third requirement is the following: each rational irreducible component of the curve must contain at least three special (marked or double) points, and each elliptic irreducible component must contain at least one special point. Note that if a node is a self-intersection point of an irreducible component, then it determines two special points on this component.

Example 4.3.2. A smooth genus g curve with n marked points is stable if and only if the pair $(g; n)$ is stable.

The points of the *Deligne–Mumford compactification* $\overline{\mathcal{M}}_{g;n}$ of the moduli space $\mathcal{M}_{g;n}$ are biholomorphic equivalence classes of genus g stable curves with n marked points. Since for a stable pair $(g; n)$ a smooth genus g curve with n marked points is stable, the Deligne–Mumford compactification procedure consists in adding to $\mathcal{M}_{g;n}$ singular stable curves.

Theorem 4.3.3 ([79]). *For each stable pair $(g; n)$ there exists the coarse moduli space $\overline{\mathcal{M}}_{g;n}$ of stable genus g complex curves with n marked points. This moduli space is a compact orbifold. It is a smooth variety if $n \geq n_0$ for some positive integer $n_0 = n_0(g)$. In particular, it is a smooth irreducible projective variety if $g = 0$. The subvariety $\mathcal{M}_{g;n} \subset \overline{\mathcal{M}}_{g;n}$ is Zariski dense. The compactified moduli space $\overline{\mathcal{M}}_{g;n}$ is endowed with the universal stable curve $\overline{\mathcal{C}}_{g;n} \to \overline{\mathcal{M}}_{g;n}$, and the marked points form n pairwise disjoint sections $\sigma_i : \overline{\mathcal{M}}_{g;n} \to \overline{\mathcal{C}}_{g;n}$.*

The set of singular stable curves $\overline{\mathcal{M}}_{g;n} \setminus \mathcal{M}_{g;n}$ is called the *boundary* of the moduli space of stable curves and is denoted by $\partial \overline{\mathcal{M}}_{g;n}$.

Example 4.3.4. The moduli space $\mathcal{M}_{0;3}$ consists of a single point and is therefore compact, whence $\overline{\mathcal{M}}_{0;3} = \mathcal{M}_{0;3}$. Indeed, each rational stable curve with three marked points is smooth.

Example 4.3.5. There are three singular stable rational curves with four marked points, see Fig. 4.5. Each of these curves consists of two irreducible rational components intersecting at a single point, and they differ by the way

236   4 Geometry of Moduli Spaces

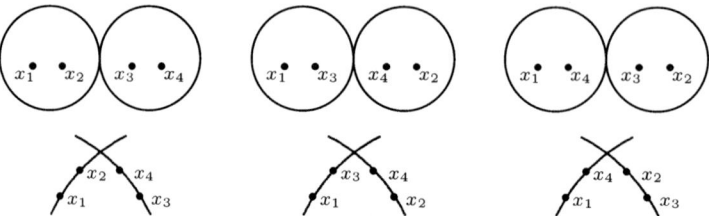

Fig. 4.5. The three singular rational curves with $n = 4$ marked points. A "complex" and a "schematic" picture

the marked points are distributed between the two components. There are exactly three possibilities:

$$\{\{x_1, x_2\}, \{x_3, x_4\}\}, \quad \{\{x_1, x_3\}, \{x_2, x_4\}\}, \quad \text{and} \quad \{\{x_1, x_4\}, \{x_2, x_3\}\}.$$

Singular curves of this form have no moduli[1], that is, each of them determines a single point in the moduli space $\overline{\mathcal{M}}_{0;4}$ of rational stable curves with 4 marked points. When added to the moduli space $\mathcal{M}_{0;4}$ of smooth curves, these three points compactify it to the projective line $\mathbb{C}P^1$. Hence, the boundary $\partial\overline{\mathcal{M}}_{0;4}$ consists of three points.

The three points of the boundary can be easily described in terms of the cross-ratio coordinate λ on $\mathcal{M}_{0;4}$ from Example 4.2.2. They have the coordinates $\lambda = \infty, 1, 0$ respectively.

Example 4.3.6. Similarly, the boundary $\partial\overline{\mathcal{M}}_{0;5}$ consists of ten projective lines corresponding to the ten ways of splitting the five marked points between the two rational components so that each component contains at least two marked points. We denote a typical component by $\{\{x_1, x_2, x_3\}, \{x_4, x_5\}\}$, which means that the marked points x_1, x_2, x_3 belong to one irreducible component of the curve, while the marked points x_4, x_5 lie on the other component. Such component of the boundary is indeed a projective line: the first irreducible component of the curve contains four special points, namely, the points x_1, x_2, x_3 and the double point, and the moduli space of configurations of four marked points on the projective line is, as we have seen above, the projective line. The configuration of the three points in the second irreducible component has no moduli.

Each irreducible component of the boundary $\partial\overline{\mathcal{M}}_{0;5}$ contains three points of further degeneration. Such a point corresponds to a stable rational curve consisting of three irreducible components, two of which intersect the third one, at a single point each. Each of the two components carries a pair of marked points, while there is one marked point on the third component. A typical curve of this kind will be denoted as $\{\{x_1, x_2\}, \{x_3\}, \{x_4, x_5\}\}$. All in all, the

---

[1] The word "moduli" usually means "continuous parameters".

ten projective lines of $\partial\overline{\mathcal{M}}_{0;5}$ intersect pairwise at 15 points corresponding to rational curves consisting of three irreducible components.

More generally, one can assign to a rational stable curve with n marked points a graph. The vertices of the graph are in one-to-one correspondence with the irreducible components of the curve. Two vertices are connected by an edge if the two irreducible components intersect. Each vertex is labelled by the marked points that lie on the corresponding irreducible component. The resulting graph is always a tree, and is called a *modular graph*. A similar construction exists for curves of higher genera as well, but in this case the graph has a more complicated structure. A modular graph describes a stratum in the moduli space $\overline{\mathcal{M}}_{g;n}$, which consists of all stable curves associated to the graph. We are not going to use modular graphs below; see [211] for further discussions.

Exercise 4.3.7. Starting from the cases $n = 3, 4$ show that the universal curve $\overline{\mathcal{C}}_{0;n}$ is isomorphic to the moduli space $\overline{\mathcal{M}}_{0;n+1}$. Verify that under the isomorphism $\overline{\mathcal{C}}_{0;4} \cong \overline{\mathcal{M}}_{0;5}$ the ten boundary lines in $\overline{\mathcal{M}}_{0;5}$ correspond to the following ten lines in $\overline{\mathcal{C}}_{0;4}$: the four sections $\sigma_1, \sigma_2, \sigma_3, \sigma_4$ and the six irreducible components of the singular fibers of the projection $\overline{\mathcal{C}}_{0;4} \to \overline{\mathcal{M}}_{0;4}$.

Example 4.3.8. The boundary $\partial\overline{\mathcal{M}}_{1;1}$ of the moduli space of stable elliptic curves consists of a single point corresponding to the curve with a single self-intersection. The normalization of this curve is the rational curve with a marked point and with two preimages of the double point. This boundary point must be attached "at the infinity" of the modular curve $\mathcal{M}_{1;1}$: see Example 4.1.5. The automorphism group of this singular elliptic curve is the cyclic group C_2, generated by the automorphism of its normalization interchanging the preimages of the double point. Thus, the automorphism behavior of the boundary point coincides with that of a generic point of the moduli space.

4.4 Combinatorial Models of the Moduli Spaces of Curves

The geometry of moduli spaces $\mathcal{M}_{g;n}$ of complex curves proved to be intimately related to the topology of graphs embedded in surfaces. Embedded graphs are used to enumerate cells in some cell decompositions of *decorated* moduli spaces. These are products of moduli spaces by the real positive octants $\mathbb{R}_+^n$. Decorated moduli spaces are no longer complex orbifolds, they carry only real orbifold structures.

In the cell decomposition of a decorated moduli space we are going to describe, a cell is associated to a genus g embedded graph with n marked faces. We suppose everywhere that $n \geq 1$. For smooth curves, only graphs without vertices of valency one and two are considered. In order to prevent confusion, the term "cell" below always refers to a cell in a moduli space, while the connected components of the complement to a graph embedded into a surface are always called "faces".

The real dimension of a cell is equal to the number of edges in the corresponding graph. A cell of the highest dimension is associated to a generic graph, i.e. to a graph with all vertices of valency three. The dimension of such a cell is $6g - 6 + 3n$. Given a link in an embedded graph, that is, an edge that is not a loop, one can obtain another embedded graph by contracting the link. The cyclic order of the darts leaving the new vertex is inherited from the two cyclic orders around the pre-contracted vertices. This operation is called the *Whitehead collapse*. A Whitehead collapse preserves the genus of an embedded graph and the number of faces in it.

Below we shall consider embedded graphs endowed with an additional structure. A *marked embedded graph* is an embedded graph with faces marked by the elements of the set $\{x_1, \ldots, x_n\}$. The Whitehead collapse operation extends to marked embedded graphs. A cell labelled by a marked embedded graph Γ' belongs to the boundary of another cell corresponding to a marked embedded graph Γ if and only if Γ' can be obtained from Γ by a sequence of Whitehead collapses. This picture gives a complete description of the combinatorial structure of the cell decomposition.

Example 4.4.1. Figure 4.6 shows the cell decomposition for the decorated moduli space $\mathcal{M}_{0;3} \times \mathbb{R}_+^3 \cong \mathbb{R}_+^3$, $\mathbb{R}_+ = \{p \in \mathbb{R} \mid p > 0\}$. This cell decomposition is invariant under the $\mathbb{R}_+$-action on $\mathbb{R}_+^3$ by multiplications. The projectivized decomposition consists of four open triangles attached to each other by three open segments. The corresponding marked embedded graphs are shown in the picture.

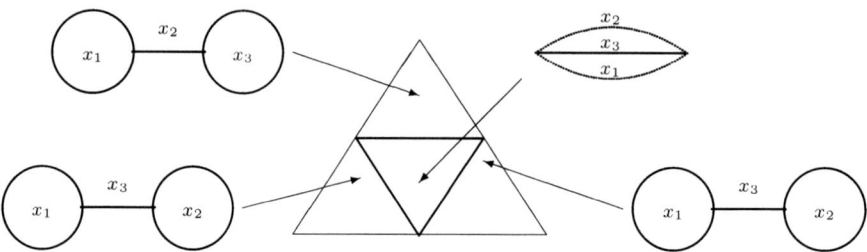

Fig. 4.6. The cell decomposition of the projectivized decorated moduli space $\mathcal{M}_{0;3} \times P\mathbb{R}_+^3$ and embedded graphs marking the cells of the highest dimension; the sides of the inner triangle correspond to the Whitehead collapses of these graphs

Such a decomposition requires associating a real positive number to each marked face of the graph. There are two natural ways to do this: that due to Mumford (see [137], [178]) exploits piecewise flat metrics on surfaces induced by Jenkins–Strebel quadratic differentials; the other one ([232]) uses hyperbolic metrics of constant negative curvature. In the first case the required number is the length of a horizontal closed curve in the punctured neighborhood of a marked point, while in the second case we take for this number the length of the corresponding horocycle. Below we describe in more detail Mumford's approach used by Kontsevich in the proof of Witten's conjecture.

Remark 4.4.2. The reader must not confuse the cell decomposition of $\mathcal{M}_{g;n} \times \mathbb{R}_+^n$ with the stratification of the compactified moduli space $\overline{\mathcal{M}}_{g;n}$ according to modular graphs described in Example 4.3.6. Note that modular graphs are abstract graphs, not embedded ones, and that the strata are generally not cells.

Mumford's model is a cell decomposition of the "decorated" moduli space of smooth stable curves, i.e., of the space $\mathcal{M}_{g;n} \times \mathbb{R}_+^n$. We consider only the case $n \geq 1$, and the stability condition implies that if $g = 0$, then $n \geq 3$. The ith coordinate p_i in the positive octant $\mathbb{R}_+^n$ is ascribed to the ith marked face x_i. Below, a point will be chosen in each marked face of an embedded graph; we mark this point with the same marking x_i and make no difference in the notation between the marked point and the marked face containing it. The value p_i is understood as the perimeter of a horizontal trajectory around this face with respect to the canonical Jenkins–Strebel quadratic differential associated to the marked curve $(X; x_1, \ldots, x_n)$ and the n-tuple $(p_1, \ldots, p_n)$; see below.

Consider a meromorphic quadratic differential ρ on a smooth curve X. Locally, such a differential has the form $\rho = \varphi(z)(dz)^2$, where φ is a meromorphic function. The set of zeroes and poles of a quadratic differential (if it is not identically zero) is finite. We restrict ourselves to quadratic differentials having poles of order not greater than 2. Outside its set of zeroes and poles, a quadratic differential determines a flat metric $dl^2 = |\varphi(z)||dz|^2$ on X. The complement to the set of zeroes and poles is endowed with two real line fields: the *horizontal* line field determined by vectors such that the value of ρ on them is positive real, and the *vertical* line field determined by vectors such that the value of ρ on them is negative real.

For a generic quadratic differential, a generic trajectory of the horizontal line field is nonclosed and can have an infinite length. However, there exist special quadratic differentials such that all horizontal trajectories except for a finite set are closed. These quadratic differentials are called *Jenkins–Strebel quadratic differentials*.

An end of a nonclosed horizontal trajectory of a Jenkins–Strebel quadratic differential is either a zero, or a first order pole of the differential. At a zero of order k the quadratic differential has, in an appropriate local coordinate z, a local presentation $z^k (dz)^2$, and there are $k + 2$ horizontal trajectories meeting at such a point. From a pole of order one issues a unique horizontal trajectory. Hence, nonclosed horizontal trajectories form a graph in X. The vertices of the graph are the zeroes of an arbitrary order and the poles of order one of the differential. The valency of a zero of order k is $k + 2$, while the valency of a pole of order one is equal to one (see Fig. 4.7). Note that this graph does

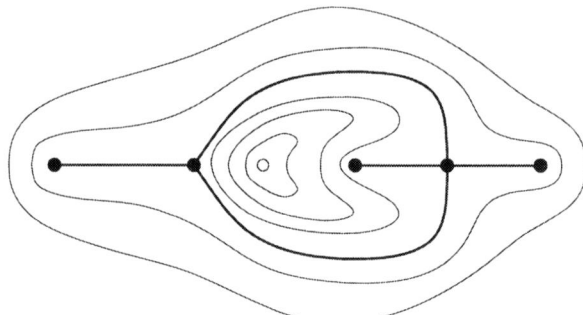

Fig. 4.7. Horizontal trajectories and the graph of the non-closed trajectories of a canonical Jenkins–Strebel quadratic differential with three poles of the first order and with two zeros, of the first and of the second order

not contain vertices of order two, and that it is not necessarily connected. Each of the edges of the graph has a positive real length with respect to the flat metric determined by the quadratic differential. This length is finite even if one or both ends of an edge are poles of order one. Indeed, at such a pole the length differential behaves as $x^{-1/2}dx$ at $x = 0$, and the length integral converges.

The connected components of the complement to the graph of nonclosed trajectories in X are open annuli and open disks. An open disk contains a single pole of order two of the differential; we call this pole the *center* of the disk. An annulus contains a connected component of the graph. Each of the connected components of the complement to the graph is composed of closed trajectories: an annulus is fibered into circles, and a disk punctured at the center is also fibered into circles. (Note that a face cannot contain two or more connected components because in this case it cannot be fibered into circles.) The Stokes theorem implies that all closed trajectories in the same connected component have the same length, whence any annulus as well as any punctured disk is isometric to a cylinder. Note that the height (the length of a vertical trajectory) of an annulus is finite, while that of a disk is infinite (because the integral of the 1-form $x^{-1}dx$ diverges at $x = 0$). Thus, we are able to associate to the ith pole of a quadratic differential a positive real number p_i, the length of a closed trajectory around this pole. If the pole is of order 2, then the quadratic differential has, in an appropriate local coordinate z, the local presentation $\rho = -(p_i dz/2\pi z)^2$, and the value p_i is uniquely determined by this property.

The following theorem is due to Strebel [269].

Theorem 4.4.3. *Let a pair $(g; n)$ be stable. Then for any genus g complex curve with n marked points $(X; x_1, \ldots, x_n)$ and any n-tuple $p_1, \ldots, p_n$ of positive real numbers there exists a unique Jenkins–Strebel quadratic differential* with poles of order two at x_i and no other poles such that the connected components of the complement to the graph of nonclosed horizontal trajectories are open disks, and the length of closed trajectories associated to the ith pole is p_i.

The Jenkins–Strebel quadratic differential ascribed to a curve $(X; x_1, \ldots, x_n)$ by the theorem will be called *canonical*. Since the complement to the graph of nonclosed horizontal trajectories for a canonical Jenkins–Strebel quadratic differential is a union of 2-disks and does not have any annuli, this graph is embedded into X. Each edge of the graph acquires a length in the metric determined by the differential. The length p_i of a closed trajectory in the ith disk is equal to the perimeter of the ith face of the graph. Recall that the valency of each vertex of the graph is not less than three.

Remark 4.4.4. In the generic case (when all the vertices are of degree 3) the graph has $6g - 6 + 3n$ edges. It might seem strange that n linear equations (the values of the perimeters) determine the lengths of the edges. In fact, $6g - 6 + 2n$ additional constraints are supplied by the complex structure of the curve: recall that the complex dimension of the moduli space is $3g - 3 + n$ (Theorem 4.2.4).

The proof of Theorem 4.4.3 is rather complicated. In [269] a special chapter is devoted to the study of quadratic differentials with closed horizontal trajectories.

Example 4.4.5. Consider an elliptic curve presented as the quotient of the complex line $\mathbb{C}$ endowed with a coordinate z modulo the lattice spanned by vectors $(\tau, 1)$, see Example 4.2.3, and let us choose 0 for the marked point. The quadratic differential dz^2 on $\mathbb{C}$ descends to a quadratic differential on the elliptic curve, but this differential has neither poles, nor zeroes. Let $\wp = \wp(z)$ be the Weierstrass $\wp$-function. This is the unique, up to a multiplicative constant, meromorphic function on the elliptic curve having a pole of order 2 at the marked point. Then $c\wp(z)(dz)^2$ is a canonical Jenkins–Strebel quadratic differential provided the constant c is chosen in such a way that the quadratic residue at 0 is strictly negative. For a general value of τ, the Weierstrass $\wp$-function has two distinct zeroes of order 1, and these zeroes are the vertices of valency three of the graph of nonclosed horizontal trajectories.

Exercise 4.4.6. Find the values of the modular parameter τ such that the Weierstrass $\wp$-function has a double zero and, therefore, the graph of nonclosed horizontal trajectories has one vertex of valency four, with two loops attached to it.

Conversely, given an embedded graph with valencies of each vertex not less than three, with faces marked by $\{x_1, \ldots, x_n\}$, and with fixed lengths of the edges, one easily constructs a flat metric on the surface punctured at the marked points. Note that knowing the lengths of the edges of the graph we uniquely recover the perimeters of the faces. In this metric, the surface consists of n cylinders of perimeters p_i and infinite height. The cylinders are attached to each other by gluing along the edges of the graph (whose lengths we already know). There is a unique complex structure on the surface such that the corresponding canonical Jenkins–Strebel differential on $(X; x_1, \ldots, x_n)$ determines the given flat metric. The construction proceeds in the same way as the construction of the complex structure in the proof of Riemann's existence theorem, and we omit the details. In the case of singular stable curves, the situation is similar, with the only exception that the graph of nonclosed trajectories is allowed to acquire vertices of valency 1 or 2 at the double points of the curve.

Denote by $\mathcal{M}_{g;n}^{\text{comb}}$ the space of genus g embedded graphs with n marked points with all vertices of valency greater than two and endowed with a metric, that is with length assigned to each edge. The cell decomposition determines the topology of $\mathcal{M}_{g;n}^{\text{comb}}$ inside each cell, and the Whitehead collapse determines the rule for gluing the cells. Thus, $\mathcal{M}_{g;n}^{\text{comb}}$ becomes a cell complex. It is also endowed with a natural real orbifold structure determined by the graph automorphisms action. The argument above yields the following statement.

Theorem 4.4.7 ([178]). *The decorated moduli space $\mathcal{M}_{g;n} \times \mathbb{R}_+^n$ and the combinatorial model $\mathcal{M}_{g;n}^{\text{comb}}$ are isomorphic as real orbifolds.*

Theorem 4.4.3 can be extended to singular curves as well. Namely,

For each genus g singular stable curve with n marked points $(X; x_1, \ldots, x_n)$ and any n-tuple $p_1, \ldots, p_n$ of positive real numbers there exists a unique Jenkins–Strebel quadratic differential possessing the following properties:

- *it has poles of order two at x_i and poles of order at most 1 at the nodes;*
- *the connected components of the complement to the graph of nonclosed horizontal trajectories are either open disks or unions of those irreducible components of the curve that do not contain marked points;*
- *the length of closed trajectories associated to the ith pole of order 2 is p_i.*

Some comments about the notion of nonclosed trajectory should be made in this case. If the quadratic differential has a pole of order one or a zero of an arbitrary order in a sheet at a double point, then at least one nonclosed trajectory in this sheet ends at this point, and the statement can be applied directly. And if there is no pole and no zero, then there is a single horizontal trajectory through this point in the given sheet. In the last case we treat this trajectory as a "nonclosed" one adding it to the graph of nonclosed trajectories.

Suppose for example that we are given a singular rational curve with four marked points split into pairs $\{x_1, x_2\}$ and $\{x_3, x_4\}$, with the horizontal trajectory lengths $\{p_1, p_2, p_3, p_4\}$. If, say, $p_1 = p_2$, then all the horizontal trajectories on the first rational irreducible component of the curve are circles of the same length p_1, and the trajectory passing through the double point is a part of the graph of nonclosed trajectories. The corresponding quadratic differential has the form $-(p_1 dz/2\pi z)^2$ in an appropriate coordinate z. The "nonclosed" trajectory splits the sphere into two discs, each consisting of closed trajectories. If $p_1 \neq p_2$, then the quadratic differential has a pole of order one at the double point and a zero somewhere else. The corresponding graph of nonclosed trajectories has two vertices, namely, the zero and the double point, an edge connecting them, and a loop attached to the zero.

As a more complicated situation, consider the case of rational curves with 8 marked points. The moduli space $\overline{\mathcal{M}}_{0;8}$ contains a particular one-dimensional subvariety consisting of curves having five irreducible components: the central rational component without marked points, and four rational components, each intersecting the central one at a single point and containing two marked points (leaf components). The graph of nonclosed trajectories in this case consists of four disjoint subgraphs (one for each leaf component). For a fixed tuple of perimeters $p_1, \ldots, p_8$ this graph is the same, independently of the mutual position of the four special points on the central rational component. The complement to this graph consists of eight disks in the leaf components and the central rational component.

Thus we see that the graph of nonclosed trajectories does not allow one to recover unambiguously the underlying stable curve: the information about the complex structure on a component without marked points is lost. This phenomenon constitutes the delicate place of Kontsevich's proof. Indeed, it shows that the combinatorial description of $\mathcal{M}_{g;n} \times \mathbb{R}_+^n$ in Theorem 4.4.7 cannot be extended to $\overline{\mathcal{M}}_{g;n} \times \mathbb{R}_+^n$. Therefore it is not obvious how to extend the 2-forms ω_i, which arise in the course of the proof (see Sec. 4.9.2), to $\overline{\mathcal{M}}_{g;n} \times \mathbb{R}_+^n$.

A possible way to resolve this difficulty is the following. Call two stable curves *equivalent* if they become isomorphic when we contract to single points all their irreducible components without marked points. Then what we need is to give a precise description of the orbifold that is obtained from $\overline{\mathcal{M}}_{g;n}$ by identifying any two points corresponding to equivalent stable curves. These identifications contract some subvarieties of the boundary of $\overline{\mathcal{M}}_{g;n}$. For the trivial case of rational curves there is an explicit description of the required space. For arbitrary g the construction is elaborated in [203], but the resulting space carries only a structure of a simplicial complex, and it is not known, whether it can be realized in the category of projective varieties.

4.5 Orbifold Euler Characteristic of the Moduli Spaces

The orbifold Euler characteristic $\chi(\mathcal{M}_{g;1})$ of the moduli space $\mathcal{M}_{g;1}$ of genus g smooth curves with one marked point was calculated by Harer and Zagier in [138]. Namely, they proved the following

244   4 Geometry of Moduli Spaces

Theorem 4.5.1. *We have*

$$\chi(\mathcal{M}_{g;1}) = \zeta(1-2g) = -\frac{B_{2g}}{2g},$$

where ζ is the Riemann zeta function and B_{2g} is the $(2g)$th Bernoulli number, that is, the coefficient of x^{2g} in the expansion

$$\frac{x}{e^x - 1} = 1 - \frac{1}{2}x + \frac{B_2}{2!}x^2 + \frac{B_4}{4!}x^4 + \cdots,$$

$B_2 = 1/6$, $B_4 = -1/30$, ...

Example 4.5.2. We have already computed in Example 4.1.7 the orbifold Euler characteristic of $\mathcal{M}_{1;1}$ (see also Example 4.2.3). It is $\chi(\mathcal{M}_{1;1}) = -\frac{1}{12}$ in accordance with the Harer–Zagier theorem above.

Now, the orbifold Euler characteristic of the moduli spaces of curves with arbitrary number of marked points is

$$\chi(\mathcal{M}_{g;n}) = (-1)^n \frac{(2g-3+n)!(2g-1)}{(2g)!} B_{2g}.$$

This is an immediate consequence of Theorem 4.5.1 and the recurrence relation

$$\chi(\mathcal{M}_{g;n+1}) = \chi(\mathcal{M}_{g;n})(2 - 2g - n) \qquad (4.1)$$

for stable (g,n). The last formula follows from the fact that, from the topological point of view, the forgetful mapping $\mathcal{M}_{g;n+1} \to \mathcal{M}_{g;n}$ is a fiber bundle whose fiber is a genus g curve punctured at n points (since the $(n+1)$th point is not allowed to coincide with either of the other marked points). This means that the Euler characteristic of the fiber is $2 - 2g - n$. Here we exploit the fact that, similarly to the smooth case, the Euler characteristic of the total space of an orbifold fiber bundle is the product of the Euler characteristics of the base and of the fiber. The recurrence formula above also yields the Euler characteristic of all moduli spaces of smooth rational curves:

$$\chi(\mathcal{M}_{0;n}) = (-1)^{n-1}(n-3)!.$$

Recall that since $\mathcal{M}_{0;n}$ is a smooth variety, its orbifold and conventional Euler characteristics coincide.

Remark 4.5.3. According to Getzler [111], the Euler characteristic of the compactified moduli space $\overline{\mathcal{M}}_{0;n}$ of rational curves is related to that of $\mathcal{M}_{0;n}$ in the following way. Consider the two exponential generating series

$$x - \sum_{n=2}^{\infty} \frac{x^n}{n!} \chi(\mathcal{M}_{0;n+1}) = x - \sum_{n=2}^{\infty} (-1)^n \frac{x^n}{n(n-1)}$$

and
$$y + \sum_{n=2}^{\infty} \frac{y^n}{n!} \chi(\overline{\mathcal{M}}_{0;n+1}).$$

Then these two generating functions are mutually inverse with respect to the composition of functions. For the first few values of n starting from $n = 3$ this gives the following values of $\chi(\overline{\mathcal{M}}_{0;n})$: 1, 2, 7, 34, 213, 1630.

A similar statement is valid for the Poincaré polynomials
$$P_{\mathcal{M}_{0;n+1}}(t) = \sum \dim H^k(\mathcal{M}_{0;n+1}, \mathbb{C}) t^k$$
and
$$P_{\overline{\mathcal{M}}_{0;n+1}}(t) = \sum \dim H^k(\overline{\mathcal{M}}_{0;n+1}, \mathbb{C}) t^k.$$

Namely, the generating function
$$y + \sum_{n=2}^{\infty} \frac{y^n}{n!} P_{\overline{\mathcal{M}}_{0;n+1}}(t)$$
is inverse to the generating function
$$x + \sum_{n=2}^{\infty} \frac{x^n}{n!} P_{\mathcal{M}_{0;n+1}}(t)$$

with respect to the composition of functions (the composition concerns the variables x and y, the parameter t being fixed). The particular case of the latter statement when $t = -1$ coincides with the statement for the Euler characteristics. The Poincaré polynomials for the moduli spaces of smooth rational curves are
$$P_{\mathcal{M}_{0;n+1}} = (t^2 - 2)(t^2 - 3) \ldots (t^2 - n + 2),$$
which follows from the study of the same forgetful mapping $\mathcal{M}_{0;n+1} \to \mathcal{M}_{0;n}$, and we can easily compute the Poincaré polynomials $P_{\overline{\mathcal{M}}_{0;n+1}}$ for small values of n (see [211]):

$$1, \ 1 + t^2, \ 1 + 5t^2 + t^4, \ 1 + 16t^2 + 16t^4 + t^6, \ 1 + 42t^2 + 127t^4 + 42t^6 + t^8.$$

Exercise 4.5.4. Compute the orbifold Euler characteristic of the compactified moduli space $\overline{\mathcal{M}}_{1;1}$.

The Harer–Zagier proof of Theorem 4.5.1 exploits the cell decomposition of the decorated moduli space $\mathcal{M}_{g;1} \times \mathbb{R}_+$ described in the previous section. Let $\mathcal{G}_{g;n}$ denote the (finite) set of genus g embedded graphs with n marked faces having no vertices of valency less than 3. The orbifold Euler characteristic of the moduli space $\mathcal{M}_{g;n}$ is equal, up to a sign, to that of the decorated moduli space $\mathcal{M}_{g;n} \times \mathbb{R}^n_+$, which, in its own turn, coincides with the sum of the orbifold Euler characteristics of the cells:

$$\chi(\mathcal{M}_{g;n} \times \mathbb{R}^n_+) = \sum_{\Gamma \in \mathcal{G}_{g;n}} \chi(M_\Gamma) = \sum_{\Gamma \in \mathcal{G}_{g;n}} \frac{(-1)^{\dim M_\Gamma}}{|\mathrm{Aut}(\Gamma)|},$$

where M_Γ is the cell corresponding to the graph Γ.

Example 4.5.5. The combinatorial moduli space $\mathcal{M}^{\mathrm{comb}}_{1;1}$ is the union of two cells of dimensions 3 and 2 respectively. The automorphism groups of these strata are the cyclic groups of order 6 and 4. Hence, the orbifold Euler characteristic of $\mathcal{M}^{\mathrm{comb}}_{1;1}$ is

$$\chi(\mathcal{M}^{\mathrm{comb}}_{1;1}) = \frac{1}{4} - \frac{1}{6} = \frac{1}{12},$$

which gives

$$\chi(\mathcal{M}_{1;1}) = -\frac{1}{12}.$$

The fact that $n = 1$ implies that we must consider embedded graphs with a single face. Therefore, the computation of $\chi(\mathcal{M}_{g;1})$ for all g is based on the calculation of the number of genus g gluings of a polygon from the previous chapter. The embedded graph produced by such a gluing marks a cell in the cell decomposition. However, not each embedded graph can appear in this way due to the fact that only graphs without vertices of valency 1 and 2 are allowed in $\mathcal{G}_{g;n}$. This restriction on gluings can be reformulated as follows:

- no edge may be identified with its neighbor;
- no adjacent pair of edges may be identified with another such pair in reverse order.

Denote the number of genus g gluings of the $2n$-gon satisfying these two conditions by $\lambda_g(n)$.

This number is closely related to the number $\varepsilon_g(n)$ computed in the previous chapter by means of the matrix model. Namely, if we denote by $\mu_g(n)$ the number of genus g gluings of the $2n$-gon with no edge identified with its neighbor (i.e., those satisfying only the first of the two conditions above), then the following statement holds.

Lemma 4.5.6. *We have*

$$\varepsilon_g(n) = \sum_{i \geq 0} \binom{2n}{i} \mu_g(n-i);$$

$$\mu_g(n) = \sum_{i \geq 0} \binom{n}{i} \lambda_g(n-i).$$

The proof is purely combinatorial and straightforward.

4.5 Orbifold Euler Characteristic

The Harer–Zagier theorem 4.5.1 follows from this lemma and the computation of $\varepsilon_g(n)$ after a rather long calculation, and we refer the reader to [138] for details.

A different derivation of Theorem 4.5.1 belongs to Kontsevich [178], Appendix D. He considers the generating function

$$F(s) = \sum_{(g,n):\, 2-2g-n<0} \frac{\chi(\mathcal{M}_{g;n})}{n!} s^{2-2g-n} = \left(\frac{\chi(\mathcal{M}_{0;3})}{3!} + \frac{\chi(\mathcal{M}_{1;1})}{1!}\right) s^{-1}$$
$$+ \left(\frac{\chi(\mathcal{M}_{0;4})}{4!} + \frac{\chi(\mathcal{M}_{1;2})}{2!} + \frac{\chi(\mathcal{M}_{2;0})}{0!}\right) s^{-2} + \dots$$

for the orbifold Euler characteristics. Note that the exponent $\chi_{g;n} = 2-2g-n$ is the Euler characteristic of a genus g curve punctured at n points, and it is negative if and only if the pair (g,n) is stable. For example, the coefficient of s^{-1} in $F(s)$ is

$$\frac{\chi(\mathcal{M}_{0;3})}{6} + \chi(\mathcal{M}_{1;1}) = \frac{1}{6} - \frac{1}{12} = \frac{1}{12}.$$

The main point here is that both the generating function $F(s)$ and the orbifold Euler characteristics $\chi(\mathcal{M}_{g;1})$ determine each other due to the recurrence relation of Eq. (4.1), which gives

$$\frac{\chi(\mathcal{M}_{g;n+1})}{(n+1)!} s^{1-2g-n} = \frac{d}{ds}\left(\frac{1}{n+1} \frac{\chi(\mathcal{M}_{g;n})}{n!} s^{2-2g-n}\right).$$

Namely, consider the generating function for the orbifold Euler characteristics,

$$A_{>0}(s) = \sum_{g \geq 1} \chi(\mathcal{M}_{g;1}) s^{\chi_{g;1}}$$

and the functions

$$A_0(s) = s \log s - s \quad \text{and} \quad A(s) = A_0(s) + A_{>0}(s).$$

Then

$$F(s) = \sum_{g \geq 1, n \geq 1} \frac{\chi(\mathcal{M}_{g;n})}{n!} s^{\chi_{g;n}} + \sum_{n \geq 3} \frac{\chi(\mathcal{M}_{0;n})}{n!} s^{\chi_{g;n}}$$
$$= \frac{A_{>0}(s)}{1!} + \frac{A'_{>0}(s)}{2!} + \frac{A''_{>0}(s)}{3!} + \dots + \frac{A''_0(s)}{3!} + \frac{A'''_0(s)}{4!} + \dots$$
$$= \frac{A(s)}{1!} + \frac{A'(s)}{2!} + \frac{A''(s)}{3!} + \dots - A_0(s) - \frac{A'_0(s)}{2}$$
$$= \int_s^{s+1} A(t)\,dt - s\log s + s - \frac{\log s}{2}.$$

Remark 4.5.7. Strange as it may seem, the function $A_0(s) = s \log s - s$ must be interpreted as the "generating function for the Euler characteristic of the moduli stack $\mathcal{M}_{0;1}$" whatever this means.

Lemma 4.5.8. *The formal power series $F(s)$ in the variable s^{-1} coincides with the formal series*

$$\log\left(\sqrt{\frac{s}{2\pi}} \int_{-\infty}^{\infty} e^{-s(\frac{x^2}{2} + \frac{x^3}{3} + \cdots)} dx\right).$$

Proof. The statement of the lemma follows from the fact that both series coincide with the formal series

$$\sum \frac{(-1)^n}{|\mathrm{Aut}(\Gamma)|} s^{|V(\Gamma)| - |E(\Gamma)|},$$

where the sum is taken over all connected embedded graphs Γ having no vertices of valency 1 and 2, and we have $|V(\Gamma)| - |E(\Gamma)| = 2 - 2g - n$. For the function $F(s)$ this is the definition, and for the integral in the lemma, this statement is the $N = 1$ version of the Hermitian matrix model statement. In order to verify it, one should make the substitution $x = ys^{-1/2}$, apply Proposition 3.3.4, and recall that the logarithm extracts connected graphs.

Now, we may consider the equality for $F(s)$ not as a formal one, but as an asymptotic expansion as $s \to \infty$, and replace the integral of the lemma with the same integral but taken over an arbitrary neighborhood $U(0)$ of 0. The independence of the asymptotic expansion of the choice of the neighborhood is obvious. Making the obvious substitution of variables, we obtain

$$\sqrt{\frac{2\pi}{s}} e^{F(s)} \sim \int_{U(0)} e^{s(x + \log(1-x))} dx = \int_{U(0)} (1-x)^s e^{sx} dx = \int_{U(1)} y^s e^{-sy} dy$$

$$\sim e^s \int_0^\infty y^s e^{-sy} dy = \frac{e^s}{s^{s+1}} \int_0^\infty z^s e^{-z} dz = \frac{e^s}{s^{s+1}} \Gamma(s+1).$$

Thus, the generating function A is a solution to the equation

$$\int_s^{s+1} A(t) dt = \log\left(\frac{\Gamma(s+1)}{\sqrt{2\pi}}\right).$$

It is easy to check that

$$A(t) = \frac{1}{2} - t + t \frac{d}{dt} \log \Gamma(t)$$

is the required solution. The Harer–Zagier formula $\chi(\mathcal{M}_{g;1}) = -B_{2g}/2g$ follows now from Stirling's formula for $\log \Gamma(t)$:

$$\frac{\Gamma'(t)}{\Gamma(t)} \sim \log t - \frac{1}{2t} - \sum_{g \geq 1} \frac{B_{2g}}{2g} t^{-2g} \quad \text{as} \quad t \to \infty.$$

4.6 Intersection Indices on Moduli Spaces and the String and Dilaton Equations

In this section we will work with the rational cohomology $H^*(\overline{\mathcal{M}}_{g;n}, \mathbb{Q})$ of the Deligne–Mumford compactifications of the moduli spaces. The cohomology is an algebra over $\mathbb{Q}$, and for two cohomology classes $\xi_1 \in H^i, \xi_2 \in H^j$ their product $\xi_1 \xi_2$ is an element in H^{i+j}. Let $\xi \in H^i(\overline{\mathcal{M}}_{g;n}, \mathbb{Q})$ be a rational cohomology class. By

$$\int_{\overline{\mathcal{M}}_{g;n}} \xi$$

we mean the result of integrating ξ over the fundamental class of the moduli space for

$$i = \dim_\mathbb{R} \overline{\mathcal{M}}_{g;n} = 2(3g - 3 + n),$$

and zero otherwise. Below we consider only rational cohomology and omit references to the coefficients in the notation.

There are numerous natural cohomology classes in $H^2(\overline{\mathcal{M}}_{g;n})$. To start with, by Poincaré duality, such a class is determined by each irreducible component in the boundary $\partial \overline{\mathcal{M}}_{g;n}$: the value of this class on a 2-cycle in $\overline{\mathcal{M}}_{g;n}$ is the number of the points of intersection of the cycle with the boundary component. Another source of such cohomology classes are the first Chern classes of line bundles over $\overline{\mathcal{M}}_{g;n}$. For example, one associates with the ith marked point the line bundle $\mathcal{L}_i$ whose fiber at a moduli point $(X; x_1, \ldots, x_n)$ is the cotangent line to X at x_i. The *first Chern class* of a holomorphic line bundle is an obstacle to the existence of a nowhere zero holomorphic section (see, e.g., [108]). By definition, it can be presented by the (Poincaré dual class to the) divisor of zeroes and poles of arbitrary meromorphic nonzero section of the bundle. Following the tradition, we use the notation $\psi_i = c_1(\mathcal{L}_i)$ for the first Chern classes of the line bundles $\mathcal{L}_i$. The classes ψ_i are sufficient for the purposes of the present chapter; other classes will also be useful in Chapter 5.

Because of the symmetry, the integral

$$\int_{\overline{\mathcal{M}}_{g;n}} \psi_1^{m_1} \ldots \psi_n^{m_n}$$

depends only on the set $\{m_1, \ldots, m_n\}$ of non-negative integers, and Witten denotes its value by

$$\langle \tau_{m_1} \ldots \tau_{m_n} \rangle = \int_{\overline{\mathcal{M}}_{g;n}} \psi_1^{m_1} \ldots \psi_n^{m_n}. \tag{4.2}$$

The order of the letters τ_i on the left-hand side is irrelevant. Informally, this number can be understood as the *number of intersection points of m_1 copies of the divisor ψ_1, $\ldots$, m_n copies of the divisor ψ_n*, although this "number of intersections" can prove to be rational, not integer. It can be nonzero only if $m_1 + \cdots + m_n = \dim_\mathbb{C} \overline{\mathcal{M}}_{g;n}$. For each set $\{m_1, \ldots, m_n\}$ of non-negative integers there is at most one value of g such that (4.2) is nonzero. For example, if we take $n = 4$, $m_1 = m_2 = m_3 = m_4 = 0$, then

$$\dim_{\mathbb{C}} \overline{\mathcal{M}}_{g;n} = 3g - 3 + n = 3g - 3 + 4 = 3g + 1$$

must be equal to zero, and the last equation has no solutions. Hence, $\langle \tau_0^4 \rangle = 0$.

Example 4.6.1. For $g = 0$, $n = 3$, there is a unique class with nonzero integral

$$\langle \tau_0 \tau_0 \tau_0 \rangle = \langle \tau_0^3 \rangle = \int_{\overline{\mathcal{M}}_{0;3}} \psi_1^0 \psi_2^0 \psi_3^0 = \int_{\overline{\mathcal{M}}_{0;3}} 1 = 1$$

since the moduli space is a single point.

Example 4.6.2. In order to compute the integral

$$\langle \tau_0 \tau_0 \tau_0 \tau_1 \rangle = \langle \tau_0^3 \tau_1 \rangle = \int_{\overline{\mathcal{M}}_{0;4}} \psi_1^0 \psi_2^0 \psi_3^0 \psi_4 = \int_{\overline{\mathcal{M}}_{0;4}} \psi_4 = \int_{\overline{\mathcal{M}}_{0;4}} \psi_1$$

let us construct a holomorphic section of the line bundle $\mathcal{L}_1$. For a smooth curve $(X; x_1, x_2, x_3, x_4)$ consider a meromorphic 1-form ω on X having poles of order one at the points x_3, x_4 and no other poles on X and such that its residue at x_3 is 1, and its residue at x_4 is -1. Such a 1-form is unique: if we choose a coordinate z on X such that the coordinates of x_3 and x_4 are 0 and ∞ respectively, then it looks like dz/z; and it has no zeroes on X. Hence, its value ω_{x_1} at the point x_1 determines a nowhere zero section of $\mathcal{L}_1$ over the moduli space $\mathcal{M}_{0;4}$ of smooth curves.

Now let us extend this section to the three boundary points (see Example 4.3.5). At the point $\{\{x_1, x_3\}, \{x_2, x_4\}\}$ the extension is determined by a pair of meromorphic 1-forms on each of the irreducible components of the curve. On the first component, this 1-form has poles of order one at the point x_3 and at the double point with residues 1 and -1 respectively, while on the second component, it has poles of order one at the point x_4 and the double point with residues -1 and 1 respectively. The value of such holomorphic 1-form at the point x_1 is nonzero. The situation with the boundary point $\{\{x_1, x_4\}, \{x_2, x_3\}\}$ is similar. The extension of the section to the boundary point $\{\{x_1, x_2\}, \{x_3, x_4\}\}$ is, however, different, since it is determined by the meromorphic 1-form, which is zero on the first irreducible component of the curve, and has poles of order one with residues $1, -1$ at the points x_3, x_4 on the second component. Indeed, there are no nontrivial meromorphic 1-forms on the first component with a single pole of order one at the double point.

Thus, we have constructed a section of the line bundle $\mathcal{L}_1$ over $\overline{\mathcal{M}}_{0;4}$ having a unique zero at the boundary point $\{\{x_1, x_2\}, \{x_3, x_4\}\}$. An easy local calculation confirms that this section intersects the zero section of the bundle transversally (see below). By definition of the first Chern class of a line bundle, this means that

$$\langle \tau_0^3 \tau_1 \rangle = \int_{\overline{\mathcal{M}}_{0;4}} \psi_1 = 1.$$

In order to check that the section constructed above does intersect the zero section transversally, consider a model example. Let C_ε be the family of rational curves $xy = \varepsilon$ on the plane $\mathbb{C}^2$ endowed with the coordinates (x, y). Let $x_1 = x_1(\varepsilon)$, $x_2 = x_2(\varepsilon)$ (resp., $x_3 = x_3(\varepsilon)$, $x_4 = x_4(\varepsilon)$) be the intersection points of the curve and the lines $x = \pm 1$ (resp., $y = \pm 1$). The 1-form $\omega(\varepsilon)$ is determined by the restriction to C_ε of the 1-form $(1/(y-1) - 1/(y+1))dy$ on the plane. The value of this 1-form at the point x_1 is given, in the coordinate x on C_ε, by the formula

$$\omega(\varepsilon)_{x_1} = -\varepsilon \left(\frac{x}{\varepsilon - x} - \frac{x}{\varepsilon + x} \right) \frac{dx}{x^2} \bigg|_{x=1}$$

$$= \varepsilon \left(\frac{1}{1-\varepsilon} + \frac{1}{1+\varepsilon} \right) dx,$$

and its order in ε at 0 is, indeed, one.

Exercise 4.6.3. 1. Verify that the 1-form $\omega(\varepsilon)$ has no poles different from x_1, x_2, x_3, x_4 (in particular, that it has no poles at infinity).

2. Find a coordinate-free way to confirm the transversality.

Example 4.6.4. The case $g = 0$, $n = 5$ can be treated similarly. Consider the section of the line bundle $\mathcal{L}_1$ determined by the meromorphic 1-forms with poles of order one at the points x_4, x_5 and fixed residues. The zero locus of this section is the union of the irreducible components

$$\{\{x_1, x_2\}, \{x_3, x_4, x_5\}\}, \quad \{\{x_1, x_3\}, \{x_2, x_4, x_5\}\}, \quad \{\{x_1, x_2, x_3\}, \{x_4, x_5\}\}$$

of the boundary divisor. The zero locus of another section of $\mathcal{L}_1$, the one determined by the family of 1-forms with poles at the points x_2, x_3, consists of the boundary lines

$$\{\{x_1, x_4\}, \{x_2, x_3, x_5\}\}, \quad \{\{x_1, x_5\}, \{x_2, x_3, x_4\}\}, \quad \{\{x_1, x_4, x_5\}, \{x_2, x_3\}\}.$$

The zero loci of these two sections intersect transversally at the single point

$$\{\{x_2, x_3\}, \{x_1\}, \{x_4, x_5\}\};$$

therefore,

$$\langle \tau_0^4 \tau_2 \rangle = \int_{\overline{\mathcal{M}}_{0;5}} \psi_1^2 = 1.$$

Similarly, the zero locus of the section of $\mathcal{L}_2$ determined by the 1-form with the poles at x_1, x_4 consists of the three boundary lines

$$\{\{x_1, x_4\}, \{x_2, x_3, x_5\}\}, \quad \{\{x_1, x_4, x_5\}, \{x_2, x_3\}\}, \quad \{\{x_1, x_3, x_4\}, \{x_2, x_5\}\}.$$

It intersects the zero locus of the first section of $\mathcal{L}_1$ transversally at two points

$$\{\{x_2, x_3\}, \{x_1\}, \{x_4, x_5\}\} \quad \text{and} \quad \{\{x_1, x_3\}, \{x_4\}, \{x_2, x_5\}\};$$

therefore,
$$\langle \tau_0^3 \tau_1^2 \rangle = \int_{\overline{\mathcal{M}}_{0;5}} \psi_1 \psi_2 = 2.$$

Exercise 4.6.5. Compute
$$\langle \tau_0^5 \tau_3 \rangle = \int_{\overline{\mathcal{M}}_{0;6}} \psi_1^3$$
by constructing appropriate sections and studying their behavior on the boundary strata.

Example 4.6.6. In the case $g = 1$, $n = 1$ the boundary divisor contains a unique component, the point corresponding to the torus with contracted nontrivial cycle (see Example 4.3.8). Consider the family of tori doubly covering the projective line by means of the projection to the x-line of the curve $y^2 = x(x-1)(x-\varepsilon)$ with x_1 at the point $x = 0$. This projection is a 2-fold covering ramified over the points $0, 1, \varepsilon, \infty$. The value of ε is defined up to a symmetry group of order 6 (interchanging the three ramification points $1, \infty, \varepsilon$), whence ε is a coordinate on the 6-fold covering of the moduli space $\overline{\mathcal{M}}_{1;1}$. The ε-family of holomorphic differentials $\omega = dx/y$ determines a holomorphic section of $\mathcal{L}_1$ (we simply take the value of ω at $x = 0$, i.e., at the point x_1).

The degeneration occurs as ε tends to either $0, 1$ or ∞. The section does not vanish at the first two points, while it degenerates at infinity, and $\omega \sim \varepsilon^{1/2}$ at this point. Therefore, we obtain a factor $1/2$. One more factor $1/2$ comes from the fact that the elliptic curve has a symmetry $(x, y) \mapsto (x, -y)$ preserving the value of ε. Finally, we conclude that

$$\langle \tau_1 \rangle = \int_{\overline{\mathcal{M}}_{1;1}} \psi_1 = \frac{1}{24}.$$

Exercise 4.6.7. Verify that the section of the line bundle $\mathcal{L}_1$ lifted to the 6-fold covering of $\overline{\mathcal{M}}_{1;1}$ in the example above has indeed no zeroes as ε tends to 0 and 1, at is has a zero of order $1/2$ as ε tends to infinity.

If the pair $(g; n)$ is stable, then there is a natural "forgetful" mapping $\pi_{g;n} : \mathcal{M}_{g;n+1} \to \mathcal{M}_{g;n}$ taking a smooth curve $(X; x_1, \ldots, x_n, x_{n+1})$ to the curve $(X; x_1, \ldots, x_n)$. This mapping extends to a mapping $\pi_{g;n} : \overline{\mathcal{M}}_{g;n+1} \to \overline{\mathcal{M}}_{g;n}$, but the definition of the extension is somewhat subtle. The subtleness is due to the fact that after forgetting a marked point on a stable curve, the latter may become unstable. There are two situations in which this happens; in both of them the irreducible component of X containing x_{n+1} is rational, and either contains a point of intersection with another component and a marked point x_j, $j \neq n+1$, or two points of intersection with other components. In both cases, the correct extension requires contracting the rational component. In the first case, the former intersection point on the other component becomes the marked point x_j; in the second case, the two former intersection points on the two neighboring components become a common point of the components (see Fig. 4.8). The proof of the fact that after these clarifying comments the forgetful mapping is well defined can be found in [211], V.4.4.

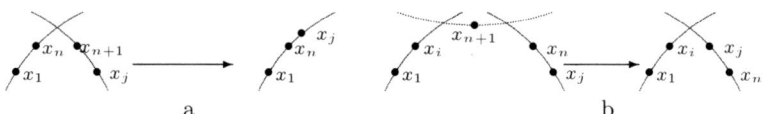

Fig. 4.8. The forgetful mapping on two types of degenerations

The forgetful mapping naturally leads to numerous recurrence relations between the intersection numbers. The most important of them are the so-called string and dilaton equations. The names came from physics, and they are related to topological gravity models.

The string equation is applicable if and only if the integrand in

$$\langle \tau_{m_1} \ldots \tau_{m_n} \tau_{m_{n+1}} \rangle = \int_{\overline{\mathcal{M}}_{g;n+1}} \psi_1^{m_1} \ldots \psi_n^{m_n} \psi_{n+1}^{m_{n+1}}$$

does not contain one of the classes ψ_i, e.g. the class ψ_{n+1}, i.e., $m_{n+1} = 0$ and $\tau_{m_{n+1}}$ is nothing else but τ_0. The *string equation* has the form

$$\langle \tau_0 \tau_{m_1} \ldots \tau_{m_n} \rangle = \langle \tau_{m_1-1} \ldots \tau_{m_n} \rangle + \cdots + \langle \tau_{m_1} \ldots \tau_{m_n-1} \rangle \quad (4.3)$$

(for j such that m_j is equal to 0 the corresponding summand vanishes). For a sketch of the proof, see below.

The string equation totally determines all intersection numbers for the ψ-classes on moduli spaces of rational curves since the dimensional restriction

$$m_1 + \cdots + m_n + m_{n+1} = (n+1) - 3$$

immediately implies that at least one of m_i equals zero (in fact, even at least three of them).

Example 4.6.8. Using the string equation we obtain the following intersection formulas for $g = 0$:

- $n = 4$:
$$\langle \tau_0^3 \tau_1 \rangle = \langle \tau_0^3 \rangle = 1;$$

- $n = 5$:
$$\langle \tau_0^3 \tau_1^2 \rangle = 2 \langle \tau_0^3 \tau_1 \rangle = 2; \quad \langle \tau_0^4 \tau_2 \rangle = \langle \tau_0^3 \tau_1 \rangle = 1,$$

in agreement with the calculations above;

- $n = 6$:
$$\langle \tau_0^3 \tau_1^3 \rangle = 3 \langle \tau_0^3 \tau_1^2 \rangle = 6;$$
$$\langle \tau_0^4 \tau_1 \tau_2 \rangle = \langle \tau_0^3 \tau_1^2 \rangle + \langle \tau_0^4 \tau_2 \rangle = 3;$$
$$\langle \tau_0^5 \tau_3 \rangle = \langle \tau_0^4 \tau_2 \rangle = 1.$$

Exercise 4.6.9. Verify the calculations in the previous example.

The following explicit formula for all intersection numbers for the genus zero moduli space is an immediate consequence of the string equation.

Proposition 4.6.10 (Intersection indices for rational curves). *For $m_1 + \cdots + m_n = n - 3$ we have*
$$\langle \tau_{m_1} \ldots \tau_{m_n} \rangle = \binom{n-3}{m_1 \ldots m_n} = \frac{(n-3)!}{m_1! \ldots m_n!}.$$

Indeed, it suffices to verify that the expression above satisfies the string equation and the initial condition $\langle \tau_0^3 \rangle = 1$. Note that we don't have to mention the genus explicitly because it is determined by the set of indices $m_1, \ldots, m_n$. This is why the parameter "genus" is absent in the generating function for intersection numbers that will be constructed in the next section.

Now let us discuss briefly the proof of the string equation. Consider the forgetful mapping $\pi : \overline{\mathcal{M}}_{g;n+1} \to \overline{\mathcal{M}}_{g;n}$. Denote by $D_j \subset \overline{\mathcal{M}}_{g;n+1}$, $j = 1, \ldots, n$, the divisor consisting of the curves containing a smooth rational irreducible component with only marked points x_j, x_{n+1} on it, intersecting other components at a single point, see Fig. 4.8a. (We consider only the first type of contracted components since the locus of nonstability degeneration of the second type, the one with two intersection points, shown in Fig. 4.8b, has codimension two and does not affect the first Chern classes we are interested in.) Note that $D_i \cap D_j = \emptyset$ for $i \neq j$ and each D_j is naturally isomorphic to $\overline{\mathcal{M}}_{g;n}$ under the forgetful mapping.

For the time being, let us denote the first Chern classes of the line bundles $\mathcal{L}_j$ on $\overline{\mathcal{M}}_{g;n}$ by ψ'_j, keeping the notation ψ_j for such classes on $\overline{\mathcal{M}}_{g;n+1}$. A nonvanishing local section of the line bundle $\mathcal{L}_j$ over $\overline{\mathcal{M}}_{g;n}$ lifts to a nonzero local section of $\mathcal{L}_j$ over $\overline{\mathcal{M}}_{g;n+1}$ having simple zero on D_j, whence
$$\psi_j = \pi^*(\psi'_j) + [D_j].$$

Therefore,
$$\langle \tau_0 \tau_{m_1} \ldots \tau_{m_n} \rangle = \int_{\overline{\mathcal{M}}_{g;n+1}} \psi_1^{m_1} \ldots \psi_n^{m_n}$$
$$= \int_{\overline{\mathcal{M}}_{g;n+1}} (\pi^*(\psi'_1) + [D_1])^{m_1} \ldots (\pi^*(\psi'_n) + [D_n])^{m_n}.$$

Now, $[D_j]\psi_j = 0$ since the restriction of the line bundle $\mathcal{L}_j$ to D_j is trivial (the marked point x_j lives, over D_j, on a separate rational irreducible component of the curve). This leads to the formula

$$(\pi^*(\psi'_j) + [D_j])^{m_j} = (\pi^*(\psi'_j))^{m_j} + [D_j](\pi^*(\psi'_j))^{m_j-1}$$

because of the decomposition $a^m - b^m = (a-b)\sum_i a^i b^{m-1-i}$ applied to $a = \psi_j$, $b = \pi^*(\psi'_j)$, $a - b = [D_j]$, $m = m_j$. Taking into account that

$$\int_{\overline{\mathcal{M}}_{g;n+1}} \pi^*(\psi'_1)^{m_1} \ldots \pi^*(\psi'_n)^{m_n} = 0$$

as an integral of a class pulled back from a variety of a smaller dimension and using the obvious equality $[D_i][D_j] = 0$, we obtain

$$\langle \tau_0 \tau_{m_1} \ldots \tau_{m_n} \rangle = \int_{\overline{\mathcal{M}}_{g;n+1}} ([D_1](\pi^*(\psi'_1))^{m_1-1} \ldots (\pi^*(\psi_n))^{m_n} \\ + \cdots + [D_n](\pi^*(\psi'_1))^{m_1} \ldots (\pi^*(\psi'_n))^{m_n-1}),$$

and the string equation follows.

The *dilaton equation* can be applied in the situation where the integrand in

$$\langle \tau_{m_1} \ldots \tau_{m_n} \tau_{m_{n+1}} \rangle = \int_{\overline{\mathcal{M}}_{g;n+1}} \psi_1^{m_1} \ldots \psi_n^{m_n} \psi_{n+1}^{m_{n+1}}$$

contains one of the classes ψ_i, e.g. the class ψ_{n+1}, in the degree one, i.e., if $m_{n+1} = 1$ and $\tau_{m_{n+1}}$ is nothing but τ_1. It looks even simpler than the string equation:

$$\langle \tau_1 \tau_{m_1} \ldots \tau_{m_n} \rangle = (2g - 2 + n)\langle \tau_{m_1} \ldots \tau_{m_n} \rangle. \tag{4.4}$$

Together with the string equation, the dilaton equation totally determines the intersection indices in genus one.

Proposition 4.6.11 (Intersection indices for elliptic curves). *For $m_1 + \cdots + m_n = n$ we have*

$$\langle \tau_{m_1} \ldots \tau_{m_n} \rangle = \frac{1}{24}\binom{n}{m_1 \ldots m_n}\left(1 - \sum_{i=2}^{n} \frac{(i-2)!(n-i)!}{n!} e_i(m_1, \ldots, m_n)\right),$$

where e_k is the kth elementary symmetric function,

$$e_k(m_1, \ldots, m_n) = \sum_{i_1 < \cdots < i_k} m_{i_1} \ldots m_{i_k}.$$

Similarly to the genus zero case, the proof of the proposition consists in verification that the expression above satisfies the string, as well as the dilaton equations and the initial condition $\langle \tau_1 \rangle = 1/24$.

The proof of the dilaton equation is similar to that of the string equation. Namely, one has

$$\langle \tau_1 \tau_{m_1} \cdots \tau_{m_n} \rangle = \int_{\overline{\mathcal{M}}_{g;n+1}} (\pi^*(\psi_1') + [D_1])^{m_1} \cdots (\pi^*(\psi_n') + [D_n])^{m_n} \psi_{n+1}$$

$$= \int_{\overline{\mathcal{M}}_{g;n+1}} (\pi^*(\psi_1'))^{m_1} \cdots (\pi^*(\psi_n'))^{m_n} \psi_{n+1},$$

since $\psi_{n+1}[D_j] = 0$ for $j = 1, \ldots, n$.

The last integral can be reduced to an integral over $\overline{\mathcal{M}}_{g;n}$ by integrating over the fibers of the mapping $\pi : \overline{\mathcal{M}}_{g;n+1} \to \overline{\mathcal{M}}_{g;n}$ forgetting the $(n+1)$th marked point. The integral of ψ_{n+1} over such a fiber is $2g - 2 + n$, that is, the number of zeroes of a meromorphic 1-form on a genus g curve X with n marked points having poles of order one at the marked points, and the dilaton equation follows.

4.7 KdV Hierarchy and Witten's Conjecture

It is convenient to unify all intersection numbers for the ψ-classes into the exponential generating function

$$F(t_0, t_1, \ldots) = \left\langle \exp\left(\sum_i t_i \tau_i\right) \right\rangle$$

$$= \sum_{n=0}^{\infty} \frac{1}{n!} \sum_{(m_1, \ldots, m_n)} \langle \tau_{m_1} \cdots \tau_{m_n} \rangle t_{m_1} \cdots t_{m_n}$$

$$= \sum_{(l_0, \ldots, l_s)} \langle \tau_0^{l_0} \cdots \tau_s^{l_s} \rangle \frac{t_0^{l_0}}{l_0!} \cdots \frac{t_s^{l_s}}{l_s!}.$$

Now Witten's conjecture (Kontsevich's theorem) states the following:

Theorem 4.7.1 (Kontsevich's theorem). *The second derivative*

$$V(t_0, t_1, \ldots) = \partial^2 F / \partial t_0^2$$

coincides with the partition function U for the universal one-matrix model (see Sec. 3.6.3).

The string and the dilaton equations can be considered as partial differential equations $L_{-1}V = 0$ and $L_0 V = 0$ on the function $V(t_0, \ldots) = \partial^2 F / \partial t_0^2$. The base for the Witten conjecture was the fact that the first of these equations $L_{-1}V = 0$ coincides with the analogous equation for the one-matrix model from Sec. 3.6.3, as well as the idea that "all models of two-dimensional quantum gravity must coincide".

The operators L_{-1} and L_0 are the first two equations in the KdV hierarchy of operators L_m.

Theorem 4.7.2. *The generating function F satisfies the system of equations $L_m(F) = 0$ for $m \geq -1$.*

One more reformulation of the Witten conjecture states that the function F in variables t_i is a τ-function for the KdV hierarchy.

Remark 4.7.3. The KdV operators satisfy the commutation relations

$$[L_n, L_m] = (n - m)L_{n+m}.$$

In particular,

- they span a Lie algebra; therefore, in order to prove Theorem 4.7.2 it is sufficient to prove that $L_2(F) = 0$ since L_{-1} and L_2 generate this Lie algebra;
- this Lie algebra is isomorphic to the Lie algebra of polynomial vector fields on the line under the isomorphism

$$L_m \mapsto -z^{m+1}\frac{d}{dz};$$

- under certain homogeneity conditions there is a unique way to extend the representation of the Lie algebra of polynomial vector fields to the algebra of differential operators in t_k starting with L_{-1} and L_0.

Note also that the string equation implies that Witten's conjecture is true for rational curves, while together with the dilaton equation it confirms the conjecture for the case of elliptic curves.

4.8 The Kontsevich Model

The Kontsevich model is an intermediate one between the one-matrix model and the model based on the intersection theory on the moduli spaces of curves. It is equivalent to both of them.

Let Λ be a diagonal $N \times N$ matrix with positive entries $\Lambda_1, \ldots, \Lambda_N$ on the diagonal. The Kontsevich model associates to the matrix Λ a new measure on the space of Hermitian matrices. This measure has the form

$$d\mu_\Lambda(H) = C_{\Lambda,N} e^{-\frac{1}{2}\operatorname{tr} H^2 \Lambda} dv(H) . \tag{4.5}$$

In fact, this measure is a family of measures parametrized by the diagonal entries of Λ. Recall that $dv(H)$ is the volume form on the space of Hermitian matrices. The constant $C_{\Lambda,N}$ is chosen so that to make the integral $\int_{\mathcal{H}_N} d\mu_\Lambda(H)$ equal to 1, see below.

The *Kontsevich model* is described by the following integral over the space of Hermitian matrices:

$$\log \int_{\mathcal{H}_N} e^{\frac{i}{6}\operatorname{tr} H^3} d\mu_\Lambda(H) . \tag{4.6}$$

258   4 Geometry of Moduli Spaces

This integral is a symmetric function of $\Lambda_1,\ldots,\Lambda_N$, but it will be more convenient for us to regard this integral as a function in a different set of variables.

We are already familiar with essential parts of integral (4.6), at least for the identity matrix Λ. The expansion of the exponent in the integrand in powers of $\operatorname{tr} H^3$ looks like

$$e^{\frac{i}{6}\operatorname{tr} H^3} = 1 + \frac{1}{1!}\frac{i}{6}\operatorname{tr} H^3 - \frac{1}{2!}\frac{1}{6^2}(\operatorname{tr} H^3)^2 - \frac{1}{3!}\frac{i}{6^3}(\operatorname{tr} H^3)^3 + \ldots,$$

and we conclude that this integral enumerates gluings of 3-stars (up to a nonessential numeric normalizing factor). Note that the number of ways to glue an odd number of 3-stars is zero, hence the imaginary part of the expansion of (4.6) vanishes, and we have

$$\int_{\mathcal{H}_N} e^{\frac{i}{6}\operatorname{tr} H^3} d\mu_\Lambda(H) = \int_{\mathcal{H}_N} \left(1 - \frac{1}{2!}\frac{1}{6^2}(\operatorname{tr} H^3)^2 + \frac{1}{4!}\frac{1}{6^4}(\operatorname{tr} H^3)^4 - \ldots\right) d\mu_\Lambda(H). \tag{4.7}$$

Taking the logarithm leads to enumeration of *connected* gluings. Thus all the innovations are due to the changing of the measure.

The changing of the measure results in appearing of the *weight* of a gluing, and therefore integral (4.7) enumerates weighted gluings, not ordinary ones.

In order to compute this weight, let us find the average $\langle h_{ij} h_{kl}\rangle$ of a monomial $h_{ij} h_{kl}$ of degree two with respect to the new measure. Recall that $H = (h_{kl})$, $h_{lk} = \bar h_{kl}$, where $h_{kl} = x_{kl} + i y_{kl}$ are the entries of the Hermitian matrix H.

In the coordinates x_{ii}, x_{ij}, y_{ij}, $1 \le i < j \le N$ the quadratic form $\operatorname{tr} H^2 \Lambda$ looks like

$$\begin{pmatrix} \Lambda_1 & & & & & & & \\ & \ddots & & & & & & \\ & & \Lambda_N & & & & & \\ & & & \Lambda_1+\Lambda_2 & & & & \\ & & & & \ddots & & & \\ & & & & & \Lambda_{N-1}+\Lambda_N & & \\ & & & & & & \Lambda_1+\Lambda_2 & \\ & & & & & & & \ddots \\ & & & & & & & & \Lambda_{N-1}+\Lambda_N \end{pmatrix}.$$

This gives the following value of the constant $C_{\Lambda,N}$ in (4.5):

$$C_{\Lambda,N} = (2\pi)^{-N^2/2}\prod_{i=1}^{N}\Lambda_i^{1/2}\prod_{i<j}(\Lambda_i+\Lambda_j).$$

The quadratic form is diagonal and we easily compute

$$\langle x_{ii}^2\rangle = \langle h_{ii}^2\rangle = \frac{1}{\Lambda_i}, \quad \langle x_{ij}^2\rangle = \langle y_{ij}^2\rangle = \frac{1}{\Lambda_i+\Lambda_j} \quad \text{for} \quad i<j.$$

The result of this computation can be summed up by the formula

$$\langle h_{ij} h_{ji} \rangle = \frac{2}{\Lambda_i + \Lambda_j} \tag{4.8}$$

and

$$\langle h_{ij} h_{kl} \rangle = 0 \quad \text{if} \quad (i,j) \neq (l,k).$$

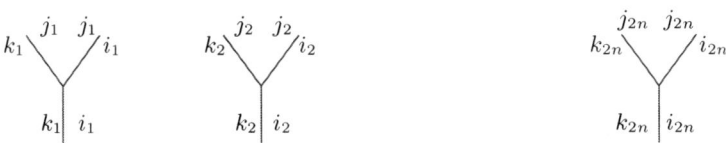

Fig. 4.9. Indexed 3-stars

Consider a set of 3-stars with darts marked, as usual, by pairs of indices (see Fig. 4.9), and consider an arbitrary gluing of these darts. Let us for the time being call such an object an *indexed gluing*. For the standard Gaussian measure the contribution of an indexed gluing depends on the indexing. However, the dependence is not too serious: the contribution of an indexed gluing equals 1 if the indexing "agrees" with the gluing, and it equals 0 otherwise. The word "agrees" means that all edge sides belonging to the same face have the same index. Thus an indexing which agrees with a gluing is simply that of the faces of the gluing. In order to compute the contribution of a gluing itself we simply summed "ones" over all indexings that agree with the gluing.

For the new measure the situation is different. The contribution of an indexed gluing, where the indexing does not agree with the gluing, is zero as well. However, if the indexing agrees with the gluing, then the contribution depends on this indexing.

Consider an indexed gluing with the indexing agreeing with the gluing. In this case an edge of the gluing corresponds to a pair $\langle h_{ij} h_{ji} \rangle$, where i and j are the indices of the faces adjacent to this edge. If an edge has the same face with index i on both sides, then it corresponds to the average $\langle h_{ii}^2 \rangle$.

Thus, the contribution of an indexed gluing is the product

$$\prod \frac{2}{\Lambda_i + \Lambda_j}$$

taken over all edges of the gluing, and the entry of the product is $2/(\Lambda_i + \Lambda_j)$ if and only if the corresponding edge is adjacent to the faces i and j.

For example, the contributions of indexed gluings in Fig. 4.10 a, b, c are

$$\frac{2}{\Lambda_i + \Lambda_j} \frac{2}{\Lambda_j + \Lambda_k} \frac{2}{\Lambda_k + \Lambda_i}, \quad \frac{1}{\Lambda_i} \frac{2}{\Lambda_i + \Lambda_j} \frac{2}{\Lambda_i + \Lambda_k}, \quad \frac{1}{\Lambda_i} \frac{1}{\Lambda_i} \frac{1}{\Lambda_i}$$

respectively.

4 Geometry of Moduli Spaces

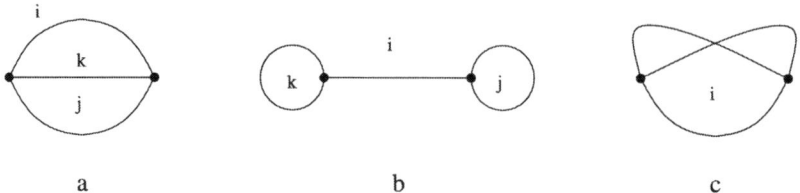

Fig. 4.10. Indexed gluings of a pair of 3-stars

The contribution of a gluing is the sum of contributions of all its possible indexings. However, it is more interesting to consider not the contribution of a separate gluing, but the contribution of a class of gluings that provide the surface of the same genus.

Example 4.8.1. There exist 15 gluings of two 3-stars. Twelve of them give a sphere, while the other three result in a torus. Those giving the sphere are of the type shown in Fig. 4.10, a and b. There are three gluings of the type a and nine gluings of the type b. Fixing the set of indices $\{i, j, k\}$ for the faces, summing over all the permutations of this set of indices and dividing by the number of these permutations we obtain the total contribution of all the spherical gluings:

$$\frac{1}{3!}\left(3 \cdot \frac{6 \cdot 8}{(\Lambda_i + \Lambda_j)(\Lambda_j + \Lambda_k)(\Lambda_k + \Lambda_i)} + 9 \cdot \left(\frac{2 \cdot 8}{2\Lambda_i(\Lambda_i + \Lambda_j)(\Lambda_i + \Lambda_k)}\right.\right.$$
$$\left.\left. + \frac{2 \cdot 8}{2\Lambda_j(\Lambda_j + \Lambda_i)(\Lambda_j + \Lambda_k)} + \frac{2 \cdot 8}{2\Lambda_k(\Lambda_k + \Lambda_i)(\Lambda_k + \Lambda_j)}\right)\right)$$
$$= \frac{12}{\Lambda_i \Lambda_j \Lambda_k} . \qquad (4.9)$$

As if by magic, all terms of the form $\Lambda_i + \Lambda_j$ with distinct i, j in the denominators are cancelled, and we are left with a denominator which is a monomial in Λ_i. Next examples also show the similar effect.

Let us explain in more detail the meaning of the coefficients on the left-hand side of (4.9). The common factor $1/3!$ corresponds to the division by the number of permutations of the three indices i, j, k. In the first summand in parentheses the coefficient 3 reflects the three different ways to obtain a gluing of the type shown in Fig. 4.10a. The coefficient 6 corresponds to the six different ways of indexing the faces of such a gluing with indices i, j, k. And the factor $8/((\Lambda_i + \Lambda_j)(\Lambda_j + \Lambda_k)(\Lambda_k + \Lambda_i))$ is the contribution of the corresponding indexed gluing.

In the second term in parentheses the coefficient 9 reflects the nine different ways to obtain a gluing of the type shown in Fig. 4.10b. Each of the three terms inside the internal parentheses corresponds to a pair of indexed gluings (all in all, there are six possibilities, similarly to the preceding case). Two gluings in a pair differ by indices, but the contribution of each gluing in a pair is the same. The remaining factor of the form $8/(2\Lambda_i(\Lambda_i + \Lambda_j)(\Lambda_i + \Lambda_k))$ is the contribution of the corresponding indexed gluing.

Now, we sum over all possible values of the indices $i, j, k = 1, \ldots, N$ and take into account the normalizing factor $1/2!6^2$ in the expansion (4.7). As a result, the total contribution of all spherical gluings of two 3-stars to integral (4.7) is:

$$c_{0;1} = \frac{1}{2!} \frac{1}{6^2} \sum_{i,j,k=1}^{N} \frac{12}{\Lambda_i \Lambda_j \Lambda_k} = \frac{1}{3!} (\operatorname{tr} \Lambda^{-1})^3 \ .$$

Here $c_{g;n}$ stands for "the contribution of all genus g gluings of $2n$ copies of a 3-star".

Example 4.8.2. The gluing in Fig. 4.10c is much easier to handle. It is the only gluing type for two 3-stars providing a torus, and there are exactly three gluings of this type. This gluing has only one face, and marking this face with the index i we obtain the resulting contribution $3/(\Lambda_i^3)$. Thus, summing over all possible indexings, we have

$$c_{1;1} = \frac{1}{2!} \frac{1}{6^2} \sum_{i=1}^{N} \frac{3}{\Lambda_i^3} = \frac{1}{24} \operatorname{tr}(\Lambda^{-3}) \ .$$

Example 4.8.3. The computational complexity grows very rapidly and we restrict ourselves with presenting only one more example, which we do not describe in detail. Figure 4.11 shows all possible types of spherical gluings of four 3-stars. For each gluing type, the number of gluings of this type is shown, as well as the contribution of a sample indexed gluing.

We attract the reader's attention to the usage of the adjective "spherical" in this example. The integral (4.7) enumerates all gluings of 3-stars, not only connected ones. For a disconnected gluing, its "Euler characteristic" χ is defined as the sum of the Euler characteristics of its connected components. For example, the Euler characteristic of the gluing in Fig. 4.11h is $\chi = 2 + 0 = 2$. The adjective "spherical" means "of Euler characteristic 2". The "genus" of a disconnected gluing can be computed from its "Euler characteristic" χ by the formula $\chi = 2 - 2g$, and exactly this value of "genus" is used in the index in $c_{g,n}$.

Summation over all indexed gluings with the set of indices $\{i, j, k, l\}$ distributed among the faces in all possible ways gives the total contribution

$$5400 \frac{\Lambda_i^2 \Lambda_j^2 \Lambda_k^2 + \Lambda_i^2 \Lambda_j^2 \Lambda_l^2 + \Lambda_i^2 \Lambda_k^2 \Lambda_l^2 + \Lambda_j^2 \Lambda_k^2 \Lambda_l^2}{\Lambda_i^3 \Lambda_j^3 \Lambda_k^3 \Lambda_l^3} \ .$$

Finally, summing over all indices $i, j, k, l = 1, \ldots, N$ and taking into account the normalizing factor $1/4!6^4$ we obtain

$$c_{0;2} = \frac{25}{144} (\operatorname{tr}(\Lambda^{-1}))^3 \operatorname{tr}(\Lambda^{-3}) \ .$$

262   4 Geometry of Moduli Spaces

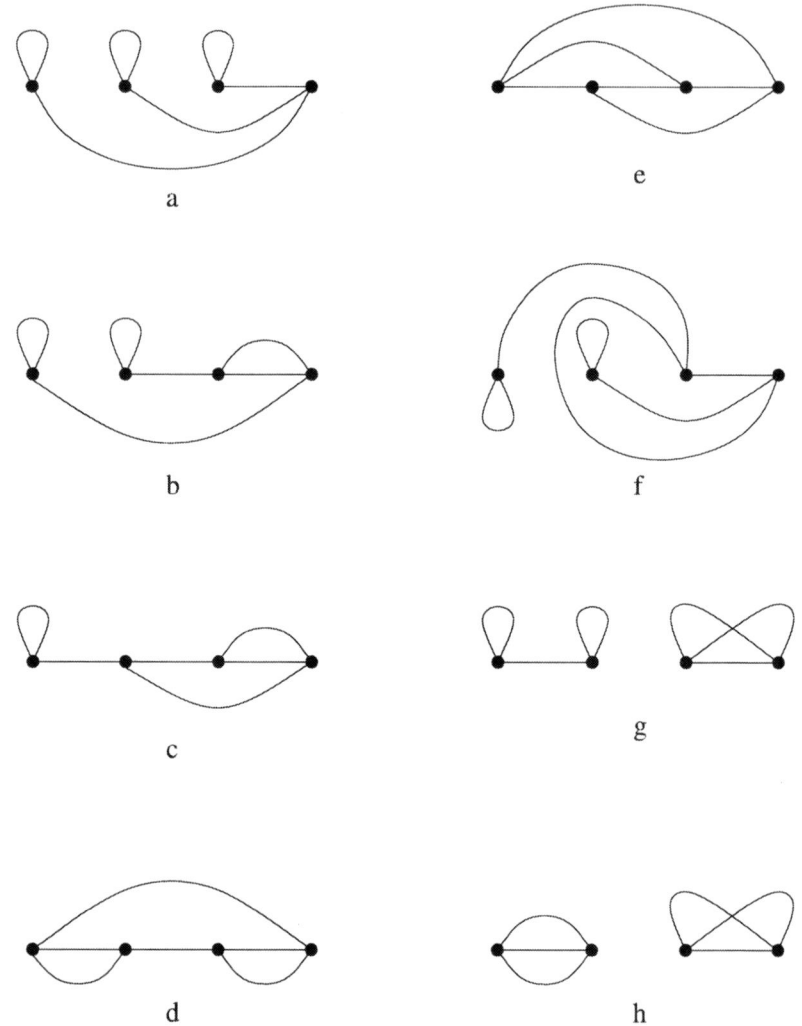

Fig. 4.11. Types of four 3-stars gluings with Euler characteristic 2. The number of gluings and a sample weight of an indexed gluing for each type are:
a) $\#=3^4\cdot 2^3$; $w = (8\Lambda_1^3(\Lambda_1+\Lambda_2)(\Lambda_1+\Lambda_3)(\Lambda_1+\Lambda_4))^{-1}$
b) $\#=3^5\cdot 2^2$; $w = (4\Lambda_1^2(\Lambda_1+\Lambda_2)^2(\Lambda_1+\Lambda_3)(\Lambda_1+\Lambda_4))^{-1}$
c) $\#=3^5\cdot 2^3$; $w = (2\Lambda_1(\Lambda_1+\Lambda_2)^2(\Lambda_1+\Lambda_3)(\Lambda_1+\Lambda_4)(\Lambda_2+\Lambda_3)^2)^{-1}$
d) $\#=3^5\cdot 2$; $w = ((\Lambda_1+\Lambda_2)(\Lambda_1+\Lambda_3)(\Lambda_1+\Lambda_4)^2(\Lambda_2+\Lambda_4)(\Lambda_3+\Lambda_4))^{-1}$
e) $\#=3^4\cdot 2$; $w = ((\Lambda_1+\Lambda_2)(\Lambda_1+\Lambda_3)(\Lambda_1+\Lambda_4)(\Lambda_2+\Lambda_3)(\Lambda_2+\Lambda_4)(\Lambda_3+\Lambda_4))^{-1}$
f) $\#=3^5\cdot 2^2$; $w = (4\Lambda_1\Lambda_3(\Lambda_1+\Lambda_2)(\Lambda_1+\Lambda_3)^2(\Lambda_3+\Lambda_4))^{-1}$
g) $\#=3^4\cdot 2$; $w = (16\Lambda_1\Lambda_4^3(\Lambda_1+\Lambda_2)(\Lambda_1+\Lambda_3))^{-1}$
h) $\#=3^3\cdot 2$; $w = (8\Lambda_4^3(\Lambda_1+\Lambda_2)(\Lambda_2+\Lambda_3)(\Lambda_1+\Lambda_3))^{-1}$
Total: $5400\,(\Lambda_1^2\Lambda_2^2\Lambda_3^2 + \Lambda_1^2\Lambda_2^2\Lambda_4^2 + \Lambda_1^2\Lambda_3^2\Lambda_4^2 + \Lambda_2^2\Lambda_3^2\Lambda_4^2)/(\Lambda_1^3\Lambda_2^3\Lambda_3^3\Lambda_4^3)$

Exercise 4.8.4. Interpret the constant $c_{-1;2}$ and compute it.
[**Answer:** $\frac{1}{72}\operatorname{tr}(\Lambda^{-1})^6$].

In the examples above, the contribution $c_{g;n}$ of all genus g gluings of $2n$ copies of 3-stars is a monomial in $\operatorname{tr}(\Lambda^{-(2k+1)})$ with rational coefficients. In the next section, we will show that this is the case for all g and n.

4.9 A Sketch of Kontsevich's Proof of Witten's Conjecture

In this section we discuss very briefly exciting connections of the Kontsevich model with the one-matrix model on the one hand, and with the intersection theory model on the other hand.

4.9.1 The Generating Function for the Kontsevich Model

The sample calculations in the previous section show that the contribution $c_{g;n}$ of all genus g gluings of $2n$ copies of 3-stars is a monomial in $\operatorname{tr}(\Lambda^{-1})$, $\operatorname{tr}(\Lambda^{-3})$, ... It is more convenient, however, to use a slightly different normalized infinite sequence of independent variables: $t_0 = -\operatorname{tr}(\Lambda^{-1})$, $t_1 = -1!!\operatorname{tr}(\Lambda^{-3})$, $t_2 = -3!!\operatorname{tr}(\Lambda^{-5})$, ..., $t_i = -(2i-1)!!\operatorname{tr}(\Lambda^{-2i-1})$, ...

Theorem 4.9.1 ([178]). *The integral* (4.6) *is a formal power series in the variables* $t_0, t_1, \ldots$ *with rational coefficients.*

Let us denote this series by $K(t_0, t_1, \ldots)$. We have already computed the first few terms:

$$K(t_0, t_1, \ldots) = \log\left(1 + \frac{1}{3!}t_0^3 + \frac{1}{24}t_1 + \frac{25}{144}t_0^3 t_1 + \frac{1}{72}t_0^6 + \ldots\right). \quad (4.10)$$

Remark 4.9.2. The integral (4.7) can be easily interpreted as a formal power series since each monomial in $\operatorname{tr}(\Lambda^{-1})$, $\operatorname{tr}(\Lambda^{-3})$, ... appears in the integral evaluation only a finite number of times. Equation (4.10) is valid for arbitrary value of the dimension N. However, if we want to compute a specific coefficient in this expansion, the value of N must be chosen sufficiently large.

Kontsevich's proof of the Witten conjecture consists of two parts. First, he shows that the coefficient of $t_0^{l_0} \ldots t_s^{l_s}/(l_0! \ldots l_s!)$ in the expansion of his integral $K(t_0, t_1, \ldots)$ in the variables t_i coincides with the intersection number $\langle \tau_0^{l_0} \ldots \tau_s^{l_s} \rangle$. This part of the proof is based on the study of the combinatorial model for the moduli space of curves. The second part consists in verification that the integral is a τ-function for the KdV hierarchy. This means, essentially, that the second derivative $\partial^2 K/\partial t_0^2$ is a solution to the KdV equation. The proof of this statement is achieved by treating the function K as a matrix Airy function.

4.9.2 The Kontsevich Model and Intersection Theory

A formal justification of the argument in this section requires the construction of a "minimal compactification" of the moduli space of smooth marked curves (elaborated by Looijenga in [203]) and an analysis of circle bundles over this compactification. The latter part is accurately written in the Ph.D. thesis of D. Zvonkine [313]. Below, we simply outline the original Kontsevich's argument.

Consider the projection

$$\pi : \mathcal{M}_{g;n}^{\text{comb}} \cong \mathcal{M}_{g;n} \times \mathbb{R}_+^n \to \mathbb{R}_+^n$$

of the combinatorial model to the second factor. This projection takes a marked graph with a metric to the n-tuple of the lengths of the perimeters of the marked points. Introduce the real 2-forms ω_i defined only on open strata of $\mathcal{M}_{g;n}^{\text{comb}}$ by the following formulas:

$$\omega_i = \sum d(l_{e'}/p_i) \wedge d(l_{e''}/p_i),$$

where p_i is the perimeter of the ith face, and e', e'' run over all pairs of distinct edges of the ith face, e' preceding e'' in some fixed order with a chosen starting vertex. The 2-form ω_i represents the class ψ_i. Indeed, fix a smooth curve $(X; x_1, \ldots, x_n)$ and take the canonical Jenkins–Strebel quadratic differential associated to the n-tuple $p_1, \ldots, p_n$. Then vertical trajectories of this quadratic differential through x_i identify the perimeter of the ith face of the corresponding embedded graph with the "spherized" cotangent line L_i considered as a real plane (that is, the fiber punctured at the origin is projected to the unit circle along the half-lines passing through the origin) at the ith point. Now it is possible to represent the intersection numbers $\langle \tau_{m_1} \ldots \tau_{m_n} \rangle$ in terms of integrals of very explicit differential forms:

$$\langle \tau_{m_1} \ldots \tau_{m_n} \rangle = \int_{\pi^{-1}(\bar{p})} \prod_{i=1}^n \omega_i^{m_i}$$

over any generic point $\bar{p} \in \mathbb{R}_+^n$.

From now on we use the notation d for the complex dimension of $\mathcal{M}_{g;n}$, $d = 3g - 3 + n$. Introduce the volume form on (the open strata of) $\mathcal{M}_{g;n}^{\text{comb}}$:

$$\text{Vol}(\lambda_1, \ldots, \lambda_n) = \frac{1}{d!} \Omega^d \times \prod_{i=1}^n e^{-\lambda_i p_i} dp_i,$$

where $\Omega = p_1^2 \omega_1 + \cdots + p_n^2 \omega_n$ and λ_i are real positive parameters.

Then the volume of $\mathcal{M}_{g;n}^{\text{comb}}$ with respect to this volume form can be computed in two ways: directly, under the projection to $\mathbb{R}_+^n$, and summing the volumes of all open cells. The first computation gives

$$\int_{\mathcal{M}_{g;n}^{\mathrm{comb}}} \mathrm{Vol}(\lambda_1,\ldots,\lambda_n) = \frac{1}{d!} \int_{\mathbb{R}_+^n} \left(\int_{\pi^{-1}(\bar{p})} \Omega^d \right) e^{-\sum \lambda_i p_i} dp_1 \wedge \cdots \wedge dp_n$$

$$= \sum_{m_1+\cdots+m_n=d} \frac{\langle \tau_{m_1} \cdots \tau_{m_n} \rangle}{m_1! \ldots m_n!} \prod_i \int_0^\infty p_i^{2m_i} e^{-\lambda_i p_i} dp_i$$

$$= \sum_{m_1+\cdots+m_n=d} \langle \tau_{m_1} \cdots \tau_{m_n} \rangle \prod_{i=1}^n \frac{(2m_i)!}{m_i!} \lambda_i^{-(2m_i+1)}$$

$$= 2^d \sum_{m_1+\cdots+m_n=d} \langle \tau_{m_1} \cdots \tau_{m_n} \rangle \prod_{i=1}^n \frac{(2m_i-1)!!}{\lambda_i^{(2m_i+1)}}.$$

The first computation is completed, and we start the second one. Consider the open cell in $\mathcal{M}_{g;n}^{\mathrm{comb}}$ corresponding to a 3-valent embedded graph Γ. The lengths $l_1,\ldots,l_{|E(\Gamma)|}$ of the edges of Γ form a set of coordinates on this cell. In these coordinates, the volume form $\mathrm{Vol}(\lambda_1,\ldots,\lambda_n)$ can be rewritten as

$$\mathrm{Vol}_\Gamma(\lambda_1,\ldots,\lambda_n) = 2^{d+|E(\Gamma)|-|V(\Gamma)|} e^{-\sum_j l_j \tilde{\lambda}_j} dl_1 \wedge \cdots \wedge dl_{|E(\Gamma)|}.$$

Here j runs over the set of all edges of Γ, and $\tilde{\lambda}_j$ is the sum

$$\tilde{\lambda}_j = \lambda_- + \lambda_+$$

of the two λ's corresponding to the two faces of Γ adjacent to the jth edge. Note that the two faces neighboring to an edge may coincide, and in this case $\lambda_- = \lambda_+$. Obtaining the correct power of 2 in the last formula (and hence showing that it is independent of the chosen cell) is a rather cumbersome task, and we refer the reader to [178] for details. An immediate calculation gives

$$\mathrm{Vol}_\Gamma(\lambda_1,\ldots,\lambda_n) = \prod_{j=1}^{|E(\Gamma)|} \frac{1}{\tilde{\lambda}_j}.$$

The contribution of a marked embedded graph to the total volume is proportional to the inverse cardinality of the automorphism group of the graph, whence summing over all 3-valent marked genus g embedded graphs with n marked faces and multiplying by 2^{-d} we obtain the *main combinatorial identity*

$$\sum_{m_1+\cdots+m_n=d} \langle \tau_{m_1} \cdots \tau_{m_n} \rangle \prod_{i=1}^n \frac{(2m_i-1)!!}{\lambda_i^{2m_i+1}} = \sum_\Gamma \frac{2^{-|V(\Gamma)|}}{|\mathrm{Aut}(\Gamma)|} \prod_{j=1}^{|E(\Gamma)|} \frac{2}{\tilde{\lambda}_j}. \quad (4.11)$$

The main combinatorial identity is an identity between two rational functions in variables λ_i. Making an arbitrary substitution of the form $\lambda_i = \Lambda_{k_i}$, $1 \leq k_i \leq N$ and summing the resulting identities over all such substitutions one gets

4 Geometry of Moduli Spaces

$$\sum_{m_1+\cdots+m_n=d} \langle \tau_{m_1}\cdots\tau_{m_n}\rangle \prod_{i=1}^{n}(2m_i-1)!!\,\mathrm{tr}(\Lambda^{-(2m_i-1)})$$

$$= \sum_{\Gamma} \frac{2^{-|V(\Gamma)|}}{|\mathrm{Aut}(\Gamma)|} \prod_{j=1}^{|E(\Gamma)|} \frac{2}{\widetilde{\Lambda}_j}. \tag{4.12}$$

Here $\widetilde{\Lambda}_j = \Lambda_- + \Lambda_+$ and the sum on the right-hand side is taken over all possible ways to color the faces of the graph Γ in N colors $\Lambda_1, \ldots, \Lambda_N$. Recall that Λ denotes the diagonal $N \times N$ matrix with positive entries $\Lambda_1, \ldots, \Lambda_N$.

The right-hand side of the last equation coincides with the matrix integral expansion in the Kontsevich model, and we obtain the first part of the Kontsevich theorem: the generating function K of the Kontsevich model coincides with the generating function F of the intersection model.

4.9.3 The Kontsevich Model and the KdV Equation

The second part of the proof consists in showing that the integral of the Kontsevich model is a τ-function for the KdV-hierarchy, in other words, that it obeys the Korteweg–de Vries equation.

Let

$$a(y) = \int_{-\infty}^{\infty} e^{i(\frac{1}{3}x^3 - yx)}\,dx$$

be the classical Airy function, i.e., the unique (up to a scalar factor) bounded solution to the linear differential equation

$$a''(y) + ya(y) = 0.$$

We are interested in the "asymptotic behavior" of this function as $y \to \infty$. An application of the stationary phase method (which must be justified in this case) gives

$$a(y) \sim e^{-\frac{2i}{3}y^{3/2}} \int_{U(y^{1/2})} e^{i(\frac{1}{3}x^3 + y^{1/2}x^2)}\,dx + e^{\frac{2i}{3}y^{3/2}} \int_{U(-y^{1/2})} e^{i(\frac{1}{3}x^3 - y^{1/2}x^2)}\,dx,$$

where the integration is carried out over arbitrary neighborhoods of the points $\pm y^{1/2}$.

Similar constructions are valid for the case of the *matrix Airy function*

$$A(Y) = \int_{\mathcal{H}_N} e^{i(\frac{1}{3}\mathrm{tr}\,H^3 - HY)}\,d\mu(H),$$

for a positive diagonal matrix Y. This function obeys the *matrix Airy equation*

$$\Delta A(Y) + \mathrm{tr}\,Y \cdot A(Y) = 0,$$

where Δ denotes the Laplace operator. Similarly to the 1-dimensional Airy function, the matrix Airy function admits an asymptotic expansion as a sum

4.9 Sketch of Kontsevich's Proof

of 2^N expressions of the form

$$e^{-i\frac{2}{3}\operatorname{tr} Y^{3/2}} \int e^{i\operatorname{tr}(\frac{1}{3}H^3 - H^2 Y^{1/2})} d\mu(H) = e^{-i\frac{2}{3}\operatorname{tr} Y^{3/2}} \int e^{i\operatorname{tr}\frac{1}{3}H^3} d\mu_{Y^{1/2}}(H).$$

The sum is taken over all 2^N quadratic roots $Y^{1/2}$ of the matrix Y, and the integral is taken over a neighborhood of the origin in $\mathcal{H}_N$. As $Y \to \infty$, the integral can be replaced with that over the entire space $\mathcal{H}_N$, i.e., it becomes the integral of the Kontsevich model for $\Lambda = Y^{1/2}$. The asymptotic expansion of the latter we already know.

Another way to compute the matrix Airy function consists in the application of formulas borrowed from [146] and [216]:

$$A(Y) = c_N \Delta(Y_i)^{-1} \int_{\mathbb{R}^N} \prod_{i=1}^n \Delta(X_i) e^{i(\frac{1}{3}X_i^3 - X_i Y_i)} dX_i$$

$$= c_N \frac{\det(a^{(j-1)}(Y_i))}{\det(Y_i^{j-1})},$$

where this time Δ denotes the Vandermonde determinant. Here we made use of the obvious identity

$$\int e^{i(x^3/3 - xy)} x^{j-1} dx = (ia(y))^{(j-1)}.$$

The derivatives of the Airy function admit natural asymptotic expansions

$$a^{(j-1)}(y) \sim \sum_{y^{1/2}} \operatorname{const} \cdot y^{-3/4} e^{-\frac{2i}{3} y^{3/2}} \cdot f_j(y^{-1/2})$$

for some Laurent series $f_j(z) = z^{-j} + \cdots \in \mathbb{Q}((z))$. Substituting the last formula into the expression for the matrix Airy function we obtain

$$A(Y) = \sum_{Y^{1/2}} \operatorname{const} \times e^{-\frac{2i}{3} \operatorname{tr} Y^{3/2}} \prod_{i=1}^N Y_i^{-3/4} \cdot \frac{\det(f_j(Y_i^{-1/2}))}{\det(Y_i^{j-1})}.$$

The last expression relates the matrix Airy function to the τ-function corresponding to the subspace $\langle f_1, f_2, \ldots \rangle \subset \mathbb{C}((z^{-1}))$, see Sec. 3.6.4. The proposition and the argument in the end of that section complete the proof of Witten's conjecture.

* * *

The main theorem established in this chapter permits to compute the intersection indices for certain classes; but the structure of the cohomology ring of the moduli spaces remains unknown. There also remains one more Witten's conjecture (it is discussed, in particular, in [178] and [202]), and though it is not apparent from its formulation, it is also related to embedded graphs.

The general idea behind the notion of a moduli space is that of "the space of parameters". In this chapter we parametrized algebraic curves. It is no less interesting to parametrize the pairs (X, f) where X is a curve and f is a meromorphic function on X. The corresponding parameter spaces are called *Hurwitz spaces*. The reader will find an introduction to this theory – from the point of view of embedded graphs, to be sure – in the next chapter.

5
Meromorphic Functions and Embedded Graphs

The starting point of the research described in this chapter is the question formulated as Problem 1.1.11 in Sec. 1.1. This question consists in enumerating constellations with given passport, and we call it the *Hurwitz problem*. In this general form, it has no satisfactory answer up to now, and the present chapter describes some partial results reflecting the best of the current knowledge. The known results are, however, rather general.

Recall that the rigid classification of complex ramified coverings is the classification up to isomorphism and corresponds to a triangular commutative diagram. The flexible (as opposed to rigid) classification is the classification up to equivalence, and is described by a quadrangular commutative diagram. These notions were introduced in Sec 1.8. Unfortunately, in the current literature these two notions are mixed, and often both of them are referred to as "topological classification".

The Hurwitz problem is equivalent to the problem of rigid topological classification of ramified coverings of the sphere. The known results concern two situations:

- enumeration of polynomial rational coverings $\overline{\mathbb{C}} \to \overline{\mathbb{C}}$;
- enumeration of genus g ramified coverings of the sphere with not more than one degenerate critical value.

In the first case, the passport of the constellation contains the partition n^1 corresponding to cyclic permutations. The answer is known in a very explicit and unexpectedly simple form. In the second case, the passport of the constellation has the form $[\kappa, 2^1 1^{n-2}, \ldots, 2^1 1^{n-2}]$ for some fixed partition $\kappa = (\kappa_1, \ldots, \kappa_p)$ of n. Hence, all the permutations except one are transpositions, and the number of transpositions is determined by the Riemann–Hurwitz formula. In this case the number of ramified coverings is expressed in terms of some intersection indices on moduli spaces of curves, whose explicit values are known only under some additional restrictions (for example, when the genus of the covering surface is at most 1).

There are different approaches to solving these enumeration problems. One of them, originating from Hurwitz and then forgotten for a century, is based on deriving, if possible, reccurrence formulas for the required numbers and on subsequent study of the resulting formulas by purely combinatorial tools of the theory of generating functions, like Lagrange inversion. I. P. Goulden, D. M. Jackson and their coauthors, see [114], [115], [116], [118], [117], reached notable success following this line.

Our presentation is based, however, on a different approach. Although the initial problem is purely combinatorial, it can be inscribed in a very rich geometric context. The geometry of moduli spaces plays a crucial role in this study, and enumeration formulas are only a part of the results obtained (it is not out of place to mention here the phrase of Hamming from his book on numerical methods: our goal is understanding, not numbers). The rigid topological classification turns out to be closely related to the behavior of the so-called Lyashko–Looijenga mapping on spaces of meromorphic functions. As a result, the enumeration problems we are interested in become related to the intersection theory on the moduli spaces of complex curves studied in the previous chapter.

We shall also touch, in Sec 5.4, the problem of flexible classification of ramified coverings. It seems to be much more difficult, and known achievements in this direction are of less generality. In the flexible classification of complex ramified coverings, the equivalence classes are orbits of the braid group action on isomorphism classes of constellations, see Sec. 1.1. For the history of the problem, see the introductory text in Sec. 5.4.

5.1 The Lyashko–Looijenga Mapping and Rigid Classification of Generic Polynomials

5.1.1 The Lyashko–Looijenga Mapping

As we have shown in Chapter 1, for a given set $\{t_1, \ldots, t_c\} \subset \mathbb{C}P^1$ of pairwise distinct ramification points and a given partition over each point, there is a finite set of isomorphism classes of meromorphic functions $f : X \to \mathbb{C}P^1$ having the prescribed ramification over the chosen points and no other ramification. Moreover, the number of elements in the set depends only on the passports and is independent of the precise positions of the points. Therefore, if we are going to approach the Hurwitz problem, then the natural idea is to consider the mapping on a space of meromorphic functions taking a function to the set of its ramification points and to try to find the *degree* of this mapping, whenever all the relative notions are well-defined. This degree coincides precisely with the required number of isomorphism classes.

Such a mapping appeared many times in related contexts, and, probably, Lyashko (see [11]) and Looijenga [201] were the first ones who understood its algebro-geometric nature in a specific situation of generic polynomials. In fact, they were interested not in topological classification of mappings, but rather in the topology of complements to some discriminants in the so-called "spaces of versal deformations of simple singularities". Following Arnold, we call the mapping the *Lyashko–Looijenga mapping*, or, for the sake of brevity, the *LL mapping*. Later, Arnold [13] gave the interpretation that we discuss here and extended the method to the case of generic "trigonometric polynomials" (rational functions having a value with only two preimages). Further developments were achieved in [190], [112], [91], [92], and we describe them below as well.

An application of the Lyashko–Looijenga method requires several steps:

- construction of a space of appropriate meromorphic functions; this space must have a natural complex structure and a "treatable" topology;
- construction of the space of values of the LL mapping;
- the proof that the LL mapping is holomorphic;
- computation of the degree of the LL mapping.

Among these steps, the first and the last one are usually the most nontrivial. In the majority of cases, the constructed space of functions requires reasonable compactification that would make easier the work with its topology. Even if such a compactification is constructed, the resulting topology remains rather complicated. In fact, these spaces rarely turn out to be simpler than the moduli spaces of curves discussed in the previous chapter. The computation of the degree is based essentially on this topology, and requires at least a minimal knowledge of it. On the contrary, the result of the second step is more or less trivial since this is simply the configuration space of points on the projective line.

We start with showing how the method works in the original case of Lyashko and Looijenga, where the topology of the moduli space of functions is simple, and then turn to various generalizations.

5.1.2 Construction of the LL Mapping on the Space of Generic Polynomials

Let $f : \mathbb{C}P^1 \to \mathbb{C}P^1$ be a polynomial. This means that there is a point in the target sphere such that its preimage is a single point. From now on we fix a coordinate t on the target sphere such that $t = \infty$ is such a point. If f is generic, i.e., all its finite critical values are distinct, then the LL mapping takes it to the unordered set of its finite critical values. A polynomial of degree n has $n - 1$ finite critical values. Of course, for $n \geq 2$ infinity also is a critical value of f, but it is convenient to forget about it.

Unordered sets of $n - 1$ distinct complex numbers are in a natural one-to-one correspondence with monic (i.e., with the leading coefficient 1) polynomials of degree $n - 1$ with distinct roots: we simply associate to a set $\{t_1, \ldots, t_{n-1}\}$ the polynomial

$$d(t) = t^{n-1} + d_1 t^{n-2} + \cdots + d_{n-1} = (t - t_1) \ldots (t - t_{n-1}). \quad (5.1)$$

Hence, in the case of polynomials the space of monic polynomials of degree $n-1$ can be chosen for the target space of the LL mapping.

The choice of the source space is more complicated. We would like to find a coordinate presentation for it. The preimage of infinity in the source sphere being fixed, we are still free in the choice of a coordinate x such that the point $x = \infty$ coincides with this preimage. There are two more degrees of freedom. For a given polynomial $f : \mathbb{C}P^1 \to \mathbb{C}P^1$ one may choose the coordinate so that f has the form

$$f = x^n + a_2 x^{n-2} + \cdots + a_n. \tag{5.2}$$

Consider the space of polynomials of the form (5.2) in a fixed coordinate x in the source sphere having simple critical values. As we have just said, each isomorphism class of a generic polynomial is represented in this space. But how many times? The answer is: exactly n times. Indeed, two polynomials of the form (5.2) are isomorphic if and only if there exists a coordinate change $x \mapsto \alpha x + \beta$ taking one of them to the other one, and the only possible form of this change is, therefore, $\beta = 0$, and α is an nth root of unity. Hence, a natural "moduli space of degree n generic polynomials" is the quotient of the space of polynomials (5.2) with distinct critical values modulo the following action of the group $C_n = \mathbb{Z}/n\mathbb{Z}$:

$$(a_2, a_3, \ldots, a_n) \mapsto (\varepsilon_n^{-2} a_2, \varepsilon_n^{-3} a_3, \ldots, \varepsilon_n^{-n+1} a_{n-1}, a_n), \tag{5.3}$$

where ε_n is a primitive nth root of unity.

It is more convenient for us, however, to deal with the space of polynomials (5.2) itself without introducing this quotient stuff. Indeed, if we manage to compute the degree of LL on this space, the calculation of the number of isomorphism types requires only dividing the answer by n. One more point is that it would be convenient to deal with the whole space of polynomials (5.2), without forgetting degenerate ones, since the topology of this larger space is rather simple. Indeed, if the LL mapping admits a reasonable extension to this space, then the degree of this extension coincides with the degree of the original mapping because the subspace of degenerate polynomials forms a subvariety of complex codimension 1 in the space (5.2). This is the main core of the Lyashko–Looijenga theorem.

Before stating it, a remark is in order. Of course, while extending the LL mapping one would expect that this extension possesses some natural properties and is related to the critical values of degenerate polynomials. This is indeed the case, and the extended mapping takes each polynomial to the set of its critical values taken with appropriate multiplicities.

Recall that a holomorphic mapping is *finite* if each point in the target space has a finite number of preimages. The *degree* of a finite mapping is the number of preimages of a generic point in the target space.

Theorem 5.1.1 (Lyashko–Looijenga). *The Lyashko–Looijenga mapping extends to a polynomial finite mapping*

$$LL : \mathbb{C}^{n-1} \to \mathbb{C}^{n-1}$$

of degree n^{n-2} from the space of polynomials (5.2) to the space of monic polynomials (5.1).

As an immediate corollary of the argument above we obtain

Corollary 5.1.2. *The number of isomorphism types of generic degree n polynomials is equal to n^{n-3}.*

Exercise 5.1.3. Show that there is a unique, up to a common conjugation, decomposition of the 3-cycle into a product of two transpositions. Write out all four pairwise nonequivalent decompositions of the 4-cycle into a product of three transpositions (cf. Example 5.5.6).

In [201] the theorem was used to prove the famous Cayley theorem [48] on enumeration of trees:

Corollary 5.1.4 (The Cayley formula for labelled trees). *The number of trees with n vertices marked by distinct marks is n^{n-2}.*

Indeed, one can associate a tree to a generic polynomial in the following way, which is a specialization of the general construction from Chapter 1 (see in particular Exercise 1.6.11). Consider a star graph in the target sphere whose center is a noncritical value t_0 and whose leaves are the *finite* critical values $t_1, \ldots, t_{n-1}$ of the polynomial. Each of the segments $[t_0, t_i]$ of the star graph has n preimages. Exactly two of these preimages have a common end, since the monodromy around the critical value t_i is a transposition. We consider these two preimages as the edge of the graph on the covering sphere whose vertices are the n preimages of the noncritical value. The standard argument from Chapter 1 shows that this graph is a tree, and there is a one-to-one correspondence between such trees and isomorphism classes of generic polynomial coverings. Note that the edges of the tree are numbered by the numbers of the corresponding critical values. And the number of trees with numbered vertices is n times the number of trees with marked edges. We leave the latter statement as an exercise for the reader. Of course, arguing in the opposite direction we can deduce the Lyashko–Looijenga theorem from Cayley's theorem.

5.1.3 Proof of the Lyashko–Looijenga Theorem

The main property, which allows one to compute the degree of the LL mapping, is quasihomogeneity. A naïve idea of establishing it is to assign weight 1 to the variable x in the space of polynomials (5.2), and to derive weights of other variables from this one ($w(a_i) = i$, $w(t) = n$, and so on). This approach does work. However, it fails in more general situations and we need some conceptual guideline for choosing the weights. We start with explaining this guideline.

The multiplication of a polynomial by a nonzero constant results in the multiplication of all its critical values by the same constant. This simple remark allows one to apply the techniques of quasihomogeneous mappings to the analysis of the LL mapping.

Let $\mathbb{C}^* = \mathbb{C} \setminus \{0\}$ denote the multiplicative group of nonzero complex numbers. This group acts linearly on the space $\mathbb{C}^m$ and, according to elementary representation theory of commutative groups, any such linear action splits into a direct sum of m one-dimensional actions of the form

$$u_\zeta : a \mapsto \zeta^w a,$$

where w is an integer called the *weight* of the action. Although the splitting may be not unique, the set of weights is well defined. If a splitting is chosen, and $a_1, \ldots, a_m$ are coordinates in $\mathbb{C}^m$ such that the action looks like

$$u_\zeta : (a_1, \ldots, a_m) \mapsto (\zeta^{w_1} a_1, \ldots, \zeta^{w_m} a_m),$$

then w_i is said to be the *weight of the coordinate* a_i, $w_i = w(a_i)$.

Suppose we have two complex spaces $\mathbb{C}^{m_1}$, $\mathbb{C}^{m_2}$ and a polynomial mapping

$$h : \mathbb{C}^{m_1} \to \mathbb{C}^{m_2}.$$

The mapping h is called *quasihomogeneous* with respect to the actions u_ζ and v_ζ on $\mathbb{C}^{m_1}$ and $\mathbb{C}^{m_2}$ respectively if $h(u_\zeta(x)) = v_\zeta(h(x))$ for any element $\zeta \in \mathbb{C}^*$ and any point $x \in \mathbb{C}^{m_1}$.

Our further considerations are based on the following well-known theorem generalizing the classical Bézout theorem to the case of quasihomogeneous polynomials.

Theorem 5.1.5. *Suppose linear actions of $\mathbb{C}^*$ on two complex spaces of the same dimension m is given, and let $w_1, \ldots, w_m$, $W_1, \ldots, W_m$ be the two sets of weights. A quasihomogeneous polynomial mapping $f : \mathbb{C}^m \to \mathbb{C}^m$ between these spaces is finite if and only if the preimage $f^{-1}(0)$ coincides with the origin. In this case the degree of f is*

$$\frac{W_1 \ldots W_m}{w_1 \ldots w_m}.$$

Note that if the formula provides a non-integral number, then the corresponding polynomial mapping cannot be finite.

Proof. The proof of this theorem is based on the fact that the space of all finite polynomial mappings with given weights of coordinates is connected.

Suppose first that all weights $w_1, \ldots, w_m$ in the preimage are equal to 1, while all weights W_i are positive integers. This means that we deal with polynomial mappings of homogeneous coordinatewise degrees W_i. All such polynomial mappings form a complex vector space. Polynomial mappings that are not finite form a proper algebraic subvariety in this vector space. Since the degree of a generic finite mapping does not vary under a small deformation of the mapping in the class of mappings of the same weight, it suffices to compute the degree of arbitrary finite mapping of given weight. The mapping $(z_1, \ldots, z_m) \mapsto (z_1^{W_1}, \ldots, z_m^{W_m})$ will do.

The case where w_i are arbitrary positive integers is reduced to the previous one if one considers the compositions of each of the mappings of our space with the mapping $(y_1, \ldots, y_m) \mapsto (y_1^{w_1}, \ldots, y_m^{w_m})$. The degree of the required mapping is then computed as the ratio of the degrees of the composite mapping and of the auxiliary mapping, which are both of the kind described earlier.

In order to understand why the LL mapping is quasihomogeneous and to compute its weights let us look more closely on the two actions of the group $\mathbb{C}^*$. The action on the space of critical values is, in fact, already introduced: we simply multiply each critical value t_i by $\zeta \in \mathbb{C}^*$, which leads to the action

$$u_\zeta : (d_1, d_2, \ldots, d_{n-1}) \mapsto (\zeta d_1, \zeta^2 d_2, \ldots, \zeta^{n-1} d_{n-1})$$

on the space of polynomials (5.1). Note that this action comes already split into one-dimensional representations. However, what does it mean to multiply a polynomial (5.2) by a constant? Indeed, we require that the leading coefficient of the polynomial be 1, and this restriction is no more valid under multiplication. The correct answer is that we consider polynomials only up to isomorphism, therefore, it suffices to find a polynomial *isomorphic* to the polynomial ζf, i.e., one of the polynomials obtained by the substitution $x \mapsto \zeta^{-1/n} x$. A new problem arises: which one of the roots $\zeta^{-1/n}$ should be chosen? The easiest and the most effective answer is that from the very beginning we should have defined the action as the multiplication by ζ^n, not by ζ, and this last action is

$$u_\zeta : \begin{cases} (a_2, \ldots, a_n) \mapsto (\zeta^2 a_2, \ldots, \zeta^n a_n) \\ (d_1, \ldots, d_{n-1}) \mapsto (\zeta^n d_1, \ldots, \zeta^{(n-1)n} d_{n-1}) \end{cases} \quad (5.4)$$

(here a_i and d_j are coefficients from (5.2) and (5.1) respectively). Note that if we delete the origins and factorize both spaces modulo this action, then we arrive at a mapping of compact varieties (weighted projective spaces).

Thus, the weights in the source space are $2, 3, \ldots, n$, while the weights in the target space are $n, 2n, \ldots, (n-1)n$. If we prove that the LL mapping is polynomial and finite, then the degree is

$$\frac{n \cdot 2n \cdots (n-1)n}{2 \cdot 3 \cdots n} = n^{n-2},$$

and we were done.

276   5 Meromorphic Functions

Following the guidelines of the present book where we try to make all things as explicit as possible, we present in the next lemma a formula from [190] for the LL mapping, which also proves that the mapping is polynomial. At the same time it shows that the extended mapping is "natural", which means that the extended mapping takes a degenerate polynomial to the set of its critical values taken with appropriate multiplicities as well: a critical value of a polynomial f with the partition $\kappa \vdash n$ is a root of multiplicity $v(\kappa)$ of LL(f) (for notation $v(\kappa)$ see Notation 1.1.9). Any computer system of symbolic computations, like Maple, using the formula in question immediately provides a coordinate presentation of the mapping for a given dimension.

Lemma 5.1.6. *The* LL *mapping takes a polynomial f of the form* (5.2) *to the discriminant with respect to x of the polynomial $f(x) - t$, up to a constant factor.*

Here the discriminant of $f(x) - t$ is a polynomial in t and in the coefficients of f, and we consider it as a polynomial in t, while the constant factor is chosen so that to make this polynomial monic.

Example 5.1.7. For $n = 4$ the LL mapping looks like

$$\text{LL}: f(x) = x^4 + a_2 x^2 + a_3 x + a_4 \mapsto -\frac{1}{256}\text{discriminant}_x(f(x) - t)$$

$$= t^3 + \left(\frac{1}{2}a_2^2 - 3a_4\right)t^2 + \left(\frac{1}{16}a_2^4 - a_2^2 a_4 + \frac{9}{16}a_2 a_3^2 + 3a_4^2\right)t$$

$$- \frac{1}{16}a_2^3 a_3^2 + \frac{1}{64}a_2^3 a_3^2 + \frac{1}{2}a_2^2 a_4^2 - \frac{9}{16}a_2 a_3^2 a_4 + \frac{27}{256}a_3^4 - a_4^3 \ .$$

Let us make a few remarks concerning the weights of the variables and parameters in this formula. (A) We take $w(x) = 1$; therefore $w(x^2) = 2$ and $w(x^4) = 4$. (B) In order for f to be homogeneous as a function of *all* the parameters involved, we take $w(a_2) = 2$, $w(a_3) = 3$, $w(a_4) = 4$. (C) We consider the polynomial $f(x) - t$; the weight of f being 4, we set $w(t) = 4$. (D) Now one can note that LL(f) is homogeneous as the polynomial in all the variables involved: the weight of each of its monomials is 12. (E) For the polynomial f the product of the weights of its coefficients is $2 \cdot 3 \cdot 4$; for the polynomial LL(f) the similar product is $4 \cdot 8 \cdot 12$; the ratio of these two numbers is $16 = 4^2$.

Proof of Lemma 5.1.6. The proof of the lemma is simple and follows more or less directly from the definition of the discriminant. A value $\bar{t}$ is a root of the discriminant of $f(x) - t$ if and only if the polynomial $f(x) - \bar{t}$ has multiple roots, i.e., if and only if $\bar{t}$ is a critical value of f. Now, the discriminant is a quasihomogeneous polynomial with respect to the above action, and its weight is $n(n-1)$. Since a generic polynomial f has $n-1$ distinct critical values, the degree of the discriminant in t is at least $n-1$. On the other hand, this degree cannot exceed $n-1$ since otherwise the weight of a monomial of higher degree would be greater than $n(n-1)$. This means that the coefficient of t^{n-1} is constant, and the lemma is proved.

5.2 Geometry of the Discriminant

The proof of the Lyashko–Looijenga theorem will be complete if we recall that $f(x) = x^n$ is the only polynomial of the form (5.2) all of whose finite critical values are 0, and it is, therefore, the unique preimage of the origin under the LL mapping.

Exercise 5.1.8 (Trigonometric polynomials [13]). Let us call *trigonometric polynomial* a function $\mathbb{C}P^1 \to \mathbb{C}P^1$ having a point in the image (infinity) with only two preimages (poles). A trigonometric polynomial is *generic* if all its finite critical values are simple. Prove that the number of isomorphism classes of trigonometric polynomials with two poles of order k_1, k_2 is

$$\frac{(k_1+k_2-1)!}{|\mathrm{Aut}(k_1,k_2)|} \cdot \frac{k_1^{k_1}}{k_1!} \cdot \frac{k_2^{k_2}}{k_2!}.$$

Here $|\mathrm{Aut}(k_1,k_2)| = 2$ if $k_1 = k_2$ and $|\mathrm{Aut}(k_1,k_2)| = 1$ otherwise. [**Hint:** make use of the following parametrization of a completed space of an appropriate covering of the space of trigonometric polynomials:

$$f(x) = x^{k_1} + a_1 x^{k_1-1} + \cdots + a_{k_1-1} x + c + b_{k_2-1}\frac{u}{x} + \cdots + b_1 \left(\frac{u}{x}\right)^{k_2-1} + \left(\frac{u}{x}\right)^{k_2},$$

and prove an analogue of Lemma 5.1.6 for rational functions.]

Generalize this statement to the case of rational mappings admitting a value with three preimages.

5.2 Rigid Classification of Nongeneric Polynomials and the Geometry of the Discriminant

5.2.1 The Discriminant in the Space of Polynomials and Its Stratification

Before we proceed further, one useful remark is in order. Besides being equivariant with respect to the action of the group $\mathbb{C}^*$ acting by multiplication of polynomials by constants, the LL mapping is also equivariant with respect to the following action of the additive group $\mathbb{C}$:

$$\mu : f \mapsto f + \mu \quad \text{for} \quad \mu \in \mathbb{C},$$

which shifts all critical values of f by the same value μ. All points in any orbit of this action in the target space of LL have the same number of preimages, and it would be convenient for us to factorize modulo this action. The corresponding quotient space of the target space is naturally identified with the subspace of polynomials with the zero sum of roots, i.e. with the zero second coefficient, and we introduce the notation

$$\mathcal{D} = \{d(t) = t^{n-1} + d_2 t^{n-3} + \cdots + d_{n-1}\} \tag{5.5}$$

for it. The source quotient space can be simply considered as the set $LL^{-1}(\mathcal{D})$ of polynomials (5.2) with the zero sum of critical values, and we denote it by $\mathcal{P}$. Both these spaces are $(n-2)$-dimensional.

A generic point in $\mathcal{D}$ is a polynomial with pairwise distinct roots, and it has n^{n-2} distinct preimages under the LL mapping. Polynomials with at least two coinciding roots form the *discriminant locus* (or, simply, the *discriminant*) $\Delta \subset \mathcal{D}$. Its preimage under the LL mapping is the discriminant $\Sigma \subset \mathcal{P}$, and it consists of polynomials having at least two coinciding critical values. In particular, polynomials with critical points of multiplicity greater than 2 also belong to the discriminant Σ.

Example 5.2.1. The equation of the discriminant locus Σ in the space of polynomials (5.2) is obtained by setting to zero the discriminant of the image polynomial under the LL mapping. For example, in the case of polynomials of degree 4 from Example 5.1.7 the discriminant locus equation is obtained by setting to zero the discriminant of the polynomial of degree 3 in t:

$$a_3^2(8a_2^3 + 27a_3^2)^3 = 0.$$

The discriminant locus is reducible and splits into two irreducible components: the "caustic" $8a_2^3 + 27a_3^2 = 0$ and the "Maxwell stratum" $a_3 = 0$, see Examples 5.2.4 and 5.2.5.

When restricting LL to the space $\mathcal{P}$ of polynomials with the zero sum of critical values we have set $d_1 = 0$, see Eq. (5.5); for polynomials (5.2) of degree 4, this gives the subvariety $a_4 = a_2^2/6$.

Since in this section we consider only polynomials, we shall not include the partition n^1 in the passport, and the passport $K = [\kappa_1, \ldots, \kappa_c]$ of a polynomial f consists of the partitions of n corresponding to each finite ramification point. We are interested in the number of permutations with cyclic type κ_i such that their product is a cyclic permutation, i.e., it has the cyclic type n^1. Recall that the *degeneracy* $v(\kappa)$ of a partition $\kappa = 1^{k_1} 2^{k_2} \ldots n^{k_n}$ of n is the number

$$v(\kappa) = 1 \cdot k_2 + 2 \cdot k_3 + \cdots + (n-1) \cdot k_n.$$

A partition is called *degenerate* if its degeneracy is greater than 1. The only nondegenerate partition is $2^1 1^{n-2}$, and its degeneracy is 1. The Riemann–Hurwitz formula implies that if $K = [\kappa_1, \ldots, \kappa_c]$ is the passport of a polynomial, then

$$v(\kappa_1) + \cdots + v(\kappa_c) = n - 1.$$

In Chapter 1 it was shown that if this requirement is satisfied, then there exists a polynomial with the passport K (see Corollary 1.6.9). Below, this statement will in fact be proved once again, by computing the degree of the closure of the variety of polynomials with this passport.

Generally speaking, a space of generic meromorphic functions admits a lot of different compactifications, and these compactifications may not contain degenerate functions of all isomorphism types. Fortunately, in the case of polynomials all types of degenerate polynomials are represented in the space of the polynomials having the form (5.2). The discriminant Σ admits a stratification according to the ramification type of polynomials in it. The *stratum* $\Sigma_K \subset \Sigma$ consists of all polynomials with the passport K. Implementing once more the Lyashko–Looijenga scheme we choose the stratum Σ_K for the space of polynomials with ramification type K. Once again, each isomorphism type of a polynomial with the passport K is represented in this space since any polynomial admits a coordinate presentation (5.2); and each isomorphism class contains exactly n representatives, except for the passport $[n^1]$. For the latter type, the representative is unique, namely, x^n.

5.2.2 Statement of the Enumeration Theorem

Fix a passport K. Let $c = c(K)$ be the number of critical values of a polynomial belonging to the stratum Σ_K, that is, the number of partitions in K. We suppose that the first r of the partitions κ_i are degenerate, while the rest $c - r$ of them are simple. Denote by $\nu(K)$ the partition of the number $n - 1$ determined by the degeneracy values of each of the partitions κ_i, $i = 1, \ldots, c$. For a degenerate partition κ of the number n, we denote by $\nu(\kappa)$ the partition of $n - 1$ corresponding to the passport, where κ is a unique degenerate partition, $\nu(\kappa) = v(\kappa)^1 1^{n-v(\kappa)-1}$. The following statement generalizes the Lyashko–Looijenga theorem to degenerate polynomials.

Let $\kappa = 1^{m_1} \ldots n^{m_n}$ be a partition of n; we denote by $|\mathrm{Aut}(\kappa)|$ the number $m_1! \ldots m_n!$ of permutations of its parts. Similarly $|\mathrm{Aut}(K)|$ is the order of the automorphism group of a passport K; if $K = [\kappa_1, \ldots, \kappa_c]$, then $|\mathrm{Aut}(K)|$ is the product of factorials of the numbers of coinciding partitions.

Theorem 5.2.2 ([190]). *The restriction of the LL mapping to each stratum Σ_K is a smooth finite mapping of degree*

$$\deg_K = n^{c-1} \frac{|\mathrm{Aut}(\nu(K))|}{|\mathrm{Aut}(K)|} \prod_{i=1}^{r} \frac{|\mathrm{Aut}(\nu(\kappa_i))|}{|\mathrm{Aut}(\kappa_i)|},$$

where the product on the right-hand side is taken over all nonsimple critical values.

The "enumerative contents" of this theorem is equivalent to the Goulden–Jackson enumerative formula for cacti (1.6). But the most interesting part of the whole picture is of course the relation between the enumerative result and the degree of the LL mapping. Sections 5.2.3, 5.2.4 are devoted to the proof of this theorem, and now we consider some of its applications.

Example 5.2.3 (Generic polynomials). The formula of the theorem remains valid for nondegenerate polynomials as well. Nondegenerate polynomials have the passport $K = [1^{n-2}2^1, \ldots, 1^{n-2}2^1]$; in this case $c = n - 1$, $|\mathrm{Aut}(K)| = |\mathrm{Aut}(\nu(K))| = (n-1)!$, $r = 0$, and we obtain

$$\deg_K = n^{n-2} \frac{(n-1)!}{(n-1)!} = n^{n-2}.$$

Example 5.2.4 (Maxwell stratum). The stratum defined by the passport $M = [1^{n-4}2^2, 1^{n-2}2^1, \ldots, 1^{n-2}2^1]$, i.e., the stratum of polynomials with two coinciding finite critical values taken at distinct points, is called the *Maxwell stratum*. We have $c(M) = n - 2$, $\nu(M) = \nu(\kappa_1) = 1^{n-3}2^1$, $|\text{Aut}(\nu(M))| = |\text{Aut}(\nu(\kappa_1))| = (n-3)!$, $r = 1$, and Theorem 5.2.2 gives the value

$$\deg_M = n^{n-3} \frac{(n-3)!}{(n-3)!} \frac{(n-3)!}{2!(n-4)!} = \frac{1}{2} n^{n-3}(n-3).$$

This result was first obtained by V. I. Arnold in [13].

Example 5.2.5 (Caustic). There is one more stratum of codimension 1, the *caustic* C. It consists of polynomials with a pair of coinciding critical points and is given by the passport $C = [1^{n-3}3^1, 1^{n-2}2^1, \ldots, 1^{n-2}2^1]$. In this case $c(C) = n - 2$, $r = 1$, $\nu(C) = \nu(\kappa) = 1^{n-3}2^1$ as well, $|\text{Aut}(\nu(C))| = |\text{Aut}(\nu(\kappa_1))| = (n-3)!$, and Theorem 5.2.2 yields the value

$$\deg_C = n^{n-3} \frac{(n-3)!}{(n-3)!} \frac{(n-3)!}{1!(n-3)!} = n^{n-3}.$$

5.2.3 Primitive Strata

We start with proving the enumeration theorem 5.2.2 for the case of primitive strata. A *primitive stratum* is a stratum of polynomials having one degenerate finite critical value, and all of whose other critical values are simple. We use the notation Σ_κ for the primitive stratum corresponding to the passport $K = [\kappa, 1^{n-2}2^1, \ldots, 1^{n-2}2^1]$. Primitive strata can be treated in a way similar to the original Lyashko–Looijenga approach.

Take, e.g., the Maxwell stratum from Example 5.2.4. Note that for $n \leq 3$ the Maxwell stratum is empty. A generic polynomial in the Maxwell stratum has the form

$$f(x) = (x^2 + a'_1 x + a'_2)^2 (x^{n-4} + a''_1 x^{n-5} + \cdots + a''_{n-4}) + a'_n.$$

Here $2a'_1 + a''_1 = 0$ and $a'_n = a'_n(a'_1, a'_2, a''_2, \ldots, a''_{n-4})$ is a polynomial chosen in such a way that it makes the sum of critical values of f equal to zero. The LL mapping restricted to the hyperplane $2a'_1 + a''_1 = 0$ in the space of the coefficients a', a'' remains polynomial, quasihomogeneous and finite. The weights of the coordinates are

$$w(a'_i) = i, \quad i = 1, 2; \qquad w(a''_j) = j, \quad j = 1, \ldots, n - 4,$$

in accordance with our naïve choice $w(x) = 1$, and the product of weights of the coordinates in the preimage is

$$2(n-4)!.$$

5.2 Geometry of the Discriminant

On the other hand, $\nu(1^{n-4}2^2) = 1^{n-3}2^1$, and a general polynomial in the image has the form

$$(t-t_1)^2(t^{n-3} + d'_1 t^{n-4} + \cdots + d'_{n-3}),$$

where $-2t_1 + d'_1 = 0$. The weights of the coordinates are

$$w(t_1) = n, \qquad w(d'_j) = nj,$$

and the product of the weights of the coordinates in the image is

$$n^{n-3}(n-3)!.$$

Dividing this by the product of the coordinates' weights in the preimage we obtain the degree of the LL mapping restricted to the Maxwell stratum:

$$\frac{n^{n-3}(n-3)!}{2(n-4)!} = n^{n-3}\frac{n-3}{2}$$

in accordance with the predictions of the main theorem.

Exercise 5.2.6. Elaborate a similar calculation for the other stratum of codimension 1, the caustic, see Example 5.2.5.

The above calculation extends easily to an arbitrary primitive stratum. Namely, suppose that the partition corresponding to the degenerate critical value has the form $\kappa = 1^{k_1} \ldots n^{k_n}$ and all other critical values are simple. Then a generic polynomial of this type with the degenerate critical level a_n has a unique representation in the form

$$f(x) = f_1(x) f_2^2(x) \ldots f_n^n(x) + a_n.$$

Here $f_i(x) = x^{k_i} + a_{i,1} x^{k_i-1} + \cdots + a_{i,k_i}$ is a polynomial of degree k_i, $k_1 + 2k_2 + \cdots + nk_n = n$, $a_{1,1} + 2a_{2,1} + \cdots + na_{n,1} = 0$, and a_n is a polynomial in $a_{i,j}$ chosen in such a way that it makes the sum of the critical values of f equal to zero. In other words, we constructed a parametrization of the closure of the primitive stratum; the space of parameters is the space of coefficients of the polynomials f_i. In particular, each primitive stratum is irreducible.

The weights of the coordinates in the preimage are $w(a_{i,j}) = j$, and their product is

$$k_1! \ldots k_n! = |\text{Aut}(\kappa)|.$$

On the other hand, a generic polynomial in the image of this stratum has the form

$$(t-t_1)^{v(\kappa)} D_1(t),$$

where $t_1 = a_n$ is the multiple critical value, and D_1 is a generic monic polynomial of degree $n - 1 - v(\kappa)$. We leave it to the reader to write explicitly the list of coefficients of D_1 and their weights; the product of the weights of the coordinates in the image is

$$n^{c-1}(n - v(\kappa) - 1)! = n^{c-1}|\mathrm{Aut}(\nu(\kappa))|,$$

and the degree of the restricted LL is

$$n^{c-1}\frac{|\mathrm{Aut}(\nu(\kappa))|}{|\mathrm{Aut}(\kappa)|}.$$

This formula coincides with the predictions of the theorem since in the case of a primitive stratum, $|\mathrm{Aut}(K)| = |\mathrm{Aut}(\nu(K))| = (n - 1 - v(\kappa))!$.

Exercise 5.2.7. Using the same technique prove the enumeration theorem for the strata where each critical value admits a unique critical preimage (but of arbitrary order). [**Hint:** make use of the auxiliary mapping A from the next section.]

5.2.4 Proof of the Enumeration Theorem

The proof of the enumeration theorem for arbitrary strata is based on the proof for primitive ones. Each stratum is the intersection of primitive strata. Unfortunately, generally speaking this intersection is not transversal, and this makes the understanding of the geometry of the intersection not an easy task. However, it is possible to overcome this difficulty by considering a suitable ramified covering space of the space $\mathcal{P}$ of polynomials, where the relevant intersections become transversal. This ramified covering space is the space of critical points of polynomials.

5.2.4.1 Auxiliary Mappings

Consider the commutative square

$$\begin{array}{ccc} \widehat{\mathcal{P}} & \xrightarrow{\widehat{\mathrm{LL}}} & \widehat{\mathcal{D}} \\ {\scriptstyle A}\downarrow & & \downarrow{\scriptstyle T} \\ \mathcal{P} & \xrightarrow{\mathrm{LL}} & \mathcal{D} \end{array} \qquad (5.6)$$

Here $\widehat{\mathcal{P}}$ is the hyperplane $\alpha_1 + \cdots + \alpha_{n-1} = 0$ in the space of critical points of polynomials, and $\widehat{\mathcal{D}}$ is the hyperplane $t_1 + \cdots + t_{n-1} = 0$ in the space of critical values. The mapping A has the form

$$A : (\alpha_1, \ldots, \alpha_{n-1}) \mapsto -\bar{t}(\alpha_1, \ldots, \alpha_{n-1}) + \int_0^x \rho,$$

where $\rho = n(\xi - \alpha_1)\ldots(\xi - \alpha_n)d\xi$, and $\bar{t}$ is a constant depending on α_i.

Exercise 5.2.8. Show that the critical values of the polynomial $\int_0^x \rho$ are $\int_0^{\alpha_i} \rho$, $i = 1, \ldots, n-1$.

This exercise shows that, since the image of A must be a polynomial with the zero sum of critical values, we must set

$$\bar{t} = \frac{1}{n}\left(\int_0^{\alpha_1} \rho + \cdots + \int_0^{\alpha_{n-1}} \rho\right).$$

The mapping T is nothing else but the Viète mapping

$$T : (t_1, \ldots, t_{n-1}) \mapsto (t - t_1) \ldots (t - t_{n-1}),$$

and the mapping $\widehat{LL}$ takes the ordered tuple of critical points to the ordered tuple of critical values.

One of the advantages of the passage to the mapping $\widehat{LL}$ is that it becomes homogeneous, not just quasihomogeneous, because the weights of all coordinates in the image are the same, and the weights of all coordinates in the preimage are also the same. The following statement is obvious.

Lemma 5.2.9. *The mappings A and T are quasihomogeneous and finite. The mapping $\widehat{LL}$ is homogeneous and finite.*

5.2.4.2 Induced Stratifications and Degrees Counting

The total preimage $A^{-1}(\Sigma_\kappa)$ of a primitive stratum Σ_κ is no longer an irreducible variety. We are free in choosing an irreducible component of this stratum, and, for a passport $K = [\kappa_1, \ldots, \kappa_c]$, we may choose these irreducible components for each κ_i so that to make them intersect transversally.

The stratification of the space $\mathcal{P}$ induces the stratification of other spaces at the vertices of the square (5.6). In order to compute the degree of the LL mapping restricted to the discriminant strata it suffices to compute the degrees of the restrictions of A, $\widehat{LL}$, and T to the corresponding strata. Let us start with the description of the induced stratification.

The strata in the space $\mathcal{D}$ are marked by partitions $\nu = 1^{m_1} \cdots (n-1)^{m_{n-1}}$ of the number $n-1$. The stratum $\Delta_\nu \subset \mathcal{D}$ consists of polynomials having m_1 roots of multiplicity 1, ..., m_{n-1} roots of multiplicity $n-1$. The strata in the space $\widehat{\mathcal{D}}$ are marked by partitions of the *set* of indices $\{1, \ldots, n-1\}$. For a partition ν of $n-1$, fix a partition $J = J_1 \sqcup J_2 \sqcup \ldots$ of the set of indices $\{1, \ldots, n-1\}$ into pairwise disjoint subsets such that there are precisely m_1 one-element subsets, m_2 two-element subsets and so on. We associate with the partition J the stratum $\widehat{\Delta}_J \subset \widehat{\mathcal{D}}$ consisting of all points $(t_1, \ldots, t_n) \in \widehat{\mathcal{D}}$ such that $t_i = t_j$ if and only if the indices i, j belong to the same set in the partition J.

The mapping T maps a stratum $\widehat{\Delta}_J$ to Δ_ν, where ν is obtained from J by forgetting the indices but preserving their numbers.

Similarly, for a passport $K = [\kappa_1, \ldots, \kappa_c]$, we fix a partition $\mathbf{I} = [I_1, \ldots, I_c]$ of the set of indices $\{1, \ldots, n-1\}$ into c tuples of pairwise disjoint subsets such that the ith tuple I_i of this partition contains precisely k_{i2} one-element subsets, k_{i3} two-element subsets and so on, where $\kappa_i = 1^{k_{i1}} 2^{k_{i2}} \ldots$. All in all, I_i consists of $v(\kappa_i)$ sets. The stratum $\widehat{\Sigma}_{\mathbf{I}} \subset \widehat{\mathcal{P}}$ consists of all points $(\alpha_1, \ldots, \alpha_{n-1}) \in \widehat{\mathcal{P}}$ such that $\alpha_i = \alpha_j$ if the indices i, j belong to the same subset of a tuple $I_l \in \mathbf{I}$, and the values of the polynomial $A(\alpha_1, \ldots, \alpha_{n-1})$ at the points α_i, α_j coincide if the indices i, j belong to subsets of the same tuple I_l.

The restrictions of the mappings in the square (5.6) to the corresponding strata form the commutative square

$$\begin{array}{ccc} \widehat{\Sigma}_{\mathbf{I}} & \xrightarrow{\widehat{\mathrm{LL}}} & \widehat{\Delta}_J \\ A \downarrow & & \downarrow T \\ \mathcal{P}_K & \xrightarrow{\mathrm{LL}} & \Delta_{\nu(K)} \end{array} \qquad (5.7)$$

where the strata $\widehat{\Sigma}_{\mathbf{I}}, \widehat{\Delta}_J$ are constructed by means of consistent partitions of the set of indices $\{1, \ldots, n-1\}$. This means that each J_i is the union of all indices from the sets of the tuple I_i.

The degrees of all mappings in this square are independent of the choice of the partition $\mathbf{I}$. We denote them by $\deg_K(\mathrm{LL})$, $\deg_K(A)$, $\deg_K(\widehat{\mathrm{LL}})$, and $\deg_{\nu(K)}(T)$ respectively. Since the square (5.7) is commutative, we obtain

$$\deg_K(\mathrm{LL}) = \frac{\deg_K(\widehat{\mathrm{LL}}) \cdot \deg_{\nu(K)}(T)}{\deg_K(A)}, \qquad (5.8)$$

and in order to compute the degree of LL it suffices to know all the degrees on the right-hand side of the last formula.

The computation of the degrees of the vertical arrows is easy. For $\nu = 1^{m_1} \ldots (n-1)^{m_{n-1}}$ we have

$$\deg_\nu(T) = m_1! \ldots m_{n-1}! = |\mathrm{Aut}(\nu)|. \qquad (5.9)$$

Indeed, if we take a polynomial belonging to the stratum Δ_ν, then the multiset of its roots is fixed. Specifying also a partition J of the set of indices $\{1, \ldots, n-1\}$ corresponding to the partition ν we arrive at $m_1!$ ways to number the roots of multiplicity one, $m_2!$ ways to number the roots of multiplicity two, and so on.

The following statement is also obvious:

$$\deg_K(A) = |\mathrm{Aut}(K)| \prod_{i=1}^{r} \frac{|\mathrm{Aut}(\kappa_i)|}{k_{i1}!}. \qquad (5.10)$$

5.2 Geometry of the Discriminant

It is proved in the same way as above: there are $|\mathrm{Aut}(K)|$ ways to permute the sets of indices corresponding to coinciding partitions κ_i and there are

$$k_{i2}! \ldots k_{in}! = \frac{|\mathrm{Aut}(\kappa_i)|}{k_{i1}!}$$

ways to permute the sets of indices inside the set I_i. For the nondegenerate partition $\kappa_i = 1^{n-2}2^1$ the last factor is 1, and we may not include it in the product.

The next section is devoted to the computation of the degree $\deg_K(\widehat{\mathrm{LL}})$.

5.2.4.3 Degree of a Projective Mapping

Let $f : \mathbb{C}^m \to \mathbb{C}^m$, $f = (f_1, \ldots, f_m)$ be a homogeneous mapping of coordinatewise constant degree W. If f is finite, then the degree of its restriction to a homogeneous subvariety in the preimage is closely related to the dimension and the degree of the subvariety. Namely, the following statement is valid.

Lemma 5.2.10. *Let $V \subset \mathbb{C}^m$ be a homogeneous affine subvariety of pure dimension k in the source space, and suppose that the image $f(V)$ is irreducible. Then*

$$\deg V \cdot W^k = \deg f(V) \cdot \deg(f|_V). \tag{5.11}$$

Moreover, if f is nondegenerate at a generic point of f, then a generic point of $f(V)$ has $\deg(f|_V)$ geometrically distinct preimages in V.

Roughly speaking, a variety is *of pure dimension k* if its every neighborhood of every point is of dimension k.

The proof is obtained by comparing the numbers of irreducible homogeneous curves in the preimage of $\mathbb{C}^{m-k+1} \cap f(V)$ belonging to V counted in two different ways: first, in terms of the degree $\deg(f_V)$, which gives us the right-hand side of the formula, and second, by considering the intersection of V with k hyperplanes, which leads to the left-hand side.

Lemma 5.2.10 allows us to compute the degree of a primitive stratum $\widehat{\Sigma}_I$ corresponding to a partition **I** with the only degenerate tuple I. Indeed, the restriction of $\widehat{\mathrm{LL}}$ to such a stratum has the degree

$$\deg_\kappa(\widehat{\mathrm{LL}}) = \frac{\deg_\kappa(A) \cdot \deg_\kappa(\mathrm{LL})}{\deg_{\nu(\kappa)}(T)}$$

$$= \frac{(n - v(\kappa) - 1)! \frac{|\mathrm{Aut}(\kappa)|}{k_1!} \cdot n^{c-1} \frac{(n-v(\kappa)-1)!}{|\mathrm{Aut}(\kappa)|}}{|\mathrm{Aut}(\nu(\kappa))|}$$

$$= n^{c-1} \frac{|\mathrm{Aut}(\nu(\kappa))|}{k_1!},$$

since $|\mathrm{Aut}(\nu(\kappa))| = (n - v(\kappa) - 1)!$ and $|\mathrm{Aut}(\kappa)| = k_1! |\mathrm{Aut}(\nu(\kappa))|$.

Using Lemma 5.2.10 we obtain the degrees of the closures of primitive strata:
$$\deg \operatorname{cl}(\widehat{\Sigma}_\kappa) = \frac{|\operatorname{Aut}(\nu(\kappa))|}{k_1!}.$$

Now, let I be an arbitrary set of pairwise disjoint subsets of the set of indices $\{1, \ldots, n-1\}$. We associate to this set the *standard plane* $\Pi_I \subset \widehat{\mathcal{P}}$ given by the equations $\alpha_j = \alpha_l$ for all pairs j, l belonging to the same element of I.

Theorem 5.2.11. *The restriction of the mapping $\widehat{\operatorname{LL}}$ to the plane Π_I is non-degenerate in the complement to the intersection of Π_I with other standard planes not containing Π_I (that is, on the set where $a_j \ne a_l$ for all pairs of indices j, l not belonging to the same element of I).*

Proof. Let $\beta_1, \beta_2, \ldots, \beta_m$ denote the coordinates of geometrically distinct critical points on the plane Π_I, and let b_i be the multiplicity of the critical point with the coordinate β_i. We have $b_1 + \cdots + b_m = n - 1$ and $b_1\beta_1 + \cdots + b_m\beta_m = 0$. Consider the subfamily ρ_I of the family of 1-forms ρ corresponding to the standard plane Π_I,
$$\rho_I = n(x - \beta_1)^{b_1} \ldots (x - \beta_m)^{b_m} dx = \Phi_I(x) dx.$$

The restriction of the mapping $\widehat{\operatorname{LL}}$ to the plane Π_I has the coordinate presentation
$$\widehat{\operatorname{LL}}_i : (\beta_1, \ldots, \beta_m) \mapsto -\bar{t}_I(\beta_1, \ldots, \beta_m) + \int_0^{\beta_i} \rho_I,$$
where $\bar{t}_I$ is the restriction of $\bar{t}$ to the plane Π_I.

Now fix the polynomial 1-forms
$$\omega_i = -b_i \frac{\rho_I}{x - \beta_i}, \qquad i = 1, \ldots, m,$$
on the line $\mathbb{C}^1$. The homology classes of these 1-forms in the space of relative cohomology $H^1(\mathbb{C}^1, \{\beta_1, \ldots, \beta_m\})$ satisfy the linear relation $[\omega_1] + \cdots + [\omega_m] = 0$ since the 1-form $d\Phi_I = \omega_1 + \cdots + \omega_m$ determines the zero relative cohomology class. If all β_i are pairwise distinct, then there are no other relations between the classes $[\omega_i]$. Indeed, if a 1-form $\omega = u_1\omega_1 + \cdots + u_m\omega_m$, $u_i \in \mathbb{C}$, has zero relative cohomology class, then the polynomial
$$\int_{\beta_1}^x \omega$$
has the same degree $n - 1$ and the same multiplicities of zeroes as the coefficient Φ_I of ρ_I, therefore, it coincides with the latter up to a nonzero constant factor. Hence, the classes $[\omega_1]$, ..., $[\omega_m]$ span the cohomology space $H^1(\mathbb{C}, \{\beta_1, \ldots, \beta_m\})$.

Now we are going to express the Jacobi matrix $(\partial \widehat{LL}_j/\partial \beta_i)$ in terms of the 1-forms ω_l. The entries of the Jacobi matrix are

$$\partial \widehat{LL}_j/\partial \beta_i = \int_0^{\beta_j} \omega_i - \partial \bar{t}/\partial \beta_i. \tag{5.12}$$

This is obvious for nondiagonal elements. In calculating diagonal elements we must also include the summand obtained by differentiating the integral along the upper integration limit. However, this summand is zero since the integrand vanishes at the point β_i, and, therefore, Eq. (5.12) is the universal formula for all entries of the Jacobi matrix, including diagonal ones. Subtracting the first row of the matrix from all other rows we arrive at the $m \times m$-matrix

$$\begin{pmatrix} \int_0^{\beta_1} \omega_1 - \partial \bar{t}/\partial \beta_1 & \cdots & \int_0^{\beta_1} \omega_m - \partial \bar{t}/\partial \beta_m \\ \int_{\beta_1}^{\beta_2} \omega_1 & \cdots & \int_{\beta_1}^{\beta_2} \omega_m \\ \vdots & \ddots & \vdots \\ \int_{\beta_1}^{\beta_m} \omega_1 & \cdots & \int_{\beta_1}^{\beta_m} \omega_m \end{pmatrix}.$$

Since the relative cohomology classes of the 1-forms ω_j span the space of relative cohomology, the rank of this matrix at points where all β_i are pairwise distinct is $m - 1$, and the theorem follows.

Now, fix a passport $K = [\kappa_1, \ldots, \kappa_c]$ and a partition $\mathbf{I} = [I_1, \ldots, I_c]$ of the set of indices $\{1, \ldots, n-1\}$ consistent with the passport. Suppose the first r partitions in the passport are degenerate, while all other $c - r$ partitions are $1^{n-2}2^1$. We call an index i *essential* for the partition $\mathbf{I}$ if the number of sets in the tuple $I_j \in \mathbf{I}$ containing i is greater than one. The critical point α_i and the corresponding critical value will also be called essential. Theorem 5.2.11 implies the following statement.

Corollary 5.2.12. *The closure* $\mathrm{cl}(\widehat{\Sigma}_\mathbf{I})$ *of a stratum* $\widehat{\Sigma}_\mathbf{I} \subset \widehat{\mathcal{D}}$ *is the intersection of the closures* $\mathrm{cl}(\widehat{\Sigma}_{I_i})$ *of the primitive strata* $\widehat{\Sigma}_{I_i}$. *These primitive strata intersect transversally almost everywhere at* $\widehat{\Sigma}_\mathbf{I}$, *i.e., everywhere outside the subvariety of complex codimension one consisting of points, where an essential critical point glues together with another critical point. In particular, the closure* $\mathrm{cl}(\widehat{\Sigma}_\mathbf{I})$ *is smooth everywhere outside the points of this subvariety, and*

$$\deg \mathrm{cl}(\widehat{\Sigma}_\mathbf{I}) = \prod_{i=1}^r \deg \mathrm{cl}(\widehat{\Sigma}_{I_i}) = \prod_{i=1}^r \frac{|\mathrm{Aut}(\nu(\kappa_i))|}{k_{i1}!}.$$

The statement of the enumeration theorem 5.2.2 follows now from this corollary and Lemma 5.2.10.

5.3 Rigid Classification of Generic Meromorphic Functions and Geometry of Moduli Spaces of Curves

5.3.1 Statement of the Enumeration Theorem

There is one more situation where the answer to the Hurwitz problem is known in a rather explicit form. This is the case where there is at most *one degenerate* ramification point, while all other ramification points are simple. We will suppose below that the point of degenerate ramification is the infinity point in the target sphere, and call its preimages *poles*. In other words, we are interested in the number of constellations $[g_1, \ldots, g_c]$, where $g_2, \ldots, g_c$ are transpositions, while g_1 is a permutation of a prescribed cycle type $\kappa = (k_1, \ldots, k_n)$. Denote this number by $h_{g;\kappa}$ and call it the *Hurwitz number*. (These numbers should not be confused with the Hurwitz numbers which are the coefficients in the expansion of the Weierstrass $\wp$-function.) Note that in this section, in contrast to the previous one, we denote by n the number of geometrically distinct preimages of the degenerate ramification point, while the degree of the functions under consideration is $k_1 + \cdots + k_n = k$.

Once again, the computation of the Hurwitz number can be achieved by the calculation of the degree of the LL mapping on suitable spaces of meromorphic functions, which are called *Hurwitz spaces*. These spaces prove to be closely related to the moduli spaces of complex curves with marked points. Namely, numbering the poles of a meromorphic function $f : X \to \mathbb{C}P^1$ we associate to such a function the curve X with n marked points, i.e., a point in the moduli space of complex marked curves. This mapping determines a mapping from the Hurwitz space (or, to be more precise, from an appropriate covering of the Hurwitz space determined by the marking) to the moduli space of curves. Now, it is not absolutely unexpected that the final result can be expressed in terms of intersection numbers on the moduli space.

Theorem 5.3.1 ([91], [92], [120]). *We have*

$$h_{g;\kappa} = \frac{(k+n+2g-2)!}{|\mathrm{Aut}(\kappa)|} \prod_{i=1}^{n} \frac{k_i^{k_i}}{k_i!} \int_{\overline{\mathcal{M}}_{g;n}} \frac{1 - \lambda_1 + \lambda_2 - \cdots + (-1)^g \lambda_g}{(1 - k_1\psi_1)\ldots(1 - k_n\psi_n)} . \quad (5.13)$$

The whole Sec. 5.3 is dedicated to the proof of this theorem.

Recall that $\overline{\mathcal{M}}_{g;n}$ denotes the moduli space of genus g stable complex curves with n marked points, and ψ_i denotes the first Chern class of the line bundle $\mathcal{L}_i$ over $\overline{\mathcal{M}}_{g;n}$; the fiber of $\mathcal{L}_i$ at a point $(X; x_1, \ldots, x_n) \in \overline{\mathcal{M}}_{g;n}$ is the cotangent line to X at x_i. Up to now the classes λ_i have not been considered. These are the Chern classes of the *Hodge bundle* $\Lambda_{g;n} \to \overline{\mathcal{M}}_{g;n}$. The fiber of the Hodge bundle at a smooth point $(X; x_1, \ldots, x_n)$ coincides with the space of all holomorphic 1-forms over X. It is, in an obvious sense, independent of the choice of the marked points. Its natural extension to a nodal curve X is

5.3 Rigid Classification and Geometry of Moduli Spaces

the space of meromorphic 1-forms over X having poles at most at the double points of X and such that the poles are of an order not greater than one, and the sum of the residues along the two sheets at a double point is zero. The rank of $\Lambda_{g;n}$ is g since the space of meromorphic 1-forms on a genus g curve is g-dimensional. We write

$$c(\Lambda_{g;n}) = c_0(\Lambda_{g;n}) + c_1(\Lambda_{g;n}) + \cdots + c_g(\Lambda_{g;n}) = 1 + \lambda_1 + \cdots + \lambda_g,$$
$$\lambda_i = c_i(\Lambda_{g;n}) \in H^{2i}(\overline{\mathcal{M}}_{g;n}).$$

As usual, we assume that the integral of a class is zero whenever the dimension of the class differs from that of the base. We are going to show in the next section how this formula works and what information about the number of meromorphic functions it produces.

It is worth mentioning that Eq. (5.13) works in both directions: it allows one to compute some intersection numbers on moduli spaces of curves provided that Hurwitz numbers are known. Basing on this remark and on the previous calculations of asymptotics of Hurwitz numbers, Okounkov and Pandharipande gave in [227] a new proof, independent of that due to Kontsevich, of the statement that the Kontsevich model coincides with the intersection theory model, which forms a part of the proof of Witten's conjecture (more exactly, they proved Theorem 4.7.2).

5.3.2 Calculations: Genus 0 and Genus 1

Before proving Theorem 5.3.1 we show how it can be applied to deduce explicit formulas for the Hurwitz numbers in genus 0 and 1

In genus 0 the theorem from the previous subsection provides an immediate answer. The rank of the Hodge bundle in this case is zero since there are no holomorphic 1-forms over rational curves, and the integral in Eq. (5.13) is

$$\int_{\overline{\mathcal{M}}_{0;n}} \frac{1}{(1 - k_1\psi_1)\ldots(1 - k_n\psi_n)} . \tag{5.14}$$

Hence, in order to compute it we need only to know the intersection numbers

$$\langle \tau_{m_1} \ldots \tau_{m_n} \rangle = \int_{\overline{\mathcal{M}}_{0;n}} \psi_1^{m_1} \ldots \psi_n^{m_n} .$$

But these indices are already known, see Proposition 4.6.10:

$$\langle \tau_{m_1} \ldots \tau_{m_n} \rangle = \frac{(n-3)!}{m_1! \ldots m_n!} .$$

Expanding the integrand in (5.14) and selecting terms of dimension $n - 3$ we obtain nothing but the expansion of the multinomial $(k_1 + \cdots + k_n)^{n-3}$:

$$\int_{\overline{\mathcal{M}}_{0;n}} \frac{1}{(1 - k_1\psi_1)\ldots(1 - k_n\psi_n)} = \sum_{m_1+\cdots+m_n=n-3} \langle \tau_{m_1} \ldots \tau_{m_n} \rangle k_1^{m_1} \ldots k_n^{m_n} .$$

Thus, we have proved the following statement.

Theorem 5.3.2. *For $g = 0$ and the partition $\kappa = (k_1, \ldots, k_n)$ of $k = k_1 + \cdots + k_n$ we have*

$$h_{0;\kappa} = \frac{(k+n-2)!}{|\mathrm{Aut}(\kappa)|} \prod_{i=1}^{n} \frac{k_i^{k_i}}{k_i!} k^{n-3} . \tag{5.15}$$

This statement was first published in [144], but without a proof. We call it the *Hurwitz formula*.

Example 5.3.3 (Generic polynomials). In the case of generic polynomials $n = 1$, $k_1 = k$, $\kappa = k^1$ and Eq. (5.15) yields

$$h_{0;\kappa} = (k-1)! \frac{k^k}{k!} k^{-2} = k^{k-3}$$

in accordance with Corollary 5.1.2.

Example 5.3.4 (Generic rational mappings). The condition of applicability of Eq. 5.15 is that all critical values except the one corresponding to κ are simple, while the one corresponding to κ may be arbitrary. In the case of generic rational mappings $\mathbb{C}P^1 \to \mathbb{C}P^1$ we may take as an additional "critical value" a generic one; then we will have $n = k$, $k_1 = \cdots = k_n = 1$, $\kappa = 1^k$, and Eq. (5.15) gives

$$h_{0;\kappa} = \frac{(2k-2)!}{k!} k^{k-3} .$$

This particular case of the Hurwitz formula was rediscovered in [72].

Exercise 5.3.5. Verify that the Hurwitz number (5.15) for $\kappa = 1^{k-2}2^1$ (that is, if κ corresponds to a simple critical value) coincides with that for generic rational mappings from the example above.

For curves of positive genus, the numerator in the integrand in (5.13) is no longer 1. The computation of the integral requires knowing the intersection numbers of the form

$$\int_{\overline{\mathcal{M}}_{g;n}} \psi_1^{m_1} \ldots \psi_n^{m_n} \lambda_j , \tag{5.16}$$

where $m_1 + \cdots + m_n + j = n + 3g - 3$. The terms corresponding to the class $\lambda_0 = 1$ have the highest homogeneous degree in k_i, and this degree is $n+3g-3$. Their value is given, at least in principle, by the Kontsevich theorem.

Much less is known about the computation of the integrals (5.16) for $j > 0$. They are called *Hodge integrals*. For a monomial containing only one λ-class (this is precisely the case we are interested in), the string and the dilaton equations remain valid. The proof proceeds similarly to the argument in the case where the integrand contains only ψ-classes. Besides, the terms including the λ-class of the highest degree, i.e. λ_g, are known due to the following statement.

5.3 Rigid Classification and Geometry of Moduli Spaces

Theorem 5.3.6 ([97]). *For* $m_1 + \cdots + m_n = n + 2g - 3$ *we have*

$$\int_{\overline{\mathcal{M}}_{g;n}} \psi_1^{m_1} \ldots \psi_n^{m_n} \lambda_g = \binom{2g + n - 3}{m_1, \ldots, m_n} b_g,$$

where $b_g = \int_{\overline{\mathcal{M}}_{g;n}} \psi_1^{2g-2} \lambda_j$ *is a constant depending only on the genus and such that*

$$1 + b_1 t^2 + b_2 t^4 + \cdots = \frac{t/2}{\sin(t/2)},$$

i.e.,

$$b_g = \frac{2^{2g-1} - 1}{2^{2g-1}} \frac{|B_{2g}|}{(2g)!},$$

where B_{2g} *is the* $(2g)$*th Bernoulli number.*

Recall that, by definition, Bernoulli numbers are coefficients of the exponential generating function

$$\frac{t}{e^t - 1} = 1 - \frac{1}{2}t + \frac{B_2}{2!}t^2 + \frac{B_4}{4!}t^4 + \cdots,$$

and the first few of them are

$$B_2 = \frac{1}{6}, \quad B_4 = -\frac{1}{30}, \quad B_6 = \frac{1}{42}, \quad B_8 = -\frac{1}{30}, \quad B_{10} = \frac{5}{66}.$$

In particular, we have

$$b_1 = \frac{1}{24}, \quad b_2 = \frac{7}{5760}, \quad b_3 = \frac{31}{967680}.$$

This theorem, together with the string and the dilaton equations, allows one to compute the genus 1 Hurwitz numbers in an explicit form. The formula below was conjectured in [118] and first proved in [116].

Theorem 5.3.7. *We have*

$$h_{1;\kappa} = \frac{(k+n)!}{24 |\mathrm{Aut}(k_1, \ldots, k_n)|} \prod_{i=1}^{n} \frac{k_i^{k_i}}{k_i!} \left(k^n - \sum_{i=2}^{n} (i-2)! e_i k^{n-i} - k^{n-1} \right),$$

where $e_i = e_i(k_1, \ldots, k_n)$ *is the* i*th elementary symmetric function in* $k_1, \ldots, k_n$ *and* $k = k_1 + \cdots + k_n = e_1$.

The reader will recognize the coefficient $b_1 = 1/24$ in front of the expression of the theorem.

Thus, a new and highly non-trivial enumerative result is obtained by using the intersection theory on the moduli spaces.

Note that the last summand in the big parentheses has degree $n - 1$ in k_i; it is the total value of the λ_1-containing integrals. All the other summands have degree n in k_i. The proof of the theorem is obtained by verifying that the numbers in the formula satisfy the string and the dilaton equations, and have the same initial values as the intersection numbers in the integrand. See Proposition 4.6.11 and [92] for details.

5.3.3 Cones and Their Segre Classes

Below, we are going to represent spaces of meromorphic functions as cones over moduli spaces of complex curves. Such cones appear naturally already in the case of nondegenerate polynomials. We have seen that the space of polynomials of the form (5.2) is not exactly the moduli space of nondegenerate polynomials of degree n: besides nondegenerate polynomials it contains also degenerate ones, and each isomorphism class of a polynomial has n representatives in it. In order to get rid of this ambiguity, we must take the quotient of this space modulo the action (5.3) of the cyclic group $\mathbb{Z}/n\mathbb{Z}$ generated by the mapping $x \mapsto \varepsilon_n x$. Here ε_n is a primitive root of unity of degree n. The resulting quotient space is no longer a vector space, but it still carries the structure of a cone.

Our treating of cones follows that of [108]. Cones are most easily defined in dual terms, that is, in terms of polynomial functions on them. By definition, a *cone* is determined by a $\mathbb{Z}_{\geq 0}$-graded $\mathbb{C}$-algebra. The cone itself is the spectrum of the algebra. A morphism of cones is determined by a graded algebra homomorphism of the defining algebras going in the opposite direction.

Example 5.3.8. Each vector space is a cone; indeed, the algebra of polynomials $\mathbb{C}[x_1, \ldots, x_n]$ is naturally graded by the degree. The $\mathbb{C}$-algebra corresponding to the quotient space of $\mathbb{C}^{n-1}$ modulo the action (5.3) is the algebra of polynomials on $\mathbb{C}^{n-1}$ invariant with respect to this action. The grading induced by setting the weights of the variables $w(a_i) = i/n$ is obviously integral. Note that each monomial (and, therefore, each polynomial) of integral weight is invariant with respect to this action.

More generally, the quotient space of a vector space modulo a linear action of a finite group is a cone. The corresponding graded algebra is the algebra of polynomials over the vector space, invariant under the group action.

Since a cone is determined by a graded algebra, an action of the multiplicative group $\mathbb{C}^*$ of nonzero complex numbers is defined on every cone.

Cones over algebraic varieties or orbifolds are analogues of vector bundles, but instead of a vector space, the fiber over each point of the underlying variety is a cone. Formally speaking, a *cone over a variety* M is a sheaf of integer-graded $\mathbb{C}$-algebras over M. Similarly to the rank of a vector bundle, the *rank of a cone over an algebraic variety* is the dimension of a fiber. We consider only locally finitely generated cones. Cones produce a natural language to describe spaces of meromorphic functions because if we fix the underlying curve and a set of points on it, then the set of meromorphic functions with poles of prescribed order at these points is not a vector space (the sum of two functions can have a pole of smaller order), but it is still a cone (multiplication by a constant preserves the orders of the poles).

A cone can be projectivized: points of the projective cone are nontrivial orbits of the $\mathbb{C}^*$-action on the cone. Projectivized cones over algebraic varieties are defined by means of the fiberwise projectivization.

5.3 Rigid Classification and Geometry of Moduli Spaces

Chern classes are not well defined for cones. However, it is possible to introduce other characteristic classes for them. Let $\mathcal{S}$ be a cone of rank r over a compact d-dimensional variety or orbifold M, and let $P\mathcal{S}$ be the corresponding projective cone. The *Segre class* is an element in the cohomology ring $H^*(M)$ (here and below cohomology is considered with rational coefficients) defined as follows. The projective cone $P\mathcal{S}$ carries a canonical line bundle usually denoted by $\mathcal{O}(1)$; this is the bundle of homogeneous rational functions of degree 1 on fibers. Its first Chern class $c_1(\mathcal{O}(1))$ is an element of the cohomology ring $H^*(P\mathcal{S})$, $c_1(\mathcal{O}(1)) \in H^2(P\mathcal{S})$. The projection mapping $\pi : P\mathcal{S} \to M$ induces the pushforward mapping $\pi_* : H_*(P\mathcal{S}) \to H_*(M)$ taking each cycle to a cycle of the same dimension. Each power $c_1^j(\mathcal{O}(1))$ of the first Chern class can be represented, by the Poincaré duality, by an element of complementary dimension in $H_*(P\mathcal{S})$. If $j \geq r - 1$, then the dimension of the cycle corresponding to $c_1^j(\mathcal{O}(1))$, that is, the number $r + d - 1 - j$, is not greater than d, the dimension of the base. Hence, we obtain a well-defined class in $H_*(M)$, and using Poincaré duality once again, a class in $H^*(M)$. The sum of all these classes for all values of j is called the Segre class of the bundle and is denoted by $s(\mathcal{S})$:

$$s(\mathcal{S}) = s_0(\mathcal{S}) + \cdots + s_d(\mathcal{S}), \qquad s_i(\mathcal{S}) \in H^{2i}(M).$$

The class $s_d(\mathcal{S}) \in H^{2d}(M)$ is called the *top Segre class* of $\mathcal{S}$ and is denoted also by $s_{\text{top}}(\mathcal{S})$. Its value $\int_M s_{\text{top}}(\mathcal{S})$ is a rational number.

Example 5.3.9. Take a cone over a point. In this case only the zero Segre class of the cone is nontrivial. Consider, for example, a k-dimensional vector space with weighted coordinates of weights $w_1, \ldots, w_k$. This means that the vector space is represented as a direct sum of one-dimensional subspaces, $\mathbb{C}^k \cong \oplus_{i=1}^k \mathbb{C}^1$, and the multiplicative group $\mathbb{C}^*$ acts on the ith summand by $u_\zeta : v \mapsto \zeta^{w_i} v$. The corresponding projective cone is the weighted projective space. The zero Segre class of such a cone is

$$s_0 = \frac{1}{w_1 \cdots w_k}.$$

If the cone $\mathcal{S}$ is a vector bundle, then the product $s(\mathcal{S})c(\mathcal{S})$ in the cohomology ring equals 1, that is, the total Segre class and the total Chern class of a vector bundle are mutually inverse. This is one of the most direct ways to introduce the notion of total Chern class provided that we know what it is for line bundles.

Both in the case of vector bundles and of general cones, the direct sum of cones corresponds to the tensor product of the sheaves of graded algebras. The total Segre class of the direct sum of cones coincides with the product of their total Segre classes. For Chern classes of vector bundles, a more general statement, called the *Whitney formula*, is true: if we have a short exact sequence

$$0 \longrightarrow \mathcal{A} \longrightarrow \mathcal{C} \longrightarrow \mathcal{B} \longrightarrow 0$$

of vector bundles over a variety M, then

$$c(\mathcal{C}) = c(\mathcal{A})\, c(\mathcal{B}).$$

Similar formulas are usually valid for the Segre classes of cones belonging to short exact sequences of cone morphisms.

5.3.4 Cones of Principal Parts

Recall that our goal is to compactify Hurwitz spaces in order to be able to compute Hurwitz numbers proceeding from geometrical information.

Fix a positive integer k. Consider two meromorphic functions f_1, f_2 defined in a neighborhood of $0 \in \mathbb{C}$ and having poles of order k at 0. We say that these functions have the same principal part at 0 if their difference $f_1 - f_2$ has no pole at 0. A *principal part* is an equivalence class of local functions with respect to this equivalence relation. Below we present a coordinate description of the space of principal parts.

If one is interested in principal parts of fixed order k of meromorphic functions at the origin of the complex line, then it seems to be natural to consider this space as a subset in the vector space of principal parts of order *at most* k. However, this direct approach to compactification of Hurwitz spaces does not work well. The reason is that when the order of the pole drops, then the degree of the function also drops, and the critical values of the function decrease in number. As a result, the LL mapping acquires discontinuity. The construction below explains how the degree of the function can be preserved under the degeneration of its pole, so that the LL mapping remains continuous. Essentially, it consists in the birth of a new rational irreducible component at the pole, so that the restriction of the degenerate principal part to this component becomes a polynomial of the same degree k.

Choose a coordinate x in a neighborhood of 0. Then a principal part can be written in the form (cf. Exercise 5.1.8)

$$\left(\frac{u}{x}\right)^k + a_1 \left(\frac{u}{x}\right)^{k-1} + \cdots + a_{k-1}\frac{u}{x} \qquad (5.17)$$

for some $u \in \mathbb{C}^*$. If $k > 1$, then this presentation is not unique; it depends on the choice of the parameter u, and there are k possibilities to make this choice. Hence, the space of expressions (5.17) covers the space of principle parts with multiplicity k. The group of the covering is $\mathbb{Z}/k\mathbb{Z}$, and it acts on the space of expressions (5.17) according to the rule

$$(u, a_1, a_2, \ldots, a_{k-1}) \mapsto (\varepsilon_k u, \varepsilon_k a_1, \varepsilon_k^2 a_2, \ldots, \varepsilon_k^{k-1} a_{k-1}), \qquad (5.18)$$

where $\varepsilon_k \in \mathbb{C}$ is a primitive root of unity of degree k.

The action of the group $\mathbb{C}^*$ by multiplications on the space of principal parts induces the following choice of weights of the parameters u and a_i:

5.3 Rigid Classification and Geometry of Moduli Spaces

$$w(u) = \frac{1}{k}, \qquad w(a_i) = \frac{i}{k}, \qquad i = 1, \ldots, k-1.$$

The algebra of polynomials in u and a_i invariant with respect to the action (5.18) is spanned, as a vector space, by monomials of integer weight. The weights of the coordinates endow this algebra with an integral grading. We denote the corresponding cone by P^k.

The cone P^k is k-dimensional, and the open dense subset $u \neq 0$ in it corresponds to principal parts with poles of order k. Its complement $A^{k-1} \subset P^k$ is the spectrum of the quotient algebra of $\mathbb{Z}/k\mathbb{Z}$-invariant polynomials modulo the ideal of polynomials divisible by u. It resembles the quotient space of polynomials (5.2) modulo the action (5.3) and carries a structure of a $\mathbb{Z}/k\mathbb{Z}$-quotient of $(k-1)$-dimensional vector space which we denote by $\widetilde{A}^{k-1}$.

Note that in spite of the fact that the value of the parameter u, and hence of the a-coordinates of a general principal part depend on the choice of the coordinate x, the a-coordinates of the elements of A^{k-1}, where $u = 0$, are unique up to the $\mathbb{Z}/k\mathbb{Z}$-action.

Now we are going to associate to a point in P^k an element of the kth tensor power of the tangent line at $0 \in \mathbb{C}$. Let L be the cotangent line to $\mathbb{C}$ at 0. Writing a point $p \in P^k \setminus A^{k-1}$ in the form (5.17) we associate to this point the principal part u/x with the pole of degree one. It is determined uniquely up to multiplication by a root of unity of degree k, and it coincides with the principal part $p^{1/k}$ (well defined up to the same ambiguity). Having a pole of degree one, this principal part determines a tangent vector at 0 as the following linear functional on L:

$$\omega \mapsto \operatorname{Res}_{x=0} \frac{u}{x}\omega,$$

ω being a local holomorphic 1-form representing a cotangent vector. The kth tensor power of the constructed tangent vector is an element in $(L^\vee)^{\otimes k}$, and this element is independent both of the choice of the coordinate x and of the root u. We call this element the *leading term* of the principal part. Associating to a principal part from A^{k-1} the zero element of $(L^\vee)^{\otimes k}$, we obtain as the result a continuous mapping

$$\phi : P^k \to (L^\vee)^{\otimes k}.$$

By abuse of language, we refer below to points of A^{k-1} as to "principal parts of order k with zero leading term".

The following statement is simple.

Lemma 5.3.10. *The dual mapping ϕ^* is a morphism of graded algebras. The multiplicity of ϕ along its zero locus A^{k-1} is k.*

The second statement of the lemma means that if we restrict ϕ to a generic curve in P^k intersecting A^{k-1} transversally, then exactly k preimages of ϕ coalesce as its image tends to 0.

Now we are able to construct cones of principal parts over the moduli spaces $\overline{\mathcal{M}}_{g;n}$. Take a stable marked curve $(X; x_1, \ldots, x_n) \in \overline{\mathcal{M}}_{g;n}$. To the ith marked point and a positive integer k_i we associate the cone of principal parts with poles of order k_i at the point x_i and the cone of principal parts with zero leading coefficients and with poles of order k_i at the point x_i. The union of these cones form the cones $\mathcal{P}_i$ and $\mathcal{A}_i$ over $\overline{\mathcal{M}}_{g;n}$. Similarly, we consider the vector bundle $\widetilde{\mathcal{A}}_i$ over $\overline{\mathcal{M}}_{g;n}$. We will also need the direct sums of cones

$$\mathcal{P} = \mathcal{P}_1 \oplus \cdots \oplus \mathcal{P}_n$$
$$\mathcal{A} = \mathcal{A}_1 \oplus \cdots \oplus \mathcal{A}_n$$
$$\widetilde{\mathcal{A}} = \widetilde{\mathcal{A}}_1 \oplus \cdots \oplus \widetilde{\mathcal{A}}_n,$$

the last of which is, in fact, a vector bundle.

The morphism ϕ above, assigned to each marked point, determines n cone morphisms $\varphi_i : \mathcal{P}_i \to (\mathcal{L}_i^\vee)^{\otimes k_i}$ over $\overline{\mathcal{M}}_{g;n}$. The ith morphism takes a principal part at the ith marked point to the k_ith power of a tangent vector at this point. The direct sum of these morphisms determines the morphism $\varphi : \mathcal{P} \to \mathcal{L}$, where

$$\mathcal{L} = (\mathcal{L}_1^\vee)^{\otimes k_1} \oplus \cdots \oplus (\mathcal{L}_n^\vee)^{\otimes k_n}.$$

The above argument justifies the following statement.

Lemma 5.3.11. *The cone $\mathcal{A}_i$ is the quotient of the vector bundle $\widetilde{\mathcal{A}}_i$ modulo the fiberwise action of the group $\mathbb{Z}/k_i\mathbb{Z}$. The cone $\mathcal{A}$ is the quotient of the vector bundle $\widetilde{\mathcal{A}}$ modulo the fiberwise action of the group $\mathbb{Z}/k_1\mathbb{Z}\oplus\cdots\oplus\mathbb{Z}/k_n\mathbb{Z}$.*

The cones $\mathcal{A}_i$, $\mathcal{A}$ are constant cones over $\overline{\mathcal{M}}_{g;n}$.

The zero locus of the morphism φ_i is the subcone $\mathcal{A}_i$ of $\mathcal{P}_i$. The multiplicity of φ_i along $\mathcal{A}_i$ equals k_i. The zero locus of the morphism φ is the subcone $\mathcal{A}$ of $\mathcal{P}$. The multiplicity of φ along $\mathcal{A}$ equals $k_1 \ldots k_n$.

This lemma immediately leads to the computation of the Segre classes of all the cones under consideration.

Lemma 5.3.12. *We have*

$$s(\mathcal{A}_i) = \frac{1}{k_i} \frac{k_i^{k_i-1}}{(k_i-1)!}, \qquad s(\mathcal{A}) = \prod_{i=1}^n \frac{1}{k_i} \frac{k_i^{k_i-1}}{(k_i-1)!}.$$

$$s(\mathcal{P}_i) = \frac{k_i!}{k_i^{k_i}} \frac{1}{1 - k_i\psi_i}, \qquad s(\mathcal{P}) = \prod_{i=1}^n \frac{k_i!}{k_i^{k_i}} \frac{1}{1 - k_i\psi_i},$$

where $\psi_i = c_1(\mathcal{L}_i)$.

First of all, since each $\mathcal{A}_i$ is a constant cone, the only nontrivial Segre class of it is the zero class $s_0(\mathcal{A}_i)$. Its value can be computed locally, at a fiber of $P\mathcal{A}_i$ over a point. This fiber coincides with the quotient of the projectivized weighted vector space $\widetilde{\mathcal{A}}_i$ modulo the action of the group $\mathbb{Z}/k_i\mathbb{Z}$. Its zero Segre class is the inverse product of the weights of the coordinates divided by the order of the group acting.

The mapping φ_i determines a short exact sequence of cones

$$0 \longrightarrow \mathcal{A}_i \longrightarrow \mathcal{P}_i \longrightarrow (\mathcal{L}_i^\vee)^{\otimes k_i} \longrightarrow 0.$$

The exactness of this sequence implies that

$$s(\mathcal{P}_i) = k_i s(\mathcal{A}_i) s((\mathcal{L}_i^\vee)^{\otimes k_i}) = k_i \frac{s(\mathcal{A}_i)}{c((\mathcal{L}_i^\vee)^{\otimes k_i})}.$$

(The factor k_i arises because φ_i has zero of multiplicity k_i along the subcone $\mathcal{A}_i$.)

The proof of the lemma is completed.

5.3.5 Hurwitz Spaces

Hurwitz spaces are spaces of meromorphic functions on complex curves. The representation of a meromorphic function by means of its ramification points and the monodromy around each point is effective not only from the point of view of representing a single function, but also if we want to describe moduli spaces of all meromorphic functions (say, on curves of given genus). The Lyashko–Looijenga mapping is simply a covering of the space of ramification loci by the moduli space of meromorphic functions, and the fiber of this covering is simply the set of all consistent monodromy data. This picture endows Hurwitz spaces with a natural topology and complex structure.

There are various reasonable ways to obtain compactifications of Hurwitz spaces. The one suggested by Harris and Mumford (see e.g. [139]) forbids ramification points to coalesce, and hence the Hurwitz space is represented as a covering of the moduli space of rational curves with marked points. Another approach introduced in [91] suggests considering the projection of a Hurwitz space to the moduli space of the underlying curves with marked poles. Then the Hurwitz space becomes fibered over the latter moduli space, and the geometry of the fiber admits a rather effective description. This description is used in the proof of Theorem 5.3.1, and we are going to present it here.

To become more precise, let $f : X \to \mathbb{C}P^1$ be a meromorphic function on a genus g smooth curve X, and let us fix some numbering $x_1, \ldots, x_n$ of its poles. Let $k_1, \ldots, k_n$ be the degrees of the poles, forming the partition κ of the degree $k = k_1 + \cdots + k_n$.

A function f determines the n-tuple of principal parts $(p_1, \ldots, p_n)$ of orders $k_1, \ldots, k_n$ at the poles. Two functions f_1, f_2 possessing the same tuple of principal parts are equal up to an additive constant. Indeed, their difference $f_1 - f_2$ has no poles on X, whence it is a constant. We fix the choice of a function with given tuple of principal parts by requiring that the sum of its critical values must be zero.

Definition 5.3.13 (Completed Hurwitz space). Let $\mathcal{P}$ denote the cone over $\overline{\mathcal{M}}_{g;n}$ of principal parts of orders k_i at the ith marked point. The closure in $\mathcal{P}$ of the set of tuples of principal parts corresponding to meromorphic functions on smooth curves is called the *completed Hurwitz space* and is denoted by $\overline{\mathcal{H}}_{g;\kappa}$.

The completed Hurwitz space is invariant under the multiplication by complex numbers, therefore it forms a subcone in $\mathcal{P}$. As well as the base moduli space, this space carries a natural complex orbifold structure.

The natural question arises whether the completed Hurwitz space coincides with the whole cone $\mathcal{P}$, or, in other words, whether any tuple of principal parts can serve as a tuple of principal parts of a meromorphic function. This is true for rational functions, i.e., for $g = 0$, but for $g \geq 1$ it is not the case. Namely, the principal parts must satisfy the following classical restriction.

Theorem 5.3.14 (see, e.g., [124]). *An n-tuple $(p_1, \ldots, p_n)$ of principal parts is the tuple of principal parts of a meromorphic function if and only if*

$$\mathrm{Res}_{x_1} p_1 \omega + \cdots + \mathrm{Res}_{x_n} p_n \omega = 0 \tag{5.19}$$

for each holomorphic 1-form ω on X.

For the rational case, requirement (5.19) is empty since there are no holomorphic 1-forms on the projective line. Note also that Theorem 5.3.14 is valid not only on smooth curves, but on stable curves as well.

Exercise 5.3.15. For $g > 0$ it can even happen that not each genus g marked curve $(X; x_1, \ldots, x_n)$ carries a meromorphic function with poles of given order at the marked points. Give an example of such a situation. Show that for each g there exists a number $n_0 = n_0(g)$ such that if $n > n_0$, then there exists a meromorphic function with poles of order one at the marked points on each smooth marked curve $(X; x_1, \ldots, x_n)$ of genus g.

Recall that we denote by $\Lambda_{g;n}$ the Hodge bundle over $\mathcal{M}_{g;n}$, that is, the vector bundle whose fibers are spaces of holomorphic 1-forms on the underlying curves. Consider the fiberwise mapping $R : \mathcal{P} \to \Lambda_{g;n}^{\vee}$ taking an n-tuple $(p_1, \ldots, p_n)$ of principal parts to the linear functional

$$\omega \mapsto \mathrm{Res}_{x_1} p_1 \omega + \cdots + \mathrm{Res}_{x_n} p_n \omega$$

on $\Lambda_{g;n}$. This mapping is a cone morphism. Theorem 5.3.14 means that the completed Hurwitz space is a part of the kernel of this morphism. Denote the whole kernel by $\mathcal{Z}_{g;\kappa}$. This kernel can be understood as the closure of tuples of principal parts associated with meromorphic functions not only on smooth curves, but on all stable curves. It can be different from the completed Hurwitz space. In other words, not each meromorphic function on a singular curve is a limit of a family of meromorphic functions on smooth curves.

5.3 Rigid Classification and Geometry of Moduli Spaces

Exercise 5.3.16. Show that for $g = 1$ and $n = 2$ there are two irreducible components in the "kernel" $\mathcal{Z}_{1;1^2}$ of φ (the morphism φ is defined before Lemma 5.3.11), one coinciding with the Hurwitz space $\overline{\mathcal{H}}_{1;1^2}$, and the other one consisting of functions over reducible curves that are constant on the elliptic component. Verify that these components are of the same dimension. Verify that their intersection consists of functions on reducible curves such that their value at the double point coincides with a critical value of their restriction to the rational component.

Having at hand a description of the completed Hurwitz space as a part of the kernel $\mathcal{Z}_{g;\kappa}$ of the mapping R, it no longer seems strange that the Hurwitz number is expressed in terms of the Chern classes of the two bundles on both sides of the mapping.

5.3.6 Completed Hurwitz Spaces and Stable Mappings

Points of the completed Hurwitz spaces $\overline{\mathcal{H}}_{g;\kappa}$ and, more generally, of the space $\mathcal{Z}_{g;\kappa}$ are stable mappings from genus g curves to $\mathbb{C}P^1$ in the sense of Kontsevich [180]. We give the definition of a stable mapping in the specific case of mappings of curves to $(\mathbb{C}P^1, \infty)$.

Definition 5.3.17. A holomorphic mapping $f : (X; x_1, \ldots, x_n) \to (\mathbb{C}P^1, \infty)$ of a nodal genus g curve X to the projective line taking marked nonsingular points $x_1, \ldots, x_n$ (and only these points) to infinity is called *stable* if its automorphism group is finite. In other words, each irreducible component of X taken by f to a single point possesses at least three singular points if its genus is zero, and at least one singular point if its genus is one.

Here by an automorphism of a mapping we mean an automorphism of the marked curve of the above definition preserving the mapping. Note that we do not allow an irreducible component to be contracted to infinity: if an irreducible component of X contains a marked point, then it cannot be contracted to a single point.

Remark 5.3.18. Any nonconstant meromorphic mapping of a stable curve is stable. However, the underlying curve of a stable mapping can well be unstable. For example, the identical mapping of $\mathbb{C}P^1$ with one marked point is stable, in contrast to the curve itself.

Now we are going to associate to a point f of the space $\mathcal{Z}_{g;\kappa}$ a stable mapping. Such a point is described by an n-tuple $(p_1, \ldots, p_n)$ of principal parts at the marked points $x_1, \ldots, x_n$ of a genus g stable curve X. If the curve X is smooth and the leading terms of all the principal parts p_i are nonzero, then the mapping f itself is stable. The same statement is true for singular curves X when all leading terms are nonzero because of the remark above. A problem arises only if one or more of the leading terms are zero.

In this last case we associate to the tuple $(p_1, \ldots, p_n)$ of principal parts on a stable curve X a meromorphic function $\tilde{f} : (\widetilde{X}; x_1, \ldots, x_n) \to (\mathbb{C}P^1, \infty)$ on a nodal curve $\widetilde{X}$ with marked points. This nodal curve is constructed as follows. For each principal part p_i with zero leading term we attach to the curve X a rational irreducible component intersecting X at x_i. The point of intersection becomes unmarked, and we mark as x_i an arbitrary nonsingular point of the attached component, all choices being equivalent. We call the added rational components *new* irreducible components, while the irreducible components of X are *old* ones. Of course, the curve $\widetilde{X}$ is unstable since the new irreducible components are rational with only two special points.

The meromorphic function $\tilde{f}$ on $\widetilde{X}$ is constructed as follows. On each old irreducible component, the restriction of $\tilde{f}$ must have principal parts p_j coinciding with those carried by this component. The restriction of $\tilde{f}$ to the ith new irreducible component is the unique polynomial of degree k_i taking x_i to ∞ and the double point to a prescribed value; the coefficients of this polynomial are determined by the corresponding point of the cone $\mathcal{A}_i$.

This construction is subject to some ambiguity since the principal parts on each irreducible component determine the restriction of the function to this component only up to an additive constant. These additive constants should be chosen so as to make the function $\tilde{f}$ on the entire curve $\widetilde{X}$ continuous, that is, its values at both sheets of each double point must coincide. After this choice is made, there is still one more degree of freedom: we may add a constant to the function as a whole. This last degree of freedom is killed by the requirement that the sum of the critical values of $\tilde{f}$ must be equal to zero. Note, however, that the definition of the set of critical values of a meromorphic function over a nodal curve requires some work, see Sec. 5.3.7 below. It is easy to verify that for each n-tuple of principal parts belonging to $\mathcal{Z}_{g;\kappa}$ the function $\tilde{f}$ thus constructed is a degree k meromorphic function on $\widetilde{X}$.

Since, conversely, any stable meromorphic function of degree k with n poles on a genus g nodal curve with zero sum of critical values determines an n-tuple of principal parts, we arrive at the following statement.

Proposition 5.3.19. *The space $\mathcal{Z}_{g;\kappa}$ coincides with the space of degree k stable meromorphic functions on genus g nodal curves, with poles of order k_i at marked points, having zero sum of critical values. The space $\overline{\mathcal{H}}_{g;\kappa}$ is the space of stable meromorphic functions that are limits of meromorphic functions on smooth curves.*

5.3.7 Extending the LL Mapping to Completed Hurwitz Spaces

The proof of Theorem 5.3.1 proceeds in the same vein as was done for polynomials. Namely, the Lyashko–Looijenga mapping associating to a meromorphic function the unordered tuple of its critical values is a well-defined mapping on the Zariski open dense subset of $\overline{\mathcal{H}}_{g;\kappa}$ consisting of functions with distinct finite critical values. The space $\overline{\mathcal{H}}_{g;\kappa}$ is no longer a vector space in our case, but it is fibered over $\overline{\mathcal{M}}_{g;n}$, and each fiber is a cone.

Theorem 5.3.20. *The* LL *mapping admits a continuous extension to the completed Hurwitz space* $\overline{\mathcal{H}}_{g;\kappa}$, *and the extended mapping is a cone morphism. Moreover, there exists an extension of* LL *to the entire kernel* $\mathcal{Z}_{g;\kappa}$ *of R possessing the same properties.*

This extension associates to a stable meromorphic function $\tilde{f}$ over a nodal curve $\tilde{X}$ described in the previous section the set of critical values, which is the union of the following sets:

- for each irreducible component of $\tilde{X}$ on which $\tilde{f}$ is not a constant, the set of critical values of the restriction of $\tilde{f}$ to this irreducible component;
- for each double point, the value of $\tilde{f}$ at this point taken twice;
- for each irreducible component of $\tilde{X}$ of genus g' on which $\tilde{f}$ is a constant, the value of $\tilde{f}$ on this component taken with multiplicity $2g' - 2$.

Note that since each rational component contracted to a point must have at least three double points, the third requirement does not cause any trouble: if g' is 0, then the corresponding negative multiplicity is compensated by the contribution of the singular points. Thus, we know what the sum of critical values of any meromorphic function is, and can choose an additive constant making it equal to 0.

Now let us verify that the LL mapping thus extended to $\overline{\mathcal{H}}_{g;n}$ is continuous.

Take a double point θ of the curve $\tilde{X}$ and consider the preimage $\tilde{f}(S^1)$ of a small circle S^1 centered at $\tilde{f}(\theta)$. We are interested in the intersection of this preimage with a small neighborhood of θ in $\tilde{X}$. If $\tilde{f}$ is nonconstant on both sheets meeting at θ, then this intersection consists of two circles, one on each sheet. If $\tilde{f}$ is nonconstant only on one sheet, then this intersection is a single circle lying on this sheet. Finally, if $\tilde{f}$ is constant on both sheets, then the intersection in question is empty; we call the corresponding double point an *inner* double point.

A natural number is assigned to each of the circles; this is the degree of the restriction of $\tilde{f}$ to this circle. This degree coincides with the order of the critical point of $\tilde{f}$ on the corresponding sheet increased by 1.

Now consider a holomorphic deformation $F : \mathcal{X} \to \mathbb{C}P^1$ of $\tilde{f}$, where $\mathcal{X} \to \mathbb{C}^1$ is a holomorphic family of marked genus g curves such that the fiber X_τ is smooth for $\tau \neq 0$ and $X_0 = \tilde{X}$, and the restriction f_τ of F to X_τ is a meromorphic function, $f_0 = \tilde{f}$. We do not require that f_τ has zero sum of critical values.

For a fixed double point θ, consider the preimage $F^{-1}(S^1)$ of the same circle S^1 in $\mathbb{C}$ centered at $\tilde{f}(\theta)$. Take the connected component of this preimage containing one of the circles on $\tilde{X}$ constructed above. For a sufficiently small $\tau \neq 0$, the intersection of this connected component with X_τ is a circle, and the degree of its mapping to S^1 under f_τ coincides with the number assigned to the circle on $\tilde{X}$.

302   5 Meromorphic Functions

Applying the procedure described above to all double points of $\widetilde{X}$ we obtain a finite set of circles on each curve X_τ for τ small enough. These circles cut each curve X_τ into connected pieces of three different types:

1. pieces containing marked points; these pieces are in one-to-one correspondence with those irreducible components of $\widetilde{X}$, on which $\tilde{f}$ is nonconstant;
2. pieces holomorphically equivalent to the annulus; these pieces are in one-to-one correspondence with the double points of $\widetilde{X}$ where $\tilde{f}$ is nonconstant on both sheets meeting at this point;
3. pieces of positive genus without marked points; these pieces are in one-to-one correspondence with the connected components of the curve $X' \subset \widetilde{X}$, the constant locus of $\tilde{f}$.

Now let us follow the behavior of the critical points and critical values of f_τ on pieces of all three types. Let the index j run over the pieces.

(1) By the Riemann–Hurwitz formula, the number of critical points of f_τ on a piece of the first type is equal to $K_j + n_j + 2g_j - 2 - D_j + c_j$, where K_j is the total order of all poles on this piece, n_j is the number of marked points, g_j is the genus of the corresponding component of $\widetilde{X}$, D_j is the total degree of the circles bounding the piece, and c_j is the number of these circles. As τ tends to 0, these critical points tend to the critical points of $\tilde{f}$ on the corresponding component of $\widetilde{X}$ with the double points excluded, and the critical values tend to that of $\tilde{f}$.

(2) The mapping f_τ takes an annulus without marked points to the small disk bounded by S^1. The degree of the mapping is the sum of the integers assigned to the boundary circles. By the Riemann–Hurwitz formula, the number of critical points of f_τ on the annulus coincides with this degree. As τ tends to 0, these critical points tend to the double point, and the critical values tend to the value of $\tilde{f}$ at the double point.

(3) The mapping f_τ takes a piece of the third type to a disk. The degree of this mapping equals D_j, the sum of the numbers assigned to the boundary circles. The number of critical values on this piece is $2g_j - 2 + D_j + c_j$, where c_j is the number of boundary circles, and g_j is the (arithmetic) genus of the corresponding connected component $X' \subset \widetilde{X}$. Further, $2g_j - 2 = 2\sum(g_{ji} - 2) + 2l$, where l is the number of inner double points on X', and g_{ji} are the genera of irreducible components of X'. As τ tends to 0, the critical values tend to the constant value of $\tilde{f}$ on the limit curve.

Now, taking the union of all limit critical values over all pieces we conclude that the set of critical values of f_τ tends precisely to the set described in the beginning of the proof.

5.3.8 Computing the Top Segre Class; End of the Proof

The image
$$\mathcal{D} = \{t^{k+n+2g-2} + d_2 t^{k+n+2g-4} + \cdots + d_{k+n+2g-2}\}$$
of the LL mapping is a cone over a point. We showed above that the LL

5.3 Rigid Classification and Geometry of Moduli Spaces

mapping LL : $\mathcal{Z}_{g;\kappa} \to \mathcal{D}$ is a cone morphism. Having the LL mapping on $\mathcal{Z}_{g;\kappa}$, how can one compute the degree of this mapping restricted to $\overline{\mathcal{H}}_{g;\kappa}$? One way to do this is to consider a line in the space of critical values as an intersection of $k + n + 2g - 4$ generic hyperplanes. The preimage of such a hyperplane under the LL mapping is a hypersurface in $\overline{\mathcal{H}}_{g;\kappa}$. The intersection of this hypersurface with a fiber of the projection $\overline{\mathcal{H}}_{g;\kappa}$ is a homogeneous affine hypersurface in this fiber, whence it defines a class in $H^2(P\overline{\mathcal{H}}_{g;\kappa})$ which is proportional to $c_1(\mathcal{O}(1))$.

Thus, we are in a situation very similar to the computation of the degree of the LL mapping for polynomials in Sec. 5.2. The only (but essential) difference consists in the fact that for polynomials the homology class of a subvariety was totally determined by the degree of the subvariety since the ambient space was simply the projective space. In the present case, however, we deal with a projectivization of a cone over the moduli space, and the structure of its (co)homology is much more complicated.

Counting the number of preimages of a generic point under the LL mapping, we must iterate the class $c_1(\mathcal{O}(1))$ over $P\overline{\mathcal{H}}_{g;\kappa}$ the maximal possible number of times, i.e., $k + n + 2g - 4$ times. In other words, we must compute the value of the top Segre class $s_{\text{top}}(\overline{\mathcal{H}}_{g;\kappa})$ on the base.

We know the total Segre class of the cone $\mathcal{Z}_{g;\kappa} \to \overline{\mathcal{M}}_{g;n}$ because of the short exact sequence of cones

$$0 \longrightarrow \mathcal{Z}_{g;\kappa} \longrightarrow \mathcal{P} \xrightarrow{R} \Lambda_{g;n}^{\vee} \longrightarrow 0,$$

which gives

$$s(\mathcal{Z}_{g;\kappa}) = s(\mathcal{P})c(\Lambda_{g;n}^{\vee}) = \prod_{i=1}^{n} \frac{k_i!}{k_i^{k_i}} \frac{1 - \lambda_1 + \cdots + (-1)^g \lambda_g}{(1 - k_1\psi_1)\dots(1 - k_n\psi_n)}.$$

(In fact, the justification of this formula requires some additional work, see the details in [92].) This formula does not allow one to compute the total Segre class of $\overline{\mathcal{H}}_{g;\kappa}$, but it yields the *top* Segre class, which coincides with that of $\mathcal{Z}_{g;\kappa}$. Indeed, all the irreducible components of $\mathcal{Z}_{g;\kappa}$ distinct from $\overline{\mathcal{H}}_{g;\kappa}$ consist of functions over singular curves; these functions inevitably have coinciding critical values, the image of these components under the LL mapping belongs to the discriminant in the space of polynomials, and, therefore, the line bundle $\mathcal{O}(1)$ over the projectivizations of these components is induced from a line bundle over a variety of dimension less than $k + n + 2g - 4$.

Now, the degree of the LL mapping can be computed as the ratio of the values of the top Segre classes of the two cones, $\overline{\mathcal{H}}_{g;\kappa}$ and $\mathcal{D}$ on the base spaces. The second value is

$$\frac{1}{(k+n+2g-2)!}$$

since the cone is just a weighted projective space with the weight of the coordinates d_i, $i = 2, \dots, k + n + 2g - 2$, being i. Therefore, the degree of the LL mapping is

$$(k+n+2g-2)! \prod_{i=1}^{n} \frac{k_i!}{k_i^{k_i}} \int_{\mathcal{M}_{g;n}} \frac{1 - \lambda_1 + \cdots + (-1)^g \lambda_g}{(1 - k_1 \psi_1) \ldots (1 - k_n \psi_n)}.$$

In order to compute the number of isomorphism classes of meromorphic functions recall that the number of ocurrences of each class in the Hurwitz space $\overline{\mathcal{H}}_{g;\kappa}$ is $|\mathrm{Aut}(\kappa)|$. Indeed, in associating a tuple of principal values to a meromorphic function we were free in marking the poles, and poles of the same order are indistinguishable. After dividing the above ratio by this number we obtain exactly the right-hand side of Eq. (5.13).

Theorem 5.3.1 is proved.

5.4 The Braid Group Action

The study of the flexible classification of complex polynomials was begun in the 19th century by Lüroth (1871) [205], Clebsch (1873) [56], and Hurwitz (1891) [144]. In the 20th century the problem reappeared after a series of short notes [77], [277], [10], and [304]. In Sec. 5.4.5 we also mention several subsequent publications, but our list probably remains incomplete. However, very few things are known today about the problem.

The notion of flexible equivalence of ramified coverings of the sphere was introduced in Definition 1.2.27 at the end of Sec. 1.2.4. The problem thus consists in describing equivalence classes of ramified coverings under this relation. In Sec. 1.2.4 we have also shown how this problem is related to the study of the orbits of braid groups action on constellations. However, concrete results are known mainly for polynomials. Therefore, in this section we will mostly study polynomials. In practical terms, until 1996 the polynomials were classified only up to degree 6. A vast computer experiment conducted in 1996–1999 (see [42], [310]), in combination with enumerative and group-theoretic methods, allowed us to achieve a classification of all polynomials up to degree 11, and also to obtain partial results for bigger degrees. These new data change entirely the global view of the problem and permit to formulate some plausible conjectures which may eventually lead to a solution of the problem.

5.4.1 Braid Groups

In Construction 1.1.17 we introduced an action of the braid group B_k on k-constellations. It is high time to introduce the group itself. A standard reference for the theory of braid groups is [33].

Let us take $\mathbb{R}^3 = \mathbb{R}^2 \times \mathbb{R}$, and fix k pairwise distinct points $p_1, \ldots, p_k$ on $\mathbb{R}^2$; let them lie, say, on the x-axis in the prescribed order. We denote the vertical axis by t; it will be convenient for us to orient it upside down. Let us fix a segment $[0, a]$, $a > 0$, on the t-axis.

5.4 The Braid Group Action

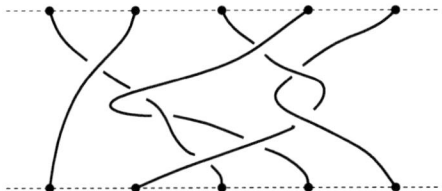

Fig. 5.1. A typical braid

Definition 5.4.1 (Braid). A *string* (the i-th string) is a continuous mapping $s_i : [0, a] \to \mathbb{R}^2$ such that $s_i(0) = p_i$, and $s_i(a) = p_j$ for some $j \in \{1, \ldots, k\}$. A *braid* β is a collection of k strings such that for every value of $t \in [0, a]$ the k values $s_1(t), \ldots, s_k(t)$ are distinct. Clearly, the mapping $\tau : i \mapsto j$ is a permutation $\tau \in S_k$; it is called the permutation *associated* to the braid β. Finally, two braids are *equivalent* if one of them may be obtained from the other one by an isotopy of the set $\mathbb{R}^2 \times [0, a]$ preserving the points $(p_i, 0)$ and (p_i, a) on the upper and lower planes, and/or by a linear "rescaling" $[0, a] \to [0, b] : t \mapsto (b/a)t$.

Remark 5.4.2. Usually the term braid is used to refer to an equivalence class of braids.

An example of a braid on 5 strings is shown in Fig. 5.1. The multiplication of two braids is carried out by attaching the starting points of the second braid to the end-points of the first one: see Fig. 5.2. (Now it is clear why the vertical axis is oriented from top to bottom: we usually draw pictures on the paper in this direction.)

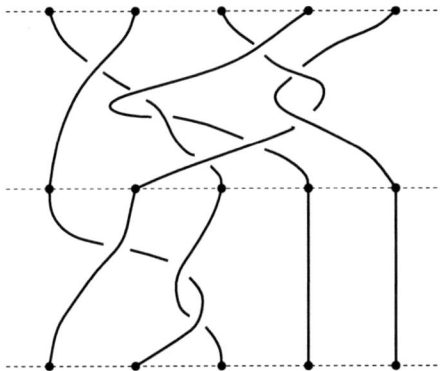

Fig. 5.2. Multiplication of braids

It is easy to see that this multiplication, indeed, creates a group, which is called the *braid group* on k strings. Fig. 5.3 shows the identity element, the (usually chosen) generators σ_i, $i = 1, \ldots, k-1$, their inverses σ_i^{-1}, and also illustrates two relations:

$$\sigma_i \sigma_j = \sigma_j \sigma_i \quad \text{for} \quad |i - j| \geq 2$$

and

$$\sigma_i \sigma_{i+1} \sigma_i = \sigma_{i+1} \sigma_i \sigma_{i+1}.$$

These relations are sufficient in order to define the braid group, though it is not at all easy to see that all other relations are consequences of these ones: see [133], Sec. 4 of Chapter 1 and Appendix 1.

Obviously, when braids are multiplied, their associated permutations are also multiplied. This leads to a homomorphism of the braid group B_k onto the permutation group S_k. (Geometrically, the group B_k "becomes" S_k if we forget the undercrossings and overcrossings of the strings; algebraically, the same effect is achieved by adding the relations $\sigma_i^2 = \text{id}$, $i = 1, \ldots, k-1$.) The preimage under this homomorphism of the identity element in S_k is called the *pure braid group*; this group is denoted by P_k. In other words, we have the following short exact sequence of groups and group homomorphisms:

$$1 \longrightarrow P_k \longrightarrow B_k \longrightarrow S_k \longrightarrow 1.$$

Being the kernel of a homomorphism, P_k is a normal subgroup of B_k.

Instead of taking k points on $\mathbb{R}^2$ we may as well take them on any 2-dimensional surface. This would give us another group, which is often called the *surface braid group*. Taking, for example, the sphere S^2 instead of the plane, we obtain the so-called *Hurwitz braid group*, or the *sphere braid group*. It is denoted by H_k. This group has the same generators and relations as above plus one additional relation:

$$\sigma_1 \sigma_2 \ldots \sigma_{k-2} \sigma_{k-1}^2 \sigma_{k-2} \ldots \sigma_2 \sigma_1 = \text{id}. \tag{5.20}$$

This relation is illustrated in Fig. 5.4: the trajectory of the first point can be contracted on the sphere (being pushed to its "opposite side"), while on the plane such a contraction would be impossible.

For the surfaces of genus $g \geq 1$ the relations in the corresponding braid groups are more complicated: see, for example, [251]. We just make an obvious but rather enlightening remark: *a surface braid group on one string is nothing else but the fundamental group of the corresponding surface*. Thus for $g \geq 1$ even the braid group on one string is not trivial.

It is difficult to convey, by this short presentation, the fundamental nature and the importance of braid groups in many branches of mathematics. What is important for us is the following interpretation of braids. The horizontal plane $\mathbb{R}^2$ or the sphere S^2 may be seen as the complex plane or the complex sphere respectively; the points $p_1, \ldots, p_k$ may be regarded as complex numbers; the

5.4 The Braid Group Action   307

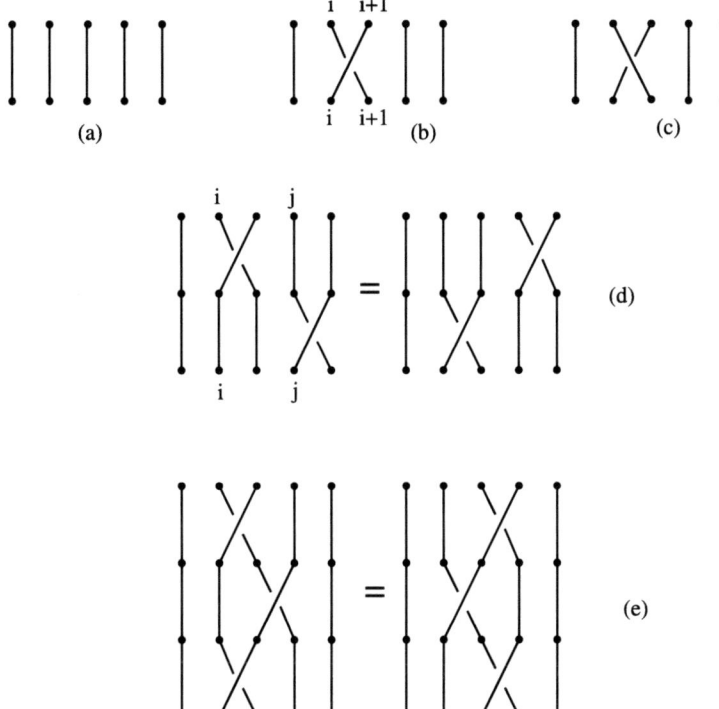

Fig. 5.3. Generators and relations in the braid group: (a) the identity; (b) the generator σ_i; (c) σ_i^{-1}; (d) relation $\sigma_i\sigma_j = \sigma_j\sigma_i$; (e) relation $\sigma_i\sigma_{i+1}\sigma_i = \sigma_{i+1}\sigma_i\sigma_{i+1}$

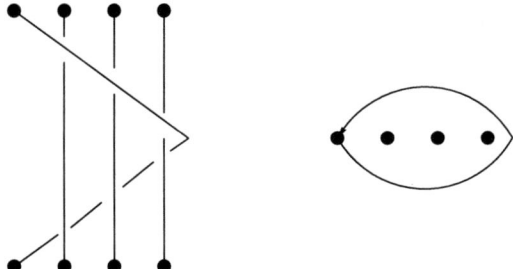

Fig. 5.4. The braid of the additional relation in H_k, and the same braid seen "from above"

t-axis may be considered as an axis representing the time; and finally, a braid itself represents a movement of the k points along the complex plane or the sphere until they arrive at the same positions *as a set* (while individual points may exchange their places). The generator σ_i makes the points p_i and p_{i+1} exchange their positions, while leaving all the other points fixed.

Now we may refer the reader once more to the end of Sec. 1.2.4: the action of the generator σ_i on constellations corresponds exactly to the exchange of places of two critical values y_i and y_{i+1}. This observation proves in fact one of the two statements of Theorem 1.2.28, namely: *if two constellations belong to the same orbit of the braid group action, then the corresponding coverings are flexibly equivalent.* This proposition was proved by Hurwitz in [144].

Remark 5.4.3. The critical values of a meromorphic function live on a complex sphere. Therefore it is rather the Hurwitz braid group H_k that acts on constellations. But this observation does not contradict the fact that the same action is also an action of the braid group B_k. An additional relation means only that we obtain the same constellations more often than we would probably suppose. But anyway our groups are infinite while their orbits are finite, and therefore quite often we obtain the same constellations without any profound reasons.

The inverse statement of Theorem 1.2.28 was proved by Zdravkovska in [304]. The proof is based on the following theorem due to H. Kneser [173]:

Theorem 5.4.4. *The space of homeomorphisms of the 2-sphere is path connected.*

According to this theorem, if $v : S^2 \to S^2$ is a homeomorphism, then there exists a family V_t of homeomorphisms of S^2 depending on a parameter $t \in [0, a]$ such that V_0 is the identity and $V_a = v$, and such that V_t is "continuous" as a function of t, whatever that means. We do not specify the topology on the space of homeomorphisms. What should, however, be true under any reasonable definition of such a topology is the following fact: taking a point $p \in S^2$, its trajectory $V_t(p)$, $t \in [0, a]$, is a continuous function of t. Applying this principle to the set of critical values $y_1, \ldots, y_k$, and taking into account that the homeomorphism v preserves this set (as a set), we conclude that the collection of the trajectories of $V_t(y_1), \ldots, V_t(y_k)$ is a braid. This finishes the proof of Theorem 1.2.28.

Remark 5.4.5. In Sec. 5.2.1 we introduced a stratification of the space of polynomials. We may now relate this stratification to the notion of flexible classification. In fact, the closure of a flexible equivalence class is an *irreducible component* of the stratification. Indeed, for algebraic varieties over $\mathbb{C}$ to be irreducible means to remain connected after the removal of singular points, and singular points correspond to sub-strata described by more degenerate flexible equivalence classes.

5.4.2 Braid Group Action on Cacti: Generalities

5.4.2.1 Polynomials and Cacti

Let us first recall that, while working with polynomials, we usually denote by k the number of *finite* critical values, while the total number of critical values, including infinity, is $k+1$. Recall also that, combinatorially, polynomials are represented by cacti (see Definition 1.6.3 in Sec. 1.6.2). A cactus is a sequence of permutations $[g_1, \ldots, g_k]$ such that their product is a cyclic permutation,

$$g_1 \ldots g_k = c = (1, 2, \ldots, n),$$

also satisfying the *planarity condition*: the total number of cycles (including the fixed points) of all the permutations g_i is equal to $(k-1)n + 1$ (or, equivalently, using the notation $v(g) = n - \#(\text{cycles in } g)$ introduced in Notation 1.1.9, we have $\sum_{i=1}^{k} v(g_i) = n - 1$). Here the cyclic permutation c corresponds to the ramification at infinity. Note that the presence of c automatically guarantees the transitivity of the action of the group generated by the permutations g_i, $i = 1, \ldots, k$, on n points.

The point at infinity plays a specific role here. In particular, while performing a homeomorphism $v : S^2 \to S^2$, we may not touch it at all: we may not put it together with other critical values. This fact permits us to act only on $g_1, \ldots, g_k$, and the action itself becomes that of B_k, not of H_{k+1}. Let us also introduce a more compact notation for a passport.

Notation 5.4.6. If the same partition $\lambda \vdash n$ is repeated in a passport p times, then, instead of writing

$$[\underbrace{\lambda, \lambda, \ldots, \lambda}_{p \text{ times}}, \mu, \ldots],$$

we will write $[p \times \lambda, \mu, \ldots]$. Also, we will not write the partition $n \vdash n$ corresponding to the cyclic permutation c. For example, instead of writing

$$[2^2 1^8, 2^2 1^8, 2^2 1^8, 2^2 1^8, 21^{10}, 21^{10}, 21^{10}, 12]$$

we will write

$$[4 \times 2^2 1^8, 3 \times 21^{10}].$$

5.4.2.2 The Action and its Invariants

The operations σ_i ($i = 1, \ldots, k-1$) act on a cactus $[g_1, \ldots, g_k]$ in the usual way:

$$\sigma_i(g_i) = g_{i+1}, \quad \sigma_i(g_{i+1}) = g_{i+1}^{-1} g_i g_{i+1} = g_i^{g_{i+1}}, \quad \sigma_i(g_j) = g_j \quad \text{for} \quad j \neq i, i+1.$$

Our goal is to study the orbits of this action. The action obviously preserves the product $g_1 \ldots g_k = c$. It is clear that the following objects are also invariant under the action:

- the cartographic group $G = \langle g_1, \ldots, g_k \rangle$ of the cactus $[g_1, \ldots, g_k]$, which is also the monodromy group of the corresponding polynomial; this group is an invariant not only as an abstract group but also as a particular subgroup $G \leq S_n$;
- the unordered refined passport $[K_1, \ldots, K_k]$ of $[g_1, \ldots, g_k]$ in G: the operation σ_i permutes the conjugacy classes K_i and K_{i+1};
- the unordered passport: the operation σ_i permutes the partitions λ_i and λ_{i+1}.

Usually we fix a passport (in terms of polynomials, we fix the "branch data", that is, the multiplicities of the preimages of the critical values), and try to find all the corresponding orbits. Recall that we call a *family* the union of all the orbits corresponding to a given passport.

5.4.2.3 General Criteria

First of all note that the case of Shabat polynomials, that is, the case of $k = 2$, from the point of view of the braid group action is trivial. Indeed, the group B_2 has only one generator σ_1, and the action of σ_1 on a tree $[g_1, g_2]$ gives $[g_2, g_1^{g_2}] = [g_2^{g_2}, g_1^{g_2}]$. Geometrically this means that we must first change black to white and white to black, taking $[g_2, g_1]$; and then move the labels one step along the border of the tree using the rotation around the former white vertices.

The situation becomes even simpler if we apply σ_1^2: taking into account that $g_1 g_2 = c$, we get

$$[g_1^{g_2}, g_2^{g_1^{g_2}}] = [g_1^{g_2}, g_2^{g_1 g_2}] = [g_1^{g_1 g_2}, g_2^{g_1 g_2}] = [g_1^c, g_2^c].$$

This tree is obviously isomorphic to the initial tree $[g_1, g_2]$. We may conclude that for $k = 2$ there are as many orbits as there are trees. (We used to color the vertices of trees in black and white; but this time after exchanging the colors we remain in the same orbit.)

In what follows we consider only the case $k \geq 3$ (if we also count infinity, then the number of critical values is ≥ 4).

The following two theorems may be found in Khovanskii and Zdravkovska [169]. For a passport $\pi = [\lambda_1, \ldots, \lambda_k]$ let us denote the parts of the partition λ_i by d_{ij}, $j = 1, \ldots, m_i$, $\sum_{j=1}^{m_i} d_{ij} = n$. The planarity condition may be rewritten as $\sum_{i,j}(d_{ij} - 1) = n - 1$. Recall that a critical value is *simple*, or *nondegenerate*, if the corresponding partition is 21^{n-2}.

Definition 5.4.7 (Defect). The *defect* of a polynomial passport is the number

$$D = \sum_{d_{ij} > 1} d_{ij},$$

where the sum is taken over all *non-simple* critical values. (For example, for generic polynomials $D = 0$.)

5.4 The Braid Group Action 311

Theorem 5.4.8. *If $D \leq n+1$, then there exists a single orbit of the braid group action (and therefore there exists a single class of flexible equivalence of polynomials).*

Example 5.4.9. Take $n = 12$ and consider the passport $[521^5, 3^2 1^6, 2 \times 21^{10}]$. We have $D = 5 + 2 + 3 + 3 = 13$ (we repeat that the two simple critical values do not count) and therefore there exists a single orbit.

Remark 5.4.10. Simple arithmetic considerations show that the maximal value of k such that there exist polynomial passports with $D > n + 1$, is equal to $k_{\max} = \lfloor 3(n-2)/4 \rfloor$: for bigger k the orbit is always unique. Taking, for example, $n = 13$, we obtain $k_{\max} = \lfloor 33/4 \rfloor = 8$, and one of the corresponding passports is $[4 \times 2^2 1^9, 4 \times 21^{11}]$ (with the defect $D = 16$). For $n = 13$ and $k = 9$ the maximal defect is attained by the passport $[3 \times 2^2 1^9, 6 \times 21^{11}]$: here D is only equal to 12.

Theorem 5.4.11. *If for every critical value there exists only one critical point, that is, if all the partitions in the passport are of the form $m1^{n-m}$, $m \geq 2$, then the braid group orbit is unique, and therefore there exists a single class of flexible equivalence of polynomials.*

Example 5.4.12. The passport $[61^6, 51^7, 31^9]$ for $n = 12$ does not satisfy the assumptions of Theorem 5.4.8: its defect is $D = 14$. Nevertheless, there exists only one orbit for this passport because it satisfies the assumptions of Theorem 5.4.11.

The proofs of both Theorems 5.4.8 and 5.4.11 are based on the same idea; it is rather topological than combinatorial. We know that the critical values of a ramified covering may be chosen arbitrarily (Riemann's existence theorem), while the critical points are subordinated to strong algebraic constraints (recall the systems of algebraic equations we had to solve in order to find Shabat polynomials in Chapter 2). But the assumptions of Theorems 5.4.8 and 5.4.11 ensure that the critical points, if not all of them then at least the "most important" ones, can also be chosen arbitrarily. This leads to the path connectedness of the space of the corresponding polynomials, and thus to the unicity of their class of flexible equivalence.

To prove Theorem 5.4.8 we will use the Lagrange interpolation procedure. Let us fix a passport π with a defect $D \leq n+1$, and look for a polynomial $P(x) = a_0 + a_1 x + \ldots + a_n x^n$ having this passport. The coefficients $a_0, \ldots, a_n$ will be $n + 1$ unknown parameters. Choose arbitrarily a critical point x_1 of multiplicity d, and a critical value y_1. These data provide us with d linear equations on parameters $a_0, \ldots, a_n$:

$$P(x_1) = y_1, \quad P'(x_1) = 0, \quad \ldots, \quad P^{d-1}(x_1) = 0.$$

We collect all such equations for all *non simple* critical values, thus obtaining D equations. To get the missing $n + 1 - D$ equations we do not use the remaining simple critical values as one might suppose, but take arbitrary $n+1-D$ regular (i.e., non critical) points and the same number of arbitrary non critical values. The necessary number of extra critical points will appear automatically just because we deal with polynomials.

We thus get a system of $n+1$ linear equations in $n+1$ variables. The uniqueness of the solution follows from the fact that if P_1 and P_2 are two such solutions, then $P_1 - P_2$ is a polynomial of degree $\leq n$ having at least $n+1$ roots, counting with multiplicities. The uniqueness implies that the matrix of the system is non degenerate, which in its turn implies the existence of a solution.

For some choices of the above data the passport of the resulting polynomial P may differ from π, because extra degeneracy might arise (extra critical points may merge, or their corresponding critical values may do so). But, in the space of parameters $a_0, \ldots, a_n$, this may happen only on a union of affine subspaces of real codimension ≥ 2. Therefore, the space of polynomials with the given passport is path connected.

To prove Theorem 5.4.11, consider a polynomial P corresponding to a passport $\pi = [m_1 1^{n-m_1}, \ldots, m_k 1^{n-m_k}]$, and let $\alpha_1, \ldots, \alpha_k$ denote the critical points of P. Then we may compute the derivative:

$$P'(x) = C \prod_{i=1}^{k} (x - \alpha_i)^{m_i - 1}.$$

This is a polynomial of degree $n-1$, as it should be, because the planarity condition in this case has the form $\sum_{i=1}^{k}(m_i - 1) = n - 1$. We see that we may choose $\alpha_1, \ldots, \alpha_k$ arbitrarily, and then compute $P(x)$ as the integral of $P'(x)$ (cf. Exercise 5.2.7 and Sec. 5.2.4).

Once more we may get some extra degeneracy, but it can take place only on a subvariety of real codimension ≥ 2. Thus the space of polynomials with the passport π is path connected.

5.4.3 Experimental Study

5.4.3.1 Preliminary Remarks

Nowadays, when the information concerning certain phenomenon is scarce, we may resort to a computer experiment. Here we give a brief account of such an experiment (see [42], [310]), which turned out to be a mixture of algorithmic, combinatorial, and group theoretic approaches.

How to compute an orbit. The principle is simple. We fix a passport, we take a cactus having this passport, and start to apply the operations σ_i. In this process we obtain new cacti, and we apply the operations to them as well. When the application of all the σ_i to all the cacti previously obtained gives nothing new, the resulting set of cacti is an orbit.

The role of an enumerative formula. Imagine that you computed an orbit, and it contains, say, 125945 cacti. Unfortunately, this information does not allow you to conclude if there are other orbits corresponding to the same passport or not. The question which remains open is, should you or should you not spend your time and energy in order to find the "remaining" cacti with this passport. And even if you find some new cacti, what is your criterion that permits you to stop the search?

The situation changes decisively if you have an enumerative formula giving you the total number of cacti with the given passport. If the formula tells you that there are, in total, 125945 cacti, then you are done: there is a single orbit, and a single equivalence class of polynomials. If, on the contrary, the formula tells you that the total number of cacti is 126000, then you know that the job is not finished. You have to find the missing 55 cacti (which may not be an easy thing to do). It goes without saying that in such a case it would also be desirable to find the reason for the splitting of this family of cacti into several orbits.

The formula we need was already given in Chapter 1: see Eq. (1.6) (it was also largely discussed in previous sections of this chapter). The concrete form of the formula is of no importance here: the important thing is *to have* such a formula. Just note that the formula enumerating the cacti with a given passport first appeared in 1992 (see [114]). This is the main reason why the experimental study of the flexible classification started only in 1995. For rational functions the enumerative formulas are known only for some particular cases (and even when such formulas are known, there are many other interesting problems!). Therefore, the experimental approach for this case is in the bud.

5.4.3.2 About the Algorithm

Algorithmic aspects of computing the orbits of a group action constitute a subject interesting in itself. It deserves a detailed exposition, but not in this book. Here we limit ourselves to a few remarks concerning some particular features of the algorithm.

Rooted vs. non-rooted cacti. The orbits as described above consist of rooted cacti. We can, however, make them consist of non-rooted cacti, and thus divide the size of an orbit by n. (The size of an orbit consisting of symmetric cacti having the symmetry of order p is divided by n/p. Anyway, symmetric and asymmetric cacti could never belong to the same orbit, because the cartographic group of a symmetric cactus contains a non-trivial center, while that of an asymmetric one does not.) Let us apply to an arbitrary cactus the operation $\sigma_1 \ldots \sigma_{k-1}$. We get

$$[g_1, g_2, g_3, \ldots, g_k] \mapsto [g_2, g_1^{g_2}, g_3, \ldots, g_k]$$
$$\mapsto [g_2, g_3, g_1^{g_2 g_3}, \ldots, g_k]$$
$$\mapsto \ldots$$
$$\mapsto [g_2, g_3, \ldots, g_k, g_1^{g_2 \cdots g_k}]$$
$$= [g_2, g_3, \ldots, g_k, g_1^{g_1 g_2 \cdots g_k}]$$
$$= [g_2, g_3, \ldots, g_k, g_1^c],$$

where $c = g_1 \ldots g_k = (1, \ldots, n)$, as was denoted before. Therefore, the operation $(\sigma_1 \ldots \sigma_{k-1})^k$, which is, by the way, the generator of the center of B_k, transforms $[g_1, g_2, \ldots, g_k]$ into $[g_1^c, g_2^c, \ldots, g_k^c]$. Thus, an orbit containing a cactus contains also all the cacti isomorphic to it.

Of course, a non-rooted cactus, which is itself an equivalence class of n isomorphic cacti, could be put in a computer memory only in the form of a representative of this class. It is a good idea to find a canonical representative: for example, the smallest one in the lexicographic order. Then to compare two cacti will mean to verify their equality and not their isomorphism.

Ordered vs. unordered passports. The operation σ_i permutes the partitions λ_i and λ_{i+1} in a passport. Therefore, all the cacti corresponding to all the differently ordered passports that may be obtained from the given one by permuting its partitions, belong to the orbit. Let us take for example the passport $\pi = [2 \times 2^2 1^8, 2 \times 31^9, 3 \times 21^{10}]$, $n = 12$. It gives rise to

$$\frac{7!}{2!2!3!} = 210$$

different orderings of the 7 partitions in question. Therefore, the corresponding orbit becomes 210 times bigger than it possibly might be if we were more clever. From the algorithmic point of view this situation is absolutely unacceptable. Not only do we need a huge amount of memory, but also the computation time increases with the growth of the orbit size.

A natural idea would be to act not by the full braid group B_k, but by the pure braid group P_k. There are, however, two unpleasant features concerning this idea.

The first one is computational. While B_k has $k-1$ generators, the minimal number of generators for P_k is $k(k-1)/2$. Even worse, the usual generators of P_k have the form

$$a_{ij} = \sigma_i \ldots \sigma_{j-1} \sigma_j^2 \sigma_{j-1}^{-1} \ldots \sigma_i^{-1}, \quad 1 \leq i \leq j \leq k-1.$$

We know only how to apply directly a "simple" generator σ_i; an application of all the a_{ij} means applying these simple generators $(k^3 - k)/2$ times. For $k = 7$, as in the example above, this means performing 112 operations instead of 6.

The second difficulty is much more profound: the action of P_k may lead to wrong results.

5.4 The Braid Group Action 315

Example 5.4.13. Let us take the passport $[2^5, 2 \times 2^2 1^6]$, $n = 10$. Acting by P_3 we obtain 5 orbits of sizes 10, 5, 5, 5, and 2. Unfortunately, while acting by B_3, we don't get 5 orbits of sizes 30, 15, 15, 15, and 6 respectively, as one might suppose: two of the orbits fuse, and we get only 4 orbits, of sizes 30, 30, 15, and 6. Needless to say that there are 4, and not 5 classes of flexible equivalence of polynomials.

We will see later that in the case of fusion a kind of "fine structure" of the two branchings of type $2^2 1^6$ is different.

What we need in fact is the subgroup of the braid group which consists of all the elements preserving a passport. Let $\mu \vdash k$ be a partition of k, $\mu = (m_1, \ldots, m_r)$. Consider a passport π of the form $[m_1 \times \lambda_1, \ldots, m_r \times \lambda_r]$, where all λ_i, $i = 1, \ldots, r$, are supposed to be distinct. While B_k projects to S_k, and P_k projects to id $\in S_k$, we need a subgroup of B_k which is the preimage, under the same homomorphism, of the subgroup $S_\mu = S_{m_1} \times \ldots \times S_{m_r} \leq S_k$. The group S_μ is often called *Young*, or *diagonal subgroup* of S_k. Let us call it a *diagonal braid group*, and denote D_μ. It is the biggest subgroup of B_k preserving the passport π considered above.

Now we need to find a set of generators of D_μ. The question may be formulated in a more general way.

Problem 5.4.14. Let $G \leq S_k$ be a permutation group. Find a set of generators of the subgroup $F \leq B_k$ which is the preimage of G under the group homomorphism $B_k \to S_k$.

The solution is given by the following procedure.

1. Take a set of permutations generating G.
2. Represent them as products of transpositions $s_i = (i, i+1)$.
3. Replace all s_i with σ_i.
4. Add the generators a_{ij} of the pure braid group P_k.

One of our goals is thus achieved: the orbits constructed by the group D_μ are correct. However, from the computational point of view we have only worsened the situation: we have added some extra generators to those of the pure braid group P_k. But quite often, after performing the above four operations, it is possible to perform the 5th one:

5. Simplify the above set of generators.

For example, if we already have σ_1 and σ_2, we don't need $\sigma_1 \sigma_2^2 \sigma_1^{-1}$ any more.

When applied to a diagonal subgroup $S_\mu = S_{m_1} \times \ldots \times S_{m_r} \leq S_k$, this procedure gives the following generators of D_μ: (1) the usual σ_i inside each segment of length m_l; (2) for any two such segments, a generator a_{ij} linking them (only one such generator for each pair of segments is needed).

For the passport $[2 \times 2^2 1^8, 2 \times 31^9, 3 \times 21^{10}]$ considered above we get the following generators:

- σ_1 (for the first segment of length 2);
- σ_3 (for the second segment of length 2);
- σ_5, σ_6 (for the segment of length 3);
- $a_{22} = \sigma_2^2$ (a link between the first two segments);
- $a_{44} = \sigma_4^2$ (a link between the second and the third segment);
- $a_{24} = \sigma_2 \sigma_3 \sigma_4^2 \sigma_3^{-1} \sigma_2^{-1}$ (a link between the first and the third segments).

As a bonus, we now have to apply the "simple" operations σ_i only 14 times instead of 112, as it was the case on page 314.

Balanced trees. Finally, we would like to mention that we represent an orbit in a computer memory using the classical data structure of balanced trees. Various species of balanced trees are described, for example, in [174], Chapter 6, or [67], Chapter 14. This data structure permits us to construct an orbit of size N in $O(N \log N)$ time, instead of the $O(N^2)$ complexity of a naïve algorithm.

Software using all the above remarks was implemented by D. Bouya (see [42]).

5.4.3.3 Results of Computations

For the degrees $n \leq 11$ there are 644 passports not covered by the general theorems of Sec. 5.4.2.3 and which therefore need a computer treatment. Among them there turned out to be 34 cases of non-uniqueness. They are summarized in Table 5.1. Note that for the degree $n = 11$ there is a single orbit for any passport, a result that is easy to state but which demanded long computational efforts.

The orbit sizes are given for *ordered* passports (that is, the orbits correspond to the action of the diagonal braid groups), and the numbers given are the numbers of *non-rooted* cacti.

Limits of the computational approach. For the degree 12, the number of passports that need to be processed is 833, which exceeds all the previous degrees taken together. The biggest putative orbit that needs to be computed corresponds to the passport $\pi = [4 \times 2^2 1^8, 3 \times 2^1 1^{10}]$ and is of size 102036672. This exceeds by 50 times the memory of the biggest computer available to us. Furthermore, the estimated computation time of this orbit is about 2 years. It is clear that we need radically new ideas in order to treat the case $n = 12$ *completely*.

However, we do have partial results for the degrees 12 and bigger.

Observations. Looking closer at the table, we see that 33 cases of non-uniqueness out of 34 are explained by symmetry and/or composition. In fact, the same is true for all the cases found up to now except for the three "exceptional" passports discussed in Sec. 5.4.4.2 and for the splitting types in A_n

5.4 The Braid Group Action   317

Table 5.1. The list of polynomial passports of degree $n \leq 11$ with a non-unique orbit (for $n = 11$ such passports don't exist)

	n	k	Passport	number of orbits	Orbit sizes	Reason of non-uniqueness
1	6	3	$[2 \times 2^2 1^2, 21^4]$	2	12, 3	Symmetry
2	7	3	$[3 \times 2^2 1^1]$	4	21, 21, 7, 7	Exceptional passport
3	8	3	$[3^2 1^2, 2^2 1^4, 21^6]$	2	28, 4	Symmetry
4	8	3	$[2^4, 2^2 1^4, 21^6]$	2	4, 2	Symmetry
5	8	3	$[41^4, 2 \times 2^2 1^4]$	2	48, 4	Symmetry
6	8	3	$[2^3 1^2, 2 \times 2^2 1^4]$	2	96, 8	Symmetry
7	8	4	$[3 \times 2^2 1^4, 21^6]$	2	992, 16	Symmetry
8	9	3	$[2 \times 2^3 1^3, 31^6]$	2	99, 3	Symmetry
9	9	3	$[421^3, 2^3 1^3, 21^7]$	2	117, 3	Composition
10	9	3	$[2^4 1, 2^3 1^3, 21^7]$	2	27, 3	Composition
11	9	3	$[2 \times 2^3 1^3, 2^2 1^5]$	2	297, 3	Composition
12	9	4	$[2 \times 2^3 1^3, 2 \times 21^7]$	2	891, 9	Composition
13	10	3	$[4^2 2^2, 2^2 1^6, 21^8]$	2	50, 5	Symmetry
14	10	3	$[3^2 2^2, 2^2 1^6, 21^8]$	2	50, 5	Symmetry
15	10	3	$[2 \times 3^2 1^4, 21^8]$	2	60, 5	Symmetry
16	10	3	$[3^2 1^4, 2^4 1^2, 21^8]$	2	60, 5	Symmetry
17	10	3	$[2 \times 2^4 1^2, 21^8]$	2	60, 5	Symmetry
18	10	3	$[3^2 1^4, 41^6, 2^2 1^6]$	2	85, 5	Symmetry
19	10	3	$[2^4 1^2, 41^6, 2^2 1^6]$	2	85, 5	Symmetry
20	10	3	$[61^4, 2 \times 2^2 1^6]$	2	120, 5	Symmetry
21	10	3	$[3^2 1^4, 2^3 1^4, 2^2 1^6]$	2	430, 15	Symmetry
22	10	3	$[2^4 1^2, 2^3 1^4, 2^2 1^6]$	2	430, 15	Symmetry
23	10	3	$[42^2 1^2, 2 \times 2^2 1^6]$	2	730, 10	Symmetry
24	10	3	$[3^2 21^2, 2 \times 2^2 1^6]$	2	730, 10	Symmetry
25	10	4	$[3^2 1^4, 2 \times 2^2 1^6, 21^8]$	2	3050, 25	Symmetry
26	10	4	$[2^4 1^2, 2 \times 2^2 1^6, 21^8]$	2	3050, 25	Symmetry
27	10	4	$[41^6, 3 \times 2^2 1^6]$	2	4275, 25	Symmetry
28	10	4	$[2^3 1^4, 3 \times 2^2 1^6]$	2	21400, 75	Symmetry
29	10	5	$[4 \times 2^2 1^6, 21^8]$	2	150000, 125	Symmetry
30	10	3	$[2^5, 31^7, 2^2 1^6]$	2	5, 2	Composition
31	10	3	$[2^5, 321^5, 21^8]$	2	10, 2	Composition
32	10	3	$[42^3, 2^2 1^6, 21^8]$	2	25, 10	Composition
33	10	4	$[2^5, 2^2 1^6, 2 \times 21^8]$	2	50, 20	Composition
34	10	3	$[2^5, 2 \times 2^2 1^6]$	4	10, 10, 5, 2	Composition and symmetry

discussed in Sec. 5.4.4.3. As we will see a few pages later, symmetry is a particular case of composition. Therefore, we must submit to a closer scrutiny the composition of polynomials and its combinatorial and group-theoretic counterparts. This is the subject of the next section.

5.4.4 Primitive and Imprimitive Monodromy Groups

5.4.4.1 Theorems of Ritt and Müller

The theorem of Ritt which was proved in Chapter 1 (Theorem 1.7.6) was in fact initially formulated in [243] (1922) only for the case of polynomials:

Theorem 5.4.15. *The monodromy group of a polynomial covering is imprimitive if and only if the polynomial is a composition of non-linear polynomials of smaller degrees.*

Remark 5.4.16. A polynomial of a prime degree obviously cannot be a composition.

Ritt's theorem suggests considering separately the primitive and imprimitive cases. The following remarkable result due to Müller[1] [221] is one of the consequences of the classification of finite groups (the reader will probably find it useful to refer to the list of definitions in Appendix 1.5.3).

Theorem 5.4.17. *Let P be a polynomial of degree n with $k \geq 3$ finite critical values and with* primitive *monodromy group not equal to A_n or S_n. Then $k = 3$, and there are only three cases possible:*

1. $n = 7$: $\pi = [3 \times 2^2 1^3] = [3 \times 2A]$, $G = \mathrm{PGL}_3(2) \cong \mathrm{PSL}_2(7)$;
2. $n = 13$: $\pi = [3 \times 2^4 1^5] = [3 \times 2A]$, $G = \mathrm{PGL}_3(3)$;
3. $n = 15$: $\pi = [2^6 1^3, 2 \times 2^4 1^7] = [2B, 2 \times 2A]$, $G = \mathrm{PGL}_4(2) \cong A_8$.

Here $2A$ and $2B$ is the notation for conjugacy classes used in the Atlas of Finite Groups [62].

We first consider these three exceptional passports. Cacti having exceptional monodromy groups are called *special*.

5.4.4.2 Exceptional Passports

Degree 7. Passport: $\pi = [3 \times 2^2 1^3]$. The total number of cacti is 56. There are four orbits:

- two orbits of size 21 with the monodromy group A_7;
- two orbits of size 7 with the monodromy group $\mathrm{PGL}_3(2)$.

---

[1] We learned this result from Adrianov who found it independently of Müller (private communication, February 1996).

Degree 13. Passport: $\pi = [3 \times 2^4 1^5]$. The total number of cacti is 35672. There are five orbits:

- one orbit of size 35620 with the monodromy group A_{13};
- four orbits of size 13 with the monodromy group $PGL_3(3)$.

Degree 15. Passport: $\pi = [2^6 1^3, 2 \times 2^4 1^7]$. The total number of cacti is 126000. There are four orbits:

- one orbit of size 125945 with the monodromy group A_{15};
- two orbits of size 5 with the monodromy group $PGL_4(2)$;
- one "imprimitive" (or composition) orbit of size 45, with the monodromy group $(S_3 \wr S_5) \cap A_{15}$ (for the notation $\wr$ see Definition 1.7.9).

One may ask why there exist several orbits having the same monodromy group. The answer is given in the following proposition. Let us recall that the monodromy group is an invariant not only as an abstract group, but also as a particular permutation group, that is, as a particular subgroup of S_n. The same group G may have several conjugate copies inside S_n. But we must not forget either that we have fixed the product $g_1 \ldots g_k = (1, 2, \ldots, n) \in G$ once and for all.

Proposition 5.4.18. *There are two conjugate copies of the group $PGL_3(2)$ inside S_7 that contain the permutation $(1, 2, \ldots, 7)$ in their intersection; four copies of $PGL_3(3)$ inside S_{13} that contain $(1, 2, \ldots, 13)$; and two copies of $PGL_4(2)$ inside S_{15} that contain $(1, 2, \ldots, 15)$.*

Remark 5.4.19. The reason why there are two A_7-orbits for the passport $[3 \times 2^2 1^3]$ remains mysterious. Up to now, this is the only case when we cannot propose a clear combinatorial invariant that would explain such a splitting. We just mention that the orbits themselves are not isomorphic: there is no bijection between them that commutes with the braid group action; see Example 5.5.8.

There exists a construction of projective linear groups containing given cyclic permutation which makes use of the properties of projective geometries over finite fields, Singer cycles, etc. Numerous details concerning all these constructions, as well as the role of projective duality and other phenomena, can be found in [162]. It is amazing that such things may influence the flexible classification of polynomials.

5.4.4.3 Splitting Types in A_n

There exists one more phenomenon which is not reflected in Table 5.1. The conjugacy classes in S_n are completely determined by the cycle structure of their elements. For the group A_n this is not the case. The following statement can be found in [152] (Lemma 1.2.10):

Lemma 5.4.20. *The set of all permutations of a given cycle structure splits into two conjugacy classes in* A_n *if and only if the lengths of all cycles in the permutations are odd and different.*

The "smallest" example that may be constructed using the idea of the above lemma is given by the following passport of degree $n = 25$: $\pi = [97531, 2 \times 31^{22}]$. And, indeed, there are two orbits for this passport, each of size 300.

5.4.4.4 Imprimitive Case

A polynomial $h : X \to Z$ is called *decomposable* if there exist two polynomials $f : X \to Y$ and $g : Y \to Z$ such that

$$h : X \xrightarrow{f} Y \xrightarrow{g} Z,$$

and the degrees of f and g are ≥ 2. Here X, Y, and Z all represent $\overline{\mathbb{C}}$, but we prefer to use three different letters in order to have a possibility to speak easily about what happens on different "levels" of the composition. We call a passport *decomposable* if there exist decomposable polynomials having this passport.

Remark 5.4.21 (How to recognize the imprimitivity). The property of a passport of being decomposable can be verified using only the passport itself, without constructing the corresponding cacti. Let us illustrate this for the case of symmetry. Consider the following passport of degree 9:

$$\pi = [\underline{3}111111, 222111, 222111].$$

We affirm that it is possible to construct a cactus with this passport having the symmetry of order 3. Indeed, there exists a vertex of degree 3 (underlined in the passport) which can serve as the center; and at the same time the remaining vertices can be equally distributed among the three branches because for any fixed color and degree the number of the corresponding vertices is divisible by 3.

For a more complicated composition the procedure is rather cumbersome and difficult to describe: we need to find not one but several "articulation points" at which a number of isomorphic smaller cacti can be glued together. We hope that the examples below will give some idea of the construction.

It is less obvious how to verify if, along with decomposable cacti, there also exist indecomposable ones having the same passport.

Composition presents a variety of different cases, and even some surprises.

Symmetry. Symmetry is a particular (in fact the simplest) case of composition. Imagine a symmetric cactus on the plane X which has p identical branches around a center. Put the center at the point $x = 0$, and apply the polynomial $f(x) = x^p$. The result will be a single branch on the plane Y. Then the mapping $g : Y \to Z$ sends this branch to an elementary polygon, as described in Sec. 1.6.

In spite of this fact, in Table 5.1 we use the term "composition" only for those cases of a composition that are more complicated than a simple symmetry.

Decomposable and indecomposable cacti may coexist. We have already many examples of this kind. First of all, the same passport may correspond to symmetric and asymmetric cacti. In all the examples marked as "compositions" in Table 5.1 there is a symmetric orbit and an asymmetric one. Finally, one more example corresponds to the exceptional passport of degree 15, see above.

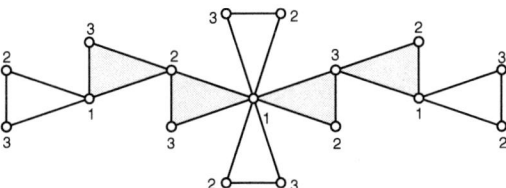

Fig. 5.5. Two blocks of imprimitivity

All cacti can be decomposable. Consider the passport $[42^2, 2 \times 21^6]$. The defect is equal to 8, and therefore, according to Theorem 5.4.8, there is a single orbit. However, if we look at Fig. 5.5, representing a cactus with this passport, then we easily find two blocks of imprimitivity; in fact, the monodromy group here is $S_4 \wr S_2$.

There is no contradiction. It turns out that the cacti with this passport are *all* decomposable; there are no indecomposable ones, and the splitting does not occur.

See however the next example.

Composition type. Important combinatorial data are supplied by the *composition type*. Without giving a formal definition, we only mention that the composition type of the polynomial $h = g \circ f$ describes how the critical values of f are related to the vertices of the cactus corresponding to g.

The following example is very instructive. Let us take the passport $\pi = [2^6, 321^7, 2^2 1^8]$ ($n = 12$). All the corresponding 72 cacti are decomposable. All of them have the same monodromy group $G = S_6 \wr S_2$. And nevertheless there are 6 orbits in this example.

322   5 Meromorphic Functions

Note that, in contrast to the situation of Proposition 5.4.18, the group $G = S_6 \wr S_2$ is also unique as a particular subgroup of S_{12}. Indeed, the presence of the long cycle $(1, 2, \ldots, 12)$ implies the unique choice of the two imprimitivity blocks of size 6, namely, $\{1, 3, 5, 7, 9, 11\}$ and $\{2, 4, 6, 8, 10, 12\}$.

What is then the reason for the splitting?

First note that the critical values of f may occupy the places that are either regular or critical points of g. It is convenient to include these places into the set of vertices of the g-cactus that lives on the plane Y, and hence to add a number of regular values to the set of the critical values of g on Z. This is what we will do. We take $z = g(y) = y^2$, and add two regular values to its only critical value $z = 0$. In this way we get the cactus of degree 2 on the plane Y, shown in Fig. 5.6.

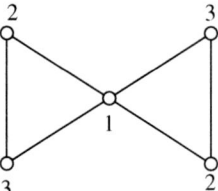

Fig. 5.6. A cactus of degree 2 on the Y plane

Now we must distribute the critical values of f among the vertices of this cactus. We denote these critical values a, b, c, d. Figure 5.7 shows the four possibilities of such a distribution, together with the corresponding partitions

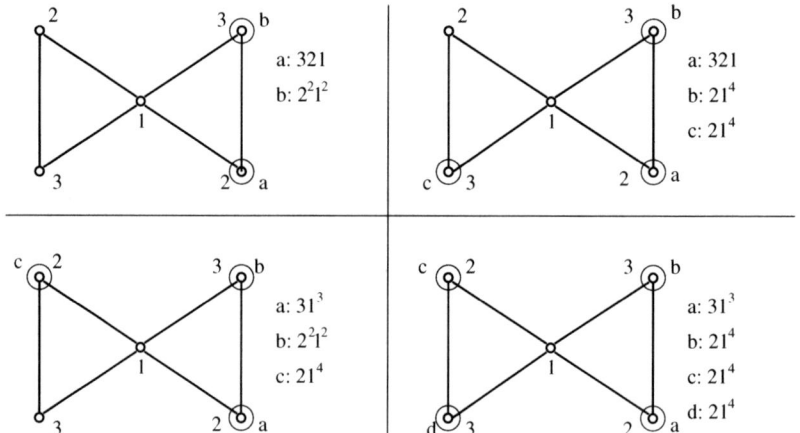

Fig. 5.7. Four composition types

of the number 6 (since deg $f = 6$). It is a trivial matter to verify that all of them give as a result our initial passport π.

Finally we remark that in the case of the composition type (a) of Fig. 5.7 the function f has only 2 critical values. The passport $[321, 2^2 1^2]$ gives rise to 3 trees. As it was already mentioned in Sec. 5.4.2.3, each tree forms a separate orbit. This leads to 3 orbits having the composition type (a). Thus we get a total of 6 orbits.

In group-theoretic language, the notion of composition type may be expressed in terms of different conjugacy classes in the wreath product of groups. We leave the details to the reader.

Fine structure of a partition. In Sec. 5.4.3.2 we already mentioned, in relation to the action of the pure braid group, the passport $[2^5, 2 \times 2^2 1^6]$. One of the possibilities to get this passport is via a composition of the type shown in Fig. 5.8. Looking closely at this covering, we discover that the partition $2^2 1^6$ for the color 2 is obtained from two critical values a and b, both corresponding to the partition 21^3, while the same partition $2^2 1^6$ for the color 3 is obtained from one critical value c corresponding to the partition $2^2 1$, and from a regular value, corresponding to the partition 1^5. This is what we mean by saying that the "fine structure" of the two partitions $2^2 1^6$ is different. This is also the reason why the pure braid group separates this type from the similar type when the roles of the colors 2 and 3 are exchanged.

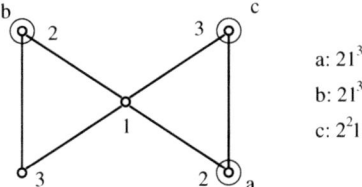

Fig. 5.8. A composition type for the passport $[2^5, 2 \times 2^2 1^6]$

A surprising example. Finally we present an example in which the equivalence class of f is unique, the same is valid for the class of g, and the composition type is the same, and nevertheless there are two equivalence classes for their composition h. The geometric meaning of this phenomenon is very simple.

Let us take f of degree 5 with the following passport: $[4 \times 21^3]$. According to all the theorems we know there is a single orbit. Two cacti of this orbit are shown in Fig. 5.9(a). We see that there is something special about the cactus on the left. Of course it is not symmetric, since only a vertex has a right to be a symmetry center. But still ... It suffices to compose it with a simple cactus of degree 2 (see Fig. 5.9(b)), and this fake symmetry becomes real!

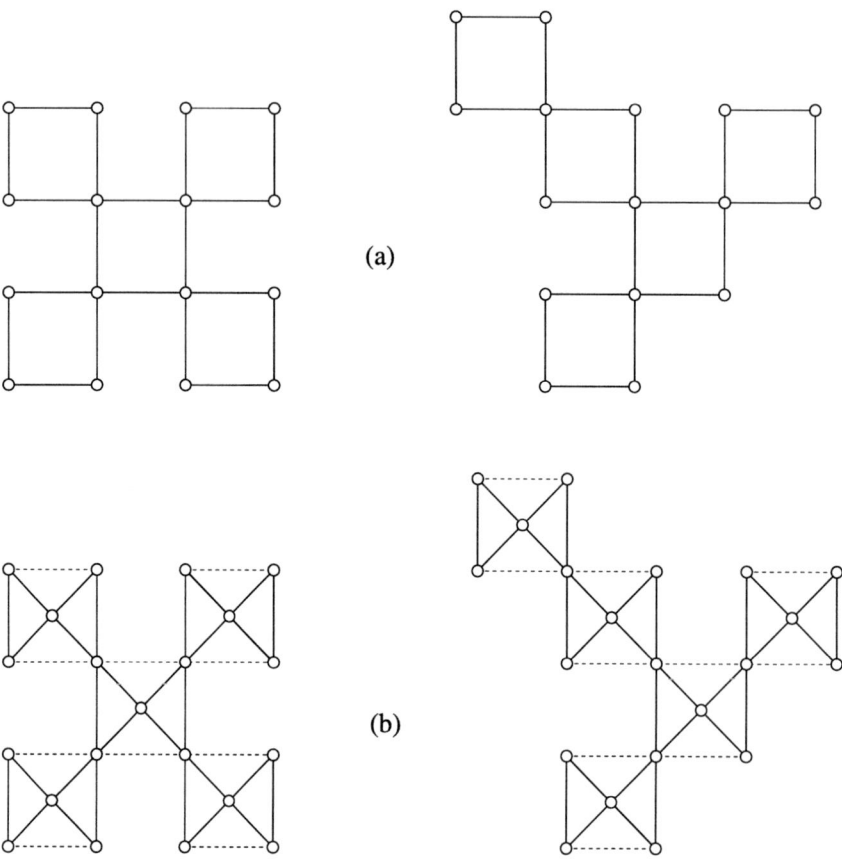

Fig. 5.9. (a) These two cacti of degree 5 with $k = 4$ belong to the same orbit; (b) These two cacti of degree 10 with $k = 3$, which are compositions of the previous ones with the same cactus of degree 2, belong to different orbits

Nevertheless, despite all the complications evoked above, we may say that the study of the imprimitive case is essentially reduced to the classification of the factors, that is, mainly to the primitive case.

5.4.5 Perspectives

5.4.5.1 Polynomial Case: Two Conjectures

The following conjectures are based on the existing experimental data. Let us say that a polynomial passport is *of type* S_n if it is indecomposable and odd (that is, if among g_i there exists at least one odd permutation). The theorems of Sec. 5.4.4.1 imply that for such a passport the only possible monodromy group is S_n. It is also easy to see that every polynomial passport with $k > (n+1)/2$ is of type S_n.

Conjecture 5.4.22. *For any passport of type S_n there exists a single orbit of the braid group action.*

Let us say that a polynomial passport is *of type* A_n if it is indecomposable, even (that is, all the permutations g_i are even) and different from the three exceptional passports of Sec. 5.4.4.2. The theorems of Sec. 5.4.4.1 imply that for such a passport the only possible monodromy group is A_n.

Conjecture 5.4.23. *For a passport of type A_n there are two possibilities:*

1. *If there are no splitting type partitions in the passport (like in Sec. 5.4.4.3), then there exists a single orbit.*
2. *If there exists a splitting type in the passport, then there exist exactly two orbits.*

Remark 5.4.24. Simple arithmetic considerations show that there can exist at most one splitting type in an even polynomial passport.

If the above conjectures turn out to be true, then the problem will be entirely reduced to the study of the imprimitive case.

5.4.5.2 General Case: Rational and Meromorphic Functions

Already the classics of the 19th century have shown that rational or meromorphic functions with only simple critical values constitute a single class of the flexible equivalence. This result implies nothing for polynomials, since each polynomial has a non-simple critical value at infinity.

Several more recent results may be found in [222], [171], [237], [290], as well as in [304] and [169] already cited previously (let us also mention an unpublished letter of P. Deligne to Looijenga [78] cited in [171]). For example, S. Natanzon [222] shows that if all critical values except one are simple, then there is a single class (generic polynomials represent a particular case of this theorem since they have only one non-simple critical value at infinity). B. Wajnryb [290] shows that if there are two non-simple critical values, then the monodromy group is the unique invariant of the braid group action. (Of course, both theorems suppose that the passport is fixed.) The results obtained by P. Kluitmann [171] concern generic polynomials, but the problem he studies is not specifically polynomial: what are the permutation groups by which the braid group permutes constellations?

We may change the notion of defect introduced in Definition 5.4.7 and also take into account infinity by adding n to the sum of the parts ≥ 2 over all the non-simple partitions. Then, the unicity criterion will be $D \leq 2n + 1$ instead of $D \leq n + 1$. This notion, and also the result of Theorem 5.4.8, extend to rational functions.

Theorem 5.4.25 (Defect for rational functions). *Let π be a planar passport, and let D be the sum of all multiplicities of the critical points corresponding to the non-simple critical values. If $D \leq 2n+1$, then there is a single braid group orbit on constellations, and therefore the corresponding flexible equivalence class of rational functions is also unique.*

The proof is similar to that of Theorem 5.4.8, with the difference that instead of the Lagrange interpolation one should use the Newton–Padé one.

Corollary 5.4.26. *If the number of critical values of a rational function of degree n is $\geq 3n/2 - 2$, then there is a single class of the flexible equivalence.*

Recall that the number of critical values of a generic rational function of degree n is $2n - 2$.

This is more or less all we can say about the general case.

5.4.5.3 Meromorphic Functions with a Single Pole

After the polynomial case, the most promising one is that of meromorphic functions of genus $g > 0$ having a single pole (an analog of polynomials for higher genera). Combinatorially, this would mean that we preserve the condition $g_1 \ldots g_k = c$, but weaken the planarity condition. There are various reasons for the hope.

(1) There are some enumerative results for the factorizations of a long cycle in a product of permutations of given cyclic structures (see, e.g., [119] or Theorem A.2.9), while the corresponding results for an arbitrary permutation concern only some particular cases.

(2) We may use the formula of Frobenius in order to count the above factorizations for an arbitrary permutation. But this method has another flaw: it counts also "non-transitive constellations" (that is, the sequences $[g_1, \ldots, g_k]$ that generate non-transitive subgroups of S_n), and therefore necessitates an inclusion-exclusion procedure, while the presence of a long cycle automatically guarantees the transitivity.

(3) Last but not least, the permutation groups containing a long cycle are classified (see below), while the "planar monodromy groups" are not [131]. The following result is due to W. Feit [98] and G. Jones [157] (for notations, see Sec. 1.5.3):

Theorem 5.4.27 (Primitive groups containing a cycle). *A primitive permutation group $G \leq S_n$ contains an n-cycle if and only if it satisfies one of the following conditions:*

1. *$G = A_n$ (n odd) or $G = S_n$, acting naturally;*
2. *$C_n \leq G \leq \mathrm{AGL}_1(n)$, acting on $\mathbb{F}_n$, where n is prime;*
3. *$\mathrm{PGL}_d(q) \leq G \leq \mathrm{P\Gamma L}_d(q)$, acting on points or hyperplanes, with $n = (q^d - 1)/(q - 1)$;*
4. *$G = \mathrm{PSL}_2(11)$, acting on the $n = 11$ cosets of a subgroup A_5;*
5. *G is a Mathieu group M_{11} or M_{23}, acting naturally with degree $n = 11$ or $n = 23$.*

In case 3, if $d \geq 3$ then G has two transitive permutation representations of degree n, on the points and hyperplanes of the projective geometry of dimension $d-1$ over the field $\mathbb{F}_q$; these two representations are equivalent under the outer automorphism of G induced by duality. There are also two representations in case 4, and in case 1 when $G = S_6$, transposed by $\mathrm{Aut}(G)$; in all other cases the representation of G is unique. In case 3 the cyclic regular subgroups are Singer subgroups, except for $G = \mathrm{P\Gamma L}_2(8)$ which has one additional conjugacy class of such subgroups.

Example 5.4.28. Let us take $n = 31$. This number is interesting, because $31 = 1 + 5 + 5^2 = 1 + 2 + 2^2 + 2^3 + 2^4$. This implies the existence of two different projective geometries with 31 points, namely: a projective plane over the field $\mathbb{F}_5$ with the automorphism group $\mathrm{PGL}_3(5)$, and a 4-dimensional projective space over the field $\mathbb{F}_2$ with the automorphism group $\mathrm{PGL}_5(2)$. The two groups are not isomorphic, but both contain an involution with the cycle structure $2^{12}1^7$. The passport $[3 \times 2^{12}1^7, 31]$ corresponds to constellations of genus $g = 3$. The total number of these constellations is equal to $3463162118054496000 = 2^8 3^5 5^3 11^3 13^3 17^3 31 \approx 3.46 \times 10^{18}$ (computations of G. Schaeffer and D. Poulalhon).

Using methods similar to those of [162], we find that there are 10 orbits of size 31 with the monodromy group $\mathrm{PGL}_3(5)$ (more exactly, their monodromy groups are the 10 conjugate copies of $\mathrm{PGL}_3(5) \leq S_{31}$ containing the fixed cycle c), plus 6 orbits of size 682 corresponding in the same manner to the group $\mathrm{PGL}_5(2)$ (computations of P. Moreau).

All the remaining constellations have the monodromy group A_{31}. We suspect that they constitute a single orbit, but we have absolutely no means to prove this fact. Just note that the size of the remaining set contains a big prime factor: $3463162118054491598 = 2 \times 31 \times 229 \times 243919010991301$. This leaves little hope that there is a nice structure on this set.

5.5 Megamaps

The term megamap is a new one. We try, by this term, to convey the idea of "a map constructed of maps". Anyway, the term hypermap is already taken, while "super" usually means anti-commutative. Possibly the term "house of maps" would be more appropriate.

There are three sources of motivation for this section. First, while studying hypermaps we usually interpret the edges geometrically as "segments of curves" drawn on a surface. However, one of the advantages of the permutation model is its abstract nature. Namely, one may take as the set E of "edges" an arbitrary finite set, and introduce the permutations acting on E by using the particular structure of the elements of this set. Concretely, in this section the elements of E will be 4-constellations with a given passport, while the permutations defining the hypermap will originate from an action of the Hurwitz braid group on these 4-constellations.

Second, we aim here at the particular goal of constructing the parameter space, or a kind of a moduli space, of the coverings of $\mathbb{C}P^1$ with four ramification points. It turns out that in this simple case the corresponding moduli space is a Riemann surface. In other words, the coverings of the Riemann sphere with four ramification points depend on one complex parameter, but this parameter is not just a complex number: it lives on a Riemann surface.

Finally, the way we represent the Riemann surfaces in question is purely combinatorial. Namely, we represent them as dessins d'enfants!

This construction is known and widely used in the inverse Galois theory, though its combinatorial aspect is sometimes difficult to extract. We cite here only few papers, mostly related to dessins d'enfants: [106], [69], [121], [311].

5.5.1 Hurwitz Spaces of Coverings with Four Ramification Points

As we already know, in order to represent a covering of the Riemann sphere $\mathbb{C}P^1 = \mathbb{C} \cup \{\infty\}$ with four ramification points we must supply the following data (see Riemann's existence theorem 1.8.14): (i) a 4-constellation $C = [g_1, g_2, g_3, g_4]$; (ii) four ramification points $y_1, y_2, y_3, y_4 \in \mathbb{C}P^1$. However, as was explained at the end of Sec. 1.8, by applying a linear fractional transformation we may take any three of the four points to 0, 1, and ∞. We prefer here to take as these three points the points y_1, y_3, and y_4. Thus, we may suppose that the four ramification points in question are $0, y, 1$, and ∞. Here y is an arbitrary complex number different from 0, 1, and ∞. We take the ramification points $0, y, 1, \infty$ in this specific order because later on it will be convenient to consider y belonging to the segment $[0, 1]$.

Of course, there is no hope that in general such a covering will be defined over $\overline{\mathbb{Q}}$, because the value of y may well be transcendental. But the object we are interested in is not a specific covering but *the space of all such coverings*. We would like to parameterize this space, or, in other words, to construct a moduli space of the coverings with four ramification points. This is a particular case of *Hurwitz space*.

Let us fix a passport π of a 4-constellation, and consider the set consisting of all the pairs $\{(C, y)\}$, where C is a constellation with the passport π and y

belongs to $\mathbb{C}P^1 \setminus \{0,1,\infty\}$. This set is endowed with a natural structure of a topological space. Let H denote a connected component of this space.

Proposition 5.5.1. *The mapping $H \to \mathbb{C}P^1 \setminus \{0,1,\infty\}$ taking a pair (C,y) to the point y is an unramified covering.*

Proof. The space H is connected, and the fiber over a point $y \in \mathbb{C}P^1 \setminus \{0,1,\infty\}$ is discrete, its points corresponding to non-isomorphic constellations with the passport π. We will see in a moment that the number of preimages of every y is the same.

In Sec. 1.8 we explained how to lift the complex structure from $\mathbb{C}P^1 \setminus \{0,1,\infty\}$ to H, and how to compactify H, thus obtaining a compact Riemann surface $\overline{H}$, and a ramified covering $f: \overline{H} \to \mathbb{C}P^1$. A remarkable fact is that $\overline{H}$ is already endowed with a Belyi function (and is therefore defined over $\overline{\mathbb{Q}}$)!

Proposition 5.5.2. *The function $f: (C,y) \mapsto y$ is a Belyi function on $\overline{H}$.*

Proof. By construction, any value of $y \neq 0,1,\infty$ is not critical because it has the same number of preimages (C,y). (We will see below what is the meaning of this number.)

5.5.2 Representation of $\overline{H}$ as a Dessin d'Enfant

Now our natural goal is to find the triple of permutations that describes the dessin d'enfant corresponding to the Belyi function

$$f: \overline{H} \to \mathbb{C}P^1 : (C,y) \mapsto y.$$

In order to do that we must understand what happens when the parameter y starts at a particular base point y_0, goes around 0, or 1, or ∞, and returns to y_0. Taking into account that our usual geometric representation of a hypermap is the preimage of the segment $[0,1]$, we will think of y_0 as of a point on $[0,1]$. In order to underline the more complicated nature of our three permutations, we change the notation from σ, α, φ to Σ, A, Φ.

We work with the Hurwitz braid group H_4 on four strings. The corresponding loops are shown in Fig. 5.10. Their representation in the form of braids is as follows:

- the loop that turns around 0 is equal to $\Sigma = \sigma_1^2$;
- the loop that turns around 1 is equal to $A = \sigma_2^2$;
- finally, the third loop, that turns around infinity, is equal to $\Phi = \sigma_2^{-1}\sigma_3^2\sigma_2$.

Lemma 5.5.3. *In the Hurwitz braid group H_4 the product $\Sigma A \Phi = \mathrm{id}$.*

Proof. We have $\Sigma A \Phi = \sigma_1^2 \sigma_2 \sigma_3^2 \sigma_2$. Conjugating this element by σ_1 we get

$$\sigma_1^{-1}(\sigma_1^2 \sigma_2 \sigma_3^2 \sigma_2)\sigma_1 = \sigma_1 \sigma_2 \sigma_3^2 \sigma_2 \sigma_1 = \mathrm{id}$$

according to the relation (5.20). One might also verify directly that the action of the product $\Sigma A \Phi$ on a constellation $[g_1, g_2, g_3, g_4]$ conjugates all the g_i by g_2 and thus produces an isomorphic constellation.

330   5 Meromorphic Functions

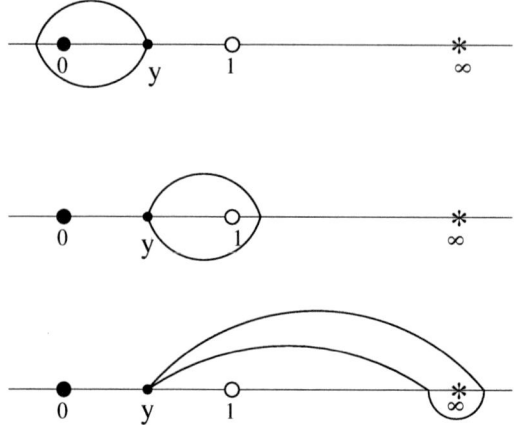

Fig. 5.10. The critical value y starts at a point y_0 in the segment $[0,1]$ and makes a complete turn around three other critical values 0, 1 and ∞ in the counter-clockwise direction

Remark 5.5.4. All the three operations Σ, A, and Φ preserve the passport of a constellation. But in order to get a connected hypermap (and a connected Riemann surface $\overline{H}$), one must take not the full set of constellations with a given passport but an *orbit* of the group $M = \langle \Sigma, A, \Phi \rangle \leq H_4 \cap P_4$. The size of the orbit is equal to the number of sheets of the covering which this time defines the connected Riemann surface $\overline{H}$, i.e., it is equal to the degree of the Belyi function f.

We summarize the above construction in the following definition.

Definition 5.5.5 (Megamap). Let us take the elements $\Sigma = \sigma_1^2$, $A = \sigma_2^2$, $\Phi = \sigma_2^{-1}\sigma_3^2\sigma_2$ of the Hurwitz braid group H_4 and consider their action on constellations defined in Construction 1.1.17. A *megamap* is an orbit of this action together with the triple of permutations which are the permutation actions of Σ, A, and Φ on this orbit.

Some remarks are in order to clarify the difference between "hyper" and "mega". A hypermap is a finite set E together with a triple of permutations $[\sigma, \alpha, \varphi]$ which act transitively on E and such that $\sigma\alpha\varphi = \text{id}$. In Chapter 2 we saw that a hypermap is one of the ways to represent a Riemann surface defined over $\overline{\mathbb{Q}}$. Thus, a megamap is a particular kind of hypermap which, in the present context, serves to represent a particular kind of Riemann surface, namely, the moduli space of coverings with four ramification points.

The black and white vertices and face centers of the resulting hypermap have their own interpretation, and some interesting phenomena take place there. Let us consider a black vertex of a megamap: it corresponds to the parameter y equal to 0. When the fourth critical value y tends to 0 and finally becomes equal to 0, what we get is a covering with three critical values $0 = y$, 1, and ∞, i.e., a dessin d'enfant. Combinatorially, it is represented by the triple of permutations $[\sigma, \alpha, \varphi] = [g_1 g_2, g_3, g_4]$. However, there is no guarantee that this dessin is connected; if not, it must be considered as a disjoint union of dessins. Note that the braid operation σ_1 does not change g_3 and g_4, and therefore the product $g_1 g_2$ is also preserved. Thus for all "edges" (i.e., 4-constellations) adjacent to a black vertex the corresponding triple $[g_1 g_2, g_3, g_4]$ will be the same. In other words, in the same way as an "edge" is in fact a 4-constellation, a "black vertex" is in fact a dessin d'enfant.

In the same way all the white vertices are labelled by the dessins of the form $[\sigma, \alpha, \varphi] = [g_1, g_2 g_3, g_4]$.

For the face centers the procedure is slightly more complicated. We would like to multiply g_2 (corresponding to y) and g_4 (corresponding to ∞). But g_2 and g_4 are not adjacent in the constellation. We must first transpose g_2 and g_3, acting by the braid operation σ_2^{-1}. (Why σ_2^{-1} and not σ_2 itself? – Because we want to preserve g_2 as it is, while σ_2 conjugates it by g_3.) We thus get the constellation $[g_1, g_2 g_3 g_2^{-1}, g_2, g_4]$. If we now act on this latter 4-constellation by Φ, and then transpose once more the second and the third permutations in it via σ_2^{-1}, the resulting triple $[g_1, g_2 g_3 g_2^{-1}, g_2 g_4]$ will be the same for all the "edges" adjacent to the face. We see that a "face center" is in fact the dessin d'enfant $[g_1, g_2 g_3 g_2^{-1}, g_2 g_4]$

5.5.3 Examples

To compute a megamap is not an easy task, at least for a "by hand" calculation. One must not only manipulate the permutations, but verify every time whether a newly obtained constellation is isomorphic to some of the previous ones. A Magma program was implemented by N. Hanusse [134]; certain calculations were also carried out by D. Bouya [41].

Example 5.5.6. Let us consider generic rational functions of degree 3 or, in other words, coverings corresponding to the passport

$$[21, 21, 21, 21].$$

There are four non-isomorphic constellations with this passport, and they are arranged into the megamap shown in Fig. 5.11.

Now consider generic polynomials of degree 4, that is, coverings with the passport

$$[21^2, 21^2, 21^2, 4].$$

It turns out that there are also four non-isomorphic constellations, and the corresponding megamap is the same as above!

We see that this example, which is probably the simplest possible, already leads to a rather non-trivial conclusion: *there exists a transformation which,*

332   5 Meromorphic Functions

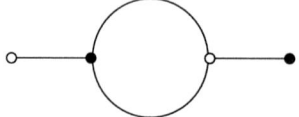

Fig. 5.11. The megamap which describes the "moduli space" of generic rational functions of degree 3, as well as that of generic polynomials of degree 4

for any given generic polynomial of degree 4, produces a generic rational function of degree 3, and which is a biholomorphic bijection (the rational function becomes non-generic when and only when the polynomial taken was also non-generic). In other words, the moduli space of generic polynomials of degree 4 and that of generic rational functions of degree 3 are isomorphic as complex curves. It would be interesting to find this isomorphism explicitly, but for the moment we have no idea of what it could look like.

Example 5.5.7. The corresponding megamap for polynomials of degree 6 with the passport $[31^3, 21^4, 2^21^2, 6]$ is shown in Fig. 5.12.

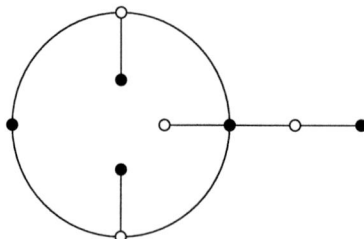

Fig. 5.12. The megamap for polynomials with the passport $[31^3, 21^4, 2^21^2, 6]$

For polynomials of degree 6 with the passport $[321, 21^4, 21^4, 6]$ the corresponding megamap is shown in Fig. 5.13.

Example 5.5.8. Let us consider polynomials of degree 7 with the passport $[2^21^3, 2^21^3, 2^21^3, 7]$. We have already seen this passport: it is one of the three exceptional passports considered in Sec. 5.4.4.2. As we know, there exist 56 non-isomorphic cacti that split into four orbits, of sizes 7, 7, 21, and 21 respectively. Therefore, we must construct four different megamaps. It turns out that the two smaller megamaps (of size 7) are isomorphic; the corresponding picture is shown in Fig. 5.14.

Concerning the bigger orbits, they are *not* isomorphic. The easiest way to see this is to look at the picture: the corresponding megamaps are shown in Fig. 5.15. For example, in the hypermap on the left there is an edge that

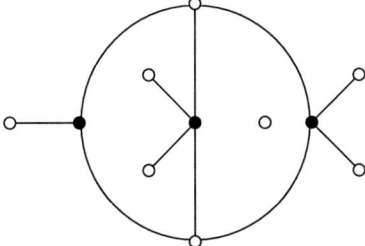

Fig. 5.13. The megamap for polynomials with the passport $[321, 21^4, 21^4, 6]$

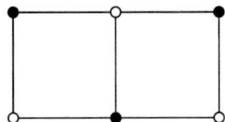

Fig. 5.14. The megamap representing each of the two orbits of size 7 of Example 5.5.8

connects two vertices of degree 2, while in the hypermap on the right such an edge does not exist.

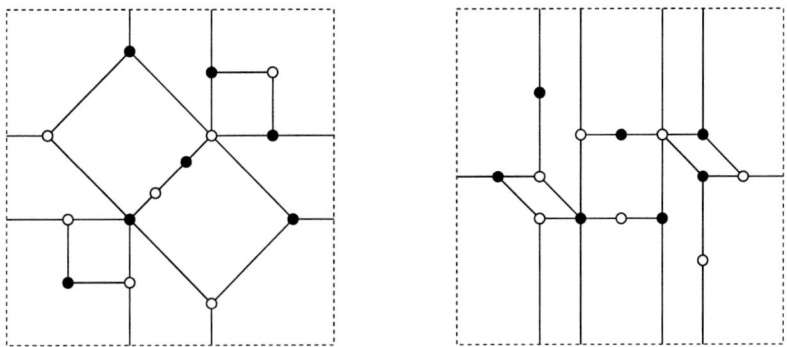

Fig. 5.15. The megamaps of genus 1 representing the two orbits of size 21 of Example 5.5.8

Both dessins are of genus 1 and thus represent elliptic curves. Note that in both cases the degrees of the black vertices, of the white vertices, and of the faces form the same partition $5^3 4 2^2$. We are unable to compute the corresponding elliptic curves and Belyi functions, but we are pretty sure that these dessins d'enfants are Galois conjugate.

Example 5.5.9. A beautiful series of examples was constructed by J.-M. Couveignes in [69]. Let us consider the following polynomial passport:

$$\pi = [31^{n-3}, (p, q, r, s), 21^{n-2}, n],$$

where p, q, r, s are pairwise distinct positive integers, $p + q + r + s = n$. According to the Goulden–Jackson formula, the number of non-rooted cacti with this passport is equal to $6n$; this will be the degree of our megamap. In Fig. 5.16 we show one of the eight faces of the megamap in question. The face is of degree $p + q + r$; one more face of the same degree is constructed by placing $p - 1$, $q - 1$, $r - 1$ "free edges" in the opposite circular order around the face. Then we prepare the other six faces by replacing p, q, r with p, q, s, then with p, r, s, and finally with q, r, s.

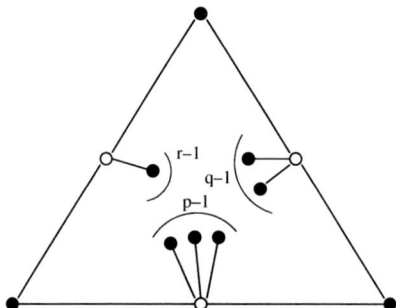

Fig. 5.16. One of the eight faces of the megamap of Example 5.5.9

Having thus prepared eight "triangles", we glue them in the form of the octahedron in such a way as to have two white vertices of degree $p + q$, two of degree $p + r$, etc., and two of degree $r + s$ (12 white vertices in total). The construction of the megamap is finished.

It is interesting to note that in this example, if all the numbers p, q, r, s are bigger than 1, then the defect of the passport π (see Definition 5.4.7) is equal to $n + 3$, and if one of them is equal to 1, then the defect is equal to $n + 2$. In both cases the defect is bigger than the bound $n + 1$ given in Theorem 5.4.8. But, having constructed the megamap explicitly, we have also shown the uniqueness of the orbit.

Example 5.5.10. It would be nice to have at least one example in which not only combinatorial, but also analytic information is supplied. Let us consider polynomials of degree n with the passport $\pi = [(n-2)1^2, 21^{n-2}, 21^{n-2}, n]$. The combinatorial computations give the megamap shown in Fig. 5.17.

This time the dessin is simple enough, and we are able to find the corresponding Belyi function explicitly: placing the black vertex of degree $n - 1$

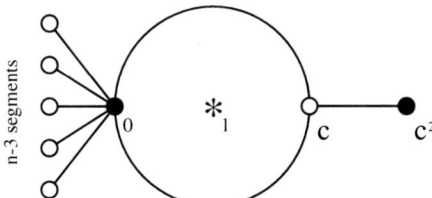

Fig. 5.17. The megamap of Example 5.5.10

to 0, the center of the smaller face to 1, and the center of the outer face at infinity, we find that the white vertex of degree 3 must be placed at the point $c = n/(n-2)$, and the black vertex of degree 1, at the point c^2. Thus we find

$$f(t) = -\frac{1}{c^n} \frac{t^{n-1}(t-c^2)}{t-1}. \tag{5.21}$$

(The only thing that needs to be verified is the fact that $f - 1$ has, indeed, a root of multiplicity 3 at c.) But this is not yet everything we want to know. We would also like to find a representation of polynomials of degree n with the passport π given above. The formula below gives the answer. Let us take the following polynomial in x depending also on a parameter t:

$$P_t(x) = \frac{1}{t-1} x^{n-2}[(n-1)x^2 - (n+(n-2)t)x + (n-1)t]. \tag{5.22}$$

Like any other polynomial, it has a critical value of multiplicity n at infinity. By construction, it also has a critical point of multiplicity $n-2$ at 0 (with the corresponding critical value also equal to 0). What are its other critical points and values? In order to answer this question let us compute its derivative:

$$P'_t(x) = \frac{n(n-1)}{t-1} x^{n-3}(x-1)(x-t/c).$$

Thus the critical points are 0, 1, and t/c, while the corresponding critical values are $P_t(0) = 0$, $P_t(1) = 1$, and

$$P_t(t/c) = -\frac{1}{c^n} \frac{t^{n-1}(t-c^2)}{t-1}.$$

The latter expression coincides with the Belyi function given in (5.21).

Let us recollect the results. The Riemann sphere of the variable t, *together with the dessin of* Fig. 5.17, is the parameter space for a family of polynomials. The polynomials in question are characterized by the fact that they have four critical values. Three of the four critical values are 0, 1, and ∞; the Belyi function for the dessin of Fig. 5.17 gives the fourth one. Therefore, we were happy to find out that for the family P_t described in (5.22) the fourth critical value, which is $P_t(t/c)$, coincides indeed with the Belyi function (5.21) of the dessin of Fig. 5.17.

* * *

Having read Chapter 3, we must not be surprised to learn that Hurwitz numbers have physical applications. Thus, for example, Gross and Taylor [126], [127] consider the free energy of the U_N Yang–Mills theory and study its asymptotics as N tends to ∞. The coefficients of the asymptotics can be expressed as sums of certain Hurwitz numbers. The behavior of these coefficients, for the group U_N as well as for other groups, permits the authors to conclude that the corresponding Yang–Mills theories are "string theories at perturbative level". Generating functions for particular classes of Hurwitz numbers often possess rather remarkable properties: Sec. A.2.5 permits to get an idea of the possibilities revealed here. Largely speaking, a general framework for the theory presented in Secs. 5.1–5.3 is provided by the mirror symmetry (see, e.g., [71]) and by the Gromov–Witten invariants, of which Hurwitz numbers represent an important particular case.

The appropriate context for the flexible classification remains to be found. There exist some obvious relations to the inverse Galois theory, and also to some algorithmic problems in group theory (the so-called "product replacement algorithm", see, e.g., [228]), but they do not provide a sufficient insight for advancing. It is not for nothing that the problem, even in the polynomial case, resists solution for more than 130 years.

6
Algebraic Structures Associated with Embedded Graphs

This chapter is devoted mainly to the description of algebraic and combinatorial constructions related to Vassiliev's theory of knot invariants [283], [284], [285]. The main combinatorial object of the theory is a chord diagram, or, which is the same, a one-vertex map. We start with a description of this notion and of the famous 4-term relation for for the chord diagrams. Our presentation follows the main lines of Bar-Natan [18], but we also pay a lot of attention to the recent development in the analysis of intersection graphs of chord diagrams. We do not present the proof of Kontsevich's theorem for Vassiliev invariants since its machinery is too far from the subject of our book. Good explanations are available in [50], [195].

6.1 The Bialgebra of Chord Diagrams

6.1.1 Chord Diagrams and Arc Diagrams

Consider a circle and put $2n$ pairwise distinct points on it. Split these points arbitrarily into n pairs and draw chords connecting the two points of each pair. The result will be a chord diagram of order n.

Fig. 6.1. Four pictures of the same chord diagram

338   6 Algebraic Structures

A rotation preserves a chord diagram, while a reflection generally yields a different chord diagram: see Figs. 6.1 and 6.2.

Fig. 6.2. Two distinct chord diagrams

Definition 6.1.1 (Chord diagram). *A chord diagram of order n (or a chord diagram with n chords) is an oriented circle with a distinguished set of n pairs of points on it considered up to orientation preserving diffeomorphisms of the circle.*

The perspicacious reader has certainly noticed that the notion of a chord diagram coincides with that of a one-face map (see the beginning of Chapter 3): both objects consist of a pair of permutations, one of them being cyclic and the other one being an involution without fixed points.

Of course, the chords may be drawn as straight lines or curve segments whose actual shape is irrelevant; what matters is the way they bind their endpoints into pairs: see Fig. 6.1. Figure 6.3 presents a list of all chord diagrams with 1, 2, and 3 chords.

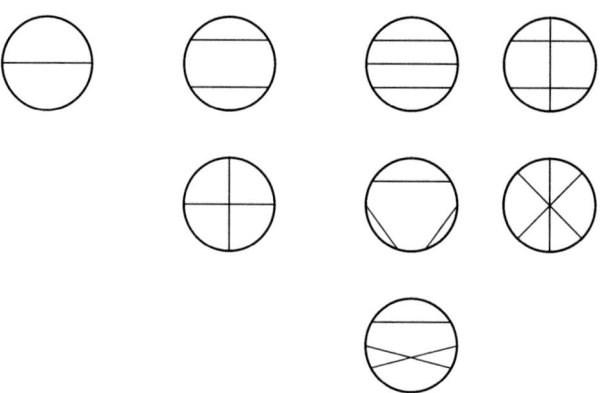

Fig. 6.3. All chord diagrams of order 1, 2, 3

Sometimes it is more convenient to represent a chord diagram as an arc diagram. To obtain an arc diagram from a chord diagram one must choose an arbitrary point on the circle (different from the ends of all chords) and "cut" the circle at this point ("push the point to infinity"). The circle becomes a straight line, and chords become arcs, see Fig. 6.4.

Of course, the representation of a chord diagram as an arc diagram depends on the choice of the point at infinity. One chord diagram usually has $2n$ different arc representations. But in the case when the chord diagram has a rotational symmetry, the number of different arc representations decreases. The cyclic group of order $2n$ acts naturally on the set of arc diagram representations of a chord diagram.

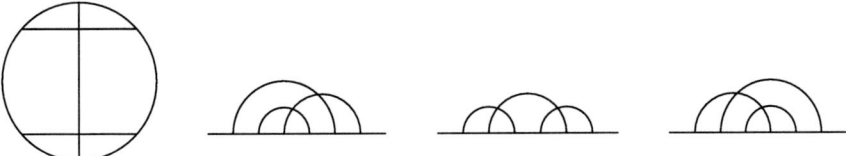

Fig. 6.4. A chord diagram and its three different arc representations

Conversely, however, an arc diagram uniquely determines the corresponding chord diagram.

Definition 6.1.2 (Arc diagram). An *arc diagram of order n* (or *an arc diagram with n arcs*) is an oriented line with a distinguished set of n pairs of points on it considered up to orientation preserving diffeomorphisms of the line.

We will usually draw an arc diagram as a horizontal line with arcs lying in the upper half-plane.

6.1.2 The 4-Term Relation

We are going to study formal linear combinations of chord diagrams. For example,

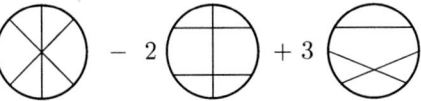

is such a formal linear combination. In this chapter we fix a commutative associative ring **k** as a ground ring. Let $\mathcal{C}_n$ denote the **k**-module spanned by all chord diagrams of order n as free generators. The above picture shows an element in $\mathcal{C}_3$. We denote by $\mathcal{A}_n$ the **k**-module generated by all arc diagrams of order n as free generators.

The 4-term relation is, in fact, a set of equivalence relations on each of the $\mathcal{C}_n$. Four diagrams take part in each such relation. In each of the four diagrams two of the chords are selected. The other chords in all four diagrams coincide.

340   6 Algebraic Structures

These common chords are irrelevant, and we will not draw them explicitly. We explain the motivation for introducing this notion in Sec. 6.2.

Let c_1, c_2, c_3, c_4 be four chord diagrams of order $n \geq 2$ such that two of their chords occupy the positions shown in Fig. 6.5, while all other chords are adjacent to the dotted circle segments and are the same in all four diagrams. The mnemonic rule for the two distinguished chords is as follows: one of the chords is fixed; one of the ends of the other chord is also fixed; the second (moving) end of this chord takes successively four places near the ends of the fixed chord.

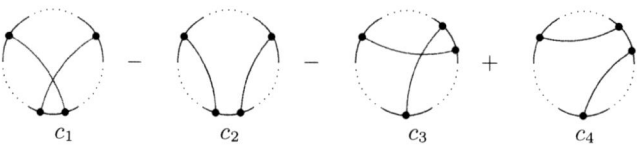

Fig. 6.5. A 4-term element in $\mathcal{C}_n$

Definition 6.1.3 (4-term element). The element

$$c_1 - c_2 - c_3 + c_4 \in \mathcal{C}_n$$

is called a *4-term element*.

A 4-term element $a_1 - a_2 - a_3 + a_4 \in \mathcal{A}_n$ has the appearance shown in Fig. 6.6.

Fig. 6.6. A four-term element in $\mathcal{A}_n$

Consider the submodule $\mathcal{C}_n^{(4)} \subset \mathcal{C}_n$ spanned by all 4-term elements in $\mathcal{C}_n$ and the submodule $\mathcal{A}_n^{(4)} \subset \mathcal{A}_n$ spanned by all 4-term elements in $\mathcal{A}_n$. We set $\mathcal{C}_0^{(4)} = \mathcal{C}_1^{(4)} = \mathcal{A}_0^{(4)} = \mathcal{A}_1^{(4)} = 0$.

Exercise 6.1.4. Verify the following statements.

1. $\mathcal{A}_2^{(4)}$ is spanned by the element ⌢⌢ − ⌢⌢ .
2. $\mathcal{C}_2^{(4)} = 0$.
3. $\mathcal{C}_3^{(4)}$ is spanned by the elements

   ⊘ − ○   and   ⊗ − 2 ⊕ + ⊠ .

Theorem 6.1.5. *There exists a natural isomorphism*

$$\mathcal{A}_n/\mathcal{A}_n^{(4)} \cong \mathcal{C}_n/\mathcal{C}_n^{(4)}$$

taking the class of an arc diagram $[a] \in \mathcal{A}_n/\mathcal{A}_n^{(4)}$ to the class $[c] \in \mathcal{C}_n/\mathcal{C}_n^{(4)}$ of the chord diagram c represented by the arc diagram a.

The proof will be given below.
The following definition introduces the main notion of the present chapter.

Definition 6.1.6 (Module of chord diagrams). The quotient module

$$\mathcal{M}_n = \mathcal{C}_n/\mathcal{C}_n^{(4)} \cong \mathcal{A}_n/\mathcal{A}_n^{(4)}$$

is said to be the *module of chord diagrams of order n*.

For example, $\mathcal{M}_3$ is a free **k**-module with 3 generators given by the equivalence classes of the elements

$$[\bigcirc],\ [\bigotimes],\ [\bigoplus],$$

while the equivalence classes of the remaining two chord diagrams can be expressed as

$$[\bigcirc\!|] = [\bigcirc] \quad \text{and} \quad [\bigotimes\!\!\!\!\times] = 2[\bigoplus] - [\bigotimes]$$

according to Exercise 6.1.4.
Thus calculations in $\mathcal{M}$ are just ordinary linear algebra calculations where chord diagrams are used instead of letters.

Exercise 6.1.7. Compute a basis for the modules $\mathcal{M}_4$ and $\mathcal{M}_5$.

From now on we will omit square brackets, thus making no distinction in the notation for chord (or arc) diagrams and their equivalence classes in the module of chord diagrams.

Proof of Theorem 6.1.5. We must prove that two arc representations of a chord diagram are equivalent modulo the 4-term relations.

Consider an arc diagram a of order n. Take the arc in a whose left end is the leftmost one. It suffices to prove that if we move the left end of this arc to the right of the rightmost end, then the equivalence class of the diagram will not change. The arc ends of other $n-1$ arcs in a split the line into $2n-1$ segments, which we number $1, 2, \ldots, 2n-1$ from left to right starting from the left semiinfinite segment. Let a_k denote the arc diagram obtained from a by moving the left end of the chosen arc to the kth segment ($k = 1, \ldots, 2n-1$).

We must prove that $a_1 - a_{2n-1} = 0$. Rewrite the left-hand side of the equality as

$$a_1 - a_{2n-1} = a_1 - a_2 + a_2 - a_3 + a_3 - \cdots - a_{2n-2} + a_{2n-2} - a_{2n-1}.$$

There are $(2n-3)\cdot 2 + 2 = 4(n-1)$ terms on the right-hand side of the previous relation. We split them into $n-1$ groups, four terms in each group. Such a group is determined by an arc in a: if the arc's ends separate segments $k, k+1$ and $m, m+1$, then the group will be $a_k - a_{k+1} + a_m - a_{m+1}$. Such a group is a 4-term element, and therefore it equals zero. The theorem is proved.

In order to find the dimensions of the spaces $\mathcal{M}_n$ over a field of characteristic zero one needs simply to encode all chord diagrams of order n, to produce all 4-term relations between them, and to solve the corresponding system of linear equations. However, this computation very soon becomes too laborious. For example, for $n = 9$ the number of chord diagrams (= the number of variables) is 644808, and the number of equations is 5056798, while the number of independent solutions is only 105 (see [18]). Kneissler [172] applied a less direct approach to compute the dimensions $\dim \mathcal{M}_n$ up to $n = 12$. The results are shown in Table 6.1 in Sec. 6.1.5 below.

6.1.3 Multiplying Chord Diagrams

Two arc diagrams may be concatenated, thus producing a new arc diagram. The concatenation of arc diagrams leads to a new operation in the module

$$\mathcal{M} = \mathcal{M}_0 \oplus \mathcal{M}_1 \oplus \mathcal{M}_2 \oplus \ldots$$

of chord diagrams. We call the operation *multiplication*. The definition proceeds as follows.

The *concatenation* $a_1 \# a_2$ of arc diagrams a_1, a_2 consists simply in drawing both diagrams one after another, see Fig. 6.7.

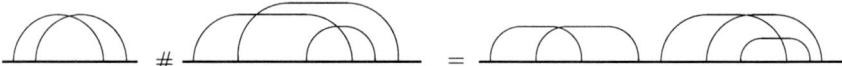

Fig. 6.7. Concatenation of arc diagrams

Definition 6.1.8 (Product of chord diagrams). The product of two chord diagrams in $\mathcal{M}$ is defined as the chord diagram represented by the concatenation of their arbitrary arc representations.

Theorem 6.1.9. *The multiplication of chord diagrams is a well-defined operation. This multiplication may be extended to a bilinear operation*

$$m : \mathcal{M} \otimes \mathcal{M} \to \mathcal{M},$$

respecting the grading

$$m : \mathcal{M}_k \otimes \mathcal{M}_l \to \mathcal{M}_{k+l}$$

making $\mathcal{M}$ into a commutative associative graded algebra over $\mathbf{k}$ with unit. The unit is given by the equivalence class of the empty diagram.

Proof. The first statement follows from Theorem 6.1.5. The commutativity follows from the fact that the arc diagrams $a_1 \# a_2$ and $a_2 \# a_1$ obviously represent the same chord diagram. We omit the purely technical verification of the other axioms of algebras.

We would like to stress that in spite of the non-commutativity of the *arc* diagram concatenation, the multiplication of *chord* diagrams is commutative.

6.1.4 A Bialgebra Structure

Besides the multiplication, one more operation may be introduced in the module of chord diagrams. This operation, called *comultiplication*, is a unary one, and it maps $\mathcal{M}$ into its tensor square:

$$\mu : \mathcal{M} \to \mathcal{M} \otimes \mathcal{M}.$$

The tensor square inherits a natural grading from $\mathcal{M}$, and the comultiplication respects this grading in the following sense:

$$\mu : \mathcal{M}_n \to \mathcal{M}_0 \otimes \mathcal{M}_n + \mathcal{M}_1 \otimes \mathcal{M}_{n-1} + \cdots + \mathcal{M}_n \otimes \mathcal{M}_0.$$

First we define an operation $\bar{\mu} : \mathcal{C} \to \mathcal{C} \otimes \mathcal{C}$.

On a chord diagram c of order n, the operation $\bar{\mu}$ acts as follows. Let $V(c)$ denote the set of all chords of the diagram c. Let $J \subset V(c)$ be a subset of $V(c)$; denote by J' its complement, $J' = V(c) \setminus J$. The subset J determines a chord diagram c_J formed by all chords belonging to the set J. We set

$$\bar{\mu}(c) = \sum_{J \subset V(c)} c_J \otimes c_{J'},$$

where the sum is taken over all 2^n subsets of the set of n chords.

For example,

The operation is extended by linearity to linear combinations of chord diagrams.

Exercise 6.1.10. Prove that the operation $\bar{\mu} : \mathcal{C} \to \mathcal{C} \otimes \mathcal{C}$ descends to a linear operation $\mu : \mathcal{M} \to \mathcal{M} \otimes \mathcal{M}$. In other words, prove that if one replaces a chord diagram by a linear combination of chord diagrams equivalent to the original one modulo the 4-term relation, then the result of comultiplication will be equivalent to the original one modulo the 4-term relation.

[**Hint:** Consider two cases: when both chords participating in the 4-term relation belong either to a set J, or to its complement J'; and when one of them belongs to J and the other one to J'.]

We attract the reader's attention to the fact that the unit in the algebra $\mathcal{M}$ over $\mathbf{k}$ may be understood as a homomorphism $e : \mathbf{k} \to \mathcal{M}$ (we simply identify the unit with $e(1)$). Since multiplication in $\mathcal{M}$ respects grading we obviously get $e(1) \in \mathcal{M}_0$. Similarly, the operation μ admits a *counit*, i.e., a homomorphism $\epsilon : \mathcal{M} \to \mathbf{k}$ given by the natural projection $\epsilon : \mathcal{M} \to \mathcal{M}_0 \cong \mathbf{k}$.

The algebra and coalgebra structures in $\mathcal{M}$ allow one to include investigation of $\mathcal{M}$ into the general framework of graded bialgebras. Let us give formal definitions.

Let $\mathbf{k}$ be a commutative associative ring with unit. The ring $\mathbf{k}$ may be regarded as a graded algebra over itself if we set $\mathbf{k} = \mathbf{k} \oplus 0 \oplus 0 \oplus \ldots$. From another point of view, $\mathbf{k}$ may be regarded as a graded coalgebra over itself with the comultiplication $\mu : a \mapsto a \otimes 1 = 1 \otimes a$ for $a \in \mathbf{k}$. The unit and the counit are represented by the identical homomorphism $\mathbf{k} \to \mathbf{k}$.

Definition 6.1.11 (Bialgebra). Let $E_0, E_1, E_2, \ldots$ be finitely generated free k-modules, $E_0 \cong \mathbf{k}$. The graded module $E = E_0 \oplus E_1 \oplus E_2 \oplus \ldots$ is called a *bialgebra* if it carries

- a graded algebra structure $m : E \otimes E \to E$, $m(E_l \otimes E_n) \subset E_{l+n}$ with a unit $e : \mathbf{k} \to E$, and
- a graded coalgebra structure $\mu : E \to E \otimes E$, $\mu(E_n) \subset E_0 \otimes E_n + E_1 \otimes E_{n-1} + \cdots + E_n \otimes E_0$ with a counit $\epsilon : E \to \mathbf{k}$

satisfying the following properties:

- μ is a graded algebra homomorphism or, equivalently, m is a graded coalgebra homomorphism;
- e is a graded coalgebra homomorphism;
- ϵ is a graded algebra homomorphism.

Theorem 6.1.12. *The module $\mathcal{M}$ of chord diagrams forms a bialgebra over $\mathbf{k}$ with respect to the operations introduced above.*

Exercise 6.1.13. We leave the proof of the theorem to the reader. It is sufficient to prove that the comultiplication is a homomorphism for the algebra structure of $\mathcal{M}$.

Definition 6.1.14 (Even commutative bialgebra). A bialgebra E is called *even commutative* if $m(a \otimes b) = m(b \otimes a)$ for any pair of elements $a, b \in E$. It is called *even cocommutative*, if for any element $a \in E$ the result of the comultiplication is a symmetric element of $E \otimes E$, i.e., $\mu(a) = \sum b_i \otimes c_i = \sum c_i \otimes b_i$.

The adjective "even" emphasizes the fact that the commutativity is understood in the usual, and not in the "super" sense, where the sign depends on the grading. It is clear from the definitions that $\mathcal{M}$ is even commutative and even cocommutative. From now on we will consider only even commutative and even cocommutative bialgebras and omit the term "even".

Example 6.1.15 (Polynomial bialgebras). Let $X = \{x_1, x_2, \ldots, x_r\}$ be a set of commuting variables. Consider the **k**-module E_n freely generated by monomials
$$x_1^{q_1} \ldots x_k^{q_k}$$
of total degree $n = q_1 + \cdots + q_k$. Setting $E_0 = \mathbf{k}$, we get a graded algebra $E = E_0 \oplus E_1 \oplus E_2 \oplus \ldots$ over **k**. The natural embedding $\mathbf{k} \to E_0$ provides E with the unit.

The coproduct for a monomial $y_1 \ldots y_m \in E$ ($y_i \in X$) is given by the formula
$$\mu(y_1 \ldots y_m) = 1 \otimes y_1 \ldots y_m + y_1 \otimes y_2 \ldots y_m + y_2 \otimes y_1 y_3 \ldots y_m + \cdots + y_1 \ldots y_m \otimes 1.$$

The bialgebra E endowed with this product and coproduct is called a *polynomial bialgebra*. We shall also make use of quasihomogeneous polynomial bialgebras. Such a bialgebra E is generated by a (possibly infinite) set $\{x_1, x_2, \ldots\}$ of pairwise commuting variables. Each variable is endowed with a positive integer weight w_i, and the grading is defined by means of the weight, not of the degree. The module E_n is spanned by all monomials
$$x_1^{q_1} \ldots x_k^{q_k}$$
of weight $n = w_1 q_1 + \cdots + w_k q_k$. We suppose that there is only finite number of variables x_i of a given weight. A polynomial bialgebra becomes a particular case of a quasihomogeneous polynomial bialgebra if one sets the weight of each independent variable equal to 1.

Quasihomogeneous polynomial bialgebras are in a sense standard bialgebras: any commutative bialgebra over a field of characteristic zero is isomorphic to a polynomial bialgebra (see below).

Exercise 6.1.16. Verify that a quasihomogeneous polynomial bialgebra is, indeed, a bialgebra.

For the sake of brevity, below we call quasihomogeneous polynomial bialgebras simply "polynomial bialgebras".

6.1.5 Structure Theorem for the Bialgebra $\mathcal{M}$

A special class of elements plays an important role in understanding the structure of bialgebras. These elements are "simpler" than the others.

Definition 6.1.17 (Primitive element). An element $p \in E$ is called *primitive* if $\mu(p) = 1 \otimes p + p \otimes 1$.

Note that a nonzero element $p \in E_0$ cannot be a primitive element. Indeed, $\mu(p) = p \cdot \mu(1) = p \cdot 1 \otimes 1$, while $1 \otimes p + p \otimes 1 = 2p \cdot 1 \otimes 1$.

Exercise 6.1.18. Prove that an element p of the polynomial bialgebra from Example 6.1.15 is primitive if and only if it is a linear polynomial.

Exercise 6.1.19. Verify that any primitive element in $\mathcal{M}_2$ is a multiple of

$$\bigotimes - \bigcirc .$$

Proposition 6.1.20. *All primitive elements of any bialgebra form a graded submodule* $\mathcal{P} \subset E$, $\mathcal{P} = \mathcal{P}_1 \oplus \mathcal{P}_2 \oplus \ldots$, *where* $\mathcal{P}_k = \mathcal{P} \cap E_k$. *In other words, all homogeneous components of a primitive element are primitive.*

The proof is a direct verification.

Exercise 6.1.21. Find a basis in the module of primitive elements in $\mathcal{M}_3$.

Answer. The module $\mathcal{P}_3$ of primitive elements is spanned, for example, by the element

$$\bigoplus - 2 \bigotimes + \bigominus .$$

Of course, primitive elements do not form either a subalgebra or a subcoalgebra in E, but they generate a subalgebra in E. In the case of a commutative and cocommutative bialgebra E over a field **k** of characteristic zero, this subalgebra is a polynomial algebra, and it coincides with E itself. This statement is a special case of the Milnor-Moore structure theorem for bialgebras.

Let us fix a basis $\{p_{k1}, \ldots\}$ in each space of primitive elements $\mathcal{P}_k \subset \mathcal{M}_k$. Denote by $S(\mathcal{P})$ the polynomial bialgebra in these elements.

Theorem 6.1.22 ([218]). *Over a field* **k** *of characteristic zero, each commutative cocommutative graded bialgebra is isomorphic to the polynomial bialgebra on its primitive elements.*

Hence, the bialgebras $\mathcal{M}$ and $S(\mathcal{P})$, if considered over a field of characteristic zero, are isomorphic.

Remark 6.1.23. For a bialgebra E, the *convolution product* $f * g : E \to E$ of two algebra endomorphisms $f, g : E \to E$ is defined by the formula

$$(f * g)(a) = m((f \otimes g)(\mu(a))), \quad a \in E.$$

A bialgebra E is called a *Hopf algebra* if it is endowed with an *antipode*, that is with an endomorphism A of E which is inverse to the identical algebra endomorphism id_E under convolution; this means that $A * \mathrm{id}_E$ is the identical operator on E_0, and it acts as zero on E_k for $k > 0$. For a polynomial bialgebra $\mathbb{C}[x_1, x_2, \ldots,]$, the unit with respect to the convolution product is the homomorphism which takes 1 to 1 and each x_i to 0. Therefore, the antipode is the a ring endomorphism acting on the generators by the formula $x_i \mapsto -x_i$. Hence, each polynomial bialgebra is endowed with a Hopf algebra structure, and all bialgebras we are going to study also carry a Hopf algebra structure. We found it more convenient to use the term "bialgebra" throughout the chapter simply because it is shorter.

The dimensions of the spaces of primitive elements $\mathcal{P}_n$ were computed by Vassiliev [283], Bar-Natan [18], and Kneissler [172] for the first few values of the grading, see Table 6.1.

Table 6.1. Dimensions of the spaces of chord diagrams and primitive elements

n	0	1	2	3	4	5	6	7	8	9	10	11	12
dim $\mathcal{P}_n$	0	1	1	1	2	3	5	8	12	18	27	39	55
dim $\mathcal{M}_n$	1	1	2	3	6	10	19	33	61	105	189	322	572

The dimensions of the spaces $\mathcal{M}_n$ can be easily calculated from that of $\mathcal{P}_n$. Indeed, the following relation holds:

Proposition 6.1.24. *We have*

$$m(t) = \frac{1}{(1-t)^{p_1}(1-t^2)^{p_2}(1-t^3)^{p_3}\cdots}$$

where $m(t) = 1 + m_1 t + m_2 t^2 + \ldots$ *is the generating function for the dimensions of the spaces* $\mathcal{M}_n$ *and* p_n *is the dimension of the space* $\mathcal{P}_n = \mathcal{P} \cap \mathcal{M}_n$.

The dimensions m_n for the first values of the grading are given in Table 6.1.

The formula in the proposition works in both directions, and it allows one to compute p_n knowing all values $m_1, \ldots, m_n$.

6.1.6 Primitive Elements of the Bialgebra of Chord Diagrams

How can one find the primitive element ⊗ − ◯ in $\mathcal{M}_2$? What must be done to construct one or more primitive elements in $\mathcal{M}_8$, for example? In this section we give an answer to these questions.

In the case of a polynomial bialgebra, the space of primitive elements coincides with that of linear polynomials in the generators. The problem of writing out primitive elements in other bialgebras is mainly related to the fact that we know a set of generators of the algebra, which are not primitive, and we do not know *a priori* their representation as polynomials in primitive elements. This is the case for the bialgebra $\mathcal{M}$.

There exists a natural projection of the bialgebra $\mathcal{M}$ onto the space of primitive elements, which allows one to construct a primitive element starting with an arbitrary element of $\mathcal{M}$. We shall write out an explicit formula for this projection.

Let $V = V(c)$ denote the set of chords of a chord diagram c. For a subset $J \subset V(c)$, let (J) denote the chord diagram formed by all chords from J. Set

$$\pi(c) = (V) - 1! \sum_{J_1 \sqcup J_2 = V} (J_1)(J_2) + 2! \sum_{J_1 \sqcup J_2 \sqcup J_3 = J} (J_1)(J_2)(J_3) - \ldots \quad (6.1)$$

where the sums are taken over all unordered disjoint partitions of the set V into *non-empty* subsets.

Theorem 6.1.25. *For each diagram c the element $\pi(c)$ is primitive.*

As an immediate consequence of the theorem we obtain the following statement.

Corollary 6.1.26. *The module of chord diagrams is generated by its primitive elements over $\mathbb{Z}$.*

Indeed, the right-hand side in (6.1) starts with the chord diagram $c = (V)$ itself. Moving all other terms to the left we obtain $c = \pi(c) + \ldots$, where, by the induction hypothesis, dots denote a polynomial in primitive elements of smaller order with integer coefficients.

In fact, the statement of the theorem can be made more precise. We call a chord diagram c *decomposable* if it can be represented as a product $c = c_1 c_2$ of two diagrams of smaller order, i.e., if the set V of its chords can be represented as a disjoint union $V = J \sqcup J'$ of non-empty subsets such that neither chord from J intersects a chord from J' and vice versa. Decomposable chord diagrams of order k span the subspace $\mathcal{D}_k \subset \mathcal{M}_k$ of *decomposable elements* in the space $\mathcal{M}_k$ of chord diagrams of order k. In other words, $\mathcal{D}_k$ is the space of order k polynomials in primitive elements of smaller order. It is easy to see that the space $\mathcal{M}_k$ is represented as the direct sum $\mathcal{M}_k = \mathcal{P}_k \oplus \mathcal{D}_k$, where $\mathcal{P}_k$ is the space of primitive elements of order k.

Theorem 6.1.27 ([185]). *The linear mapping π given by the formula (6.1) determines a projection $\pi : \mathcal{M}_k \to \mathcal{P}_k$ along $\mathcal{D}_k$, i.e., it takes each primitive element into itself, and it takes all decomposable elements to zero.*

Example 6.1.28. For the chord diagram 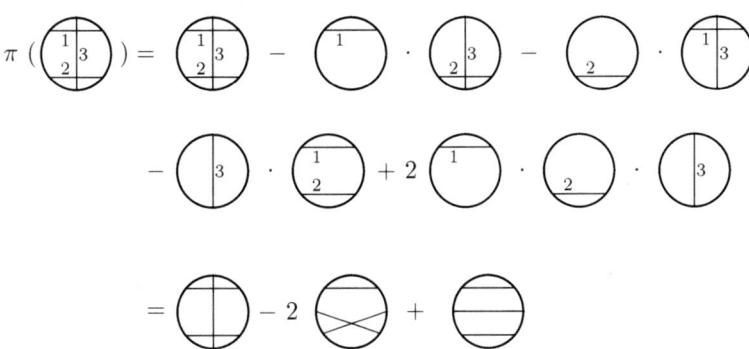, we have

which is a primitive element in $\mathcal{M}_3$.

We do not present a proof of Theorem 6.1.25. A direct proof can be found in [185]. A different approach looks like follows. The projection formula is a special case of the following more general construction (see [230], [246], [193]). Let $E = E_0 \oplus E_1 \oplus E_2 \oplus \ldots$ be a graded cocommutative coalgebra with a comultiplication $\mu : E \to E \otimes E$. Consider the space $\mathrm{Hom}(E, K)$ of linear mappings $E \to K$, where K is an arbitrary commutative algebra over $\mathbb{C}$. Then this space can be endowed with the *convolution product*

$$\varphi_1 \varphi_2(a) = m_K(\phi_1 \otimes \phi_2)\mu(a),$$

where $m_K : K \otimes K \to K$ is the multiplication in K. Since a product of mappings exists, we can consider power series of linear mappings. In particular, if E is a bialgebra, then any coalgebra homomorphism $\varphi : E \to E$ can be represented in the form $\varphi = 1 + \varphi_0$, where 1 is the identity mapping when restricted to E_0, and zero otherwise, and φ_0 vanishes on E_0. Therefore, the logarithm

$$\log \varphi = \log(1 + \varphi_0) = \varphi_0 - \frac{1}{2}\varphi_0^2 + \frac{1}{3}\varphi_0^3 - \ldots$$

is well defined. Indeed, φ_0 takes value 0 on E_0, hence φ_0^2 takes value 0 on $E_0 \oplus E_1$, φ_0^3 takes value 0 on $E_0 \oplus E_1 \oplus E_2$, and so on. As a result, only finite number of summands differ from zero on any value $a \in E$ in the expansion of $\log \varphi$.

Proposition 6.1.29 ([246]). *For any linear mapping $\varphi : E \to E$ which is a coalgebra homomorphism, the logarithm $\log \varphi$ maps E to the subspace of primitive elements $P(E) \subset E$. The logarithm of the identity mapping is the projection onto the subspace of primitive elements.*

6.2 Knot Invariants and Origins of Chord Diagrams

Knot invariants of finite order were introduced in [283] (see also [284], [285]). In this paper Vassiliev associated chord diagrams to singular knots and deduced the 4-term relations.

6.2.1 Knot Invariants and their Extension to Singular Knots

Consider an embedding of the oriented circle S^1 into $\mathbb{R}^3$, i.e., a smooth mapping $u : S^1 \to \mathbb{R}^3$ without self-intersections and with a non-vanishing tangent vector. We also suppose that the space $\mathbb{R}^3$ is endowed with an orientation. Two embeddings $u_0, u_1 : S^1 \to \mathbb{R}^3$ are called *isotopy equivalent* if there exists an orientation preserving diffeomorphism $g : \mathbb{R}^3 \to \mathbb{R}^3$ such that the diagram

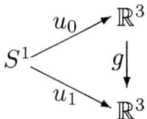

is commutative. This notion is equivalent (because the group of diffeomorphisms of $\mathbb{R}^3$ is contractible) to the perhaps more intuitive notion of homotopy, which would be that $u_0, u_1 : S^1 \to \mathbb{R}^3$ are equivalent if they can be joined by a continuous path of embeddings $u_t : S^1 \to \mathbb{R}^3$, $t \in [0, 1]$.

Definition 6.2.1 (Knot). A *knot* is an isotopy class of embeddings $S^1 \to \mathbb{R}^3$.

In the sequel we will make no distinction between a knot and an embedding $S^1 \to \mathbb{R}^3$ representing this knot.

Definition 6.2.2 (Knot invariant). A *knot invariant* is a function on knots.

We usually consider knot invariants with values in an arbitrary commutative associative algebra K with unit over the ground ring **k**.

Knots are usually presented by their diagrams – generic plane projections of the knots. At double points of the projection overpasses and underpasses of the two branches are shown, and an orientation of the knot is given.

The main problem in the theory of knots is "How to distinguish knots". Of course, if we manage to find a knot invariant taking distinct values on two presentation of two knots, then the corresponding knots are also distinct. Various types of knot invariants have been introduced since the beginning of the study of knots. Some of them are explained later in this section.

The main idea of constructing Vassiliev knot invariants is to consider "singular knots", not only smooth ones, see Fig. 6.9.

Fig. 6.8. Diagrams of various knots

Fig. 6.9. Singular knots

Definition 6.2.3 (Singular knot). A *singular knot* is an isotopy class of mappings $u : S^1 \to \mathbb{R}^3$ such that

- $u'(t) \neq 0$ for all $t \in S^1$;
- all points of self-intersection of the image $u(S^1)$ are ordinary double points with transversal self-intersection (i.e., the two tangent vectors to the knot at a double point are not proportional).

Consider a small neighborhood of a double point of a singular knot. In such a neighborhood each singular knot looks like two intersecting segments:

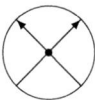

The arrows denote the orientation of the knot corresponding to the orientation of the initial circle S^1.

A singular point can be "resolved" by a small deformation of the knot in two different directions – the *positive* and the *negative* one (see Fig. 6.10). The sign of a resolution coincides with the orientation of the triple of vectors (the upper vector, the lower vector, the vector from the lower image of the resolved intersection point to its upper image). Recall that the ambient space $\mathbb{R}^3$ is supposed to be oriented. The main point here is that, as the reader can easily check, the sign of a resolution does not change if one looks at the same picture from the "opposite side of the page".

Fig. 6.10. Positive (on the left) and negative (on the right) resolutions of a singular point

If two (possibly singular) knots differ from each other only inside a small ball, where they look like on the left- and on the right-hand side of Fig. 6.10, then we say that they can be obtained one from the other by an *elementary transformation*.

Theorem 6.2.4. *Let v be a knot invariant. Then it has a (unique) extension to singular knots satisfying the following relation:*

$$v\left(\vcenter{\hbox{\includegraphics{}}}\right) = v\left(\vcenter{\hbox{\includegraphics{}}}\right) - v\left(\vcenter{\hbox{\includegraphics{}}}\right)$$

The three pictures are meant to show three (singular) knots, coinciding everywhere outside a small ball, where they look as in the pictures.

Proof. We will proceed by induction over the number of double points in a singular knot.

Consider a knot with one singular point. The two resolutions of the point are uniquely determined, and we just determine the value of a knot invariant on the singular knot according to the defining rule, which provides the base of induction.

Suppose a knot invariant v is extended to all singular knots with less than n double points. Let us prove that it may be extended to any singular knot u with n double points.

Let $x, y \in \mathbb{R}^3$ be two double points of the singular knot u. We are going to show that the extension of the invariant v to u does not depend on which of the points x or y is resolved.

Consider four possible ways of resolving both singular points x and y (see Fig. 6.11).

Then the extension of the invariant v to the singular knot u should be equal to $v(u_1) - v(u_2) - v(u_3) + v(u_4)$ independently of what the order of resolutions is.

The theorem is proved.

6.2 Knot Invariants   353

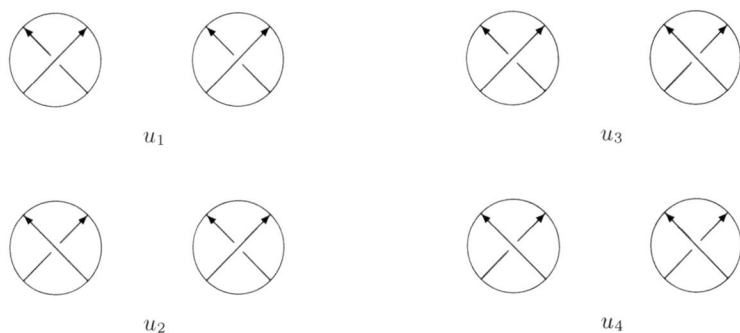

Fig. 6.11. Four possible resolutions of two singular points

6.2.2 Invariants of Finite Order

We are now able to introduce the main notion of the theory.

Definition 6.2.5 (Vassiliev invariant). A knot invariant is called a *Vassiliev invariant, of order not greater than* n, if it vanishes on all singular knots with more than n double points.

Vassiliev invariants of order not greater than n with values in a **k**-algebra K form a **k**-module which we denote by $\mathcal{V}_n$. The union $\mathcal{V} = \mathcal{V}_0 \cup \mathcal{V}_1 \cup \mathcal{V}_2 \cup \ldots$ forms a **k**-module $\mathcal{V}$ of *finite order invariants* endowed with the filtration $\mathcal{V}_0 \subset \mathcal{V}_1 \subset \mathcal{V}_2 \subset \cdots \subset \mathcal{V}$.

Example 6.2.6. Constant functions on nonsingular knots provide Vassiliev invariants of order ≤ 0. On the other hand, each invariant from $\mathcal{V}_0$ must be a constant on nonsingular knots, i.e., $\mathcal{V}_0 \cong K$.

Indeed, an elementary transformation of a knot does not change the value of any invariant of order ≤ 0, and any knot may be obtained from any other knot by a series of elementary transformations.

Example 6.2.7. Vassiliev invariants of order not greater than 1 also come from constant functions on knots. Indeed, consider a singular knot with one double point. After a series of elementary transformations it may be transformed to a "decomposed" knot with one singular point (see Fig. 6.12).

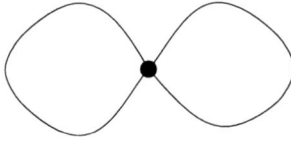

Fig. 6.12. A decomposed knot with one singular point

These elementary transformations do not change the value of an invariant of order ≤ 1, and for the final knot, shown in Fig. 6.12, it should be equal to zero: both resolutions give the same knot which is the unknot. Thus, $\mathcal{V}_1 = \mathcal{V}_0 \cong K$.

Now we are going to associate a chord diagram to any singular knot. A Vassiliev invariant of order not greater than n determines, in fact, a function on chord diagrams with n chords.

Definition 6.2.8 (Chord diagram of a singular knot). Let $u : S^1 \to \mathbb{R}^3$ be a singular knot with n double points $x_1, \ldots, x_n \in \mathbb{R}^3$. The *chord diagram of the singular knot* u is the circle S^1 with n pairs of points, which are the preimages $\{u^{-1}(x_1), \ldots, u^{-1}(x_n)\}$.

Obviously, any chord diagram can be realized as a chord diagram of a singular knot.

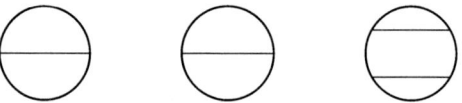

Fig. 6.13. Chord diagrams of singular knots from Fig. 6.9

Theorem 6.2.9. *The value of a Vassiliev invariant of order not greater than n on a singular knot with exactly n double points depends only on the chord diagram of the knot, and not on the knot itself.*

The proof is based on the following statement.

Lemma 6.2.10. *Let two singular knots have the same chord diagram. Then one of them can be obtained from the other one by a series of elementary transformations.*

Proof. Indeed, let u_1, u_2 be two singular knots with n double points, that have identical chord diagrams. By an appropriate isotopy the knot u_2 can be transformed so that

- each of the double points of u_2 coincides with the corresponding double point of u_1;
- each of the arcs of u_2 coincides with the corresponding arc of u_1 in a small neighborhood of each double point;
- the arcs of u_1 and u_2 have no common points outside these small neighborhoods.

Now take two successive double points of u_2 and the arc of u_2 connecting these points. This arc can be transformed into the corresponding arc of u_1 by a series of elementary transformations preserving the small neighborhoods of the ends of the arcs. Due to the defining condition, the value of v does not change under such deformations, whence the required assertion.

Exercise 6.2.11. Consider a singular knot u with n double points. Resolving these double points in all possible ways one obtains 2^n distinct (nonsingular) knots $u_{\varepsilon_1,\ldots,\varepsilon_n}$. Here $\varepsilon_i = 1$ if the i-th double point is resolved in the positive direction, $\varepsilon_i = -1$ otherwise.

Verify that a knot invariant v is of order not greater than $n-1$ if and only if
$$\sum_{(\varepsilon_1,\ldots,\varepsilon_n)} \varepsilon_1 \ldots \varepsilon_n v(u_{\varepsilon_1,\ldots,\varepsilon_n}) = 0$$
for any singular knot u with n double points.

Theorem 6.2.9 implies that values of the invariant do not depend on the specific behavior of the knotted curve, being completely determined by the corresponding chord diagram, i.e., by the order in which double points are encountered while moving along the curve. Moreover, if the restrictions of two invariants of order not greater than n to the set of all singular knots with exactly n double points coincide, then their difference is an invariant of order not greater than $n-1$. This means that an element of $\mathcal{V}_n/\mathcal{V}_{n-1}$ determines a function on chord diagrams of order n, and this mapping is a monomorphism.

As an immediate corollary we obtain the following important statement, valid over a field **k** of characteristic zero.

Proposition 6.2.12. *The vector space $\mathcal{V}_n$ is finite-dimensional for each $n = 0, 1, 2, \ldots$.*

Indeed, the dimension of the quotient space $\mathcal{V}_n/\mathcal{V}_{n-1}$ does not exceed the number of chord diagrams of order n. This dimension is, in fact, significantly smaller, as we will see below.

6.2.3 Deducing 1-Term and 4-Term Relations for Invariants

Although each invariant of order $\leq n$ determines a function on chord diagrams, not every function can be obtained in this way. Indeed, we have seen, for example, that each invariant of order ≤ 1 is a constant function on knots. It follows that each function on the chord diagram with one chord coming from invariants of order ≤ 1 must be equal to zero. Thus, there are some restrictions on functions on chord diagrams coming from finite order knot invariants.

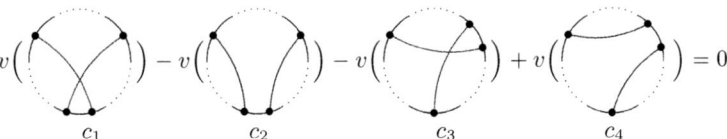

Fig. 6.14. The 4-term relation

Theorem 6.2.13. *Any function v on chord diagrams of order n coming from a knot invariant of order not greater than n satisfies two types of relations:*

- *1-term relations: if a chord diagram contains a chord that does not intersect any other chord, then v equals zero on this diagram;*
- *4-term relations, see Fig. 6.14.*

Proof. We start by deducing the 1-term relation. Similarly to the proof of Lemma 6.2.10, a series of elementary transformations transforms a singular knot with a chord diagram having a chord which does not intersect any other chord, into a knot of the following type:

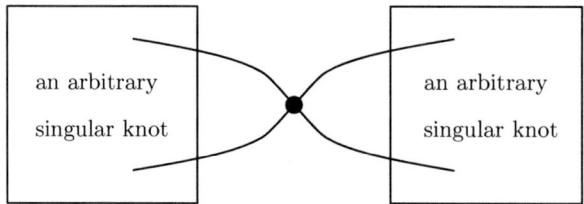

It is obvious that the type of the singular knot resolved at the distinguished double point does not depend on the resolution, and the 1-term relation follows.

Now, consider two chords taking part in a 4-term relation. One of them is a "moving" chord, we denote it by A, the other one is "stable", we denote it by B.

Each of the chords A and B corresponds to a double point of the knot. Let us draw four pictures of the knots corresponding to the four terms of the 4-term relation: see Fig. 6.15. The right-hand side presents two resolutions of the double point A in each knot.

Summing all the four relations we see that the terms on the right-hand side pairwise annihilate each other, and thus we obtain the 4-term relation.

We denote the module of order n chord diagrams considered modulo the 1-term and the 4-term relations by $\mathcal{N}_n$ and set

$$\mathcal{N} = \mathcal{N}_0 \oplus \mathcal{N}_1 \oplus \mathcal{N}_2 \oplus \dots.$$

This module is obviously the quotient module of $\mathcal{M}$ modulo the 1-term relations. The following statement is obvious.

Theorem 6.2.14. *The module $\mathcal{N}$ inherits a commutative cocommutative graded bialgebra structure from $\mathcal{M}$.*

A natural question arises: whether there exist other restrictions for knot invariants of finite order that do not follow from 1-term and 4-term relations? In the case when **k** is a field of characteristic zero the negative answer is given by the following theorem due to M. Kontsevich:

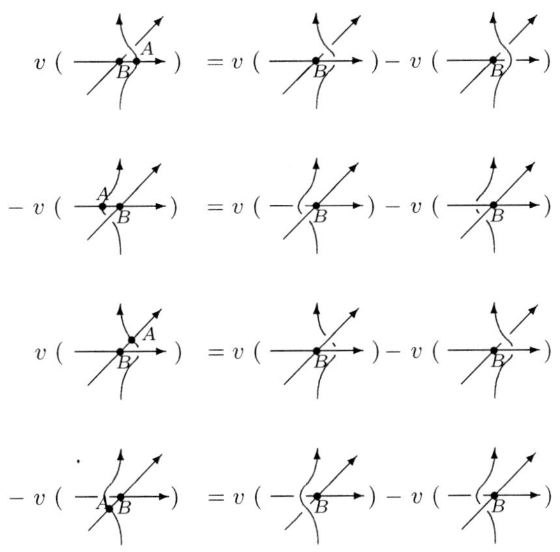

$$v(\ \cdot\ ) = v(\ \cdot\ ) - v(\ \cdot\ )$$

$$-v(\ \cdot\ ) = v(\ \cdot\ ) - v(\ \cdot\ )$$

$$v(\ \cdot\ ) = v(\ \cdot\ ) - v(\ \cdot\ )$$

$$-v(\ \cdot\ ) = v(\ \cdot\ ) - v(\ \cdot\ )$$

Fig. 6.15. Resolutions of the four singular knots corresponding to the terms of the 4-term relation

Theorem 6.2.15 ([179], [18]). *Every function on chord diagrams of order n with values in an algebra over a field $\mathbf{k}$ of characteristic zero satisfying 1-term and 4-term relations can be obtained from a knot invariant of order not greater than n.*

Remark 6.2.16. Note that if we are working with framed knots instead of simple knots, then the 1-term relation is no longer valid, and the Hopf algebra of chord diagrams provides the correct picture for finite order invariants of framed knots. (A *framed knot* is a smooth mapping of a narrow cylindric band to $\mathbb{R}^3$; the restriction of the mapping to one side of the band is the knot, and the image of the band is the *framing*.) It is worth mentioning, however, that working with framed knots is somewhat subtle, see the discussion in [164]. Anyway, the renormalization described in Sec. 6.3.2 allows one to construct a function satisfying both the 1-term and the 4-term relation starting from a function satisfying only the 4-term relation.

6.2.4 Chord Diagrams of Singular Links

As we have already mentioned, a chord diagram is nothing else but a one-face map, or, by duality, a one-vertex map (we have studied them in detail in Chapter 3). Indeed, a chord diagram can be represented by a pair of permutations on the set of chord ends. One of the permutations, which represents the cyclic order of the points around the circle, is cyclic. The other one, which represents the chords, is an involution without fixed points. It is not difficult to show that the genus of the corresponding one-face map is a "weight system", see Exercise 6.3.13 below.

One may ask if it is possible to generalize the theory of Vassiliev invariants to general maps. The answer is yes. In order to do that we must consider links instead of knots.

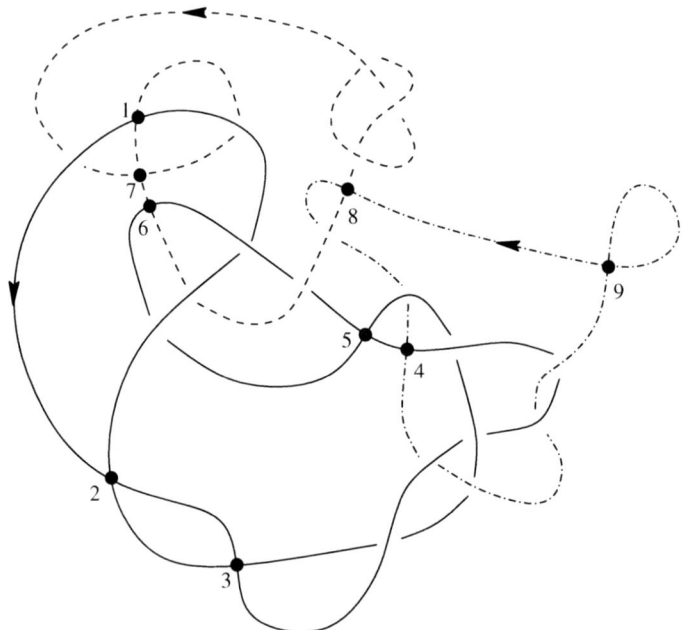

Fig. 6.16. A singular link

Fig. 6.16 shows a singular link, which is obtained by gluing certain (non-singular) points of three (singular) knots. Following our general scheme, we take the preimage of each component of the link, and mark the two preimages of a singular point by the marks i and $\bar{i}$. We thus obtain three circles shown in Fig. 6.17. The result may be expressed as the permutation

$$\sigma = (1, 2, 3, 4, 5, 6, \bar{5}, \bar{3}, \bar{2})(8, 7, \bar{1}, \bar{7}, \bar{6})(\bar{8}, \bar{4}, 9, \bar{9})$$

on the set
$$D = \{1, 2, \ldots, 9, \bar{1}, \bar{2}, \ldots, \bar{9}\}.$$

In order not to encumber the picture with unnecessary details, we did not draw the chords in Fig. 6.17. But their representation as a permutation is clear:

$$\alpha = (1, \bar{1})(2, \bar{2})(3, \bar{3})(4, \bar{4})(5, \bar{5})(6, \bar{6})(7, \bar{7})(8, \bar{8})(9, \bar{9}).$$

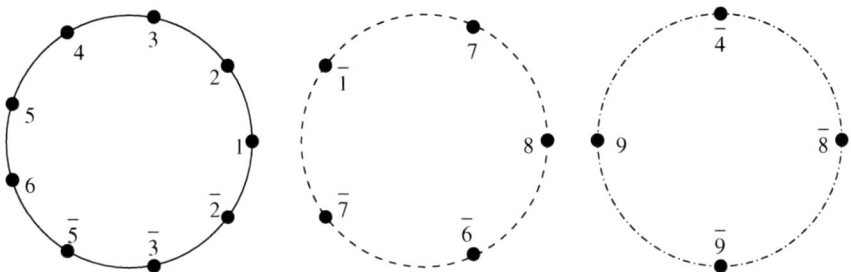

Fig. 6.17. Three circles that form the base for the chord diagram corresponding to Fig. 6.16. The chords are not shown: they connect 1 to $\bar{1}$, 2 to $\bar{2}$, etc.

Thus, the "chords" may join points of different circles.

The result of the whole construction is a map, represented by the pair of permutations $[\sigma, \alpha]$. Vassiliev invariants of links are described in terms of maps modulo the one- and the four-term relation. We just have to keep in mind that the corresponding maps may well be non-connected.

6.3 Weight Systems

6.3.1 A Bialgebra Structure on the Module $\mathcal{V}$ of Vassiliev Knot Invariants

The multiplication and the comultiplication in the module of chord diagrams do not appear by accident. They just mimic corresponding (dual) operations on knot invariants.

Consider the module $\mathcal{V}$ of finite-order (or Vassiliev's) invariants with values in an algebra K over the ground ring **k**. This module is, in fact, a **k**-algebra with respect to the usual multiplication of invariants as functions on knots:

$$m(v_1, v_2) = v_1 v_2, \text{ where } v_1 v_2(u) = v_1(u) v_2(u).$$

This algebra is a filtered algebra:

Proposition 6.3.1. $\mathcal{V}_m \mathcal{V}_n \subset \mathcal{V}_{m+n}$.

Proof. Exercise for the reader. [**Hint:** Use Exercise 6.2.11.]

The *concatenation* (or the *connected sum*) of knots is a natural operation of gluing two knots together respecting their orientation: see Fig. 6.18.

We denote the concatenation of two knots u_1, u_2 by $u_1 \# u_2$.

The concatenation supplies the algebra $\mathcal{V}$ with a *comultiplication*. For $v \in \mathcal{V}$ we set

$$\mu(v)(u_1 \otimes u_2) = v(u_1 \# u_2).$$

360   6 Algebraic Structures

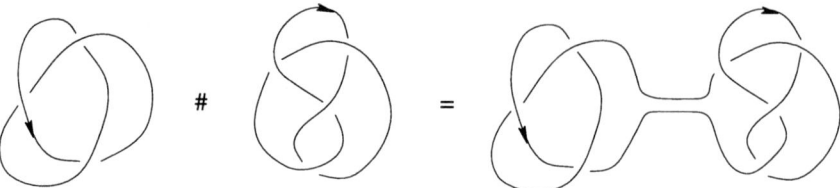

Fig. 6.18. Concatenation of knots

Here the tensor product of knots is understood as an element of the tensor square of the **k**-module freely spanned by knots.

The comultiplication makes $\mathcal{V}$ into a filtered coalgebra:

Proposition 6.3.2.

$$\mu(\mathcal{V}_n) \subset \mathcal{V}_0 \otimes \mathcal{V}_n + \mathcal{V}_1 \otimes \mathcal{V}_{n-1} + \cdots + \mathcal{V}_n \otimes \mathcal{V}_0.$$

Proof. Exercise for the reader.

Now let us associate with the filtered module $\mathcal{V}$ the corresponding graded module

$$\operatorname{gr} \mathcal{V} = \mathcal{V}_0 \oplus \mathcal{V}_1/\mathcal{V}_0 \oplus \mathcal{V}_2/\mathcal{V}_1 \oplus \ldots.$$

Theorem 6.3.3. *The operations of multiplication $m : \mathcal{V} \otimes \mathcal{V} \to \mathcal{V}$ and comultiplication $\mu : \mathcal{V} \to \mathcal{V} \otimes \mathcal{V}$ descend to multiplication $m : \operatorname{gr} \mathcal{V} \otimes \operatorname{gr} \mathcal{V} \to \operatorname{gr} \mathcal{V}$ and comultiplication $\mu : \operatorname{gr} \mathcal{V} \to \operatorname{gr} \mathcal{V} \otimes \operatorname{gr} \mathcal{V}$ making the **k**-module $\operatorname{gr} \mathcal{V}$ a graded bialgebra with the unit $e: \mathbf{k} \to \mathcal{V}_0$ and the counit $\epsilon : \mathcal{V}_0 \to \mathbf{k}$ given by the natural isomorphism mappings.*

This is a straightforward calculation, and we leave it to the reader.

Now Kontsevich's theorem 6.2.15 can be reformulated as follows:

Theorem 6.3.4. *Over a field **k** of characteristic zero, the bialgebra $\operatorname{gr} \mathcal{V}$ is isomorphic to the dual bialgebra $\mathcal{N}^* = \mathcal{N}_0^* \oplus \mathcal{N}_1^* \oplus \mathcal{N}_2^* \oplus \ldots$ (having the natural structure of a graded bialgebra).*

6.3.2 Renormalization

In this section we suppose that the ground ring **k** is a field of characteristic zero.

Functions on chord diagrams originating from Vassiliev knot invariants satisfy two types of relations, 1-term relations and 4-term relations. But in Sec. 6.1 we considered chord diagrams modulo only 4-term relations. The reason is that the 1-term relations are much less important than the 4-term ones.

6.3 Weight Systems

To be more precise, let $\mathcal{M}^*$ be the space of linear functions on $\mathcal{M}$, and let $\mathcal{N}^* \subset \mathcal{M}^*$ be the subspace of functions satisfying also 1-term relations. Let $X_n : \mathcal{M}_n \to \mathcal{M}_n$ be the *chord separating operator* defined, on a diagram c of order n, by the formula

$$ X_n : c \mapsto \ominus \cdot \sum_{i=1}^{n} c_i. $$

Here c_i is the chord diagram of order $n-1$ obtained from c by removing the i-th chord.

Introduce the operator $\mathrm{Rn} : \mathcal{M} \to \mathcal{M}$ by the following formula:

$$ \mathrm{Rn}_n = \prod_{i=1}^{n} \left(1 - \frac{X_n}{i}\right) = (-1)^n \binom{X_n - 1}{n}, \quad \mathrm{Rn}_n : \mathcal{M}_n \to \mathcal{M}_n. \quad (6.2) $$

The bialgebra $\mathcal{M}$ is isomorphic to the polynomial algebra in primitive elements. We shall see that, in fact, the application of Rn is equivalent to setting $\ominus = 0$ in each polynomial.

Theorem 6.3.5. *The operator Rn^* is a projection $\mathrm{Rn}^* : \mathcal{M}^* \to \mathcal{N}^*$, i.e., $(\mathrm{Rn}^*)^2 = \mathrm{Rn}^*$ and $\mathrm{Rn}^*|_{\mathcal{N}^*} = \mathrm{id}$.*

This projection is called the *renormalization operator*. Having the renormalization operator in mind, the study of functions satisfying only 4-term relations is equivalent to the study of functions satisfying both 4-term and 1-term relations.

Proof. In order to prove the theorem it is sufficient to prove two lemmas.

Lemma 6.3.6. *The kernel of Rn contains the submodule $\ominus \cdot \mathcal{M} \subset \mathcal{M}$.*

Proof. The factor $1 - X_n/k$ in the product kills each chord diagram with exactly k isolated chords. But this factor can produce a linear combination of chord diagrams with more than k isolated chords. Since these new chord diagrams are killed by other factors corresponding to greater values of k, we can proceed by induction.

Lemma 6.3.7. *The operator Rn_n is the identity on $\mathcal{P}_n \subset \mathcal{M}_n$ for $n \geq 2$.*

Indeed, applying the chord separating operator X_n to the right-hand side of Eq. (6.1) (see page 348) we conclude that it maps $\mathcal{P}_n$ to zero.

We finish this section with a proposition describing properties of the renormalization operator which can be verified directly. Let $X : \mathcal{M} \to \mathcal{M}$ denote the linear operator whose restriction to each $\mathcal{M}_n$ coincides with X_n.

Proposition 6.3.8. 1. *The operator* $X\colon \mathcal{M} \to \mathcal{M}$ *is a derivation, i.e.,*

$$X(c_1 c_2) = X(c_1)c_2 + c_1 X(c_2).$$

2. *The operator* $\mathrm{Rn}\colon \mathcal{M} \to \mathcal{M}$ *is a bialgebra homomorphism.*
3. *The operator* $\mathrm{Rn}_n \colon \mathcal{M}_n \to \mathcal{M}_n$ *can also be given by the formula*

$$\mathrm{Rn}\colon c \mapsto \sum_J \left(-\bigcirc\right)^{n-|J|} \cdot (J)$$

where $J \subset V(c)$ runs over all subsets of chords of the diagram c, the diagram (J) being formed by all chords from J, and $|J|$ being the number of chords in J.

6.3.3 Weight Systems

As we have seen in the previous section, the investigation of finite order knot invariants is equivalent, in a way, to the investigation of linear functions on $\mathcal{M}$ with values in an algebra K over the ground ring **k**, i.e., of the module $\mathcal{M}^*$.

Definition 6.3.9 (Weight systems). Elements of $\mathcal{M}^* = \mathrm{Hom}(\mathcal{M}, K)$ are called *weight systems* (or *preinvariants*, as in [52]) with values in K. A weight system is called *multiplicative* if it is a homomorphism of ring structures in $\mathcal{M}$ and K.

Modules $\mathcal{M}_k^*$ for $k = 0, 1, 2, \ldots$ of weight systems of order k are naturally embedded in $\mathcal{M}^*$ which consists of linear functionals on $\mathcal{M}$ vanishing on all $\mathcal{M}_k$ for k sufficiently large. Indeed, each weight system of order k can be extended to an element in $\mathcal{M}^*$ simply by setting it equal to zero on $\mathcal{M}_i$ for $i \neq k$. (More formally, $\mathcal{M}^*$ does not coincide with $\mathcal{M}_0^* \oplus \mathcal{M}_1^* \oplus \mathcal{M}_2^* \ldots$, but it is equal to the projective limit of the finite sums $\mathcal{M}_0^* \oplus \mathcal{M}_1^* \oplus \cdots \oplus \mathcal{M}_n^*$.)

We are now ready to present some simple examples of weight systems. Large series of examples are described in Secs. 6.4 and 6.5. Using renormalization, it is possible to construct an invariant out of each weight system.

We usually define values of a weight system on either chord or arc diagrams, meaning that it can be extended to a linear function on $\mathcal{M}$.

Example 6.3.10 (Number of chords). The number of chords in the diagram is obviously a weight system. It is a constant function on $\mathcal{M}_n$, and thus is of no interest.

Example 6.3.11 (Chords intersections number). We say that two chords in a chord diagram intersect if their ends alternate along the circle. The number of chord intersections (i.e., the number of pairs of intersecting chords) is a weight system. Indeed, it is obvious that interchanging the ends of two neighboring chords we change the number of intersections by one, and therefore the 4-term relation is trivially satisfied.

Exercise 6.3.12. Prove that the renormalized number of chords intersections is identically equal to zero.

Example 6.3.13. This example shows a deep and rather nontrivial interrelation between chord diagrams and graphs on surfaces. As we have mentioned in the beginning of the chapter, a chord diagram c can be considered as a one-face map. Denote by $f(c)$ the number of faces in the *dual* map (that is, the number of vertices in the original map). Recall that, for a chord diagram with n chords, the dual map is obtained by gluing darts of the $2n$-star. The number of faces of this map coincides with the number of boundary components of the two-surface obtained by thickening each chord and by attaching the disk to the circle. The number of faces for all chord diagrams with up to 3 chords are shown in Fig. 6.19.

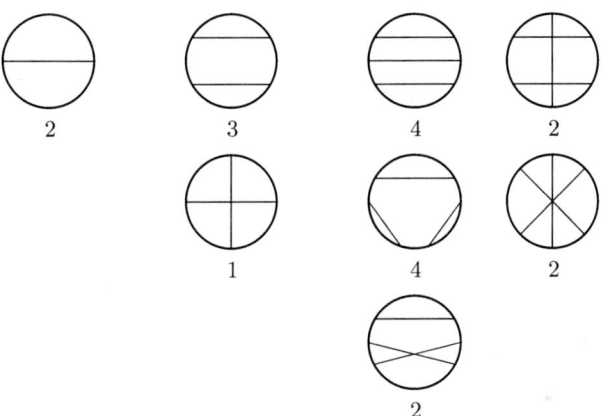

Fig. 6.19. Number of faces for all chord diagrams of order 1, 2, 3

Let $F : \mathbb{N} \to K$ be any fixed function with values in an arbitrary ring K. Then the mapping $F \circ f$ provides a weight system with values in K.

Indeed, consider the four terms of a 4-term element with doubled chords (see Fig. 6.20). In the chord diagram c_1 with doubled chords the tails (a, d),

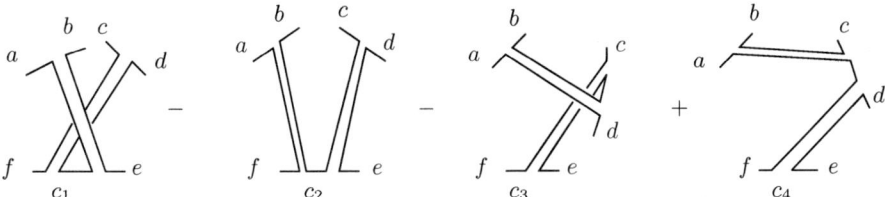

Fig. 6.20. Number of faces in the 4-term element

(b, e), (c, f) are glued pairwise. The same pairs of tails are glued in the diagram c_3. Therefore the numbers of faces in the diagrams c_1 and c_3 coincide. Similar argument is valid for the pair c_2 and c_4. We see that the 4-term relation splits into a pair of 2-*term relations* $f(c_1) = f(c_3)$ and $f(c_2) = f(c_4)$. Composition with an arbitrary function does not change the situation.

Exercise 6.3.14 ([19]). Prove that two chord diagrams are equivalent modulo the 2-*term relation* $c_1 = c_3$, $c_2 = c_4$ if and only if the numbers of faces of these chord diagrams coincide. In other words, consider the subalgebra in the algebra of chord diagrams generated by the unique chord diagram with one chord and the unique indecomposable chord diagram with two chords. Then the mapping taking each chord diagram to the chord diagram in this algebra with the same number of faces descends to an algebra homomorphism.

Example 6.3.15. Associate to any chord diagram c its *intersection graph* $\Gamma(c)$. The vertices of the graph correspond to the chords of the diagram, and two vertices are connected by an edge if the corresponding chords intersect (see Definition 6.4.1). Let $\chi : G \mapsto \chi_G \in \mathbb{Z}[t]$ be the mapping that takes each graph G to its chromatic polynomial. Then the mapping $c \mapsto \chi_{\Gamma(c)}$ is a weight system with values in $\mathbb{Z}[t]$. The proof is given in Sec. 6.4.

6.3.4 Vassiliev Knot Invariants and Other Knot Invariants

This subsection is dedicated to examples of some classical knot invariants that were known long before Vassiliev introduced finite order invariants. All these invariants are functions on nonsingular knots.

Knot invariants are usually defined for plane projections of knots, and then one must prove that they are well defined. For such proofs we refer the reader to [74]. But sometimes the invariance of the function follows directly from the definition, as in the following two examples.

Example 6.3.16. The *genus* of a knot u is the minimal genus of a surface with boundary embedded in $\mathbb{R}^3$ and such that the boundary of the surface coincides with u. (Making several holes in a surface of genus g we obtain a surface with boundary and we agree that the genus of this surface also is g. If the number of holes (= the number of boundary components) is m, then the Euler characteristic of the surface with boundary is $2 - 2g - m$: the boundary components are considered as "missing faces".) The genus of a knot is obviously a knot invariant.

Example 6.3.17. The *fundamental group* of the complement of a knot in $\mathbb{R}^3$ is a knot invariant.

As it is usual in knot theory, these two invariants, which are easily defined, are very complicated to handle. For example, the complexity of known algorithms for computing the genus of a knot from a knot diagram exceeds the capacity of contemporary computers even for diagrams with a small number of self-intersections. Concerning the fundamental group, it is in fact not difficult to find a presentation of the fundamental group from a knot diagram, but comparing two presentations obtained from two distinct diagrams is quite a job. To the best of our knowledge, nobody has proved that the genus of a knot is not a finite order invariant, but it has no reason to be one. And if it is not, this would not mean that finite order invariants do not distinguish knots. On the other hand, if anybody proved that any two knots with distinct fundamental group of the complement are distinguished by a Vassiliev invariant, this would be serious evidence confirming that finite order invariants distinguish knots.

For large classes of knot invariants it has already been proved that they can be expressed in terms of Vassiliev invariants. The words "can be expressed" should be specified in each case separately. We hope, however, that the two following examples clearly explain the meaning of these words.

Example 6.3.18 (Conway polynomial [18]). The *Conway polynomial* is a knot invariant with values in the ring $\mathbb{Z}[x]$ of polynomials in one variable x. In fact, the Conway polynomial is defined on *links*, not only on knots. A *link* is an isotopy class of nonsingular embeddings of a finite set of circles in $\mathbb{R}^3$.

Consider a generic plane projection of a knot (or a link), i.e., a projection with only transversal double points of self-intersection. We define the Conway polynomial by the recurrence rule:

$$x \cdot \mathrm{Con}\!\left(\!\!\begin{array}{c}\includegraphics{}\end{array}\!\!\right) = \mathrm{Con}\!\left(\!\!\begin{array}{c}\includegraphics{}\end{array}\!\!\right) - \mathrm{Con}\!\left(\!\!\begin{array}{c}\includegraphics{}\end{array}\!\!\right)$$

and the initial value $\mathrm{Con}(\text{unknot}) = 1$.

This definition allows one to calculate the value of the Conway polynomial knowing a generic projection of the knot. The definition is very close to that of Vassiliev invariants so it is not surprising that the *coefficients* of the Conway polynomial are Vassiliev invariants.

Proposition 6.3.19. *Let C_n be the knot invariant with values in $\mathbb{Z}$ equal to the coefficient of x^n in the Conway polynomial. Then C_n is a Vassiliev knot invariant of order $\leq n$.*

Indeed, extend the Conway polynomial to singular knots according to the rule

$$\mathrm{Con}\!\left(\!\!\begin{array}{c}\includegraphics{}\end{array}\!\!\right) = \mathrm{Con}\!\left(\!\!\begin{array}{c}\includegraphics{}\end{array}\!\!\right) - \mathrm{Con}\!\left(\!\!\begin{array}{c}\includegraphics{}\end{array}\!\!\right).$$

Then, by the defining recurrence rule we have

$$x \cdot \mathrm{Con}\left(\smile\hspace{-0.3em}\frown\right) = \mathrm{Con}\left(\times\right).$$

Therefore, the Conway polynomial of a singular knot with k double points is divisible by at least x^k, and the coefficient C_n obviously vanishes on singular knots with more than n double points.

Thus each coefficient C_n determines a weight system of order n.

Remark 6.3.20. Although the coefficients of the Conway polynomial are Vassiliev invariants, the polynomial itself is not a Vassiliev invariant. Indeed, the sequence of its coefficients C_n has increasing order.

Exercise 6.3.21. According to Example 6.2.11 the number of faces in the map determined by a chord diagram is a weight system. Prove that the value of C_n on a chord diagram is equal to 1 if there is only one face, and that it vanishes otherwise.

Remark 6.3.22. The reader should not conclude that the coefficients of the Conway polynomial themselves can only take values 0 and 1. This is true only for their extensions C_n to singular knots with n double points.

Example 6.3.23 (HOMFLY polynomial). The *HOMFLY polynomial* $H = H(N, q)$ (see [34], [18]) is a knot (and link) invariant. It depends on two variables, denoted usually by N and $q^{1/2}$. Similarly to the Conway polynomial, the HOMFLY polynomial is defined by the relation

$$(q^{1/2} - q^{-1/2})H\left(\smile\hspace{-0.3em}\frown\right) = q^{N/2}H\left(\times\right) - q^{-N/2}H\left(\times\right)$$

and the initial value $H(\text{unknot}) = 1$.

Substitute $q = q(x) = 1 + q_1 x + q_2 x^2 + \ldots$ into the HOMFLY polynomial and consider it as a power series in the variable x. Then we have

Proposition 6.3.24. *The coefficient $h_n = h_n(N) \in \mathbb{Z}[N]$ of the power series $H = H(N, q(x))$ is an invariant of order $\leq n$.*

Proof. Indeed, the extension of the HOMFLY polynomial to singular knots is given by the equation

$$H\left(\times\right) = H\left(\times\right) - H\left(\times\right).$$

The leading term in the expansion of the right-hand side of the last equation in powers of x coincides with the leading term of the expansion of the right-hand side in the defining equation for the HOMFLY polynomial, and is therefore divisible by x. Further we can argue as in the case of the Conway polynomial.

Similarly to the Conway polynomial, the HOMFLY polynomial itself is not an invariant of finite order.

6.4 Constructing Weight Systems via Intersection Graphs

6.4.1 The Intersection Graph of a Chord Diagram

Definition 6.4.1 (Intersection graph). The *intersection graph* $\Gamma(c)$ of a chord diagram c is the graph whose vertices correspond to the chords of c, with two vertices connected by an edge if the corresponding chords intersect. (We say that two chords, A and B, intersect if their endpoints A_1, A_2 and B_1, B_2 appear in the alternating order A_1, B_1, A_2, B_2 along the circle.)

For example,

$$\Gamma\!\left(\!\!\begin{array}{c}\includegraphics{}\end{array}\!\!\right) \;=\; \Gamma\!\left(\!\!\begin{array}{c}\includegraphics{}\end{array}\!\!\right) \;=\; \bullet\!\!-\!\!\bullet\!\!-\!\!\bullet$$

(we see that the fake intersections do not count: cf. Fig. 6.1). Note that not every graph can be obtained as the intersection graph of a diagram. The simplest counterexample appears in degree 6 and is given by the graph

A complete description of obstructions for a graph to be an intersection graph is given in [37].

Distinct diagrams can have the same intersection graph. For example, consider the chain graph

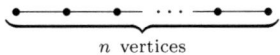

For $n = 5$ there are three diagrams having it as their intersection graph:

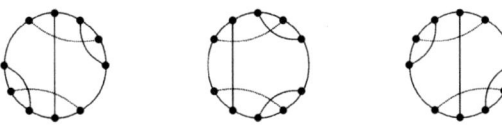

The intersection graph of a chord diagram carries a lot of information about the diagram. Many chord diagram invariants depend, in fact, on the intersection graph of the diagram. We will show also how it can be used for constructing a large family of chord diagram invariants.

6.4.2 Tutte Functions for Graphs

The theory of graph invariants we are going to present in Sec. 6.4.3 is in a certain sense close to the classical theory of Tutte functions. Here we give a brief description of the theory of Tutte functions for graphs. A complete description can be found in [279].

In the paper [279] W. T. Tutte introduced a ring of graphs. In Tutte's ring, graphs are allowed to have loops, multiple edges, and to be disconnected. The underlying module of this ring is spanned, over $\mathbb{Z}$, by finite graphs modulo the following relation:

$$\Gamma - \Gamma'_e - \Gamma''_e = 0, \tag{6.3}$$

which we call the *Tutte relation*. Here Γ is a graph, e is a link, i.e., an edge in Γ which is not a loop, the graph Γ'_e is obtained from Γ by deletion of the edge e, and the graph Γ''_e is the result of contraction of e. Of course, Eq. (6.3) is in fact a huge set of relations, one for each pair (a graph, an edge in the graph). The multiplication in the ring is induced by the disjoint union of graphs.

The time Tutte wrote his paper, a number of interesting graph invariants were known satisfying the property

$$f(\Gamma) = f(\Gamma'_e) + f(\Gamma''_e). \tag{6.4}$$

Tutte called such invariants W-functions, and they are precisely the linear functions on his module. Below, we call Tutte's W-functions *Tutte invariants*. The *graph complexity*, i.e., the number of spanning trees of the graph, is an example of a Tutte invariant.

He also introduced V-functions, that is, W-functions satisfying the additional multiplicativity requirement

$$f(\Gamma_1 \cdot \Gamma_2) = f(\Gamma_1)f(\Gamma_2) \tag{6.5}$$

for any two graphs Γ_1, Γ_2. In other words, V-functions are ring homomorphisms. The chromatic polynomial (taken with the appropriate sign) is an example of a V-function.

Tutte gave a complete description of all V-functions, which is as follows. Eq. (6.3) allows one to present each graph as an equivalent linear combination of graphs without links (since both graphs Γ'_e and Γ''_e have fewer links than the graph Γ). A graph without links is a disjoint union of graphs consisting of one vertex and a number of loops, i.e., a product of such graphs. Hence, one-vertex graphs are multiplicative generators of the ring of graphs, and any V-function is uniquely determined by its values on the one-vertex graphs with $0, 1, 2, \ldots$ loops.

Theorem 6.4.2 ([279], [281]). *A V-function can take arbitrary values on one-vertex graphs with n loops. In other words, any graph is equivalent to a unique linear combination of linkless graphs.*

One more point of view is that Tutte's ring is isomorphic to the polynomial ring in infinite number of variables, $\mathbb{Z}[s_0, s_1, s_2, \ldots]$, where the variable s_n corresponds to the one-vertex graph with n loops. A W-function can take arbitrary values on disjoint unions of one-vertex graphs and is uniquely determined by these values.

In the context of this chapter, the Tutte theorem states that graphs with one vertex and a number of loops "generate a bialgebra of graphs", and this bialgebra is isomorphic to the filtered polynomial bialgebra $\mathbb{Z}[s_0, s_1, s_2, \ldots]$ having one generator of each order. We are not going to make these statements more precise, preferring to explain all constructions for another bialgebra of graphs, which is more closely related to chord diagrams.

6.4.3 The 4-Bialgebra of Graphs

The 4-bialgebra of graphs was introduced in [187], and below we follow the approach of this paper.

6.4.3.1 The 4-Term Relation

Let Γ be a graph. We restrict our consideration to graphs without loops and multiple edges (*simple* graphs).

A *graph invariant* is simply a function on (isomorphism classes of) graphs. A graph invariant can take values in an arbitrary Abelian group or in a commutative algebra K over the ground ring **k**, although it is usually sufficient to keep in mind either the ring $\mathbb{Z}$ of integers, or the field $\mathbb{C}$ of complex numbers, or a polynomial ring.

Denote by $V(\Gamma)$ the set of vertices of Γ and by $E(\Gamma)$ the set of its edges. Let us associate to each ordered pair of (distinct) vertices $A, B \in V(\Gamma)$ of a graph Γ two other graphs Γ'_{AB} and $\widetilde{\Gamma}_{AB}$.

The graph Γ'_{AB} is obtained from Γ by erasing the edge $AB \in E(\Gamma)$ in the case that this edge exists, and by adding the edge otherwise. In other words, we simply change the adjacency between the vertices A and B in Γ. This operation is an analogue of edge deletion, but we prefer to formulate it in a slightly more symmetric way.

The graph $\widetilde{\Gamma}_{AB}$ is obtained from Γ in the following way. For any vertex $C \in V(\Gamma) \setminus \{A, B\}$ we switch its adjacency with A to the opposite one if C is joined with B, and we do nothing otherwise. All other edges do not change. Note that the graph $\widetilde{\Gamma}_{AB}$ depends not only on the pair (A, B), but on the order of vertices in the pair as well.

Exercise 6.4.3. Show that, for a given ordered pair A, B of vertices, the two operations $\Gamma \mapsto \Gamma'_{AB}$ and $\Gamma \mapsto \widetilde{\Gamma}_{AB}$ commute.

Definition 6.4.4 (4-invariant). A graph invariant is a *4-invariant* if it satisfies the *4-term relation*

$$f(\Gamma) - f(\Gamma'_{AB}) = f(\widetilde{\Gamma}_{AB}) - f(\widetilde{\Gamma}'_{AB}) \qquad (6.6)$$

for each graph Γ and for any pair $A, B \in V(\Gamma)$ of its vertices.

In contrast to the Tutte relation (6.3), all four graphs entering Eq. (6.6) have the same set of vertices, and therefore the number of vertices is also the same.

As an immediate consequence of the definition we obtain the following statement. Let $\tilde{\chi}(\Gamma) = (-1)^{|V(\Gamma)|}\chi(\Gamma)$ denote the *modified chromatic polynomial* of a graph Γ, where $\chi(\Gamma)$ is the usual chromatic polynomial and $|V(\Gamma)|$ is the number of vertices in Γ. Recall that the *chromatic polynomial* $\chi(\Gamma) \in \mathbb{Z}[t]$ is the polynomial whose value at an integer point $t = t_0$ is equal to the number of colorings of the vertices of Γ in t_0 colors so that any two neighboring vertices have distinct colors.

Proposition 6.4.5. *The modified chromatic polynomial of a graph is a 4-invariant (with values in the algebra of polynomials in one variable).*

Proof. Suppose that Γ contains an edge $e = AB \in E(\Gamma)$. Introduce the short notation $\widetilde{\Gamma} = \widetilde{\Gamma}_{AB}$. For the modified chromatic polynomial we have

$$\tilde{\chi}(\Gamma) - \tilde{\chi}(\Gamma'_e) = \tilde{\chi}(\Gamma''_e), \quad \tilde{\chi}(\widetilde{\Gamma}) - \tilde{\chi}(\widetilde{\Gamma}'_e) = \tilde{\chi}(\widetilde{\Gamma}''_e).$$

For the chromatic polynomial, the contraction of an edge requires eliminating multiple edges in the resulting graph. After eliminating possible multiple edges we obtain $\Gamma''_e = \widetilde{\Gamma}''_e$, which is obvious from the definitions. The proposition is proved.

There are a lot of 4-invariants that are not Tutte invariants, and not every Tutte invariant is a 4-invariant. Probably the most obvious 4-invariant that is not Tutte is the number $|E(\Gamma)|$ of edges in Γ. The verification of Eq. (6.6) is easy. Indeed, the difference in both sides of Eq. (6.6) is either 1 or -1 for any graph and any pair of its vertices. Other examples of 4-invariants will be presented later in this section.

6.4.3.2 Bialgebra Structure

Consider the (infinitely generated) module over $\mathbf{k}$ spanned by all graphs as free generators. In his original paper [279] Tutte suggested treating the disjoint union of graphs as a multiplication of the generators. This multiplication can be extended by linearity to linear combinations of graphs making the space into a commutative algebra. We denote this algebra by $\mathcal{G}$. The empty graph plays the role of the unit in this algebra. The number of graph vertices induces a grading in $\mathcal{G}$,

$$\mathcal{G} = \mathcal{G}_0 \oplus \mathcal{G}_1 \oplus \mathcal{G}_2 \oplus \ldots,$$

where $\mathcal{G}_k$ is the finite dimensional linear space freely spanned by graphs with k vertices, $k = 0, 1, \ldots$. The multiplication $m : \mathcal{G} \otimes \mathcal{G} \to \mathcal{G}$ preserves this grading:

$$m : \mathcal{G}_l \otimes \mathcal{G}_n \to \mathcal{G}_{l+n}.$$

Let us define a second operation, the comultiplication $\mu : \mathcal{G} \to \mathcal{G} \otimes \mathcal{G}$, as follows. For a set $J \subset V(\Gamma)$ of vertices of a graph Γ let us denote by (J) the induced subgraph of Γ with the set of vertices J, i.e., J is the set of vertices of (J), and $e \in E(\Gamma)$ is an edge in (J) if both ends of e belong to J. We set

$$\mu(\Gamma) = \sum_{J \sqcup J' = V(\Gamma)} (J) \otimes (J'), \qquad (6.7)$$

where the sum is taken over all subsets J of the set of vertices $V(\Gamma)$ and $J' = V(\Gamma) \setminus J$. There are $2^{|V(G)|}$ summands on the right-hand side of Eq. 6.7. The comultiplication is extended to linear combinations of graphs by linearity.

Fig. 6.21 demonstrates an example of comultiplication (compare with the example of comultiplication of a chord diagram in Sec. 6.1.4).

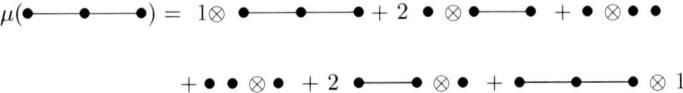

Fig. 6.21. Coproduct of a graph

The comultiplication, as well as the multiplication, respects the grading:

$$\mu : \mathcal{G}_n \to \mathcal{G}_0 \otimes \mathcal{G}_n \oplus \mathcal{G}_1 \otimes \mathcal{G}_{n-1} \oplus \cdots \oplus \mathcal{G}_n \otimes \mathcal{G}_0.$$

Theorem 6.4.6. *The multiplication and the comultiplication defined above make the algebra of graphs a commutative cocommutative bialgebra.*

We omit purely technical verification of the bialgebra axioms. This example of a bialgebra arising in graph theory appeared in [246].

The bialgebra of graphs is too complicated: it carries the same information as graphs themselves. It is reasonable, therefore, to look for some quotient bialgebra, which is easier to handle.

Similarly to Tutte relation (6.3), one can consider the quotient space of graphs modulo the 4-term relations,

$$\Gamma - \Gamma'_{AB} - \widetilde{\Gamma}_{AB} + \widetilde{\Gamma}'_{AB} = 0 \qquad (6.8)$$

for all graphs Γ and all ordered pairs A, B of their vertices. We denote this quotient space by $\mathcal{F}$. This space inherits a bialgebra structure from the bialgebra of graphs, and we call it the 4-*bialgebra* of graphs. The dual space $\mathcal{F}^*$ also carries a bialgebra structure, that of the dual bialgebra, and it will be called the *bialgebra of* 4-*invariants*, and its elements will be called 4-*invariants*. A 4-invariant is *multiplicative* if its value on a disjoint union of graphs is the product of its values on the components.

By a slight abuse of language, below we shall make no difference between a graph and the equivalence class it represents in $\mathcal{F}$.

In contrast to the case of Tutte relations, all terms in the 4-term relation have the same number of vertices. Therefore, the number of vertices induces a grading in $\mathcal{F}$,

$$\mathcal{F} = \mathcal{F}_0 \oplus \mathcal{F}_1 \oplus \mathcal{F}_2 \oplus \ldots,$$

where $\mathcal{F}_n \subset \mathcal{F}$, $n = 0, 1, 2, \ldots$ is the subspace spanned by graphs with n vertices modulo the 4-term relations.

Theorem 6.4.7. *The multiplication and the comultiplication defined above induce a bialgebra structure on the space $\mathcal{F}$ of graphs modulo 4-term relations.*

Proof. The only thing we need to verify is that the multiplication and the comultiplication both respect the 4-term relation (6.8). For the disjoint union of graphs this statement is obvious. In order to verify it for the comultiplication, it is sufficient to consider two different cases. Namely, let $A, B \in V(\Gamma)$ be two distinct vertices of a graph Γ. The right-hand side summands in the comultiplication formula (6.7) split into two groups: those where both vertices A and B belong either to the subset $J \subset V(\Gamma)$, or to its complement $J' = V(\Gamma) \setminus J$; and those where A and B belong to distinct subsets. Summing up terms of the first kind for the coproduct $\mu(\Gamma - \Gamma'_{AB} - \widetilde{\Gamma}_{AB} + \widetilde{\Gamma}'_{AB})$ gives zero as the result, while the sum of the terms of the second kind is already equal to zero for each of the coproducts $\mu(\Gamma - \Gamma'_{AB})$ and $\mu(\widetilde{\Gamma}_{AB} - \widetilde{\Gamma}'_{AB})$.

The theorem is proved.

According to the structure theorem 6.1.22, the 4-bialgebra is isomorphic to a polynomial bialgebra.

6.4.3.3 Examples of 4-Invariants

Example 6.4.8 (Vertex quadrangles). Probably the first non-trivial example of a 4-invariant that is not a Tutte invariant was suggested by Chmutov and Varchenko in [55] (although the notion of 4-invariant did not yet exist at the time). Let a *vertex quadrangle* in a graph Γ be a subset $\{A_1, A_2, A_3, A_4\} \subseteq V(\Gamma)$ of the set of vertices consisting of four vertices satisfying the following requirement: there is a cyclic order (i, j, k, l) on the set of the indices $\{i, j, k, l\} = \{1, 2, 3, 4\}$ such that $E(\Gamma)$ contains four edges $A_i A_j$,

A_jA_k, A_kA_l, A_lA_i. Then the number $Q^v(\Gamma)$ of vertex quadrangles in Γ is a 4-invariant.

For example, one can check this statement for the 4-term relation in Fig. 6.22: each of the graphs on the left-hand side of the equality contains one vertex quadrangle, while there are no vertex quadrangles in both graphs on the right-hand side.

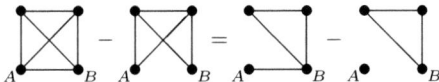

Fig. 6.22. A 4-term relation for graphs with 4 vertices

It is easy to see that the number of vertex quadrangles is not a Tutte invariant. Indeed, for the 4-cycle it equals 1, while for the other two terms in (6.3) it is zero.

In order to prove that the number of vertex quadrangles is indeed a 4-invariant, one can verify this statement first for the graphs with 4 vertices, which can be done by a direct computation. Denote by Q_4^v the 4-invariant equal on all 4-vertex graphs to the number of vertex quadrangles, and equal to zero on all graphs with other number of vertices. Now let U denote the function on graphs, identically equal to 1. This function is obviously a 4-invariant since it satisfies the 4-term relation. Then we have $Q^v = Q_4^v U$, where the product is understood as the convolution product, see Sec. 6.1.6. In other words, the vertex quadrangle invariant is the product of two invariants. A direct verification of the fact that the number of vertex quadrangles satisfies the 4-term relation is also simple.

The above example exploits the following general fact.

Proposition 6.4.9. *If we have some 4-invariant $f \in \mathcal{F}_n^*$ for some n, then the function*
$$F(\Gamma) = \sum_J f((J)),$$
where the sum is taken over all n-element subsets J of $V(\Gamma)$ and (J) is the induced subgraph, is a 4-invariant on graphs with arbitrary number of vertices.

Example 6.4.10 (Edge polygons mod 2). We define an *edge quadrangle* as a subset $\{e_1, e_2, e_3, e_4\} \subseteq E(\Gamma)$ of the set of edges of Γ forming a quadrangle after an appropriate ordering. The number $Q^e(\Gamma)$ of edge quadrangles is not a 4-invariant. However, it becomes one if considered modulo 2. For example, the first graph on the left-hand side of Fig. 6.22 contains 3 edge quadrangles, the number of edge quadrangles in the second graph equals 1, while there are no edge quadrangles in both graphs on the right-hand side. Either of the proofs described above for the number of vertex quadrangles works in this case as well.

Similar statement is valid for the number of edge k-gons modulo 2 for each $k \geq 4$. Indeed, let $E_k(\Gamma)$ be the number of edge k-gons in Γ having k pairwise distinct vertices. Suppose Γ contains an edge AB. Then for the two terms on the left-hand side of the 4-term relation (6.6) we have that $E_k(\Gamma) - E_k(\Gamma'_{AB})$ is the number of edge k-gons in Γ passing through the edge AB. Similarly, for the right-hand side, $E_k(\widetilde{\Gamma}_{AB}) - E_k(\widetilde{\Gamma}'_{AB})$ is the number of edge k-gons in $\widetilde{\Gamma}_{AB}$ passing through AB.

All the k-gons in Γ passing through AB contain a chain $CABD$ and split into three disjoint classes according to the adjacency of the vertices C and D to A and B:

- the vertex C is adjacent to B and D is adjacent to A;
- the vertex C is adjacent to B, but D is not adjacent to A;
- the vertex C is not adjacent to B.

All edge k-gons in $\widetilde{\Gamma}_{AB}$ passing through the four points A, B, C, D admit a similar classification.

Now, the k-gons in Γ belonging to the second class are in one-to-one correspondence with the k-gons in $\widetilde{\Gamma}_{AB}$ containing the path $CBAD$. The edge k-gons of the third kind are the same in both graphs Γ and $\widetilde{\Gamma}_{AB}$. And, finally, the edge k-gons of the first kind in each of the two graphs come in pairs: the chain $CABD$ can be replaced with the chain $CBAD$. Hence, the number of edge k-gons of the first type is even for each of the two graphs, and the required assertion follows.

Example 6.4.11 (Perfect matchings). A *perfect matching* (or a *1-factor*) in a graph Γ with $2n$ vertices is a union of n edges in Γ passing through each vertex. The number of perfect matchings $m_1(\Gamma)$ is a 4-invariant (we set $m_1(\Gamma) = 0$ if the number of vertices in Γ is odd). For example, in Fig. 6.22 the numbers of perfect matchings are respectively $3 - 2 = 1 - 0$.

To verify this statement, similarly to the previous example, we note that $m_1(\Gamma) - m_1(\Gamma'_{AB})$ is the number of perfect matchings in Γ containing the edge AB, as well as $m_1(\widetilde{\Gamma}_{AB}) - m_1(\widetilde{\Gamma}'_{AB})$ is the number of perfect matchings in $\widetilde{\Gamma}_{AB}$ containing the edge AB. But each set of edges containing the edge AB and forming a perfect matching in Γ forms a perfect matching in $\widetilde{\Gamma}_{AB}$, and vice versa.

It is obvious that m_1 is multiplicative: the number of perfect matchings in a disjoint union of graphs is the product of these numbers in each of the graphs.

Note, however, that this 4-invariant is not too powerful. Indeed, consider the 4-invariant $\bar{m}_1$ with values in $\mathbb{Z}$, whose value equals 1 on the graph with two vertices and one edge, and 0 on all other graphs. Then it is obvious that

$$m_1 = e^{\bar{m}_1} = 1 + \frac{\bar{m}_1}{1!} + \frac{\bar{m}_1^2}{2!} + \ldots,$$

i.e., the number of perfect matchings is the exponent of a very simple 4-invariant $\bar{m}_1$. As above, the multiplication in the space of invariants is understood not as a pointwise multiplication, but as the convolution described in Sec 6.1.6.

Exercise 6.4.12. Define the *matching polynomial* $M_1(\Gamma) \in \mathbb{Z}[t]$ of a graph Γ as the sum
$$M_1(\Gamma) = \sum_{E_1 \subset E(\Gamma)} t^{|E_1|},$$
where the sum is taken over all the subsets E_1 in the set $E(\Gamma)$ of the edges of Γ such that each two edges in the subset have distinct vertices. Show that the matching polynomial is a multiplicative 4-invariant. Show that it coincides with $e^{\overline{M}_1}$, where $\overline{M}_1$ is the 4-invariant equal to 1 on the graph with one vertex, equal to t on the graph with two vertices and one edge, and equal to zero otherwise.

Example 6.4.13 (Corank of the adjacency matrix). A slightly more sophisticated example of a 4-invariant, due to Yu. Vol'vovskii, is given by the following proposition.

Let us number the vertices of a graph Γ and let $A(\Gamma)$ denote the *adjacency matrix* of Γ. This means that the entry a_{ij} of $A(\Gamma)$ is 1 if the ith and the jth vertices are connected by an edge, and $a_{ij} = 0$ otherwise. Since the graph is simple, all diagonal elements of the adjacency matrix are 0. Note that the adjacency matrix is symmetric.

Proposition 6.4.14. *The corank of the adjacency matrix considered as a matrix over $\mathbb{Z}_2$ is a $\mathbb{Z}$-valued 4-invariant.*

Of course, the same statement is valid for the *rank* of the adjacency matrix, but we prefer to choose the corank because in this form the relation to a weight system is more apparent.

Proof. We are going to prove a stronger statement, namely that the corank of the adjacency matrix is invariant under the mapping $\Gamma \mapsto \widetilde{\Gamma}_{AB}$ for any graph Γ and any pair of vertices $A, B \in V(\Gamma)$. In other words, the corank satisfies the *2-term relation*.

Indeed, let the vertex A be numbered by 1 and the vertex B, by 2. Then the mapping $\Gamma \mapsto \widetilde{\Gamma}_{AB}$ results, in the language of adjacency matrices, in the conjugation $A(\widetilde{\Gamma}_{AB}) \equiv C^T A(\Gamma) C \mod 2$, where C is the matrix

$$C = \begin{pmatrix} 1 & 1 & 0 & \ldots \\ 0 & 1 & 0 & \ldots \\ 0 & 0 & 1 & \ldots \\ \ldots & \ldots & \ldots & \ldots \end{pmatrix}$$

and C^T is the transpose matrix. In particular, this transformation does not change the corank of $A(\Gamma)$.

376   6 Algebraic Structures

As we are going to see in Example 6.5.7, the weight system of the present example coincides with the one constructed from the standard representation of the Lie algebra $gl(N)$. It would be interesting to know *what graph invariants can be constructed from Lie algebras and their representations* (see Sec. 6.5). At the moment, however, even the number of available examples is small.

Exercise 6.4.15. Let N be a formal variable. Show that the mapping $\Gamma \mapsto N^{\operatorname{corank} A(\Gamma)}$ is a multiplicative 4-invariant with values in $\mathbb{Z}[N]$ satisfying the 2-term relation.

Exercise 6.4.16. Let us call a *framed graph* a graph with vertices marked by elements of $\mathbb{Z}_2$. The adjacency matrix of a framed graph coincides with that of the underlying graph, but its diagonal elements coincide with the framing of the corresponding vertices. Let $|f|$ denote the number of ones in the framing f. Show that the mapping

$$\Gamma \mapsto \sum_f (-1)^{|f|} N^{\operatorname{corank} A(\Gamma_f)},$$

where the sum is taken over all framings f of the vertices of Γ, is a multiplicative 4-invariant with values in $\mathbb{Z}[N]$ satisfying the 2-term relation.

Remark 6.4.17. Recall that by attaching a disk to the circle and thickening each chord we obtain from a chord diagram a two-surface with boundary; for a chord diagram c, denote this surface by $S(c)$. The adjacency matrix of the intersection graph of a chord diagram c is nothing else but the intersection matrix in the homology $H_1(S(c), \mathbb{Z}_2)$ written in a specific basis. The elements of the basis are in one-to-one correspondence with the chords: a representative of a basic vector is obtained by connecting the ends of the corresponding chord in the disk attached to the circle. For a framed chord diagram (that is, a chord diagram with each chord marked by an element of $\mathbb{Z}_2$), the surface $S(c)$ is constructed by twisting all one-handles corresponding to the chords with framing 1. Note that the self-intersection index of such basic element is 1.

We hope the reader will fill in all missing details.

6.4.3.4 Primitive Elements

Primitive elements form a vector space $\mathcal{P} \subset \mathcal{F}$ which is also graded,

$$\mathcal{P} = \mathcal{P}_1 \oplus \mathcal{P}_2 \oplus \mathcal{P}_3 \oplus \dots.$$

Let P_k be the dimension of the space $\mathcal{P}_k$. After picking a basis s_{kj}, $j = 1, \dots, P_k$, in each $\mathcal{P}_k$ we make the 4-bialgebra into the polynomial bialgebra in variables s_{ij}. Therefore, the crucial point in the description of the structure of the 4-bialgebra is the study of its primitive elements; in particular, it is most important to find the sequence $P_1, P_2, P_3, \dots$.

Let us present some examples of primitive elements in $\mathcal{F}$. The space $\mathcal{F}_1$ is spanned by the unique graph with one vertex. We denote this graph by s_{11}. This graph is obviously primitive, whence $P_1 = 1$.

The space $\mathcal{F}_2$ is spanned by the two two-vertex graphs: the segment and the graph consisting of two disjoint vertices. The 4-term relations for two-vertex graphs are empty, whence $\dim \mathcal{F}_2 = 2$. There is a unique (up to a multiplicative constant) primitive element in $\mathcal{F}_2$ (see Fig. 6.23 (a)), cf. Exercise 6.1.19.

(a)   (b)

Fig. 6.23. Basic primitive elements (a) in $\mathcal{P}_2$ and (b) in $\mathcal{P}_3$

We denote this element by s_{21}. Its primitiveness can be verified directly from the definition. Therefore, $P_2 = 1$ as well. The space $\mathcal{F}_2$ is spanned by two elements s_{21} and s_{11}^2.

In order 3 we meet the first non-trivial example of the 4-term relation. There are 4 different graphs with three vertices, and the only essential 4-term relation is shown in Fig. 6.24.

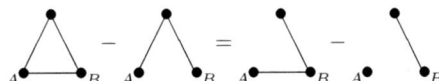

Fig. 6.24. A 4-term relation for graphs with 3 vertices

Hence, $\dim \mathcal{F}_3 = 3$. Note that, in contrast to the case of Tutte invariants, the graph K_3, the triangle, can be expressed as a linear combination of forests. The subspace $\mathcal{P}_3$ of primitive elements is spanned by the linear combination shown in Fig. 6.23 (b), cf. Exercise 6.1.21.

We denote this element by s_{31}. Hence, $P_3 = 1$. The space $\mathcal{F}_3$ is thus spanned by the elements $s_{11}^3, s_{21}s_{11}, s_{31}$. The results of similar computations in higher orders due to E. Soboleva and A. Kaishev [264] are presented in

Table 6.2. Dimensions of subspaces in the 4-bialgebra

n	0	1	2	3	4	5	6	7
$\dim \mathcal{P}_n$	0	1	1	1	2	3	5	7
$\dim \mathcal{F}_n$	1	1	2	3	6	10	19	32

378   6 Algebraic Structures

Table 6.2. Note that the computational complexity grows very rapidly because of the growth both of the number of graphs and of the number of 4-term relations, and because of the necessity to verify the graph isomorphisms.

There is a projection $\pi_n : \mathcal{F}_n \to \mathcal{P}_n$ of the module of n-vertex graphs onto the submodule of primitive elements similar to that for chord diagrams, see Sec. 6.1.6.

Theorem 6.4.18. *For any graph $\Gamma \in \mathcal{F}_n$ the element*

$$\pi_n(\Gamma) = \Gamma - 1! \sum_{J_1 \sqcup J_2 = V(\Gamma)} (J_1)(J_2) + 2! \sum_{J_1 \sqcup J_2 \sqcup J_3 = V(\Gamma)} (J_1)(J_2)(J_3) - \dots, \quad (6.9)$$

where the kth sum is taken over all unordered partitions of the set $V(\Gamma)$ of vertices of Γ into a disjoint union of nonempty subsets $J_1, \dots, J_k$ and (J) is the induced subgraph on the set J, is primitive. The mapping π_n extends to a projection $\pi_n : \mathcal{F}_n \to \mathcal{P}_n$ along the submodule of decomposable elements.

As an application of the above theorem let us construct the primitive element $\pi_4(C_4)$, where C_4 is a cycle on four vertices. The construction is shown in Fig. 6.25. Let us clarify how, for example, the coefficient -3 in the edgeless graph on the right-hand side of Fig. 6.25 can be obtained from Eq. (6.9). There is a unique way to split the vertices $V(C_4)$ into two nonempty subsets J_1, J_2 such that both graphs $(J_1), (J_2)$ are edgeless: each of the sets J_i must consist of opposite vertices of the square. There are two ways to split $V(C_4)$ into three non-empty subsets so that all the three induced graphs are edgeless: the two vertices contained in one subset must be opposite vertices of the square. Finally, the unique way to split $V(C_4)$ into four non-empty subsets leads to an edgeless graph. Hence, the coefficient at the edgeless graph in the projection must be

$$-1! \cdot 1 + 2! \cdot 2 - 3! \cdot 1 = -3.$$

Other coefficients are obtained in a similar way.

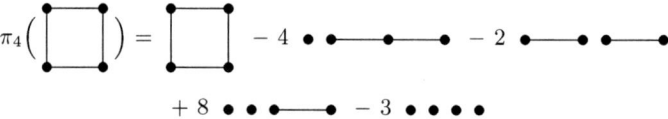

Fig. 6.25. Projecting the 4-cycle to the space of primitive elements

Remark 6.4.19. We can consider the 4-bialgebra over $\mathbb{Z}$, not over a field of characteristic zero. In this case the structure theorem 6.1.22 cannot be applied. However, since all the coefficients in (6.9) are integers, primitive elements of the 4-bialgebra generate it over $\mathbb{Z}$ as well. What we do not know is whether there is torsion in the submodule of primitive elements. Up to now calculations show no torsion, but it may appear in higher orders.

6.4.4 The Bialgebra of Weighted Graphs

6.4.4.1 Definitions

The 4-bialgebra of graphs is not an easy one to understand. Already the dimensions of the spaces of primitive elements in small orders show that it is a complicated one: it has many primitive elements even in small orders. This algebra possesses, however, a rather simple, but rich, quotient bialgebra whose structure admits a very explicit description. This is the bialgebra of weighted graphs introduced in [54]. Its structure is similar to that of the Tutte's ring in the sense that it has exactly one generator in each dimension.

Definition 6.4.20 (Weighted graph). A *weighted graph* is a graph Γ without loops and multiple edges endowed with a mapping $w : V(\Gamma) \to \mathbb{N}$ called the *weight* which assigns a positive integer to each vertex of the graph. The *weight $w(\Gamma)$ of a graph* Γ is the sum of weights of all its vertices, $w(\Gamma) = \sum_{A \in V(\Gamma)} w(A)$.

Ordinary graphs without loops and multiple edges can be treated as weighted graphs with the weights of all vertices equal to 1.

We will use two natural operations on weighted graphs with a distinguished edge: *deletion* and *contraction*.

If e is an edge of the graph Γ, then the new graph Γ'_e is obtained from Γ by removing the edge e. The weights of the vertices do not change. This is what we call *deletion*.

The *contraction* $\Gamma \mapsto \Gamma''_e$ of an edge e is defined as follows:

- the edge e is contracted into a vertex A of the new graph Γ''_e;
- if multiple edges appear, then each of them is replaced by a single edge;
- the weight $w(A)$ of the vertex A is set equal to the sum of weights of the two ends of the edge e in Γ; the weights of the other vertices do not change.

Definition 6.4.21. A *weighted 3-term element* is an expression of the form

$$\Gamma - \Gamma'_e - \Gamma''_e,$$

where Γ is an arbitrary weighted graph and e is its arbitrary edge.

We set $\mathcal{W}_n$ to be the **k**-module generated by all weighted graphs of weight n modulo all weighted 3-term elements. We set $\mathcal{W} = \mathcal{W}_0 \oplus \mathcal{W}_1 \oplus \mathcal{W}_2 \oplus \ldots$ with $\mathcal{W}_0 = \mathbf{k}$.

Each weighted graph is equivalent modulo weighted 3-term elements to a linear combination of weighted graphs without edges.

Example 6.4.22. Successive deletion and contraction of edges provide the following decomposition for a given graph, see Fig. 6.26.

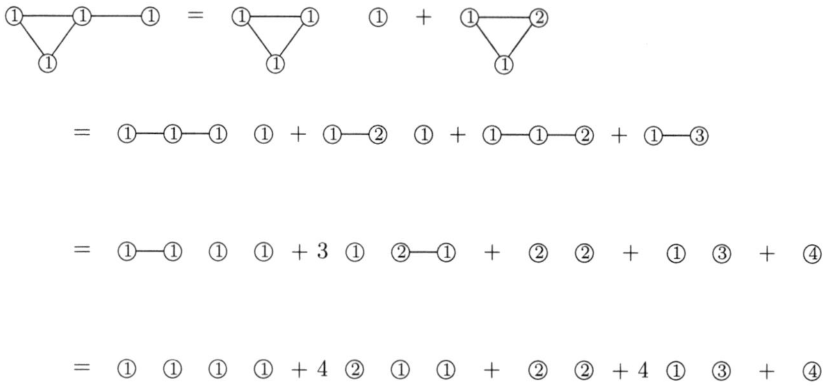

380   6 Algebraic Structures

[Fig. 6.26 graph decomposition diagrams]

Fig. 6.26. A decomposition of a graph

The question is, whether this decomposition depends on the sequence of edges chosen? It may be checked for any given graph that the answer is negative. The complete proof for all weighted graphs is given by the structure theorem, see below.

Definition 6.4.23 (Bialgebra of weighted graphs). By definition, the *bialgebra of weighted graphs* is the module $\mathcal{W}$ over $\mathbf{k}$ with the following operations of multiplication and comultiplication:

- the multiplication
$$m : \mathcal{W}_n \otimes \mathcal{W}_m \to \mathcal{W}_{m+n}$$
comes from the disjoint union of graphs;
- the comultiplication
$$\mu : \mathcal{W}_n \to \mathcal{W}_0 \otimes \mathcal{W}_n \oplus \mathcal{W}_1 \otimes \mathcal{W}_{n-1} \oplus \ldots \oplus \mathcal{W}_n \otimes \mathcal{W}_0$$
is defined on generators as follows: consider a weighted graph Γ; let $J \subset V(\Gamma)$ be a subset in the vertices of Γ and let $J' = V(\Gamma) \setminus J$ be its complement; we set
$$\mu(\Gamma) = \sum_J (J) \otimes (J').$$

The sum is taken over all subsets J, and (J) is the induced subgraph of Γ with vertices in J;
- the unit is represented by the empty graph;
- the counit is the mapping taking each element (a linear combination of graphs) to the coefficient of the empty graph in the canonical expansion of an element as a linear combination of edgeless graphs (this coefficient is not equal to zero only if the initial linear combination contains a non-zero constant times the empty graph).

We omit purely technical verification of the bialgebra axioms.

Definition 6.4.24 (Weighted graph invariant). A *weighted graph invariant* is a linear function on $\mathcal{W}$ with values in an algebra K over $\mathbf{k}$. A weighted graph invariant is called *multiplicative* if it is a homomorphism of ring structures.

We are going to prove further that each weighted graph invariant generates a weight system.

6.4.4.2 Structure Theorem for Weighted Graphs

Theorem 6.4.25 ([54]). *The submodule of primitive elements of $\mathcal{W}_n$ is isomorphic to $\mathbf{k}$ and freely generated by the graph $\textcircled{n}$ with one vertex of weight n. The mapping*

$$\iota : \textcircled{n} \mapsto s_n$$

extends to an isomorphism of graded bialgebras

$$\iota : \mathcal{W} \to \mathbf{k}[s_1, s_2, \ldots],$$

where the grading of the variable s_n is set to n.

Remark 6.4.26. The theorem gives a description of all weighted invariants since ι is a universal weighted invariant. This means that an arbitrary weighted invariant can be obtained from ι by substituting appropriate values for the variables s_k.

Proof. Put

$$\iota(\Gamma) = \sum_{\gamma} (-1)^{\beta(\gamma)} \prod_{\gamma_i} s_{w(\gamma_i)}, \qquad (6.10)$$

where the sum is taken over all spanning subgraphs γ of the graph Γ and the product is taken over all connected components γ_i of γ. Here $\beta(\gamma)$ is the first Betti number of the graph γ, i.e., the number of independent circuits in γ. (A *spanning* subgraph is a subgraph whose set of vertices coincides with that of the whole graph. Thus, the spanning subgraphs of a graph Γ are in one-to-one correspondence with subsets of the set of edges of Γ.)

Now we extend ι by linearity to the **k**-module freely generated by all weighted graphs. We claim that this mapping descends to the quotient algebra $\mathcal{W}$ and defines an isomorphism between $\mathcal{W}$ and $\mathbf{k}[s_1, s_2, \ldots]$ (also denoted by ι in the sequel).

In order to prove the theorem we must verify that the mapping ι

- vanishes on 3-term elements;
- is a homomorphism of graded bialgebras; and
- is one-to-one.

Let Γ be an arbitrary weighted graph and let e be an arbitrary edge of Γ. We are going to prove that

$$\iota(\Gamma) = \iota(\Gamma'_e) + \iota(\Gamma''_e).$$

There exists a natural one-to-one correspondence between subgraphs of Γ'_e and those subgraphs of the graph Γ that do not contain the edge e. Thus we have

$$\sum_{\gamma'_e} t(\gamma'_e) = \iota(\Gamma'_e),$$

where the sum on the left-hand side is taken over all spanning subgraphs γ'_e of the graph Γ that do not contain the edge e, and $t(\gamma'_e)$ denotes the corresponding product in (6.10).

We are going to prove that the other part of the sum for $\iota(\Gamma)$ is equal to $\iota(\Gamma''_e)$. Let γ'' be a spanning subgraph of the graph Γ''_e. Let b be an edge of the graph γ'' that is twice covered by the edges of the graph Γ in the process of contraction. The preimage of b under the contraction thus consists of two edges which, together with the edge e, form a triangle. A spanning subgraph γ of Γ which is contracted to γ'' and contains the edge e may either contain any of the two preimages of the edge b or both of them (see Fig. 6.27).

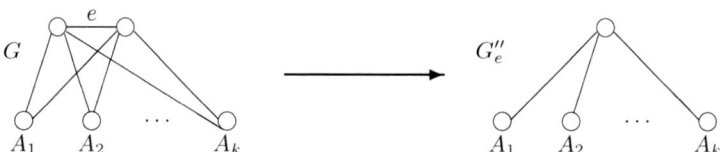

Fig. 6.27. Triangles double covering edges in the graph with a contracted edge

Thus, the spanning subgraph γ'' of Γ''_e corresponds to 3^k spanning subgraphs of G containing e, where k is the number of twice covered edges in γ''.

Mark an edge in each triangle in Γ that contains e. We have three possibilities for the preimage of each twice covered edge. Two of these possibilities correspond to the same first Betti number of the covering graph. The third one gives a graph whose first Betti number differs from this value by one. Thus the products for two of the three possibilities mutually cancel, and we obtain a one-to-one correspondence between spanning subgraphs of Γ''_e and spanning subgraphs of Γ that contain e and a marked edge in each triangle containing e. Therefore, we have

$$\sum_{\gamma_e} t(\gamma_e) = \iota(\Gamma''_e),$$

where the sum is taken over all spanning subgraphs of G containing the edge e.

The multiplicativity of ι follows from a straightforward calculation.

Both bialgebras $\mathcal{W}$ and $\mathbf{k}[s_1, s_2, \dots]$ are freely generated by their generators. Therefore, in order to establish their isomorphism it suffices to establish an isomorphism between their generators. The assignment $s_n \mapsto \textcircled{n}$ defines a homomorphism $\mathbf{k}[s_1, s_2, s_3, \dots] \to \mathcal{W}$ which is obviously inverse to ι.

Remark 6.4.27. It is proved in [199] that the weight system coming from weighted graphs coincides with a weight system constructed from a direct sum of several copies of the Lie algebras $\operatorname{gl}(N)$ and $\operatorname{so}(N)$ (cf. Sec. 6.5).

6.4.4.3 The Bialgebra of Weighted Graphs and the 4-Bialgebra

Proposition 6.4.28. *The mapping assigning weight 1 to all vertices of a graph extends to a homomorphism of bialgebras $\mathcal{F} \to \mathcal{W}$. This homomorphism is surjective.*

Proof. In the bialgebra of weighted graphs, both differences on the left- and on the right-hand side of the 4-term relation

$$\Gamma - \Gamma'_{AB} = \widetilde{\Gamma}_{AB} - \widetilde{\Gamma}'_{AB}$$

are equivalent in $\mathcal{W}$ to graphs with contracted edges. The weight of the new vertex in both cases is 2, and it is obvious that the underlying graphs Γ''_{AB} and $\widetilde{\Gamma}''_{AB}$ of the two weighted graphs coincide. In order to verify that each weighted graph $\textcircled{n}$ can be obtained in this way, it is sufficient to compute the value of the mapping on the chain with n vertices.

6.4.5 Constructing Vassiliev Invariants from 4-Invariants

Explicit calculations show that up to order 8 the equivalence class of a chord diagram modulo the 4-term relation is completely determined by its intersection graph. This evidence led to the conjecture [52] that this was the case for diagrams of all orders. However, this conjecture proved to be false (T. Le, unpublished, see [50]).

The 4-bialgebra of graphs from Sec. 6.4.3 was invented to identify properties of chord diagrams that are determined by their intersection graphs.

Theorem 6.4.29. *The mapping $\Gamma : c \mapsto \Gamma(c)$ associating to each chord diagram c its intersection graph $\Gamma(c)$ extends to a homomorphism $\Gamma : \mathcal{M} \to \mathcal{F}$.*

Note that the mapping Γ preserves the grading, and is therefore the direct sum of linear mappings $\Gamma_n : \mathcal{M}_n \to \mathcal{F}_n$.

As an immediate consequence of the theorem we obtain the following statement.

Corollary 6.4.30. *Any 4-invariant $f : \mathcal{F}_n \to K$ determines a weight system $f \circ \Gamma_n : \mathcal{M}_n \to K$.*

Here K is an arbitrary algebra over **k**. In particular, each 4-invariant from the examples given in Sec. 6.4.3.3 determines a weight system.

Example 6.4.31. The corank of the adjacency matrix $A(\Gamma)$ of a graph Γ satisfies the 2-term relation $\Gamma \sim \widetilde{\Gamma}_{AB}$. A similar statement is true for the number of faces of a chord diagram (see Example 6.3.13). There is nothing strange, therefore, that these two functions are very close to each other.

Proposition 6.4.32. *For any chord diagram c the number $f(c)$ of faces of the diagram coincides with* $\mathrm{corank}(A(\Gamma(c))) + 1$.

This statement has been rediscovered several times by various people, see [217], [220], [264].

The proof of Theorem 6.4.29 is obvious. The only thing we need to verify is that the mapping Γ respects the 4-term relation. But the rule for the 4-term relation for graphs (6.8) shows precisely what happens with the intersection of chords in all four chord diagrams of Fig. 6.5.

To conclude, let us state a conjecture about the properties of the mapping $\Gamma: \mathcal{M} \to \mathcal{F}$. The statement of Le cited above shows that Γ cannot be a monomorphism. We also know that not each graph can be obtained as the intersection graph of a chord diagram. Nevertheless, for all n the image of Γ_n contains the complete graph K_n (the graph with n vertices with each pair of vertices joined by an edge). The complete graph is, in a sense, "the most complicated" graph, all other graphs are "simpler". Therefore, one can expect that each graph is equivalent, modulo the 4-term relation, to a linear combination of intersection graphs.

Conjecture 6.4.33. *The mapping $\Gamma_n : \mathcal{M}_n \to \mathcal{F}_n$ is surjective for all n.*

Calculations up to order 7 confirm this conjecture. Besides, one would expect that the kernel of Γ_n is "thin" in $\mathcal{D}_n$. In order 7 A. Kaishev found that the kernel of the mapping $\Gamma_7 : \mathcal{M}_7 \to \mathcal{F}_7$ was one-dimensional and computed its generator. However, up to now the evidence is too small to make a precise conjecture. Note that the study of this kernel is, probably, the most interesting part of the work. For example, only elements of this kernel (if any) may distinguish a chord diagram from its mirror image.

6.5 Constructing Weight Systems via Lie Algebras

In this section we describe the construction of weight systems coming from Lie algebras due to Bar-Natan and Kontsevich [18], [179]. Lie algebras endowed with an invariant nondegenerate bilinear form provide a powerful source of weight systems. Bar-Natan even conjectured that each weight system could be obtained in this way. In [287] Vogel showed that for semisimple Lie (super)algebras this statement was false by giving an explicit description of a primitive element in the bialgebra of chord diagrams with 32 chords such that all semisimple Lie algebras weight systems vanished on this element. Other examples of such elements are given in [198]. Lie algebra weight systems are extremely nontrivial for calculations. They also proved to be a useful tool to find lower bounds for the number of independent primitive elements in the bialgebra of chord diagrams.

6.5.1 Free Associative Algebras

Let $\mathcal{B}_m = \langle x_1, \ldots, x_m \rangle$ be the free associative algebra generated by $x_1, \ldots, x_m$ as free generators. We are going to associate with each positive integer m a weight system b_m. The ring of values of the weight system will be described later.

Consider the mapping $\hat{b}_m : \mathcal{A}_n \to \mathcal{B}_m$ of the module of arc diagrams of order n into the algebra $\mathcal{B}_m$ constructed as follows.

Let $a \in \mathcal{A}_n$ be an arc diagram. Mark each arc of the diagram a with one of the marks $1, \ldots, m$. Denote such a marking by $\nu : V(a) \to \{1, \ldots, m\}$ (where $V(a)$ is the set of arcs in a). We associate with the diagram a and the marking ν an element $\bar{b}_m(a, \nu)$ by writing the letter $x_{\nu(v)}$ on both ends of each arc $v \in V(a)$ and reading these letters from left to right. The word thus obtained will be the element $\bar{b}_m(a, \nu)$.

We set

$$\hat{b}_m(a) = \sum_\nu \bar{b}_m(a, \nu).$$

The sum is taken over all m^n possible markings $\nu : V(a) \to \{1, \ldots, m\}$. For example, for $m = 2$ and the diagram

one has

$$\hat{b}_2(a) = x_1^6 + x_1^2 x_2 x_1 x_2 x_1 + x_1 x_2 x_1^3 x_2 + x_2 x_1^2 x_2 x_1^2$$
$$+ x_1 x_2^2 x_1 x_2^2 + x_2 x_1 x_2^3 x_1 + x_2^2 x_1 x_2 x_1 x_2 + x_2^6.$$

Remark 6.5.1. Obviously, the image of $\hat{b}_m$ lies in the symmetric part of $\mathcal{B}_m$, that is, the value of $\hat{b}_m$ on an arc diagram is invariant under the action of the symmetric group S_m by permuting the coordinates x_i.

Introduce a scalar product on the linear part of the algebra $\mathcal{B}_m$ by setting $(x_i, x_j) = \delta_{ij}$. Thus, the orthogonal group $O(m)$ acts on the linear part of the algebra $\mathcal{B}_m$. This action may be naturally extended to an action of $O(m)$ on the algebra $\mathcal{B}_m$ itself. The action preserves the grading of elements in $\mathcal{B}_m$.

Proposition 6.5.2. *The mapping $\hat{b}_m$ is invariant under the action of the orthogonal group $O(m)$ on $\mathcal{B}_m$.*

Proof. The group $O(m)$ is generated by subgroups $O(2)$ acting by rotating coordinate planes. Therefore, it is sufficient to prove the proposition for the action of the group $O(2)$ on a pair of letters x_1, x_2.

Now, let a be an arc diagram. The image $\hat{b}_m(a)$ is a sum of monomials. Consider the $O(2)$-change of variables $x_1 = ax_1' + bx_2'$, $x_2 = cx_1' + dx_2'$. We need to check that each monomial of the type $\ldots (x_1')^{\alpha_1} \ldots (x_2')^{\beta_1} \ldots$ appears in the transformed image if and only if the monomial $\ldots (x_1)^{\alpha_1} \ldots (x_2)^{\beta_1} \ldots$ appears in $\hat{b}_m(a)$. Otherwise the monomial must have coefficient 0.

This is indeed the case. If there is no monomial of the above type in $\hat{b}_m(a)$, then there exists a pair $\ldots x_1' \ldots x_2' \ldots$ on two ends of the same arc. Such a monomial comes from two types of monomials

$$\ldots x_1 \ldots x_1 \ldots \quad \text{and} \quad \ldots x_2 \ldots x_2 \ldots$$

after substitutions, and we take ax_1' times dx_2' from the first monomial and cx_1' times bx_2' from the second monomial. Thus the coefficient is a multiple of $ad + bc = 0$ since our matrix is orthogonal.

Similarly, if there exists a monomial of the above type in the original image $\hat{b}_2(a)$, then all letters x_1' as well as x_2' are split into pairs corresponding to the ends of the arc. Each pair $\ldots x_1 \ldots x_1 \ldots$ introduces a factor in the coefficient of the type $(a^2 + b^2) = 1$, while the pair $\ldots x_2 \ldots x_2 \ldots$ provides the factor $(c^2 + d^2) = 1$ and the product equals 1.

The mapping $\hat{b}_m$ descends to a linear mapping (which we also denote by the same letter) $\hat{b}_m : \mathcal{A} \to \mathcal{B}_m$. The image $\hat{b}_m(\mathcal{A})$ forms a subalgebra in $\mathcal{B}_m$. The image $\hat{b}_m(\mathcal{A}^{(4)}) \subset \hat{b}_m(\mathcal{A})$ of the subspace of all 4-term elements forms a two-sided ideal in this subalgebra.

Theorem 6.5.3. *The quotient mapping $b_m: \mathcal{M} = \mathcal{A}/\mathcal{A}^{(4)} \to \hat{b}_m(\mathcal{A})/\hat{b}_m(\mathcal{A}^{(4)})$ is a weight system.*

The proof follows directly from the definition.

Of course, having an arbitrary associative algebra B with a system $x_1, \ldots, x_m$ of generators and a quotient mapping $\mathcal{B}_m \to B$ one may construct a composite weight system with values in an appropriate quotient of a subalgebra in B.

The weight system b_m is difficult for calculations. Each of the weight systems b_m is more complicated than the previous one due to the natural embedding $\mathcal{B}_m \subset \mathcal{B}_{m+1}$. The sequence of mappings b_m carries important information about a chord diagram.

6.5.2 Universal Enveloping Algebras of Lie Algebras

A special (and the most important) case of the above construction is given by the universal enveloping algebra of a Lie algebra.

Let H be a Lie algebra. For simplicity we restrict ourselves to complex Lie algebras. Consider a basis $\{x_1, \ldots, x_m\} \subset H$. The *universal enveloping algebra* $U(H)$ of H is the quotient algebra of the algebra of noncommutative polynomials in variables x_i modulo the ideal generated by elements of the form $x_i x_j - x_j x_i - [x_i, x_j]$. The mapping $\hat{b}_m : \mathcal{A} \to \mathcal{B}_m$ descends to a mapping $b_m^H : \mathcal{A} \to U(H)$. We denote the last mapping simply by b^H since the value of m is encoded in the Lie algebra H as well. The mapping b^H depends, of course, on the choice of the basis in H.

Suppose now that H admits a nondegenerate bilinear invariant form $(\cdot, \cdot)$. Here the word "invariant" means that $([x, y], z) = (x, [y, z])$ for each triple $x, y, z \in H$. Let $\{x_1, \ldots, x_m\}$ be an orthonormal basis with respect to this form, i.e., $(x_i, x_j) = \delta_{ij}$.

Theorem 6.5.4. *If $\{x_1, \ldots, x_m\}$ is an orthonormal basis in H, then we have*

- $\mathcal{A}^{(4)} \subset \operatorname{Ker} b^H$;
- $\operatorname{Im} b^H$ *is a subset of the center* $Z(U(H)) \subset U(H)$;
- b^H *does not depend on the choice of the orthonormal basis.*

The proof will be given on page 389.

Thus, having a Lie algebra H endowed with a nondegenerate bilinear invariant form, the mapping b^H determines a weight system with values in the center of the universal enveloping algebra of H.

Example 6.5.5. For the only chord diagram with one chord, the value of the weight system b^H is $x_1^2 + x_2^2 + \cdots + x_m^2$. The latter element in $Z(U(H))$ is called *the Casimir element* for arbitrary Lie algebra H; we will denote it by C_H.

There are two chord diagrams of order 2. One has

$$b^H\left(\bigcirc\!\!\!\!\!\ominus\right) = C_H^2$$

since this diagram is the square of the one-chord diagram.

This chord diagram has two distinct arc representations. These two different representations give the relation:

$$x_1^2 x_2^2 + x_1^2 x_3^2 + \cdots = x_1 x_2^2 x_1 + x_1 x_3^2 x_1 + \ldots \qquad (6.11)$$

Since we have

6 Algebraic Structures

$$x_i^2 x_j^2 - x_i x_j^2 x_i + x_j^2 x_i^2 - x_j x_i^2 x_j = x_i^2 x_j^2 - x_i x_j x_i x_j + x_i x_j x_i x_j - x_i x_j^2 x_i +$$
$$x_j^2 x_i^2 - x_j x_i x_j x_i + x_j x_i x_j x_i - x_j x_i^2 x_j$$
$$= x_i[x_i, x_j]x_j - x_j[x_i, x_j]x_i +$$
$$x_i x_j[x_i, x_j] - x_j x_i[x_i, x_j],$$
(6.12)

it follows that relation (6.11) can be rewritten as

$$\sum_{i \neq j}(x_i[x_i, x_j]x_j - x_j[x_i, x_j]x_i) = -\sum_{i \neq j}[x_i, x_j]^2.$$

The last equality allows one to calculate the value of the invariant b^H on the second diagram of order 2 for arbitrary Lie algebra H:

$$b^H\left(\bigotimes\right) = \sum x_i x_j x_i x_j$$

and

$$\sum x_i x_j x_i x_j = \sum x_i x_j x_i x_j - \sum x_i^2 x_j^2 + \sum x_i^2 x_j^2$$
$$= -\sum_{i \neq j} x_i[x_i, x_j]x_j + C_H^2$$
$$= \sum_{i \neq j}[x_i, x_j]^2 + C_H^2$$
(6.13)

The simplest noncommutative Lie algebra with a nondegenerate bilinear form is the simple Lie algebra $sl_2(\mathbb{C})$. It is generated by three elements x, y, z such that

$$[x, y] = z, \quad [y, z] = x, \quad [z, x] = y.$$

The bilinear form is given by the relations

$$(x, x) = (y, y) = (z, z) = 1, \quad (x, y) = (y, z) = (z, x) = 0.$$

The center of the universal enveloping algebra coincides with the polynomial algebra in the Casimir element

$$C_{sl_2} = x^2 + y^2 + z^2.$$

Thus we obtain a weight system with values in the ring of polynomials in one variable. The value of (6.13) becomes $C_{sl_2}^2 + 2C_{sl_2}$.

Corollary 6.5.6. *Let $R : H \to GL_N$ be a representation of the Lie algebra H, let $\hat{R} : U(H) \to GL_N$ be the corresponding representation of the universal enveloping algebra, and let $\mathrm{Tr} : GL_N \to \mathbb{C}$ be the trace mapping. Then the composition mapping*

$$\frac{1}{N}\mathrm{Tr} \circ \hat{R} \circ b^H : \mathcal{A} \to \mathbb{C}$$

provides a multiplicative weight system with values in $\mathbb{C}$.

The coefficient $1/N$ results from the requirement that the unit matrix must go to $1 \in \mathbb{C}$ under this homomorphism.

Proof of Theorem 6.5.4. Fix a numbering on the set of arcs that do not participate in the 4-term relation. The part of the 4-term element value corresponding to this numbering looks like

$$\Delta = \sum_{i,j} w_0 x_i x_j w_1 x_i w_2 x_j w_3 - \sum_{i,j} w_0 x_j x_i w_1 x_i w_2 x_j w_3$$
$$- \sum_{i,j} w_0 x_i w_1 x_j x_i w_2 x_j w_3 + \sum_{i,j} w_0 x_i w_1 x_i x_j w_2 x_j w_3, \qquad (6.14)$$

where w_0, w_1, w_2, w_3 are fixed words in the alphabet $\{x_1, \ldots, x_m\}$ corresponding to the given numbering of the arcs. Transforming the expression (6.14) one obtains

$$\Delta = \sum_{i,j} w_0 [x_i, x_j] w_1 x_i w_2 x_j w_3 + \sum_{i,j} w_0 x_i w_1 [x_i, x_j] w_2 x_j w_3$$
$$= \sum_{i,j,k} C_{i,j}^k w_0 x_k w_1 x_i w_2 x_j w_3 + \sum_{i,j,l} C_{i,j}^l w_0 x_i w_1 x_l w_2 x_j w_3,$$

or, changing indices $i \to l, k \to i$ in the first sum,

$$\Delta = \sum_{i,j,l} C_{l,j}^i w_0 x_i w_1 x_l w_2 x_j w_3 + \sum_{i,j,l} C_{i,j}^l w_0 x_i w_1 x_l w_2 x_j w_3$$
$$= \sum_{i,j,l} (C_{l,j}^i + C_{i,j}^l) w_0 x_i w_1 x_l w_2 x_j w_3$$
$$= 0.$$

Here $C_{i,j}^k$ are the structure constants for the basis $\{x_1, \ldots, x_m\}$, i.e.,

$$[x_i, x_j] = \sum_k C_{ij}^k x_k$$

in the Lie algebra H. The last sum in the expression for Δ equals zero since for an orthonormal basis the structure constants are anticommutative under the exchange of the upper and the left lower indices: $C_{l,j}^i = -C_{i,j}^l$.

In order to prove that the image of b^H belongs to the center of $U(H)$ it suffices to show that $x_j b^H(a) = b^H(a) x_j$ for each j, say for $j = 1$, and for each arc diagram a. We will proceed step by step, moving the letter x_1 through the element $b^H(a)$. Fix an arc A and consider "the contribution" of this arc:

$$\Delta(A) = \sum_i w_0 x_1 x_i w_1 x_i w_2 - \sum_i w_0 x_i x_1 w_1 x_i w_2$$
$$- \sum_i w_0 x_i w_1 x_1 x_i w_2 + \sum_i w_0 x_i w_1 x_i x_1 w_2$$
$$= \sum_{i,k} C_{i,1}^k w_0 x_k w_1 x_i w_2 + \sum_{i,k} C_{k,1}^i w_0 x_i w_1 x_k w_2$$
$$= 0.$$

Here we made use of the orthonormality of the basis $\{x_i\}$ once again.

Thus, the contribution of each arc equals zero, as required.

The last statement of the theorem follows from Theorem 6.1.22.

6.5.3 Examples

In these examples we use the term *complete weight system* to refer to a Lie algebra weight system with values in the center of the enveloping algebra, and we specify the representation otherwise.

Example 6.5.7 (The standard representation of gl(N)). Elaborating the above construction for the standard representation of the Lie algebra $\mathrm{gl}(N)$ one obtains the following picture. It will be more convenient for us to change our convention and to choose not an orthonormal basis of $\mathrm{gl}(N)$, but a basis e_{ij} such that $(e_{ij}, e_{ji}) = (e_{ii}, e_{ii}) = 1$, $i, j = 1, \ldots, N$, and we set all other scalar products equal to zero. Under this choice, when associating an element in $U(\mathrm{gl}(N))$ to a chord (or arc) diagram, we write on the ends of each chord (or arc) not the same element x_i, but the pair of elements e_{ij}, e_{ji}. For example, the chord diagram of order 2 with two intersecting chords is taken to
$$\sum_{i,j,k,l=1}^N e_{ij} e_{kl} e_{ji} e_{lk}.$$

Obviously, this mapping coincides with $b^{\mathrm{gl}(N)}$.

The basis e_{ij} can be chosen in such a way that in the standard representation these elements are taken to the $N \times N$-matrices having 1 in the (i, j)-cell, and 0 in all other places. (Recall that the standard representation of $\mathrm{gl}(N)$ is its representation as the Lie algebra of all $N \times N$-matrices.) The nondegenerate invariant scalar product is $(A, B) = \mathrm{tr}(A\overline{B}^t)$.

Proposition 6.5.8. *The weight system associated to the standard representation of the Lie algebra* $\mathrm{gl}(N)$ *takes each chord diagram c to $N^{f(c)-1}$, where $f(c)$ is the number of faces of the diagram.*

In order to prove this statement let us consider the one-vertex map dual to the chord diagram in question. Marking the edges of the map (= chords) by a pair of elements e_{ij}, e_{ji} is equivalent to marking the two sides of the edge by the indices i and j. Similarly to the case of Hermitian matrix integrals (see Sec. 3.3.2), the contribution of an indexed gluing is 1 if the sides of each face are marked with the same index; otherwise the contribution is zero. Indeed, the product of matrices corresponding to an indexed gluing is zero if the indexing does not agree with the gluing, and it contains a single unit on the diagonal otherwise. The final step consists in summing over all possible markings, and the required assertion follows.

We have seen already (see Example 6.3.13) that not only $N^{f(c)-1}$, but the value $f(c)$ itself is a weight system, and it satisfies the "2-term relation". The relationship of this weight system with the corank of the adjacency matrix is discussed in Example 6.4.13 and Proposition 6.4.32.

Example 6.5.9 (The standard representation of $\mathrm{so}(N)$).
The similar construction is valid for the representation of the Lie algebra $\mathrm{so}(N)$ as the Lie algebra of $N \times N$ skew symmetric matrices.

Proposition 6.5.10. *The weight system associated to the standard representation of the Lie algebra $\mathrm{so}(N)$ takes each chord diagram c to $\sum_I (-1)^{|I|} N^{f(c_I)-1}$, where the sum is taken over all subsets $I \subset V(c)$ of the chords of c, c_I is the surface obtained from c by thickening all chords, with chords belonging to I twisted, and $f(c_I)$ is the number of faces of the surface c_I.*

This weight system is related to matrix integrals over the space of symmetric matrices (see Exercise 3.2.12) in the same way as the $\mathrm{gl}(N)$ standard representation weight system is related to the Hermitian matrix integration. We leave the proof to the reader.

Example 6.5.11 (The complete $\mathrm{sl}(2)$ weight system). The complete $\mathrm{sl}(2)$ weight system was studied by Chmutov and Varchenko [55]. The center of the universal enveloping algebra of the Lie algebra $\mathrm{sl}(2)$ is naturally identified with the algebra $\mathbb{C}[C]$ of polynomials in a single variable C, the Casimir element. Chmutov and Varchenko found recurrence relations allowing one to compute the value of the complete $\mathrm{sl}(2)$ weight system on chord diagrams with $n+1$ chords knowing its values on diagrams with fewer chords or with fewer chord intersections.

The value of $b^{\mathrm{sl}(2)}$ is equal to C on the chord diagram with one chord, and it equals $C(C-1)$ on the chord diagram of order two with two intersecting chords. More generally, if a chord diagram contains a chord intersecting a single other chord (a leaf), then the leaf can be eliminated, and the value of $b^{\mathrm{sl}(2)}$ on the initial chord diagram equals the one on the chord diagram with the eliminated leaf times $C-1$. In particular, the value of $b^{\mathrm{sl}(2)}$ on a chord diagram of order n whose intersection graph is a tree is equal to $C(C-1)^{n-1}$.

If a chord diagram does not contain a leaf, then the recurrence relation of Chmutov and Varchenko is more complicated. It is presented in Fig. 6.28. In fact, the relation shown in the figure differs slightly from the one from [55].

392 6 Algebraic Structures

Firstly, we use a different normalization of the weight system; this normalization is more convenient and leads to smaller coefficients in the invariant values. Secondly, we give only the two first of the four Chmutov–Varchenko recurrence relations. The reason is that they are, in fact, sufficient: each indecomposable chord diagram of order greater than two without leaves contains a triple of chords looking like the triple in the leftmost chord diagram in the two equations in Fig. 6.28. It suffices to take for the "vertical" chord the "shortest" chord of the diagram (the "length" of a chord is the minimum of the number of chord ends in the two parts in which the chord splits the circle).

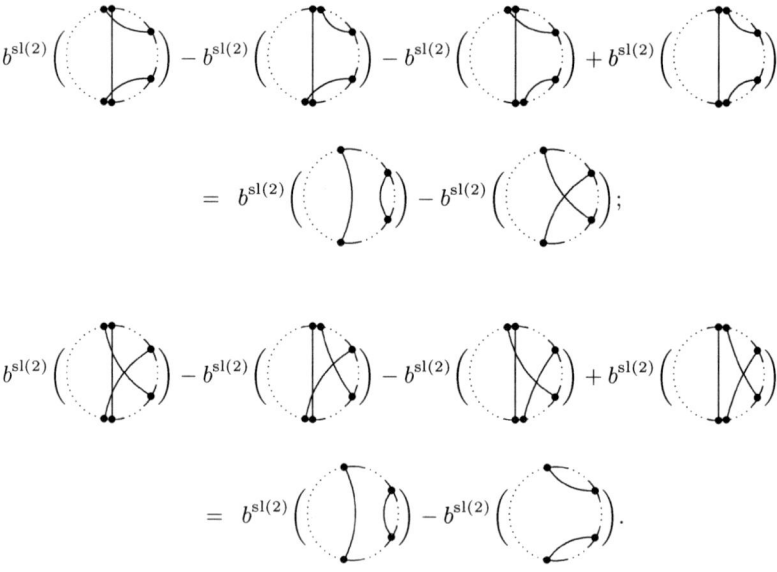

Fig. 6.28. The recurrence relation for the complete sl(2) weight system

Example 6.5.12 (The complete gl(1|1) weight system). In [282] Vaintrob extended the construction of weight systems to Lie superalgebras. In [100] his construction was elaborated in the special case of the Lie superalgebra gl(1|1). In this case the center of the universal enveloping algebra is graded by a weight and is generated by two elements: the Casimir element C of weight 1 and an element y of weight 2. It was shown that the value of the gl(1|1)-invariant on a chord diagram can be expressed as the following linear combination of its values on chord diagrams of smaller order:

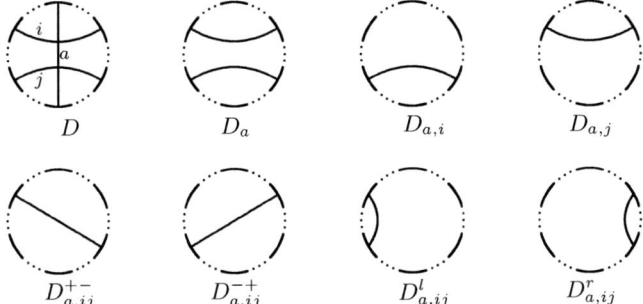

Fig. 6.29. Chord diagrams entering the recurrence relation for the complete gl(1|1) weight system. Besides the chords indicated in the figure, each diagram may contain many other chords, but they are common to all of them and do not change from one diagram to another

$$W(D) = C \cdot W(D_a) - y \sum_i W(D_{a,i})$$
$$+ y \sum_{i<j} (W(D_{a,ij}^{+-}) + W(D_{a,ij}^{-+}) - W(D_{a,ij}^l) - W(D_{a,ij}^r)),$$

where a is a fixed chord in the chord diagram D, i, j run over the set of chords in D intersecting the chord a, and the diagrams participating in the recurrence relation look like those shown in Fig. 6.29.

6.6 Some Other Algebras of Embedded Graphs

In this section we describe briefly other algebras appearing in relation to embedded graphs.

6.6.1 Circle Diagrams and Open Diagrams

In the original construction of Bar-Natan [18] the Hopf algebra of chord diagrams comes together with the Hopf algebra of circle diagrams, which is isomorphic to the first one. A *circle diagram* is a connected graph with all vertices of order 3 and with a distinguished oriented simple cycle. (In [18] these objects were called "Chinese characters".) The vertices lying on the distinguished cycle are called the *external* vertices of the diagram, while all the other vertices are *internal*. At each internal vertex, a cyclic order of darts is chosen. If we remove the distinguished circle from a diagram, then the underlying graph splits into a finite number of connected components, and we require that each connected component contains at least one external vertex. Chord diagrams are circle diagrams without internal vertices; the number of connected components in the graph obtained by removing the distinguished circle from a chord diagram coincides with the number of chords in it.

The *STU-relation* for circle diagrams is shown in Fig. 6.30. Modulo the STU-relation each circle diagram is equivalent to a linear combination of chord diagrams. The module spanned by all circle diagrams is graded by half the number of the vertices in the diagram. It is endowed with a natural product, and with the coproduct shown in Fig. 6.31.

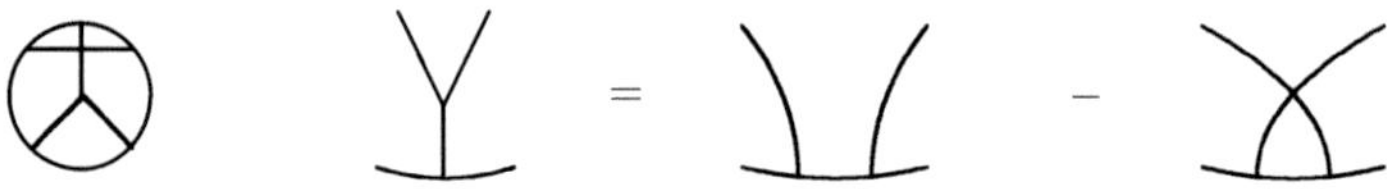

Fig. 6.30. (a) A circle diagram, and (b) the STU-relation. All internal vertices are presumed to be oriented counterclockwise, and points of chords' intersection are phantoms

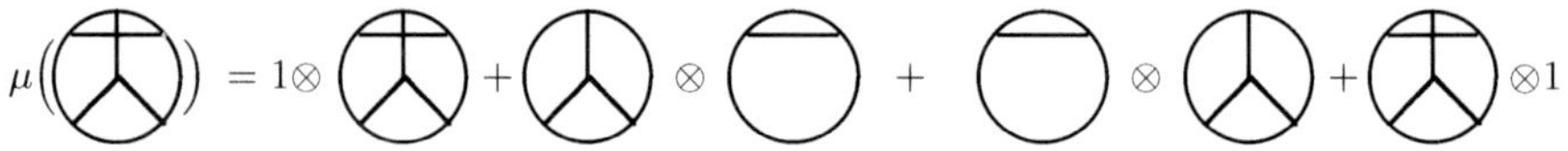

Fig. 6.31. The coproduct of a circle diagram

Proposition 6.6.1 ([18]). *The module of circle diagrams endowed with the multiplication and the comultiplication defined above is a commutative cocommutative bialgebra. The mapping which associates to a circle diagram the corresponding linear combination of chord diagrams is well defined, and it induces the bialgebra isomorphism.*

The graph with oriented vertices obtained by removing the distinguished circle from a circle diagram is called an *open diagram*. It has vertices of valency only 3 and 1, and each its connected component contains at least one vertex of valency 1. Consider the module spanned by open diagrams modulo the antisymmetry and the *IHX-relations* (see Fig. 6.32). This module is endowed with the bialgebra structure with respect to the multiplication induced by the disjoint union of open diagrams, and the comultiplication for which connected

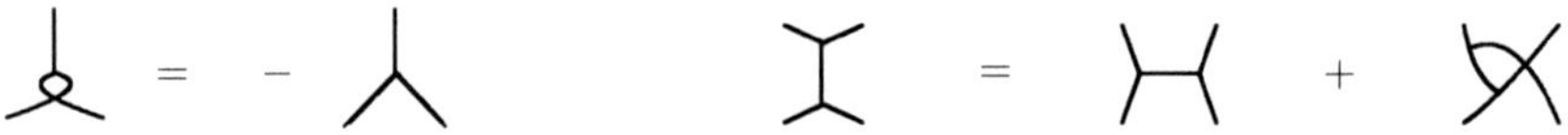

Fig. 6.32. (a) The antisymmetry, and (b) the IHX relations

open diagrams are primitive elements. It is graded by half the number of vertices of an open diagram.

Let k be the number of vertices of valency 1 of an open diagram. Fix k distinct points at the circle and consider all possible ways to attach the 1-valency vertices to the circle (in a generic case, there is $k!$ ways of such an attachment). Summing over all these ways we obtain a linear combination of circle diagrams.

Theorem 6.6.2 ([18]). *Both the mapping from the space of circle diagrams to the space of open diagrams and the mapping from the space of open diagrams to the space of circle diagrams described above identify them as graded vector spaces. The first of the two mappings is an isomorphism of the two bialgebra structures, while the second one induces a new multiplication on the space of open diagrams, compatible with the comultiplication.*

6.6.2 The Algebra of 3-Graphs

The algebra of 3-graphs introduced in [51] is obtained from the algebra of circle diagrams by forgetting the distinguished circle (the Wilson loop). The resulting object carries a well-defined multiplication, but considered as an algebra it admits nontrivial relations between generators, and therefore carries no cocommutative coalgebra structure compatible with the algebra structure.

By definition, the *algebra of 3-graphs* is the **k**-module freely generated by connected graphs whose all vertices are oriented and of valency 3, modulo the antisymmetry relation and the IHX-relation shown in Fig. 6.32.

The product of two graphs is defined as their connected sum obtained by choosing an arbitrary edge in the first graph and an arbitrary edge in the second graph, cutting them and gluing the ends pairwise. The algebra of 3-graphs is graded by half the number of vertices in the graph. The multiplication respects this grading.

Table 6.3 presents the dimensions of the spaces of 3-graphs for small values of the grading.

Table 6.3. Dimensions of homogeneous spaces in the algebra of 3-graphs

n	0	1	2	3	4	5	6	7	8	9	10	11
dim	1	1	1	1	2	2	3	4	5	6	8	9

6.6.3 The Temperley–Lieb Algebra

Here we follow the approach of [82]. The *Temperley–Lieb algebra* $\mathrm{TL}_n(q)$ (introduced in [276], see also [236]) on n strings with the parameter q is the associative algebra over $\mathbb{Z}[q]$ in n generators $1, e_1, \ldots, e_{n-1}$ satisfying the relations

396   6 Algebraic Structures

- $e_i^2 = qe_i,$ $i = 1, 2, \ldots, n-1;$
- $[e_i, e_j] = 0$ if $|i - j| > 1;$
- $e_i e_{i\pm 1} e_i = e_i,$ $i = 1, 2, \ldots, n-1.$

The generator e_i of $\mathrm{TL}_n(q)$ admits the following pictorial presentation:

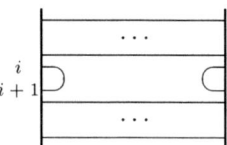

The multiplication of two monomials consists in attaching the right-hand side of the pictorial representation of the first monomial to the left-hand of the second one, which implies the relations above:

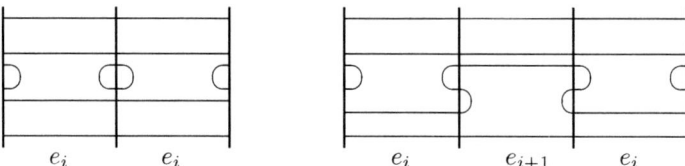

The relations show that closed loops can be eliminated from a product of generators causing the factor q each. Hence, each product of generators is equal to an element without loops (a *reduced element*) times some power of q. Now, there is a one-to-one correspondence between reduced elements in $\mathrm{TL}_n(q)$ and arc diagrams on n arcs, see Fig. 6.33.

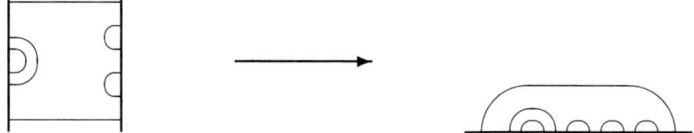

Fig. 6.33. Associating an arc diagram to a reduced element in the Temperley–Lieb algebra: the vertical strip opens as a fan; its vertical sides form together the horizontal line; and the strings of the algebra element become arcs of the arc diagram

In particular, the algebra $\mathrm{TL}_n(q)$ is finitely generated over $\mathbb{Z}[q]$ and has a basis consisting of $\mathrm{Cat}_n = (2n)!/n!(n+1)!$ elements. For two basis elements a_1, a_2, it is possible to define their scalar product (a_1, a_2) according to the following rule. Attach the arc diagram corresponding to the basis element a to the rotated arc diagram corresponding to b and obtain a system of meanders (see Fig. 6.34, cf. Sec. 3.4.3).

Fig. 6.34. A system of meanders with three connected components

Then, by definition, $(a_1, a_2) = q^c$, where c is the number of connected components in the meander system. For example, for the system in Fig. 6.34 we have $c = 3$. All pairwise scalar products of the basis elements form the *Lickorish bilinear form*. This bilinear form was introduced in [197] in relation with the calculation of the Jones polynomial. Lickorish also posed the problem of computing the determinant of this matrix: such a computation would give a new, purely combinatorial approach to construction of certain invariants of 3-manifolds. This determinant is a polynomial in q, and it suffices to find its zeroes together with the multiplicities. The zeroes were found in [175], and in [82] the multiplicities were also computed. The determinant in question is expressed in terms of the Chebyshev polynomials of the second kind in q.

* * *

Vassiliev knot invariants constitute a part of his far-reaching theory of the complements to discriminants and their topology (see [286]). The combinatorics of chord diagram bialgebras and that of their bases is not well understood. It may seem strange that while Hopf algebras play an important role in the group representation theory (see, e.g., [109], [305]), the permutational representation of embedded graphs does not play any role at all in the theory presented in this chapter. The future research must shed more light on all these intricate questions.

Algebraic structures associated to embedded graphs did not yet reveal all their potential power. See, for instance, their applications to the renormalization in quantum field theory [61], [103].

A

Applications of the Representation Theory of Finite Groups

by Don Zagier

This appendix consists of two sections. In the first we give a self-contained and fairly complete introduction to the representation and character theory of finite groups, including Frobenius's formula and a higher genus generalization. In the second we give several applications related to topics treated in this book.

A.1 Representation Theory of Finite Groups

A.1.1 Irreducible Representations and Characters

Let G be a finite group. A (finite-dimensional; we will always assume this) *representation* (V, π) of G is a finite-dimensional complex vector space V and a homomorphism $\pi : G \to \mathrm{GL}(V)$. Thus each $g \in G$ defines a linear map $v \mapsto \pi(g)v$ from V to V, with $\pi(g_1 g_2) = \pi(g_1)\pi(g_2)$. One often drops the "$\pi$" and writes the action simply as $v \mapsto gv$. Alternatively, one often drops the "V" and simply denotes the representation itself by π. The definition given here corresponds to left representations; one also has right representations (where $v \mapsto vg$ with $v(g_1 g_2) = (vg_1)g_2$); this leads to an isomorphic theory, by replacing g by g^{-1}.

We call two representations V and V' *isomorphic*, denoted $V \simeq V'$, if there is a G-equivariant isomorphism from V to V', and write $V \cong V'$ if such an isomorphism has been fixed. If V is a (left) representation of G and A a complex vector space, then both $V \otimes_{\mathbb{C}} A$ and $\mathrm{Hom}_{\mathbb{C}}(A, V)$ are (left) representations in the obvious way ($g(v \otimes a) = (gv) \otimes a$, $(g\phi)(a) = g\phi(a)$). If $\dim_{\mathbb{C}} A = k$, then both of these representations are isomorphic to $V \oplus \cdots \oplus V$ (k copies), the isomorphisms being canonical if one has chosen a basis of A. Similarly, the dual space $V^* = \mathrm{Hom}_{\mathbb{C}}(V, \mathbb{C})$ is in a natural way a right representation of G via $(\phi g)(v) = \phi(gv)$ and the space $\mathrm{Hom}_{\mathbb{C}}(V, A)$ is also a right representation, isomorphic to k copies of V^*. Finally, if V and V' are two representations of G, then we write $V \otimes_G V'$ for the quotient of $V \otimes_{\mathbb{C}} V'$ by the relation $gv \otimes v' = v \otimes gv'$ and $\mathrm{Hom}_G(V, V')$ for the set of G-equivariant linear maps from V to V'; these are simply vector spaces, without any natural G-action.

A representation V of G is called *irreducible* if it contains no proper subspace which is invariant under the action of G. As a simple example, let $G = S_n$ and $V = \mathbb{C}^n$ with the obvious action of G by permutation of the coordinates. Then V is not irreducible, since it contains the two subspaces $W_1 = \{(x,\ldots,x) \mid x \in \mathbb{C}\}$ and $W_2 = \{(x_1,\ldots,x_n) \in \mathbb{C}^n \mid x_1 + \cdots + x_n = 0\}$, of dimensions 1 and $n-1$ respectively, which are obviously invariant under the action of G. On the other hand, these two representations are irreducible and V is their direct sum. More generally, one has:

Lemma A.1.1. *Any representation of G is a direct sum of irreducible ones.*

Proof. Pick a G-invariant non-degenerate scalar product on V. (To obtain one, start with any positive-definite scalar product and replace it by the obvious average over G.) If V is not already irreducible, it contains a proper G-invariant subspace W. But then the orthogonal complement $W^\perp$ of W is also G-invariant, and $V = W \oplus W^\perp$. The result now follows by induction on the dimension. □

This will be used in conjunction with the following property of irreducible representations:

Lemma A.1.2 (Schur's Lemma). *Let V and V' be two irreducible representations of G. Then the complex vector space $\mathrm{Hom}_G(V, V')$ is 0-dimensional if $V \not\simeq V'$ and 1-dimensional if $V \simeq V'$. The space $\mathrm{Hom}_G(V, V)$ is canonically isomorphic to $\mathbb{C}$.*

Proof. Since neither V nor V' has a non-trivial G-invariant subspace, any non-zero G-equivariant map $\phi : V \to V'$ has trivial kernel and cokernel. Hence $\mathrm{Hom}_G(V, V') = \{0\}$ if V and V' are not isomorphic. If they are, then we may assume that $V' = V$. Then for any eigenvalue λ of ϕ, the map $\phi - \lambda$ has a kernel and therefore is zero. Hence $\mathrm{Hom}_G(V, V) \cong \mathbb{C}$ canonically. □

These two lemmas already suffice to prove one of the first basic facts of the theory, the "first orthogonality relation for characters." If (V, π) is an irreducible representation, we define its *character* as the function $\chi_\pi(g) = \mathrm{tr}\,(\pi(g), V)$ from G to $\mathbb{C}$. Then we have:

Corollary A.1.3 (First orthogonality relation). *Let (V, π) and (V', π') be two irreducible representations of G. Then*

$$\frac{1}{|G|} \sum_{g \in G} \chi_\pi(g) \overline{\chi_{\pi'}(g)} = \begin{cases} 1 & \text{if } \pi \simeq \pi', \\ 0 & \text{otherwise.} \end{cases} \tag{A.1}$$

Proof. The dimension of the space V^G of G-invariant vectors in any representation V of G is given by $\dim(V^G) = |G|^{-1} \sum_{g \in G} \operatorname{tr}(g, V)$. (*Proof.* The linear map $v \mapsto |G|^{-1} \sum_{g \in G} gv$ is a projection from V to V^G, so its trace equals $\dim V^G$.) Apply this to the G-representation $\operatorname{Hom}_{\mathbb{C}}(V, V') \cong V \otimes_{\mathbb{C}} V'^*$, with G acting by $g(v, \phi) = (gv, \phi \circ g^{-1})$. (Recall that V'^* is naturally a right representation of G, so that we have to replace g by g^{-1} when we act from the left.) Then the trace of $g \in G$ equals $\chi_\pi(g) \overline{\chi_{\pi'}(g)}$, and the dimension of $\bigl(\operatorname{Hom}(V, V')\bigr)^G = \operatorname{Hom}_G(V, V')$ is 1 or 0 by Schur's Lemma. $\square$

Now let $\{(V_i, \pi_i)\}_{i \in I}$ be a full set of non-isomorphic irreducible representations of G. Lemma A.1.1 tells us that any representation V of G is isomorphic to a direct sum $\bigoplus_{i \in I} \underbrace{V_i \oplus \cdots \oplus V_i}_{k_i}$ of the representations V_i or equivalently, by what was said above, that

$$V \cong \bigoplus_i V_i \otimes_{\mathbb{C}} A_i \cong \bigoplus_i \operatorname{Hom}_{\mathbb{C}}(B_i, V_i) \tag{A.2}$$

for some k_i-dimensional vector spaces A_i and B_i over $\mathbb{C}$, but we do not yet know that these spaces, or even the multiplicities k_i, are independent of the decomposition chosen. The following lemma shows that this is true and gives a canonical description of the spaces A_i and B_i.

Lemma A.1.4. *Let V be an arbitrary representation of G. Then we have canonical G-equivariant isomorphisms*

$$\bigoplus_{i \in I} V_i \otimes_{\mathbb{C}} \operatorname{Hom}_G(V_i, V) \xrightarrow{\sim} V, \quad V \xrightarrow{\sim} \bigoplus_{i \in I} \operatorname{Hom}_{\mathbb{C}}(\operatorname{Hom}_G(V, V_i), V_i) \tag{A.3}$$

given by sending $x \otimes \phi \in V_i \otimes \operatorname{Hom}_G(V_i, V)$ to $\phi(x)$ and $v \in V$ to the homomorphism $\phi \mapsto \phi(v)$ from $\operatorname{Hom}_G(V, V_i)$ to V_i. Conversely, given any decompositions of V of the form (A.2), there are canonical isomorphisms $A_i \cong \operatorname{Hom}_G(V_i, V)$ and $B_i \cong \operatorname{Hom}_G(V, V_i)$ as complex vector spaces.

Proof. Since both isomorphisms in (A.2) are additive under direct sums, we may assume by Lemma A.1.1 that V is irreducible, say $V = V_j$ for some $j \in I$. Then both statements in (A.3) follow immediately from Lemma A.1.2, since $V_j \otimes_{\mathbb{C}} \mathbb{C} \cong \operatorname{Hom}_{\mathbb{C}}(\mathbb{C}, V_j) \cong V_j$. The proof of the last statement, which will not be used in what follows, is similar and will be left to the reader. $\square$

We next introduce the *group algebra* $\mathbb{C}[G]$. This is the set of linear combinations $\sum_{g \in G} \alpha_g[g]$ ($\alpha_g \in \mathbb{C}$) of formal symbols $[g]$ ($g \in G$), with the obvious addition and multiplication. It can be identified with $\operatorname{Maps}(G, \mathbb{C})$ via $\alpha(g) = \alpha_g$. The group algebra is a left *and* right representation of G via $g_1[g]g_2 = [g_1 g g_2]$, or equivalently $(g_1 \alpha g_2)(g) = \alpha(g_1^{-1} g g_2^{-1})$ if α is a map from G to $\mathbb{C}$. The central result of the representation theory of finite groups is the following assertion.

Theorem A.1.5. *Let G be a finite group. Then there is a canonical $(G \times G)$-equivariant algebra isomorphism*

$$\mathbb{C}[G] \cong \bigoplus_{i \in I} \mathrm{End}_{\mathbb{C}}(V_i) \tag{A.4}$$

sending $[g]$ to the collection of linear maps $\pi_i(g) : V_i \to V_i$.

Proof. For any representation V of G, $\mathrm{Hom}_G(\mathbb{C}[G], V) \cong V$ as G-representations, since an equivariant map $\phi : \mathbb{C}[G] \to V$ is uniquely determined by $\phi([1]) \in V$, which is arbitrary. Applying this to $V = V_i$ ($i \in I$) and then applying the second isomorphism in (A.3) to $V = \mathbb{C}[G]$, we obtain the assertion of the theorem. We can also obtain the isomorphism (A.4), in the reverse direction, by applying the first isomorphism in (A.3) to $V = \mathbb{C}[G]$ and using the canonical isomorphisms $\mathrm{Hom}_G(V_i, \mathbb{C}[G]) \cong V_i^*$ and $V_i \otimes_{\mathbb{C}} V_i^* \cong \mathrm{Hom}_{\mathbb{C}}(V_i, V_i) \cong \mathrm{End}_{\mathbb{C}}(V_i)$. □

Essentially all important general facts about representations of finite groups are corollaries of this theorem. To state them, let us introduce the notation $\mathcal{C}$ for the set of conjugacy classes in G and $\mathcal{R}$ for the set of isomorphism classes of irreducible representations. (Of course $\mathcal{R}$ and the index set I used above are in canonical bijection, but we will no longer need to have picked representatives for the elements of $\mathcal{R}$.) Since the value of the character $\chi_\pi(g)$ depends only on the isomorphism class of π and the conjugacy class of g, we can write, with some abuse of notation, $\chi_\pi(C)$ for any $\pi \in \mathcal{R}$ and $C \in \mathcal{C}$. For instance, with these notations the first orthogonality relation (A.1) becomes

$$\sum_{C \in \mathcal{C}} |C| \chi_\pi(C) \overline{\chi_{\pi'}(C)} = |G| \delta_{\pi,\pi'} \qquad (\pi, \pi' \in \mathcal{R}). \tag{A.5}$$

Then Theorem A.1.5 has the following consequences.

Corollary A.1.6. *The cardinality of $\mathcal{R}$ is finite and*

$$\sum_{\pi \in \mathcal{R}} (\dim \pi)^2 = |G|. \tag{A.6}$$

Proof. Compare the dimensions on both sides of (A.4). □

Corollary A.1.7. *The sets $\mathcal{C}$ and $\mathcal{R}$ have the same cardinality: there are as many irreducible representations of G as there are conjugacy classes in G.*

Proof. A basis for the center $Z(\mathbb{C}[G])$ of $\mathbb{C}[G]$ is clearly given by the elements $e_C = \sum_{g \in C} [g]$ ($C \in \mathcal{C}$). On the other hand, $\mathrm{End}_{\mathbb{C}}(V_i)$ is the matrix algebra $M_{\dim V_i}(\mathbb{C})$, with 1-dimensional center. Hence the algebra isomorphism (A.4) tells us that $|\mathcal{C}| = \dim_{\mathbb{C}} Z(\mathbb{C}[G]) = |\mathcal{R}|$. □

Corollary A.1.8 (Second orthogonality relation). *Let* $C_1, C_2 \in \mathcal{C}$. *Then*

$$\sum_{\pi \in \mathcal{R}} \chi_\pi(C_1)\overline{\chi_\pi(C_2)} = \begin{cases} |G|/|C_1| & \text{if } C_1 = C_2, \\ 0 & \text{otherwise.} \end{cases} \quad (A.7)$$

Notice that this formula agrees with (A.6) when $C_1 = C_2 = \{1\}$, since $\chi_\pi(1) = \dim \pi$.

Proof. This follows from (A.5) and Corollary A.1.7, since these imply that the matrix $\left(|C|^{1/2}|G|^{-1/2}\chi_\pi(C)\right)_{\pi \in \mathcal{R}, C \in \mathcal{C}}$ is square and unitary, and the inverse of a unitary matrix is also unitary. But we can also obtain (A.7) directly (and then, if we wish, deduce (A.5) from it) by computing the trace of the action of $(g_1, g_2) \in C_1 \times C_2$ on both sides of (A.4). The action of (g_1, g_2) on the basis $\{[g]\}_{g \in G}$ of $\mathbb{C}[G]$ is given by the permutation $[g] \mapsto [g_1 g g_2^{-1}]$ (as before, we have to invert g_2 to turn the right action into a left one), so its trace is the number of fixed points of this permutation, which is clearly $|G|/|C_1|$ if g_1 and g_2 are conjugate and 0 otherwise. On the other hand, the trace of (g_1, g_2) on $\text{End}_\mathbb{C}(\pi) = \pi \otimes_\mathbb{C} \pi^*$ equals $\chi_\pi(g_1)\overline{\chi_\pi(g_2)}$. □

A.1.2 Examples

We illustrate the theory explained in Sec. A.1.1 in a few important special cases.

Abelian groups. If G is a cyclic group of order n, with generator γ, then there are n obvious 1-dimensional (and hence irreducible!) representations of G given by $V = \mathbb{C}$ and $\gamma v = \zeta v$ with $\zeta \in \mathbb{C}$ a (not necessarily primitive) nth root of unity. By the dimension formula (A.6), these are the only irreducible representations. More generally, for any finite abelian group one sees easily that all irreducible representations of G are 1-dimensional (because the commuting operators $\pi(g)$ on any representation V have a common eigenvector, or alternatively by reducing to the cyclic case) and that the corresponding characters are simply the homomorphisms from G to $\mathbb{C}^*$.

Symmetric groups of small order. The symmetric group S_n has two 1-dimensional representations **1** (the trivial representation, $V = \mathbb{C}$ with all elements of G acting as $+1$) and ε_n (the *sign* representation, $V = \mathbb{C}$ with odd permutations acting as -1) and an $(n-1)$-dimensional irreducible representation **St**$_n$ which is the space $W_2 \subset \mathbb{C}^n$ mentioned at the beginning of this section. For $n = 2$ and $n = 3$, the dimension formula (A.6) shows that these are the only irreducible representations (and **St**$_n \simeq \varepsilon_n$ for $n = 2$), with character tables given by

C	Id	(1,2)
$\|C\|$	1	1
1	1	1
ε_2	1	−1

C	Id	(1,2)	(1,2,3)
$\|C\|$	1	3	2
1	1	1	1
ε_3	1	−1	1
St_3	2	0	−1

(Here the numbers in the second row show the size of the conjugacy classes, needed as weights to make the rows of the table orthogonal; the columns are orthogonal as they stand.) For $n = 4$, the orthogonality relations again have a unique solution and the character table must take the form

C	Id	(1,2)	(1,2,3)	(1,2)(3,4)	(1,2,3,4)
$\|C\|$	1	6	8	3	6
1	1	1	1	1	1
ε_4	1	−1	1	1	−1
A	2	0	−1	2	0
St_4	3	1	0	−1	−1
$\mathrm{St}_4 \otimes \varepsilon_4$	3	−1	0	−1	1

for some 2-dimensional irreducible representation A of S_4. We can construct A explicitly as $\{(x_s)_{s \in S} \mid \sum x_s = 0\}$, where S is the 3-element set of decompositions of $\{1, 2, 3, 4\}$ into two (unordered) subsets of cardinality 2. The reader may wish to attempt constructing "by hand" the 7×7 character table for the group S_5, where the dimensions of the irreducible representations are 1, 1, 4, 4, 5, 5 and 6.

Observe that in the above tables for S_n the character values $\chi_\pi(g)$ are all integers. This is a general fact. For arbitrary finite group representations, $\chi_\pi(g)$ is a sum of roots of unity and hence an algebraic integer, and its Galois conjugates are simply the values of $\chi_\pi(g^\ell)$ with $\ell \in \mathbb{Z}$ prime to the order of g. (This is because the action of $\mathrm{Gal}(\overline{\mathbb{Q}}/\mathbb{Q})$ on roots of unity is given by $\zeta \mapsto \zeta^\ell$ with ℓ prime to the order of ζ.) For $G = \mathrm{S}_n$, however, g^ℓ and g are conjugate since they have the same cycle structure, so $\chi_\pi(g)$ is Galois invariant and hence belongs to $\mathbb{Z}$.

Symmetric groups of arbitrary order. We will not give a complete account of the general representation theory of S_n here, since it is a little complicated and there are many good accounts, but will only mention some highlights, following the approach given in the beautiful paper [OV], which we highly recommend to the reader. The statements given here will be used only in Sec. A.2.3. We denote by $\mathcal{R}_n$ the sets of isomorphism classes of irreducible representations of S_n. We will also consider S_{n-1} as a subgroup of S_n (namely, the set of elements fixing n), and similarly for all S_i, $i < n$.

A. The first basic fact is that each irreducible representation of S_n, when restricted to the subgroup S_{n-1}, splits into the direct sum of distinct irreducible representations of S_{n-1}. For $\pi \in \mathcal{R}_n$ and $\pi' \in \mathcal{R}_{n-1}$ we write $\pi' \prec \pi$ if π' occurs in $\pi|_{S_{n-1}}$, so $\pi|_{S_{n-1}} = \oplus_{\pi' \prec \pi} \pi'$. Similarly, each $\pi' \prec \pi$ when restricted to S_{n-2} splits into a sum of irreducible rerpresentations π'' of S_{n-2}, and continuing this process, we see that π splits canonically into a direct sum of 1-dimensional spaces V_ξ indexed by all possible chains $\xi : \pi_1 \prec \cdots \prec \pi_n = \pi$ with $\pi_i \in \mathcal{R}_i$. (Specifically, we have $V_\xi = V_1 \subset \cdots \subset V_n = \pi$ where each V_i is S_i-invariant and isomorphic to π_i.) Notice how unusual this behavior is: for general finite groups G it is more modern, and of course better, to think of representations as actions of G on abstract vector spaces, without any choice of basis, rather than as collections of matrices satisfying the same relations as the elements of G, but for the symmetric groups the irreducible representations come equipped with their own nearly (i.e., up to scalar multiples) canonical bases, and we have matrices after all!

B. The next key idea is to introduce the so-called *Jucys-Murphy element*

$$X_n = (1,n) + (2,n) + \cdots + (n-1,n)$$

of the group algebra $\mathbb{Z}[S_n]$. We can write this element as $e_{[T]_n} - e_{[T]_{n-1}}$, where $e_{[T]_n} \in Z(\mathbb{Z}[S_n])$ as in the proof of Corollary A.1.7 above is the sum of all elements in the conjugacy class $[T]_n$ of $T = (1,2) \in S_n$ and $e_{[T]_{n-1}}$ is the corresponding element for $n-1$. Since $e_{[T]_n}$ is central, by Schur's lemma it acts on each $\pi \in \mathcal{R}_n$, and hence on each subrepresentation $\pi' \prec \pi$, as multiplication by a scalar $\nu_\pi(T)$ (which belongs to $\mathbb{Z}$ by the remark on integrality made above), and similarly $e_{[T]_{n-1}}$ acts on π' as a scalar $\nu_{\pi'}(T)$, so X_n acts on π' as multiplication by the number $a_n = \nu_\pi(T) - \nu_{\pi'}(T)$. By induction on i it follows that for each chain $\xi = (\pi_1, \ldots, \pi_n)$ each element $X_i \in \mathbb{Z}[S_i] \subset \mathbb{Z}[S_n]$ acts on π_{i-1} as multiplication by some integer $a_i(\xi)$, so we can associate to the chain ξ a *weight vector* $a(\xi) = (a_1(\xi), \ldots, a_n(\xi)) \in \mathbb{Z}^n$.

C. Conversely, the weight vector $a(\xi)$ determines ξ (and hence also π) completely, and there is a complete description of which vectors $a = (a_1, \ldots, a_n) \in \mathbb{Z}^n$ occur as weight vectors. (The conditions are (i) $a_1 = 0$, (ii) for each $j > 1$, we have $|a_j - a_i| = 1$ for some $i < j$, and (iii) if $a_i = a_j$ for some $i < j$ then both $a_i - 1$ and $a_i + 1$ occur among the a_k with $i < k < j$.) Furthermore, two weight vectors $a, a' \in \mathbb{Z}^n$ correspond to the same representation π if and only if they are permutations of one another, so that π is uniquely characterized by the function $f : \mathbb{Z} \to \mathbb{Z}_{\geq 0}$ defined by $f(r) = \#\{i \mid a_i = r\}$, and this sets up a bijection between $\mathcal{R}_n$ and the set of finitely supported functions $f : \mathbb{Z} \to \mathbb{Z}_{\geq 0}$ satisfying $f(r+1) - f(r) \in \{0, |r| - |r+1|\}$ for all r and $\sum_{r \in \mathbb{Z}} f(r) = n$. These functions in turn correspond bijectively to the elements of the set $\mathcal{Y}_n$ of *Young diagrams* (= subsets $Y \subset \mathbb{N}^2$ such that $(x,y) \in Y \Rightarrow (x',y') \in Y$ whenever $1 \leq x' \leq x$, $1 \leq y' \leq y$) of cardinality n, the correspondence $f \leftrightarrow Y$ being given by $f(r) = \#\{(x,y) \in Y \mid x - y = r\}$ and $Y = \{(x,y) \in \mathbb{N}^2 \mid \min(x,y) \leq f(x-y)\}$. In the bijection between $\mathcal{R}_n$

and $\mathcal{Y}_n$ obtained in this way one has $\pi' \prec \pi$ if and only if the corresponding Young diagrams $Y' \in \mathcal{Y}_{n-1}$ and $Y \in \mathcal{Y}_n$ satisfy $Y' \subset Y$.

D. Finally, there is an explicit inductive procedure, given by the so-called *Murnaghan-Nakayama rule*, to compute the value of the character $\chi_\pi(C)$ for any $\pi \in \mathcal{R}_n$ and conjugacy class $C \subset \mathsf{S}_n$ in terms of the Young diagram associated to π and the partition of n associated to C. However, we will not use this and do not give the details here, noting only that the case $C = [T]_n$, which will be needed in Sec. A.2.3, follows from **B.** and **C.** above.

A.1.3 Frobenius's Formula

We end Sec. A.1 by describing a formula which has applications to combinatorial problems in many parts of mathematics and in particular to several of the topics treated in this book; some of these applications will be discussed in Sec. A.2. This formula computes the number

$$\mathcal{N}(G; C_1, \ldots, C_k) := \#\{(c_1, \ldots, c_k) \in C_1 \times \cdots \times C_k \mid c_1 \cdots c_k = 1\}$$

for arbitrary conjugacy classes $C_1, \ldots, C_k \in \mathcal{C}$ in terms of the characters of the irreducible representations of G. Note that $\mathcal{N}(G; C_1, \ldots, C_k)$ is independent of the order of the arguments, because the identity $c_i c_{i+1} = c_{i+1}(c_{i+1}^{-1} c_i c_{i+1})$ lets us interchange C_i and C_{i+1}.

Theorem A.1.9 (Frobenius's formula). *Let G be a finite group and $C_1, \ldots, C_k$ conjugacy classes in G. Then*

$$\mathcal{N}(G; C_1, \ldots, C_k) = \frac{|C_1| \cdots |C_k|}{|G|} \sum_\chi \frac{\chi(C_1) \cdots \chi(C_k)}{\chi(1)^{k-2}}, \qquad (A.8)$$

where the sum is over all characters of irreducible representations of G.

Before giving the proof, we note three special cases. If $k = 1$ or $k = 2$, then (A.8) reduces to the orthogonality relation (A.7), applied to $(C_1, 1)$ or (C_1, C_2^{-1}), respectively. For $k = 3$, we write $(C_1, C_2, C_3) = (A, B, C^{-1})$ with $A, B, C \in \mathcal{C}$. Then $\mathcal{N}(G; C_1, C_2, C_3) = n_{AB}^C$, where

$$n_{AB}^C = \#\{(a, b) \in A \times B \mid ab \in C\}.$$

The integers n_{AB}^C are nothing but the structure constants (i.e., the numbers defined by $e_A e_B = \sum_C n_{AB}^C e_C$) of the center of the group ring $\mathbb{Z}[G]$ with respect to the basis $\{e_C\}$ defined in the proof of Corollary A.1.7 above. Formula (A.8) therefore describes the ring structure of this commutative ring in terms of the character theory of G. This formula plays a role in mathematical physics in connection with the so-called "fusion algebras."

Proof. If C is any conjugacy class of G, then the element $e_C = \sum_{g \in C} [g]$ is central and hence, by Schur's lemma, acts on any irreducible representation

A.1 Representation Theory of Finite Groups

π of G as multiplication by a scalar $\nu_\pi(C)$. Since each element $g \in C$ has the same trace $\chi_\pi(g) = \chi_\pi(C)$, we find

$$|C|\chi_\pi(C) = \sum_{g \in C} \chi_\pi(g) = \operatorname{tr}(\pi(e_C), V) = \operatorname{tr}(\nu_\pi(C) \cdot \operatorname{Id}, V) = \nu_\pi(C) \dim \pi$$

and hence

$$\nu_\pi(C) = \frac{|C|}{\dim \pi} \chi_\pi(C) = \frac{\chi_\pi(C)}{\chi_\pi(1)} |C|. \tag{A.9}$$

Now we compute the trace of the action by left multiplication of the product of the elements $e_{C_1}, \ldots, e_{C_k}$ on both sides of (A.4). On the one hand, this product is the sum of the elements $[c_1 \cdots c_k]$ with $c_i \in C_i$ for all i, and since the trace of left multiplication by $[g]$ on $\mathbb{C}[G]$ is clearly $|G|$ for $g = 1$ and 0 otherwise, the trace equals $|G|\mathcal{N}(G; C_1, \ldots, C_k)$. On the other hand, the product of the e_{C_i} acts as scalar multiplication by $\prod \nu_\pi(C_i)$ on π and hence also on the $(\dim \pi)^2$-dimensional space $\operatorname{End}_\mathbb{C}(\pi)$. Formula (A.8) follows immediately. □

Theorem A.1.9 has a clear topological interpretation. Let X be the 2-sphere with k (numbered) points $P_1, \ldots, P_k$ removed. Then $\pi_1(X)$ is a (free) group on k generators $x_1, \ldots, x_k$ with $x_1 \cdots x_k = 1$, and $\mathcal{N}(G; C_1, \ldots, C_k)$ simply counts the number of homomorphisms ρ from $\pi_1(X)$ to G with $\rho(x_i) \in C_i$ for each i. As explained in the text of the book, if G acts faithfully on some finite set F, then each such homomorphism corresponds to a (not necessarily connected) Galois covering of X with fibre F, Galois group G, and ramification points P_i, such that for each i the permutation of the elements of a fixed fibre induced by the local monodromy at P_i belongs to the conjugacy class C_i. Hence $\mathcal{N}(G; C_1, \ldots, C_k)$ counts the coverings with these properties. Observe that the above-mentioned invariance of $\mathcal{N}(G; C_1, \ldots, C_k)$ under permutation of its arguments is clear from this topological point of view.

A natural idea is now to generalize this to ramified coverings of Riemann surfaces of arbitrary genus $g \geq 0$. In view of the structure of the fundamental group of a k-fold punctured surface of genus g, this boils down to computing the number

$$\mathcal{N}_g(G; C_1, \ldots, C_k) = \#\{(a_1, \ldots, a_g, b_1, \ldots, b_g, c_1, \ldots, c_k) \in$$
$$G^{2g} \times C_1 \times \cdots \times C_k \mid [a_1, b_1] \cdots [a_g, b_g] c_1 \cdots c_k = 1\}.$$

The result, given in the theorem below, turns out to be surprisingly simple, but was apparently not given in the classical group-theoretical literature and was discovered first in the context of mathematical physics ([DW], [FQ]). The proof is an almost immediate consequence of the special case $g = 0$.

Theorem A.1.10. *With the same notations as above, we have for all $g \geq 0$*

$$\mathcal{N}_g(G; C_1, \ldots, C_k) = |G|^{2g-1} |C_1| \cdots |C_k| \sum_\chi \frac{\chi(C_1) \cdots \chi(C_k)}{\chi(1)^{k+2g-2}}. \tag{A.10}$$

Proof. Note that $[a_1, b_1] \cdots [a_g, b_g] = a_1 (b_1 a_1 b_1^{-1})^{-1} \cdots a_g (b_g a_g b_g^{-1})^{-1}$ and that, for a and a' in the same conjugacy class A, there are $|G|/|A|$ elements $b \in G$ with $bab^{-1} = a'$. Hence

$$\mathcal{N}_g(G; C_1, \ldots, C_k) = \sum_{A_1, \ldots, A_g \in \mathcal{C}} \frac{|G|}{|A_1|} \cdots \frac{|G|}{|A_g|} \times \mathcal{N}(G; A_1, A_1^{-1}, \ldots, A_g, A_g^{-1}, C_1, \ldots, C_k).$$

Now applying Theorem A.1.9 we find

$$\mathcal{N}_g(G; C_1, \ldots, C_k) = |G|^{g-1} |C_1| \cdots |C_k| \sum_{A_1, \ldots, A_g \in \mathcal{C}} |A_1| \cdots |A_g| \times$$
$$\left(\sum_\chi \frac{\chi(A_1)\overline{\chi(A_1)} \cdots \chi(A_g)\overline{\chi(A_g)} \chi(C_1) \cdots \chi(C_k)}{\chi(1)^{k+2g-2}} \right)$$
$$= |G|^{g-1} |C_1| \cdots |C_k| \sum_\chi \frac{\chi(C_1) \cdots \chi(C_k)}{\chi(1)^{k+2g-2}} \left(\sum_{A \in \mathcal{C}} |A| \chi(A) \overline{\chi(A)} \right)^g.$$

The theorem now follows from the case $\pi = \pi'$ of the orthogonality relation (A.5). □

We observe that, by formula (A.9), both Frobenius's formula and its higher genus generalization can be written more naturally in terms of the eigenvalues $\nu_\pi(C)$ than the character values $\chi_\pi(C)$. In particular, equation (A.10) when expressed in terms of the $\nu_\pi(C)$ takes the simple form

$$\mathcal{N}_g(G; C_1, \ldots, C_k) = |G|^{2g-1} \sum_{\pi \in \mathcal{R}} \frac{\nu_\pi(C_1) \cdots \nu_\pi(C_k)}{(\dim \pi)^{2g-2}}. \tag{A.11}$$

A.2 Applications

In this section we give several applications of the character theory of finite groups in the case when $G = S_n$, the symmetric group on n letters. In this case the conjugacy classes are described simply by the partitions of n, the conjugacy class of an element $g \in S_n$ with d_1 one-cycles, d_2 two-cycles, etc., corresponding to the partition $\lambda = 1^{d_1} 2^{d_2} \cdots$ of n. It turns out that much of the part of the information which we are interested in can be captured by any of three polynomials $\Phi_g(X)$, $P_g(r)$ or $Q_g(k)$ which mutually determine each other but which encode the interesting combinatorial information about g in different ways.

A.2.1 Representations of S_n and Canonical Polynomials Associated to Partitions

In this subsection we define the three polynomials $\Phi_g(X)$, $P_g(r)$ or $Q_g(k)$ mentioned above and prove their main properties. Applications to several of the topics treated in the main text of the book will then be given in Sec. A.2.2–A.2.4. Since each of these three polynomials depends only on the conjugacy class C of $g \in S_n$ or equivalently only on the partition $\lambda \vdash n$ corresponding to C, we will also use the notations $\Phi_C(X)$, $P_C(r)$ or $Q_C(k)$ and $\Phi_\lambda(X)$, $P_\lambda(r)$ or $Q_\lambda(k)$.

We recall from Sec. A.1.2 the standard irreducible representation $\mathbf{St}_n = \mathbb{C}^n/\mathbb{C}$ of S_n of dimension $n-1$. We define $\Phi_g(X)$ as the characteristic polynomial of g on $\mathbf{St}_n$:

$$\Phi_g(X) = \det(1 - gX, \mathbf{St}_n). \tag{A.12}$$

It is easy to see that, if g has the cycle structure $1^{d_1} 2^{d_2} \cdots$, then the characteristic polynomial of g on $\mathbb{C}^n = \mathbf{St}_n \oplus \mathbf{1}$ is simply $\prod (1 - X^i)^{d_i}$, so in terms of λ we have

$$\Phi_\lambda(X) = \frac{(1-X)^{d_1}(1-X^2)^{d_2}\cdots}{1-X} \qquad (\lambda = 1^{d_1} 2^{d_2} \cdots). \tag{A.13}$$

This is the first of our three polynomials.

From linear algebra, we know that the coefficient of X^r in the characteristic polynomial of any endomorphism ϕ of a finite-dimensional vector space V equals $(-1)^r$ times the trace of the endomorphism induced by ϕ on the rth exterior power $\bigwedge^r(V)$. We therefore have

$$\Phi_g(X) = \sum_{r=0}^{n-1} (-1)^r \chi_r(g) X^r, \tag{A.14}$$

where we have abbreviated

$$\chi_r(g) := \operatorname{tr}(g, \pi_r), \qquad \pi_r := \bigwedge^r(\mathbf{St}_n) \qquad (0 \le r \le n-1).$$

We shall show below that the representations π_r are irreducible and distinct, so the χ_r are distinct characters of S_n. We have $\chi_r(1) = \dim \pi_r = \binom{n-1}{r}$. Now, since there is a unique polynomial of degree $n-1$ having specified values at any n specified points, we can define a polynomial $P_g(r)$ by the requirements that $\deg P_g \le n-1$ and that

$$P_g(r) = \frac{\chi_r(g)}{\chi_r(1)} = \binom{n-1}{r}^{-1} \chi_r(g) \qquad (0 \le r \le n-1). \tag{A.15}$$

This is our second polynomial attached to (the conjugacy class of) g.

Finally, we recall the notations $\ell(g) = \sum d_i$ and $v(g) = \sum (i-1) d_i$ for the number of cycles of an element g of S_n and for its complement $v(g) = n - \ell(g)$.

410   A Representation Theory

(Again, since both depend only on the conjugacy class of g, we will use the alternative notations $\ell(C)$, $v(C)$ or $\ell(\lambda)$, $v(\lambda)$ as the situation requires.) Let σ denote the cyclic element $(1\,2\,\ldots\,n)$ of S_n. We then define

$$Q_C(k) = \frac{1}{|C|} \sum_{g \in C} k^{\ell(g\sigma)} \tag{A.16}$$

for any conjugacy class C of S_n. This is our third polynomial.

Before stating the main result, which describes how each of the three polynomials just defined determines the other two, we need two simple (and well-known) lemmas concerning the representations π_r.

Lemma A.2.1. *The representations $\pi_r = \bigwedge^r(\mathbf{St}_n)$ $(0 \leq r \leq n-1)$ of S_n are irreducible and distinct.*

Proof. Set $V = \pi_0 \oplus \cdots \oplus \pi_{n-1}$, a representation of $G = S_n$ of dimension 2^{n-1}. By the results of Sec. A.1, we know that V can be decomposed uniquely as $\bigoplus_{\pi \in \mathcal{R}} m(\pi)\pi$, where $\mathcal{R}$ is a set of representatives of the isomorphism classes of irreducible representations of G and the multiplicities $m(\pi)$ are non-negative integers. Since V is by construction a sum of n non-trivial representations, we have $\sum m(\pi) \geq n$. Define $\chi_V : S_n \to \mathbb{C}$ by $\chi_V(g) = \text{tr}(g, V)$. Then χ_V decomposes as $\chi_V = \sum m(\pi)\chi_\pi$. We define a scalar product on the vector space of conjugacy-invariant functions $f : S_n \to \mathbb{C}$ by $(f_1, f_2) = |G|^{-1} \sum_{g \in G} f_1(g)\overline{f_2(g)}$. By equation (A.1) (first orthogonality relation) we know that the characters χ_π of the irreducible representations of G form an orthogonal basis for this vector space with respect to this scalar product, so $(\chi_V, \chi_V) = \sum m(\pi)^2$. If we can show that $(\chi_V, \chi_V) = n$, it will follow that $\sum(m(\pi)^2 - m(\pi)) \leq 0$ and hence that each multiplicity $m(\pi)$ equals 0 or 1, with $m(\pi) = 1$ for exactly n distinct irreducible representations of G, proving the lemma.

To calculate (χ_V, χ_V), we first use equations (A.14) and (A.13) to obtain that

$$\chi_V(g) = \sum_{r=0}^{n-1} \chi_r(g) = \Phi_g(-1) = \begin{cases} 2^{(d_1+d_3+\cdots)-1} & \text{if } d_2 = d_4 = \cdots = 0, \\ 0 & \text{otherwise} \end{cases}$$

for g with the cycle structure $\prod i^{d_i}$. Since the number of elements with this cycle structure is $n!/\prod(d_i!\, i^{d_i})$, we obtain

$$(\chi_V, \chi_V) = \frac{1}{n!} \sum_{g \in S_n} |\chi_V(g)|^2$$

$$= \frac{1}{4} \sum_{\substack{d_1, d_3, d_5, \ldots \geq 0 \\ d_1 + 3d_3 + 5d_5 + \cdots = n}} \frac{4^{d_1+d_3+d_5+\cdots}}{d_1!\, d_3!\, d_5! \cdots 1^{d_1}\, 3^{d_3}\, 5^{d_5} \cdots}$$

$$= \frac{1}{4} \cdot \text{coefficient of } x^n \text{ in} \prod_{\substack{i \geq 1 \\ i \text{ odd}}} e^{4x^i/i} = n,$$

since $4 \sum_{i \text{ odd}} \dfrac{x^i}{i} = 2 \log \dfrac{1+x}{1-x}$ and $\left(\dfrac{1+x}{1-x}\right)^2 = 1 + 4 \sum_{n \geq 1} nx^n$. □

Lemma A.2.2. *The value of the character of an irreducible representation π of S_n on the conjugacy class of the cyclic element $\sigma \in S_n$ is given by*

$$\chi_\pi(\sigma) = \begin{cases} (-1)^r & \text{if } \pi \simeq \pi_r \text{ for some } r, \ 0 \leq r \leq n-1, \\ 0 & \text{otherwise.} \end{cases}$$

Proof. The first statement follows immediately from formulas (A.14) and (A.13), since

$$\sum_{r=0}^{n-1} (-1)^r \chi_r(\sigma) X^r = \Phi_\sigma(X) = \dfrac{1 - X^n}{1 - X} = \sum_{r=0}^{n-1} X^r.$$

The second then follows immediately from the second orthogonality relation for characters, since $\sum_{\pi \in \mathcal{R}} |\chi_\pi(\sigma)|^2 = n!/(n-1)! = n = \sum_{r=0}^{n-1} |\chi_r(\sigma)|^2$. □

Theorem A.2.3. *For any integer $n \geq 1$ there are canonical linear isomorphisms between the three n-dimensional vector spaces*

$$\langle 1, X, \ldots, X^{n-1} \rangle, \quad \langle 1, r, \ldots, r^{n-1} \rangle, \quad \langle k, k^2, \ldots, k^n \rangle,$$

such that the three polynomials $\Phi_C(X)$, $P_C(r)$ and $Q_C(k)$ correspond to one another for every conjugacy class C in S_n. These isomorphisms are given by $\Phi \leftrightarrow P \leftrightarrow Q$, where $\Phi(X)$, $P(r)$ and $Q(k)$ are related by the generating function identities

$$\Phi(X) = \sum_{r=0}^{n-1} (-1)^r \binom{n-1}{r} P(r) X^r = (1-X)^{n+1} \sum_{k=1}^{\infty} Q(k) X^{k-1}, \quad (A.17)$$

or alternatively in terms of bases of the three vector spaces by

$$(1-X)^{n-1-s} \leftrightarrow \binom{n-1-r}{s} \Big/ \binom{n-1}{s} \leftrightarrow \binom{s+k}{s+1} \quad (A.18)$$

or

$$X^s(1-X)^{n-1-s} \leftrightarrow (-1)^s \binom{r}{s} \Big/ \binom{n-1}{s} \leftrightarrow \binom{k}{s+1} \quad (A.19)$$

for $0 \leq s \leq n-1$. These isomorphisms are equivariant with respect to the three involutions $*$ defined by

$$\Phi^*(X) = (-X)^{n-1} \Phi(1/X), \quad P^*(r) = P(n-1-r),$$
$$Q^*(k) = -Q(-k), \quad (A.20)$$

under which the polynomials $\Phi_C(X)$, $P_C(r)$ and $Q_C(k)$ are invariant or antiinvariant according to whether C is a conjugacy class of even or odd permutations.

Proof. The fact that the formulas (A.17) give isomorphisms between the spaces in question can be checked either directly or by using the binomial theorem to check that the collections of functions listed in (A.18) or in (A.19), each of which forms a basis for the relevant vector space, satisfy the identities in (A.17). The equivariance of the isomorphisms with respect to the involutions (A.20) follows most easily by noting that these involutions exchange the bases given in (A.18) and (A.19) up to a factor $(-1)^s$. The fact that $\Phi_g(X)$ and $P_g(r)$ are related by the first formula in (A.17) follows immediately from equations (A.13) and (A.14) above. The statement about the invariance or anti-invariance of Φ_g, P_g and Q_g under the involutions $*$ is easy in each case (although of course it would suffice to verify only one of them and then use the rest of the theorem): for Φ_g it follows immediately from formula (A.13) and the fact that $\mathrm{sgn}(g) = (-1)^{v(g)}$ with $v(g) = \sum(i-1)d_i$; for P_g it follows from (A.15) and the fact that $\bigwedge^{n-1-r}(\mathbf{St}_n)$ is dual to the tensor product of $\bigwedge^{r}(\mathbf{St}_n)$ with the sign representation ε_n (because $\bigwedge^{d-r} V \otimes \bigwedge^r(V) \xrightarrow{\wedge} \bigwedge^d(V)$ is a perfect pairing for any d-dimensional vector space V and $\bigwedge^{n-1}(\mathbf{St}_n) = \varepsilon_n$); and for Q_g it follows from the fact that $\ell(g\sigma) = n - v(g\sigma) \equiv v(g) + 1 \pmod{2}$ for any $g \in S_n$. The only thing we have to show is therefore that the polynomials $\Phi_C(X)$ and $Q_C(k)$ are related by the generating series identity

$$\sum_{k=1}^{\infty} Q_C(k) X^{k-1} = \frac{\Phi_C(X)}{(1-X)^{n+1}}, \qquad (A.21)$$

for any conjugacy class C in S_n. For this we use the following lemma.

Lemma A.2.4. *For any integers n, r and k with $0 \le r \le n-1$ we have*

$$\frac{1}{n!} \sum_{g \in S_n} P_g(r) k^{\ell(g)} = \text{coefficient of } X^{k-1} \text{ in } \frac{X^r}{(1-X)^{n+1}}.$$

Proof. By a calculation similar to the one in the proof of Lemma A.2.1, we have

$$\frac{1}{n!} \sum_{g \in S_n} \Phi_g(Y) k^{\ell(g)} = \sum_{\substack{d_1, d_2, \dots \\ d_1 + 2d_2 + \dots = n}} \frac{k^{d_1 + d_2 + \cdots}}{d_1! d_2! \cdots 1^{d_1} 2^{d_2} \cdots} \times$$

$$\frac{(1-Y)^{d_1}(1-Y^2)^{d_2}\cdots}{1-Y}$$

$$= \frac{1}{1-Y} \cdot \text{coefficient of } u^n \text{ in } \prod_{i=1}^{\infty} \exp\left(\frac{k(1-Y^i)u^i}{i}\right)$$

$$= \frac{1}{1-Y} \cdot \text{coefficient of } u^n \text{ in } \left(\frac{1-uY}{1-u}\right)^k$$

$$= \text{coefficient of } X^{k-1} \text{ in } \frac{(1-XY)^{n-1}}{(1-X)^{n+1}}, \qquad (A.22)$$

where the last equality follows either by using residue calculus to get

$$\frac{1}{1-Y}\operatorname{Res}_{u=0}\left(\left(\frac{1-uY}{1-u}\right)^k \frac{du}{u^{n+1}}\right) = -\operatorname{Res}_{X=1}\left(\frac{(1-XY)^{n-1}}{(1-X)^{n+1}}\frac{dX}{X^k}\right)$$

$$= +\operatorname{Res}_{X=0}\left(\frac{(1-XY)^{n-1}}{(1-X)^{n+1}}\frac{dX}{X^k}\right)$$

(here $u = \frac{1-X}{1-XY}$) or else from the geometric series identity

$$\sum_{k=1}^{\infty} X^{k-1}\left(\frac{1-uY}{1-u}\right)^k = \frac{1-uY}{1-u-X+uXY}$$

$$= \frac{1}{1-X} + (1-Y)\sum_{n=1}^{\infty}\frac{(1-XY)^{n-1}}{(1-X)^{n+1}}u^n.$$

The lemma follows by comparing the coefficients of $(-1)^r\binom{n-1}{r}Y^r$ on both sides of (A.22). □

Remark A.2.5. Lemma A.2.4 implies that

$$\frac{1}{n!}\sum_{g\in S_n} P_g(r)\,k^{\ell(g)} = \binom{k+n-r-1}{n} - \binom{n-r-1}{n}$$

as polynomials in r and n, since both sides of this identity are polynomials in k and r, both have degree $\leq n-1$ in r, and they agree for $k \in \mathbb{N}$ and for $r = 0, 1, \ldots, n-1$.

We can now complete the proof of equation (A.21) and hence of Theorem A.2.3. Frobenius's formula (Theorem A.1.9) in the case $k = 3$ can be rewritten in the form

$$\frac{1}{|C|}\sum_{c\in C} F(bc) = \sum_{\pi} \chi_\pi(b)\chi_\pi(C)\left(\frac{1}{|G|}\sum_{A\in G}|A|\frac{\chi_\pi(A)}{\chi_\pi(1)}F(A^{-1})\right)$$

for any finite group G, conjugacy classes A and C of G, and class function (= conjugacy-invariant function) $F : G \to \mathbb{C}$, where the sum is over all irreducible representations of G. Specializing to $G = S_n$ and $b = \sigma$ and using Lemma A.2.2, we find

$$\frac{1}{|C|}\sum_{c\in C} F(\sigma c) = \sum_{r=0}^{n-1}(-1)^r \chi_r(C)\left(\frac{1}{n!}\sum_{g\in G} P_g(r) F(g^{-1})\right).$$

Now specializing further to $F(g) = k^{\ell(g)}$ and using Lemma A.2.4 and equation (A.14) gives

$$Q_C(k) = \text{coefficient of } X^{k-1} \text{ in } \frac{\Phi_C(X)}{(1-X)^{n+1}}. \qquad \square$$

We mention two simple consequences of Theorem A.2.3 which will be used later.

Corollary A.2.6. *Let $\lambda = 1^{d_1} 2^{d_2} \cdots$ be a partition of n. Then the polynomial $P_\lambda(r)$ has degree $v(\lambda) = \sum (i-1) d_i$ and leading coefficient $(-1)^{v(\lambda)} K(\lambda)$, where*

$$K(\lambda) = \frac{(n-1-v(\lambda))!}{(n-1)!} 1^{d_1} 2^{d_2} \cdots n^{d_n}. \tag{A.23}$$

The polynomial $Q_\lambda(k)$ has degree $v(\lambda) + 1$ and leading coefficient

$$\frac{\prod i^{d_i}}{(v(\lambda)+1)!}.$$

Proof. From equation (A.13) we see that $\Phi_\lambda(X) \sim \frac{(n-1)!}{(n-1-v(\lambda))!} K(\lambda) \times (1-X)^{n-1-v(\lambda)}$ as $X \to 1$. But this means that $\Phi_\lambda(X)$ is a linear combination of the basis elements (A.18) with $s \leq v(\lambda)$, with the coefficient for $s = v(\lambda)$ being $\frac{(n-1)!}{(n-1-v(\lambda))!} K(\lambda)$. The correspondence (A.18) then tells us that P_λ and Q_λ have the degrees and leading coefficients stated. □

Corollary A.2.7. *The polynomial $Q_\lambda(k)$ takes integer values for $k \in \mathbb{Z}$. Moreover, the value of $Q_\lambda(k)$ for $k \in \mathbb{N}$ depends only on the numbers $n - d_1$, $d_2, \ldots, d_{k-1}$, the first few being*

$$Q_\lambda(0) = 0, \quad Q_\lambda(1) = 1, \quad Q_\lambda(2) = n+2-d_1, \quad Q_\lambda(3) = \binom{n+3-d_1}{2} - d_2.$$

Proof. Equations (A.21) and (A.13) give

$$Q_C(k) = \text{coefficient of } X^{k-1} \text{ in}$$
$$(1-X)^{d_1-n-2} (1-X^2)^{d_2} (1-X^3)^{d_3} \cdots (1-X^{k-1})^{d_{k-1}}. \quad \square$$

Remark A.2.8. The fact that $\deg Q_\lambda \leq v(\lambda)$ says that $\ell(g\sigma) \leq v(g)$ for all $g \in S_n$. Equivalently, $v(g) + v(g\sigma) \geq n - 1 = v(\sigma)$. This is a special case of the general fact that $v(g_1) + v(g_2) \geq v(g_3)$ for any three elements $g_1, g_2, g_3 \in S_n$ with product 1, which can be seen most easily by noticing that $v(g_i)$ is the codimension of the fixed point set of g_i acting on $\mathbb{C}^n$ (or on its irreducible subspace $\mathbf{St}_n$) and that codimensions of subspaces behave subadditively. The statement that the polynomial $Q_\lambda(k)$ depends only on $n - d_1$ and the d_i with $i \geq 2$ says that it is stable under the inclusions $S_n \subset S_{n+1}$ and hence depends only on the class of g in S_∞. This is not difficult to see directly from the definition, but there does not seem to be any obvious reason why the value of $Q_\lambda(k)$ for $k \in \mathbb{N}$ depends only on the d_i with $i < k$ or why it is an integer.

A.2.2 Examples

We give a number of examples of the polynomials introduced in Sec. A.2.1 for special conjugacy classes and for small values of n.

Trivial element: $g = 1$. Here the cycle structure is simply 1^n, so equation (A.13) gives $\Phi_1(X) = (1 - X)^{n-1}$, while from equations (A.4) and (A.5) we immediately get $P_1(r) = 1$ and $Q_1(k) = k$.

Transposition: $T = (1, 2)$. Now the cycle structure is $1^{n-2}2$, so equation (A.13) gives

$$\Phi_T(X) = (1 - X)^{n-3}(1 - X^2) = (1 - X)^{n-2}(1 + X). \tag{A.24}$$

From equation (A.14) we therefore obtain

$$\chi_r(T) = \binom{n-2}{r} - \binom{n-2}{r-1} = \binom{n-3}{r} - \binom{n-3}{r-2}$$

(with the obvious conventions when $r < 2$), so from (A.15) and the formula $\chi_r(1) = \binom{n-1}{r}$ we obtain

$$P_T(r) = 1 - \frac{2r}{n-1} \tag{A.25}$$

while equation (A.17) or (A.18) or (A.19) gives

$$Q_T(k) = k^2.$$

Note that this last formula, unlike the formulas for $\Phi_T(X)$ and $P_T(r)$, is independent of n. This is a special case of the above-mentioned fact that $Q_g(k)$ is stable under $S_n \subset S_{n+1}$.

Cyclic permutation: $\sigma = (1, 2, \cdots, n)$. Here the cycle structure is simply n^1, so, as we have already seen, $\Phi_\sigma(X) = 1 + X + \cdots + X^{n-1}$ and $\chi_r(\sigma) = (-1)^r$. However, the explicit forms of the polynomials P_σ and Q_σ are quite complicated, as can be seen from the examples for $n = 5$ and 6 below.

Free involution: $\tau = (1, 2)(3, 4) \cdots (n-1, n)$. Here n must be even, say $n = 2m$. The cycle structure of the conjugacy class of τ is 2^m, so (A.13) gives

$$\Phi_\tau(X) = \frac{(1 - X^2)^m}{1 - X} = (1 + X)(1 - X^2)^{m-1} \tag{A.26}$$

and hence

$$\chi_r(\tau) = (-1)^{[(r+1)/2]} \binom{m-1}{[r/2]},$$

but here again the polynomials P_τ and Q_τ have no simple closed form.

Small n. Finally, to give the reader a better feel for the $\Phi_\lambda \leftrightarrow P_\lambda \leftrightarrow Q_\lambda$ correspondence, we give a complete table of these three polynomials for partitions of n with $1 \leq n \leq 6$ (see the next page). For convenience we tabulate the polynomial $\widetilde{P}_\lambda(t) := P\left(\frac{n-1}{2} - t\right)$ instead of $P_\lambda(r)$, since by the last statement of Theorem A.2.3 this is an even or an odd polynomial in t for all λ. The stability of Q_λ mentioned in the remark at the end of the last section is visible in this table: for example, the first seven entries in the column giving $Q_\lambda(k)$ for $n = 6$ agree with the values for $n = 5$.

A.2.3 First Application: Enumeration of Polygon Gluings

This is the combinatorial question which was discussed in Chapter 3 in connection with the evaluation of the (orbifold) Euler characteristic of the moduli space of curves of genus g. The problem was to count the number $\varepsilon_g(m)$ of ways to identify in pairs (with reverse orientation) the sides of a $(2m)$-gon such that the closed oriented surface obtained has a given genus g. Clearly this is the same as the number of free involutions in S_{2m} whose product with the standard cyclic permutation σ has $m + 1 - 2g$ cycles, so formula (A.16) gives

$$Q_\tau(k) = \frac{1}{(2m-1)!!} \sum_{0 \leq g \leq m/2} \varepsilon_g(m)\, k^{m+1-2g}, \qquad (A.27)$$

where τ denotes any free involution (and where we have used that the conjugacy class of τ has cardinality $(2m-1)!!$). On the other hand, from equations (A.21) and (A.26) we get

$$\sum_{k=1}^\infty Q_\tau(k)\, X^{k-1} = \frac{\Phi_\tau(X)}{(1-X)^{2m+1}} = \frac{(1+X)^m}{(1-X)^{m+2}},$$

which we can rewrite equivalently by using the same calculation as in the last line of (A.22) (with n replaced by $m + 1$ and Y by -1) as

$$Q_\tau(k) = \frac{1}{2} \cdot \text{coefficient of } u^{m+1} \text{ in } \left(\frac{1+u}{1-u}\right)^k.$$

Since the sum on the right-hand side of (A.27) is the polynomial denoted $T_m(k)$ in Sec. 3.1, this reproduces the evaluation of $\varepsilon_g(m)$ in terms of generating functions given in Theorem 3.1.5.

n	λ	$\widetilde{P}_\lambda(t)$	$Q_\lambda(k)$	$\Phi_\lambda(X)$
1	1^1	1	k	1
2	1^2	1	k	$1-X$
	2^1	$2t$	k^2	$1+X$
3	1^3	1	k	$1-2X+X^2$
	$1^1 2^1$	t	k^2	$1-X^2$
	3^1	$\frac{1}{2}(3t^2-1)$	$\frac{1}{2}(k^3+k)$	$1+X+X^2$
4	1^4	1	k	$1-3X+3X^2-X^3$
	$1^2 2^1$	$\frac{2}{3}t$	k^2	$1-X-X^2+X^3$
	$1^1 3^1$	$\frac{1}{8}(4t^2-1)$	$\frac{1}{2}(k^3+k)$	$1-X^3$
	2^2	$\frac{1}{6}(4t^2-3)$	$\frac{1}{3}(2k^3+k)$	$1+X-X^2-X^3$
	4^1	$\frac{1}{6}(4t^3-5t)$	$\frac{1}{6}(k^4+5k^2)$	$1+X+X^2+X^3$
5	1^5	1	k	$1-4X+6X^2-4X^3+X^4$
	$1^3 2^1$	$\frac{1}{2}t$	k^2	$1-2X+2X^3-X^4$
	$1^2 3^1$	$\frac{1}{4}t^2$	$\frac{1}{2}(k^3+k)$	$1-X-X^3+X^4$
	$1^1 2^2$	$\frac{1}{3}(t^2-1)$	$\frac{1}{3}(2k^3+k)$	$1-2X^2+X^4$
	$1^1 4^1$	$\frac{1}{6}(t^3-t)$	$\frac{1}{6}(k^4+5k^2)$	$1-X^4$
	$2^1 3^1$	$\frac{1}{4}(t^3-2t)$	$\frac{1}{4}(k^4+3k^2)$	$1+X-X^3-X^4$
	5^1	$\frac{1}{24}(5t^4-15t^2+4)$	$\frac{1}{24}(k^5+15k^3+8k)$	$1+X+X^2+X^3+X^4$
6	1^6	1	k	$1-5X+10X^2-10X^3+5X^4-X^5$
	$1^4 2^1$	$\frac{2}{5}t$	k^2	$1-3X+2X^2+2X^3-3X^4+X^5$
	$1^3 3^1$	$\frac{1}{80}(12t^2+5)$	$\frac{1}{2}(k^3+k)$	$1-2X+X^2-X^3+2X^4-X^5$
	$1^2 2^2$	$\frac{1}{20}(4t^2-5)$	$\frac{1}{3}(2k^3+k)$	$1-X-2X^2+2X^3+X^4-X^5$
	$1^2 4^1$	$\frac{1}{60}(4t^3-t)$	$\frac{1}{6}(k^4+5k^2)$	$1-X-X^4+X^5$
	$1^1 2^1 3^1$	$\frac{1}{90}(4t^3-9t)$	$\frac{1}{4}(k^4+3k^2)$	$1-X^2-X^3+X^5$
	$1^1 5^1$	$\frac{1}{384}(16t^4-40t^2+9)$	$\frac{1}{24}(k^5+15k^3+8k)$	$1-X^5$
	2^3	$\frac{1}{30}(4t^3-13t)$	$\frac{1}{3}(k^4+2k^2)$	$1+X-2X^2-2X^3+X^4+X^5$
	$2^1 4^1$	$\frac{1}{240}(16t^4-64t^2+15)$	$\frac{1}{15}(k^5+10k^3+4k)$	$1+X-X^4-X^5$
	3^2	$\frac{1}{640}(48t^4-216t^2+115)$	$\frac{1}{40}(3k^5+25k^3+12k)$	$1+X+X^2-X^3-X^4-X^5$
	6^1	$\frac{1}{960}(48t^5-280t^3+259t)$	$\frac{1}{120}(k^6+35k^4+84k^2)$	$1+X+X^2+X^3+X^4+X^5$

The proof just given follows the exposition in [Z], where the identity (A.21) was proved and a few other applications were given. We mention two briefly:

- the probability that the product of two random cyclic permutations in S_n has exactly ℓ cycles is $(1+(-1)^{n-\ell})$ times the coefficient of x^ℓ in $\binom{x+n}{n+1}$. In particular, for n odd the probability that such a product is cyclic equals $\frac{2}{n+1}$, as opposed to the probability that a random even element of S_n is cyclic, which equals $\frac{2}{n}$.

- the number of representations of an arbitrary even element of S_n as a product of two cyclic permutations is $\geq \frac{2(n-1)!}{n+2}$, as opposed to the average number of such representations for even permutations, which equals $\frac{2(n-1)!}{n}$.

A.2.4 Second Application: the Goulden–Jackson Formula

Our second application, again reproducing a result proved by a different method within the main text of the book, is the formula of Goulden-Jackson given in various forms in Theorems 1.5.12, 1.5.15, 1.6.6 and 5.2.2. The problem, in our present language, is to count the "Frobenius number" $\mathcal{N}(S_n; C_0, C_1, \cdots, C_k)$ of $(k+1)$-tuples $(c_0, \ldots, c_k) \in C_0 \times \cdots \times C_k$ when C_0 is the class of the cyclic element $\sigma \in S_n$ and $C_1, \ldots, C_k$ are arbitrary conjugacy classes in S_n. The Goulden-Jackson formula says that this number is 0 if $v = v(C_1) + \cdots + v(C_k)$ is less than $n-1$ and gives an explicit formula for it when $v(C_1) + \cdots + v(C_k) = n - 1$. We give a somewhat more general result.

Theorem A.2.9. *Let $C_1, \ldots, C_k$ be arbitrary conjugacy classes in S_n and C_0 the class of the cyclic element σ. Then*

$$\mathcal{N}(S_n; C_0, C_1, \ldots, C_k) = \frac{(-1)^{n-1}}{n} |C_1| \cdots |C_k| \cdot \Delta^{n-1}(P_{C_1} \cdots P_{C_k})(0), \tag{A.28}$$

where $P_{C_i}(r)$ is the polynomial of degree $v(C_i)$ associated to the conjugacy class C_i as in Sec. A.2.1 and Δ denotes the forward differencing operator $\Delta P(r) = P(r+1) - P(r)$.

For the conjugacy class C of S_n corresponding to a partition $\lambda = 1^{d_1} \cdots n^{d_n}$ of n, let us write $N(C)$ for the number $N(\lambda) = \frac{(d_1 + \cdots + d_n - 1)!}{d_1! \cdots d_n!}$ defined in 1.5.11. Then we have:

Corollary A.2.10 (Goulden-Jackson formula). *Let the C_i be as in the theorem. Then the number $\mathcal{N}(S_n; C_0, C_1, \ldots, C_k)$ vanishes if the number $v := v(C_1) + \cdots + v(C_k)$ is less than $n-1$ and is given by*

$$\frac{1}{(n-1)!} \mathcal{N}(S_n; C_0, C_1, \ldots, C_k) = n^{k-1} N(C_1) \cdots N(C_k) \tag{A.29}$$

if $v = n - 1$.

Proof. The theorem is an immediate consequence of Frobenius's formula (Theorem A.1.9) and Lemma A.2.2, which says that $\chi(C_0)$ equals $(-1)^r$ if $\chi = \chi_r$ and vanishes for all other irreducible characters, together with definition (A.15), which together give

$$\mathcal{N}(S_n; C_0, C_1, \ldots, C_k) = \frac{1}{n} |C_1| \cdots |C_k| \sum_{r=0}^{n-1} (-1)^r \binom{n-1}{r} P_{C_1}(r) \cdots P_{C_k}(r).$$

The corollary follows immediately from the theorem together with Corollary A.2.6 above and the formula $|C_i|K(\lambda_i) = nN(\lambda_i)$ (compare equation (A.23) and the definition of $N(\lambda)$), since the polynomial $P_{C_1}(r) \cdots P_{C_k}(r)$ has degree v and the mth difference of a polynomial of degree v is 0 for $m > v$ and equal to $m!$ times the leading coefficient of the polynomial if $m = v$. □

Remark A.2.11. The fact that $\mathcal{N}(S_n; C_0, C_1, \ldots, C_k) = 0$ unless $v \geq n - 1$ is also obvious from the remark at the end of Sec. A.2.1, which says that the function v satisfies the triangle inequality and hence necessarily $v(c_1) + \cdots + v(c_k) \geq v(c_0)$ if $c_0 c_1 \cdots c_k = 1$. It also follows from the topological interpretation of $\mathcal{N}(S_n; C_0, C_1, \ldots, C_k)$ as the counting function for ramified coverings of S^2 with ramification types C_i, since $v - n - 1$ equals the Euler characteristic of the total space of the covering, which has the form $2 - 2g \leq 2$ because a covering one of whose ramification types is cyclic is necessarily connected. This shows also that the number $\mathcal{N}(S_n; C_0, C_1, \ldots, C_k)$ vanishes unless $v \equiv n - 1 \pmod{2}$. To see this from Theorem A.2.9, it is convenient to rewrite formula (A.28) in the form

$$\frac{n\mathcal{N}(S_n; C_0, C_1, \ldots, C_k)}{|C_1| \cdots |C_k|} = \Delta_+^{n-1}(\widetilde{P}_{C_1} \cdots \widetilde{P}_{C_k})(0), \qquad (A.30)$$

where $\widetilde{P}_C(t) = P_C(\frac{n-1}{2} - t)$ is the shifted version of $P_C(r)$ mentioned in Sec. A.2.2 and Δ_+ is the symmetric difference operator $\Delta_+ f(t) = f(t + \frac{1}{2}) - f(t - \frac{1}{2})$. This expression obviously vanishes if $v \not\equiv n - 1 \pmod{2}$ because of the symmetry property $\widetilde{P}_C(-t) = (-1)^{v(C)} \widetilde{P}_C(t)$ and the fact that Δ_+ reverses the parity of an even or odd function.

Remark A.2.12. As a special case, we can take $C_1 = \cdots = C_k = [T]$, the class of transpositions in S_n. Then $\widetilde{P}_{C_i}(t) = \frac{2t}{n-1}$ by formula (A.25), so equation (A.30) gives

$$\mathcal{N}(S_n; \sigma, \underbrace{T, \ldots, T}_{k}) = n^{k-1} \Delta_+^{n-1}(t^k)\big|_{t=0}.$$

Using the identity

$$\Delta_+^{n-1}(t^k)\big|_{t=0} = \sum_{r=0}^{n-1} (-1)^r \binom{n-1}{r} \left(\frac{n-1}{2} - r\right)^k = \frac{d^k}{du^k}\left[\left(e^{u/2} - e^{-u/2}\right)^{n-1}\right]\big|_{u=0}$$

and the residue theorem, we can write this equivalently as

$$\mathcal{N}(G; \sigma, \underbrace{T, \ldots, T}_{k}) = k! \cdot n^{k-1} \cdot S_g(n) \quad \text{for } k = n - 1 + 2g \quad (g \geq 0), \quad (A.31)$$

where $S_g(n)$ is the polynomial of degree g in n given by

420 A Representation Theory

$$S_g(n) = \text{coefficient of } u^{2g} \text{ in } \left(\frac{\sinh u/2}{u/2}\right)^{n-1}$$

$$= \frac{n-1}{n-1+2g} \cdot \text{coefficient of } s^{2g} \text{ in } \left(\frac{s/2}{\operatorname{asinh} s/2}\right)^{n+2g-1}, \quad \text{(A.32)}$$

the first values of which are given by

$$S_0(n) = 1, \quad S_1(n) = \frac{n-1}{24}, \quad S_2(n) = \frac{(n-1)(5n-7)}{5760}. \quad \text{(A.33)}$$

Formula (A.31) for $\mathcal{N}(S_n; \sigma, T, \ldots, T)$ was given in [SSV].

Remark A.2.13. Another observation is that, from formula (A.10) together with the fact that $\chi(\sigma) = \chi(\sigma)^3$ for all irreducible characters χ of S_n (Lemma A.2.2), we have the formula

$$\mathcal{N}_g(S_n; \sigma, C_1, \ldots, C_k) = n^{2g} \mathcal{N}(S_n; \underbrace{\sigma, \ldots, \sigma}_{2g+1}, C_1, \ldots, C_k) \quad \text{(A.34)}$$

for the "generalized Frobenius number" of Theorem A.1.10, for any integer $g \geq 0$ and any conjugacy classes $C_1, \ldots, C_k$ in S_n. This allows one to generalize Theorem A.2.9 to the case of ramified coverings of a Riemann surface of arbitrary genus with cyclic ramification at at least one point.

In view of the very simple form of equation (A.34), it might be of interest to give a direct combinatorial or topological proof, without using character theory.

In the rest of this subsection, we show how one can use Theorem A.2.9 to give explicit formulas for $\mathcal{N}(S_n; C_0, C_1, \ldots, C_k)$ for arbitrary values of v. The calculation and the results are considerably simplified if we use the symmetric formula (A.30) instead of (A.28), but still rapidly become complicated as the number $2g = v - n + 1$ grows.

The main step in the calculation is a refinement of Corollary A.2.6 of Theorem A.2.3 giving more leading terms of $P_\lambda(r)$ (or equivalently, of $\widetilde{P}_\lambda(t)$) for an arbitrary conjugacy class λ of S_n. Write $\lambda = n_1 + \cdots + n_s = 1^{d_1} 2^{d_2} \cdots$ and define invariants $\ell_h(\lambda)$ (higher moments) of λ by

$$\ell_h(\lambda) = n_1^h + \cdots + n_s^h = \sum_{i \geq 1} i^h d_i \quad (h = 0, 1, \ldots), \quad \text{(A.35)}$$

so that $\ell_0(\lambda) = \ell(\lambda)$, $\ell_1(\lambda) = n$. Assign to $\ell_h(\lambda)$ the weight h. Then we have:

Lemma A.2.14. *The polynomial* $\widetilde{P}_\lambda(t) = P_\lambda\left(\frac{n-1}{2} - t\right)$ *has the form*

$$\widetilde{P}_\lambda(t) = K(\lambda) \left(1 + a_1(\lambda) t^{v(\lambda)-2} + a_2(\lambda) t^{v(\lambda)-4} + \cdots \right),$$

where $K(\lambda)$ is given by (A.23) and each $a_j(\lambda)$ is a universal polynomial of weighted degree $\leq 2j$ in n and the invariants $\ell_h = \ell_h(\lambda)$ with h even, the first two values being

$$a_1(\lambda) = (\ell_0 + 1)\ell_0 \frac{\ell_2 - 1}{24} - \frac{(n-1)n(n+1)}{24},$$

$$a_2(\lambda) = (\ell_0 + 3)(\ell_0 + 2)(\ell_0 + 1)\ell_0 \frac{5(\ell_2 - 1)^2 - 2(\ell_4 - 1)}{5760}$$
$$- \frac{(n-1)n(n+1)}{24}(\ell_0 + 1)\ell_0 \frac{\ell_2 - 1}{24}$$
$$+ \frac{(n-3)(n-2)(n-1)n(n+1)(5n+7)}{5760}.$$

Proof. We already know that $\widetilde{P}_\lambda(t)$ is a polynomial of degree and parity $v(\lambda)$ and hence has the form $K(\lambda) \sum_j a_j(\lambda) t^{v(\lambda) - 2j}$ for some numbers $a_j(\lambda)$. Define a power series $\widetilde{\Phi}_\lambda(u) \in \mathbb{Q}[[u]]$ by $\widetilde{\Phi}_\lambda(u) = e^{(n-1)u/2} \Phi_\lambda(e^{-u})$. Then equations (A.14) and (A.15) give

$$\widetilde{\Phi}_\lambda(u) = \sum_{r=0}^{n-1} (-1)^r \binom{n-1}{r} \widetilde{P}_\lambda\left(\frac{n-1}{2} - r\right) e^{(\frac{n-1}{2} - r)u}$$

$$= K(\lambda) \sum_{j \geq 0} a_j(\lambda) \frac{d^{v(\lambda) - 2j}}{du^{v(\lambda) - 2j}} \left[\sum_{r=0}^{n-1} (-1)^r \binom{n-1}{r} e^{(\frac{n-1}{2} - r)u}\right]$$

$$= K(\lambda) \sum_{j \geq 0} a_j(\lambda) \frac{d^{v(\lambda) - 2j}}{du^{v(\lambda) - 2j}} \left[(e^{u/2} - e^{-u/2})^{n-1}\right]$$

$$= K(\lambda) \sum_{j \geq 0} a_j(\lambda) \frac{d^{v(\lambda) - 2j}}{du^{v(\lambda) - 2j}} \left[\sum_{g \geq 0} S_g(n) u^{n-1+2g}\right]$$

$$= (n-1)! K(\lambda) \sum_{j,g \geq 0} a_j(\lambda) \widehat{S}_g(n) \frac{u^{n-v(\lambda) - 1 + 2j + 2g}}{(n - v(\lambda) - 1 + 2j + 2g)!},$$

where $S_g(n)$ is as in (A.32) and

$$\widehat{S}_g(n) = \frac{(n-1+2g)!}{(n-1)!} S_g(n) = \frac{1}{(n-1)!} \Delta_+^{n-1}(t^{n-1+2g})\big|_{t=0}, \quad (A.36)$$

a polynomial of degree $3g$ in n. On the other hand, from formula (A.13) and the expansion

$$e^{x/2} - e^{-x/2} = x \exp\left(\sum_{h=2}^{\infty} \frac{B_h}{h} \frac{x^h}{h!}\right) = x \exp\left(\frac{x^2}{24} - \frac{x^4}{2880} + \cdots\right),$$

where B_h denotes the hth Bernoulli number, we have

$$\widetilde{\Phi}_\lambda(u) = \prod_{i\geq 1}\left(e^{iu/2}-e^{-iu/2}\right)^{d_i^*}$$

$$= \left(\prod_i i^{d_i}\right)\cdot u^{\ell(\lambda)-1}\cdot\exp\left(\sum_{h=2}^{\infty}\frac{B_h}{h}\ell_h^*(\lambda)\frac{x^h}{h!}\right)$$

$$= \frac{(n-1)!\,K(\lambda)}{(n-1-v(\lambda))!}\,u^{n-v(\lambda)-1}$$

$$\times\left(1+\frac{\ell_2^*(\lambda)}{24}u^2+\frac{5\ell_2^*(\lambda)^2-2\ell_4^*(\lambda)}{5760}u^4+\cdots\right),$$

where $d_i^* = d_i - \delta_{i,1}$ and $\ell_h^*(\lambda) = \sum_{i\geq 1} i^h d_i^* = \ell_h(\lambda) - 1$. Comparing with the previous formula for $\widetilde{\Phi}_\lambda(u)$, we find

$$a_0(\lambda) = 1,$$

$$a_1(\lambda) + \widehat{S}_1(n) = \frac{(n+1-v(\lambda))!}{(n-1-v(\lambda))!}\,\frac{\ell_2^*(\lambda)}{24},$$

$$a_2(\lambda) + \widehat{S}_1(n)\,a_2(\lambda) + \widehat{S}_2(n) = \frac{(n+3-v(\lambda))!}{(n-1-v(\lambda))!}\,\frac{5\ell_2^*(\lambda)^2-2\ell_4^*(\lambda)}{5760},$$

etc. The lemma now follows by induction on j. □

Returning to the situation of Theorem A.2.9, we consider arbitrary conjugacy classes $C_i \leftrightarrow \lambda_i$ $(i=1,\ldots,k)$ of S_n. From Lemma A.2.14 we get

$$\prod_{i=1}^k \widetilde{P}_{\lambda_i}(t) = \left(\prod_{i=1}^k K(\lambda_i)\right)\left(t^v + a_1(\lambda_1,\ldots,\lambda_k)t^{v-2} + a_2(\lambda_1,\ldots,\lambda_k)\,t^{v-4}+\cdots\right)$$

where $v = \sum v(\lambda_i)$ as before and

$$a_1(\lambda_1,\ldots,\lambda_k) = \sum_{1\leq i\leq k} a_1(\lambda_i),$$

$$a_2(\lambda_1,\ldots,\lambda_k) = \sum_{1\leq i<j\leq k} a_1(\lambda_i)a_2(\lambda_j) + \sum_{1\leq i\leq k} a_2(\lambda_i),$$

etc. Substituting this into equation (A.30) and recalling that $|C_i|K(C_i) = nN(C_i)$, we find

$$\frac{1}{(n-1)!}\,\frac{\mathcal{N}(S_n;C_0,C_1,\ldots,C_k)}{n^{k-1}N(\lambda_1)\cdots N(\lambda_k)}$$

$$= \frac{1}{(n-1)!}\,\Delta_+^{n-1}\left(\sum_{j\geq 0} a_j(\lambda_1,\ldots,\lambda_k)\,t^{v-2j}\right)\bigg|_{t=0}$$

$$= \sum_{j=0}^g \widehat{S}_{g-j}(n)\,a_j(\lambda_1,\ldots,\lambda_k) \qquad (v = n-1+2g),$$

where the polynomials $\widehat{S}_{g-j}(n)$ are given by (A.36) and (A.32). This is the desired generalization of the Goulden-Jackson formula (A.29) to the case of ramified coverings of S^2 by surfaces of arbitrary genus g with at least one cyclic ramification point. We note the case $g = 1$ separately since it is not too complicated:

Corollary A.2.15. *With the notations of Theorem A.2.9 and Corollary A.2.10, we have*

$$\frac{1}{(n-1)!} \frac{\mathcal{N}(\mathsf{S}_n; C_0, C_1, \ldots, C_k)}{n^{k-1} N(C_1) \cdots N(C_k)}$$
$$= \sum_{i=1}^{k} \ell(C_i)(\ell(C_i)+1) \frac{\ell_2(C_i) - 1}{24} - (k-1) \frac{n^3 - n}{24}$$

if $v(C_1) + \cdots + v(C_k) = n + 1$, where $\ell_2(C_i)$ denotes the sum of the squares of the lengths of the cycles in the conjugacy class C_i.

A.2.5 Third Application: "Mirror Symmetry" in Dimension One

"Mirror symmetry", originally discovered in the context of mathematical physics (string theory) and intensively studied during recent years, is a predicted duality between certain families of Calabi–Yau manifolds. It manifests itself on several levels, one of which has to do with the counting functions (Gromov–Witten invariants) that enumerate the holomorphic mappings $C \to X$ of complex curves C into a Calabi–Yau manifold X representing a given class in the second homology group of X.

A Calabi–Yau manifold of dimension n is a complex (projective) n-manifold X such that the space of holomorphic i-forms on X is 0-dimensional for
$0 < i < n$ and 1-dimensional for $i = n$. The original mirror symmetry phenomenon concerned the case $n = 3$, but it was observed by Dijkgraaf [D] that it also occurs in the far simpler case $n = 1$. Here "Calabi–Yau" is just a synonym of "genus one", and the coefficients of the counting functions merely enumerate the generic mappings $Y \to X$, appropriately weighted, from curves of genus g to a given complex curve X of genus 1, where both g and the degree n of the mapping are fixed. Here "generic" means that exactly two sheets of the covering come together over every ramification point, so that each point of X has either n or $n-1$ preimages, and "appropriately weighted" means, as usual, that each covering is counted with a weight equal to the reciprocal of the number of its automorphisms over X. Note that, by the Riemann–Hurwitz formula, a covering of the above type will be ramified over precisely $2g - 2$ points of X.

The problem is therefore to compute $h^0(2g - 2, n)$, where $h^0(k, n)$ denotes the number of (weighted) isomorphism classes of connected n-sheeted coverings of a given Riemann surface X of genus 1 with generic ramification over k

given points of X. If we denote by $h(k,n)$ the corresponding number without the connectedness condition, then a standard argument gives the relation

$$\sum_{n>0}\sum_{k\geq 0} h^0(k,n) \frac{X^k}{k!} q^n = \log\left(1 + \sum_{n>0}\sum_{k\geq 0} h(k,n) \frac{X^k}{k!} q^n\right), \quad \text{(A.37)}$$

while another standard argument gives the formula

$$h(k,n) = \frac{1}{n!} \mathcal{N}_1\big(S_n; \underbrace{T,\ldots,T}_{k}\big) \quad \text{(A.38)}$$

for $h(k,n)$ in terms of the generalized Frobenius number $\mathcal{N}_1(S_n; T,\ldots,T)$ of $(k+2)$-tuples $(a,b,c_1,\ldots,c_k) \in (S_n)^2 \times [T]^k$ with $[a,b]c_1\cdots c_k = 1$ as introduced in Sec. A.1.3; here $[T]$ denotes the conjugacy class of transpositions, see Sec. A.2.2. (The number $h^0(n,k)$ counts the tuples which generate a subgroup of S_n acting transitively on $\{1,\ldots,n\}$, but since we have no closed formula for this we must use (A.37) and (A.38) instead.)

From formulas (A.38) and (A.11) with $g = 1$ we find

$$h(k,n) = \sum_{\pi \in \mathcal{R}_n} \nu_\pi(T)^k$$

where $\mathcal{R}_n$ as in Sec. A.1.2 denotes the set of irreducible representations of S_n and $\nu_\pi(T)$ ($\pi \in \mathcal{R}_n$) as in Sec. A.1.2 and A.1.3 is the eigenvalue of $\sum_{g\in[T]_n} g$ on the representation π. From this we obtain

$$\sum_{k\geq 0} h(k,n) \frac{X^k}{k!} = \sum_{\pi \in \mathcal{R}_n} e^{\nu_\pi(T)X} = H_n(e^X), \quad \text{(A.39)}$$

where

$$H_n(u) = \sum_{\pi \in \mathcal{R}_n} u^{\nu_\pi(T)} \in \mathbb{Z}[u, u^{-1}],$$

a symmetric Laurent polynomial in u which for small values of n can be computed directly from the information given in Sec. A.1.2:

$H_2(u) = u + u^{-1}$,
$H_3(u) = u^3 + 1 + u^{-3}$,
$H_4(u) = u^6 + u^2 + 1 + u^{-2} + u^{-6}$,
$H_5(u) = u^{10} + u^5 + u^2 + 1 + u^{-2} + u^{-5} + u^{-10}$,
$H_6(u) = u^{15} + u^9 + u^5 + 2u^3 + 1 + 2u^{-3} + u^{-5} + u^{-9} + u^{-15}$.

In general, we see from **B.** and **C.** of Sec. A.1.2 that, if $\pi \in \mathcal{R}_n$ corresponds to the function $f : \mathbb{Z} \to \mathbb{Z}_{\geq 0}$ and to the Young diagram $Y \in \mathcal{Y}_n$, then

$$\nu_\pi(T) = \sum_{r \in \mathbb{Z}} rf(r) = \sum_{(x,y)\in Y} (x-y).$$

Either of these descriptions leads to a formula for $H_n(u)$, but a more convenient formula is obtained by using a third parametrization of the irreducible representations of S_n: if π corresponds to the function f and the Young diagram Y, denote by $s = f(0) = \max\{i \mid (i,i) \in Y\}$ the number of diagonal elements in Y and let

$$a_i = \max\{j \mid (i+j-1,i) \in Y\}, \qquad b_i = \max\{j \mid (i,i+j-1) \in Y\}$$

be the number of elements of Y to the right of or above the point $(i,i) \in Y$, respectively, so that Y is described as $\bigcup_{i=1}^{s} ([i, i+a_i-1] \times \{i\} \cup \{i\} \times [i, i+b_i-1])$. Then we have

$$n = \sum_{i=1}^{s}(a_i + b_i - 1), \qquad \nu_\pi(T) = \sum_{i=1}^{s}\left(\binom{a_i}{2} - \binom{b_i}{2}\right), \qquad (A.40)$$

and $\mathcal{R}_n$ is parametrized by the set of tuples $(s, (a_1, \ldots, a_s), (b_1, \ldots, b_s))$ satisfying the inequalities $a_1 > \cdots > a_s > 0$, $b_1 > \cdots > b_s > 0$ and the first of equations (A.40). This gives the generating function identity

$$\sum_{n \geq 0} H_n(u) q^n = \text{coefficient of } \zeta^0$$

$$\text{in } \prod_{a \geq 1}\left(1 + u^{\binom{a}{2}} q^{a-1} \zeta\right) \cdot \prod_{b \geq 1}\left(1 + u^{-\binom{b}{2}} q^{b-1} \zeta^{-1}\right).$$

This can be written more symmetrically by shifting a and b by $\frac{1}{2}$ to get

$$\sum_{n \geq 0} H_n(u) q^n = \text{coefficient of } \zeta^0 \text{ in}$$

$$\prod_{m \in \{\frac{1}{2}, \frac{3}{2}, \frac{5}{2}, \ldots\}} \left(1 - u^{m^2/2} q^m \zeta\right)\left(1 - u^{-m^2/2} q^m \zeta^{-1}\right).$$

(Here we have multiplied ζ by $-q^{1/2}u^{1/8}$, which does not affect the coefficient of ζ^0.) Combining this with the previous formulas, we obtain the following theorem, due to Douglas [Dou] and Dijkgraaf [D].

Theorem A.2.16. *For $g \geq 1$, let $F_g(q) = \sum_{n \geq 1} h^0(2g-2, n) q^n \in \mathbb{Q}[[q]]$ denote the counting function of generically ramified coverings of a genus 1 Riemann surface by Riemann surfaces of genus g. Then*

$$\sum_{g=1}^{\infty} F_g(q) \frac{X^{2g-2}}{(2g-2)!} = \log\left(\sum_{n=0}^{\infty} H_n(e^X) q^n\right)$$

$$= \log\left(\text{coefficient of } \zeta^0 \text{ in} \prod_{m \in \mathbb{Z}_{\geq 0}+\frac{1}{2}} \left(1 - u^{m^2/2} q^m \zeta\right)\left(1 - u^{-m^2/2} q^m \zeta^{-1}\right)\right).$$

This theorem has an interesting corollary, which was discovered and proved in the language of mathematical physics by Dijkgraaf [Dij] and Rudd [Ru] and proved from a purely mathematical point of view in [KZ]. Recall that a *modular form* of weight k on the full modular group $\Gamma = \mathrm{SL}(2,\mathbb{Z})$ is a function $F(z)$, defined for complex numbers z with $\Im(z) > 0$, which has a Fourier expansion of the form $F(z) = \sum_{n=0}^{\infty} a(n) e^{2\pi i n z}$ with coefficients $a(n)$ of polynomial growth and which satisfies the functional equation $F\left(\frac{az+b}{cz+d}\right) = (cz+d)^k F(z)$ for all $\left(\begin{smallmatrix} a & b \\ c & d \end{smallmatrix}\right) \in \Gamma$. A *quasimodular form* of weight k on Γ is a function having a Fourier expansion with the same growth condition and such that for each value of z the function $(cz+d)^{-k} F\left(\frac{az+b}{cz+d}\right)$ is a polynomial in $(cz+d)^{-1}$ as $\left(\begin{smallmatrix} a & b \\ c & d \end{smallmatrix}\right) \in \Gamma$ varies. The ring of modular forms is generated by the two functions

$$E_4(z) = 1 + 240 \sum_{n=1}^{\infty} \frac{n^3 q^n}{1-q^n} = 1 + 240 q + 2160 q^2 + 6720 q^3 + \cdots$$

and

$$E_6(z) = 1 - 504 \sum_{n=1}^{\infty} \frac{n^5 q^n}{1-q^n} = 1 - 504 q - 16632 q^2 - 122976 q^3 - \cdots$$

of weight 4 and 6, respectively, where $q = e^{2\pi i z}$, while the ring of quasimodular forms is generated by these two functions together with the quasimodular form

$$E_2(z) = 1 - 24 \sum_{n=1}^{\infty} \frac{n q^n}{1-q^n} = 1 - 24 q - 72 q^2 - 96 q^3 - \cdots$$

of weight 2. Theorem A.2.16 then implies:

Corollary A.2.17. *For all $g \geq 2$, $F_g(q)$ is the q-expansion of a quasimodular form of weight $6g - 6$.*

We omit the proof, referring the reader to the papers cited above. The first example is

$$F_2(q) = \frac{1}{2^7 3^4 5}\left(5 E_2^3 - 3 E_2 E_4 - 2 E_6\right) = q^2 + 8 q^3 + 30 q^4 + 80 q^5 + 180 q^6 + \cdots,$$

which leads easily to the amusing closed formula $h^0(2,n) = \frac{n}{6} \sum_{d \mid n} (d^3 - nd)$ for the number of coverings (generically ramified, with fixed ramification points) of an elliptic curve by Riemann surfaces of genus 2. In general, however, the generating functions $F_g(q)$ do not have integral coefficients, e.g., $F_3(q)$ begins

$$\frac{1}{12} q^2 + \frac{20}{3} q^3 + 102 q^4 + \frac{2288}{3} q^5 + \cdots.$$

References

[D] R. Dijkgraaf, Mirror symmetry and elliptic curves, in: The Moduli Spaces of Curves, R. Dijkgraaf, C. Faber, G. van der Geer eds., Progress in Mathematics, **vol. 129**, Birkhäuser, Boston, 149–164 (1995)

[Dij] R. Dijkgraaf, Chiral deformations of conformal field theories, Nuclear Phys. B, **493**, 588–612 (1997)

[Dou] M. R. Douglas, Conformal string theory techniques in large N Yang–Mills theory, in: Quantum field theory and string theory (Cargese 1993), 119–135, NATO Adv. Sci. Inst. Ser. B Phys., **328**, Plenum, New York (1995)

[DW] R. Dijkgraaf and E. Witten, Topological gauge theories and group cohomology, Commun. Math. Phys., **129**, 393–429 (1990)

[FQ] D. Freed and F. Quinn, Chern–Simons gauge theory with a finite gauge group, Commun. Math. Phys., **156**, 435–472 (1993)

[KZ] M. Kaneko and D. Zagier, A generalized Jacobi theta function and quasimodular forms, in: The Moduli Spaces of Curves, R. Dijkgraaf, C. Faber, G. van der Geer eds., Progress in Mathematics, **vol. 129**, Birkhäuser, Boston, 165–172 (1995)

[OV] A. Okounkov and A. Vershik, A new approach to representation theory of symmetric groups, Selecta Math. (N.S.), **2**, 581–605 (1996)

[Ru] R. Rudd, The string partition function for QCD on the torus, hep-th/9407176

[SSV] B. Shapiro, M. Shapiro and A. Vainshtein, Ramified coverings of S^2 with one degenerate branch point and enumeration of edge-ordered graphs, in: Topics in singularity theory, Amer. Math. Soc. Transl. Ser. 2, 180, AMS. Providence RI, 219–227 (1997)

[Z] D. Zagier, On the distribution of the number of cycles of elements in symmetric groups, Nieuw Arch. Wiskd., IV Ser., **13**, 489–495 (1995)

References

1. **Abel N. H.** Über die Integration der Differetial-Formen $\frac{\rho dx}{\sqrt{R}}$ wenn ρ und R ganze Funktionen sind. – *J. Reine Angew. Math.*, 1826, vol. **1**, 185–221. See also: Sur l'intégration de la formule différentielle $\frac{\rho dx}{\sqrt{R}}$, R et ρ étant des fonctions entières. – In "Oeuvres Complètes de Niels Henrik Abel" (L. Sylow, S. Lie eds.), Christiania, 1881, t. **1**, 104–144.
2. **Abramovitz M., Stegun I. A.** Handbook of Mathematical Functions, with Formulas, Graphs and Mathematical Tables. – Dover, 1965.
3. **Adrianov N. M.** Classification of primitive edge rotation groups of plane trees. – *Fundamentalnaya i Prikladnaya Matematika*, 1997, vol. **3**, no. 4, 1069–1083 (in Russian).
4. **Adrianov N. M.** Arithmetic theory of graphs on the surfaces. – Ph.D. thesis, Moscow State University, 1997, 116 pp. (in Russian).
5. **Adrianov N. M., Kochetkov Yu. Yu., Shabat G. B., Suvorov A. D.** Mathieu groups and plane trees. – *Fundamentalnaya i Prikladnaya Matematika*, 1995, vol. **1**, no. 2, 377–384 (in Russian).
6. **Adrianov N. M., Kochetkov Yu. Yu., Suvorov A. D.** Plane trees with special primitive edge rotation groups. – *Fundamentalnaya i Prikladnaya Matematika*, 1997, vol. **3**, no. 4, 1085–1092 (in Russian).
7. **Adrianov N. M., Shabat G. B.** Unicellular cartography and Galois orbits of plane trees. – In: "Geometric Galois Action" (L. Schneps, P. Lochak eds.), Vol. 2: "The Inverse Galois Problem, Moduli Spaces and Mapping Class Groups", London Math. Soc. Lecture Notes Series, vol. **243**, 1997, 13–24.
8. **Adrianov N. M., Zvonkin A. K.** Composition of plane trees. – *Acta Applicandae Mathematicae*, 1998, vo. **52**, no. 1–3, 239–245.
9. **Appel K., Haken W.** Every Planar Map Is Four-Colorable. – AMS, "Contemporary Mathematics", 1989, vol. **98**.
10. **Arnold V. I.** Braids of algebraic functions and cohomology of swallowtails. – *Russian Mathematical Surveys*, 1968, vol. **23**, 247–248.
11. **Arnold V. I.** Critical points of functions and the classification of caustics. – *Russ. Math. Surveys*, 1974, vol. **29**, 243–244
12. **Arnold V. I.** Plane curves, their invariants, perestroikas and classification. – In: Adv. Soviet Math., AMS, Providence RI, 1994, vol. **21**, 33-91.

13. **Arnold V. I.** Topological classification of trigonometric polynomials and combinatorics of graphs with an equal number of vertices and edges. – *Funct. Anal. and its Appl.*, 1996, vol. **30**, no. 1, 1–14.
14. **Artin E.** Theorie der Zöpfe. – *Hamburger Abhandlungen*, 1925, vol. **4**, 47–72.
15. **Ashkinuze V. G.** On the number of semi-regular polyhedra. – *Matematicheskoe Prosveshchenie*, 1957, vol. **1**, 107–118 (in Russian).
16. **Atkinson M. D.** An algorithm for finding the blocks of a permutation group. – *Math. Comp.*, 1975, vol. **29**, 911–913.
17. **Babai L.** The probability of generating the symmetric group. – *J. of Combinat. Theory A*, 1989, vol. **52**, 148–153.
18. **Bar-Natan D.** On Vassiliev knot invariants. – *Topology*, 1995, vol. **34**, no. 2, 423–472.
19. **Bar-Natan D., Garoufalidis S.** On the Melvin–Morton–Rozansky conjecture. – *Invent. Math.*, 1996, vol. **125**, 103–133.
20. **Bauer M., Itzykson C.** Triangulations. – In: "Formal Power Series and Algebraic Combinatorics 93" (A. Barlotti, M. Delest, R. Pinzani eds.), Florence, June 1993, 1–45. Also in: "The Grothendieck Theory of Dessins d'Enfant" (L. Schneps ed.), London Math. Soc. Lecture Notes series, vol. **200**, Cambridge Univ. Press, 1994, 179–236.
21. **Belyĭ G. V.** On Galois extensions of a maximal cyclotomic field. – *Math. USSR Izvestija*, 1980, vol. **14**, no. 2, 247–256. (Original Russian paper: *Izv. Acad. Nauk SSSR, cer. mat.*, 1979, vol. **43**, no. 2, 269–276.)
22. **Belyĭ G. V.** A new proof of the three-point theorem. – *Matematicheskii Sbornik*, 2002, vol. **193**, no. 3, 21–24.
23. **Bender E. A., Richmond L. B.** A survey of the asymptotic behaviour of maps. – *J. of Combinat. Theory B*, 1986, vol. **40**, 297–329.
24. **Bergeron F., Labelle G., Leroux P.** Combinatorial Species and Tree-like Structures. – Cambridge University Press, Cambridge, 1998.
25. **Bessis D., Itzykson C., Zuber J.-B.** Quantum field theory techniques in graphical enumeration. – *Adv. Appl. Math.*, 1980, vol. **1**, no. 2, 109–157.
26. **Bétréma J., Péré D., Zvonkin A. K.** Plane trees and their Shabat polynomials. Catalog. – Rapport interne du LaBRI, Bordeaux, 1992, 119 pp.
27. **Bétréma J., Zvonkin A. K.** La vraie forme d'un arbre. – In: "TAPSOFT '93: Theory and Practice of Software Development. 4-th International Joint Conference CAAP/FASE, Orsay, April 1993. Proceedings" (M.-C. Gaudel, J.-P. Jouannaud eds.), Lecture Notes in Computer Science, Springer, 1993, vol. **668**, 599–612.
28. **Bétréma J., Zvonkin A. K.** Plane trees and Shabat polynomials. – *Discrete Mathematics*, 1996, vol. **153**, 47–58.
29. **Biane P., Capitaine M., Guionnet A.** Large deviation bounds for matrix Brownian motion. – *Invent. Math.*, 2003, vol. **152**, no. 2, 433–459.
30. **Birch B. J.** Noncongruence subgroups, covers and drawings. – In: "The Grothendieck Theory of Dessins d'Enfants" (L. Schneps ed.), London Math. Soc. Lecture Notes Series, Cambridge Univ. Press, 1994, vol. **200**, 25–46.
31. **Birch B. J.** Shabat trees of diameter 4: Appendix to a paper of Zvonkin. – 1995, unpublished.
32. **Birch B. J., Chowla S., Hall M., Jr., Schinzel A.** On the difference $x^3 - y^2$. – *Norske Vid. Selsk. Forh. (Trondheim)*, 1965, vol. **38**, 65–69.
33. **Birman J. S.** Braids, Links and Mapping Class Groups. – Princeton University Press, 1974 (Annals of Math. Studies, vol. **82**).

34. **Birman J., Lin X.-S.** Knot polynomials and Vassiliev invariants. – *Invent. Math.*, 1993, vol. **111**, 225–270.
35. **Bóna M., Bousquet M., Labelle G., Leroux P.** Enumeration of m-ary cacti. – *Advances in Applied Math.*, 2000, vol. **24**, 22–56.
36. **Bonnington C. P., Little C. H.** The Foundation of Topological Graph Theory. – Springer, 1995, 180 pp.
37. **Bouchet A.** Circle graph obstructions. – *J. of Combinat. Theory B*, 1994, vol. **60**, 107–144.
38. **Boulatov D. V., Kazakov V. A., Kostov I. K., Migdal A. A.** Analytical and numerical study of a model of dynamically triangulated random surfaces. – *Nuclear Physics B*, 1986, vol. **275** [FS17], 641–686.
39. **Boulatov D. V., Kazakov V. A.** The Ising model on a random planar lattice: the structure of the phase transition and the exact critical exponents. – *Phys. Lett. B*, 1987, vol. **186**, no. 3-4, 379–384.
40. **Bouttier J., Di Francesco P., Guitter E.** Critical and tricritical hard objects on bicolourable random lattices: exact solutions. – *J. Phys. A*, 2002, vol **35**, no. 17, 3821–3854.
41. **Bouya D.** Mégacartes. – Mémoire de DEA, Université Bordeaux I, 1997.
42. **Bouya D., Zvonkin A. K.** Topological classification of complex polynomials: New experimental results. – Rapport interne du LaBRI no. 1219-99, May 1999, 25 pp. (http://dept-info.labri.u-bordeaux.fr/~zvonkin)
43. **Brezin E., Byrne R., Levy J., Pilgrim K., Plummer K.** A census of rational maps. – *Conformal Geometry and Dynamics. An Electronic Journal of the AMS*, 2000, vol. **4**, 35–74.
(http://www.ams.org/ecgd/home-2000.html)
44. **Brézin E., Itzykson C., Parisi G., Zuber J.-B.** Planar diagrams. – *Commun. Math. Phys.*, 1978, vol. **59**, 35–51.
45. **Brézin E., Kazakov V. A.** Exactly solvable field theories of closed strings. – *Physics Letters B*, 1990, vol. **236**, no. 2, 144–150.
46. **Browkin J., Brzezinski J.** Some remarks on the abc-conjecture. – *Math. Comp.*, 1994, vol. **62**, 931–939.
47. **Butler G., McKay J.** The transitive groups of degree up to eleven. – *Commun. in Algebra*, 1983, vol. **11**, no. 8, 863–911.
48. **Cayley A.** A theorem on trees. – *Quarterly J. of Pure and Appl. Math.*, 1889, vol. **23**, 376–378. (Reprinted in: "The Collected Mathematical Papers of A. Cayley", in 13 volumes, Cambridge Univ. Press, 1889–1897, vol. **13**, 26–28.)
49. **Chmutov S. V., Duzhin S. V.** An upper bound for the number of Vassiliev knot invariants. – *J. of Knot Theory and its Ramifications*, 1994, vol. **3**, no. 2, 141–151.
50. **Chmutov S. V., Duzhin S. V.** The Kontsevich integral. – *Acta Appl. Math.*, 2001, vol. **66**, no. 2, 155–190.
51. **Chmutov S. V., Duzhin S. V., Kaishev A. I.** The algebra of 3-graphs. – In: *Proc. Steklov. Inst. Math.*, 1998, **vol. 221**, 157–186.
52. **Chmutov S. V., Duzhin S. V., Lando S. K.** Vassiliev knot invariants I. Introduction. – In: Advances in Sov. Math., 1994, vol. **21**, 117–126.
53. **Chmutov S. V., Duzhin S. V., Lando S. K.** Vassiliev knot invariants II. Intersection graph conjecture for trees. – In: Advances in Sov. Math., 1994, vol. **21**, 127–134.

54. **Chmutov S. V., Duzhin S. V., Lando S. K.** Vassiliev knot invariants III. Forest algebra and weighted graphs. – In: Advances in Sov. Math., 1994, vol. **21**, 135–146.
55. **Chmutov S.V., Varchenko A.N.** Remarks on the Vassiliev knot invariants coming from sl$_2$. – *Topology*, 1997, vol. **36**, 153–178.
56. **Clebsch A.** Zur Theorie der Riemann'schen Fläche. – *Mathematische Annalen*, 1873, vol. **6**, 216–230.
57. **Cohen H.** A Course in Computational Algebraic Number Theory. – Springer (Graduate Texts in Mathematics, vol. 138), 1993.
58. **Cohen P. B., Itzykson C., Wolfart J.** Fuchsian triangle groups and Grothendieck dessins: Variations on a theme of Belyi. – *Comm. Math. Phys.*, 1994, vol. **163**, no. 3, 605–627.
59. **Cohen P. B., Wolfart J.** Dessins de Grothendieck et variétés de Shimura. – *C. R. Acad. Sci. Paris, Sér. I*, 1992, t. **315**, 1025–1028.
60. **Colin de Verdière Y.** Spectres de Graphes. – Société Mathématique de France, 1998 (Cours Spécialisés, vol. **4**).
61. **Connes A., Kreimer D.** Renormalization in quantum field theory and the Riemann–Hilbert problem I. The Hopf algebra of graphs and the main theorem. – *Comm. Math. Phys.*, 2000, vol. **210**, no. 1, 249–273.
62. **Conway J. H., Curtis R. T., Norton S. P., Parker R. A., Wilson R. A.** (with computational assistance from **Thackray J. G.**) ATLAS of Finite Groups: Maximal Subgroups and Ordinary Characters for Simple Groups. – Clarendon Press, Oxford, 1985.
63. **Cori R.** Un Code pour les Graphes Planaires et ses Applications. – Astérisque, 1975, vol. **27**.
64. **Cori R., Machì A.** Construction of maps with prescribed automorphism group. – *Theoret. Comput. Sci.*, 1982, vol. **21**, 91–98.
65. **Cori R., Machì A.** Maps, hypermaps and their automorphisms: a survey, I, II, III. — *Expositiones Mathematicae*, 1992, vol. **10**, 403–427, 429–447, 449-467.
66. **Cori R., Penaud J.-G.** The complexity of a planar hypermap and that of its dual. – *Ann. of Discr. Math.*, 1980, vol. **9**, 53–62.
67. **Cormen T. H., Leiserson C. E., Rivest R. L.** Introduction to Algorithms. – The MIT Press, 1990.
68. **Couveignes J.-M.** Calcul et rationalité de fonctions de Belyi en genre 0. – *Ann. de l'Inst. Fourier*, 1994, vol. **44**, no. 1, 1–38.
69. **Couveignes J.-M.** Quelques revêtements définis sur $\mathbb{Q}$. – *Manuscripta Math.*, 1997, vol. **92**, no. 4, 409–445.
70. **Couveignes J.-M., Granboulan L.** Dessins from a geometric point of view. – In: "The Grothendieck Theory of Dessins d'Enfant" (L. Schneps ed.), Cambridge Univ. Press, 1994, vol. **200**, 79–113.
71. **Cox D. A., Katz S.** Mirror Symmetry and Algebraic Geometry. – AMS, 1999.
72. **Crescimanno M., Taylor W.** Large N phases of chiral QCD$_2$. – *Nuclear Phys. B*, vol. **437**, 1995, 3–24.
73. **Cromwell P. R.** Kepler's work on polyhedra. – *The Math. Intelligencer*, 1995, vol. **17**, no. 3, 23–33.
74. **Crowell R. H., Fox R. H.** Introduction to Knot Theory. – Ginn, 1963.
75. **Davenport H.** On $f^3(t) - g^2(t)$. – *Norske Vid. Selsk. Forh. (Trondheim)*, 1965, vol. **38**, 86–87.

76. **Davenport J. H.** On the Integration of Algebraic Functions. – Springer, 1981.
77. **Davis C.** Extrema of a polynomial. – *Amer. Math. Monthly*, 1957, vol. **64**, 679–680.
78. **Deligne P.** Letter to Looijenga of March 9, 1974. – Unpublished (cited in [171]).
79. **Deligne P., Mumford D.** The irreducibility of the space of curves of given genus. – *Inst. Hautes Études Sci., Publ. Math.*, 1969, vol. **36**, 75–109.
80. **Di Francesco P.** 2-D quantum gravity and topological gravities, matrix models, and integrable differential systems. – In: "The Painlevé Property", CRM Ser. Math. Phys., Springer, New York, 1999, 229–285.
81. **Di Francesco P., Ginsparg P., Zinn-Justin J.** $2D$ gravity and random matrices. – *Phys. Rep.*, 1995, vol. **254**, 1–133.
82. **Di Francesco P., Golinelli O., Guitter E.** Meanders and the Temperley–Lieb algebra. – *Comm. in Math. Phys.*, 1997, vol. **186**, no. 1, 1–59.
83. **Di Francesco P., Golinelli O., Guitter E.** Meanders: exact asymptotics. – *Nuclear Physics B*, 2000, vol. **570** [**FS**], 699–712.
84. **Di Francesco P., Itzykson C., Zuber J.-B.** Polynomial averages in the Kontsevich model. – *Commun. Math. Phys.*, 1993, vol. **151**, 193–219.
85. **Dixon J. D.** The probability of generating the symmetric group. – *Mathematische Zeitschrift*, 1969, vol. **110**, 199–205.
86. **Douady A., Hubbard J.** A proof of Thurston's characterization of rational functions. – *Acta Mathematica*, 1993, vol. **171**, 263–297.
87. **Douglas M. R., Shenker S. H.** Strings in less than one dimension. – *Nuclear Physics B*, 1990, vol. **335**, 635–654.
88. **Dyck W.** Über das Problem der Nachbargebiete. – *Math. Ann.*, 1888, vol. **32**, 457–512.
89. **Edmonds A. L., Kulkarni R. S., Stong R. E.** Realizability of branched coverings of surfaces. – *Trans. Amer. Math. Soc.*, 1984, vol. **282**, no. 2, 773–790.
90. **Edmonds J. R.** A combinatorial representation for polyhedral surfaces. – *Notices Amer. Math. Soc.*, 1960, vol. **7**, p. 646.
91. **Ekedahl T., Lando S.K., Shapiro M., Vainshtein A.** On Hurwitz numbers and Hodge integrals. – *C. R. Acad. Sci. Paris Sér. I Math.*, 1999, vol. **328**, 1175–1180.
92. **Ekedahl T., Lando S.K., Shapiro M., Vainshtein A.** Hurwitz numbers and intersections on moduli spaces of curves. – *Invent. Math.*, 2001, vol. **146**, no. 2, 297–327.
93. **Elkies N.** *ABC* implies Mordell. – *Intern. Math. Res. Notes*, 1991, no. 7, 99–109.
94. **El Marraki M., Hanusse N., Zipperer J., Zvonkin A.** Cacti, braids and complex polynomials. – *Séminaire Lotharingien de Combinatoire*, 1997, vol. **37**, 36 pp. (http://www.mat.univie.ac.at/~slc)
95. **El Marraki M., Zvonkin A.** Composition des cartes et hypercartes. – *Séminaire Lotharingien de Combinatoire*, 1995, vol. **34**, 18 pp. (http://www.mat.univie.ac.at/~slc)
96. **Euler L.** Elementa doctrinae solidorum. – *Novi Comment. Acad. Sci. Petrop.*, 1752, vol. **4**, 109–140.
97. **Faber C., Pandharipande R.** Hodge integrals and Gromov-Witten theory. – *Invent. Math.*, 2000, vol. **139**, no. 1, 173–199.

98. **Feit W.** Some consequences of the classification of finite simple groups. – In: B. Cooperstein, G. Mason eds. "The Santa Cruz Conference on Finite Groups" (Proceedings of Symposia in Pure Mathematics, vol. **37**), AMS, 1980, 175–181.
99. **Feller W.** An Introduction to Probability Theory and its Application. – John Wiley & Sons, 1971, vol. **2**.
100. **Figueroa-O'Farrill J. M., Kimura T., Vaintrob A.** The universal Vassiliev invariant for the Lie superalgebra gl(1|1). – *Commun. Math. Phys.*, 1997, vol. **185**, 93–127.
101. **Filimonenkov V. O., Shabat G. B.** Fields of definition of Belyi function and Galois cohomology. – *Fundamentalnaya i Prikladnaya Matematika*, 1995, vol. **1**, no. 3, 781–799 (in Russian).
102. **Flajolet Ph.** Elements of a general theory of combinatorial structures. – In: Lect. Notes in Comp. Sci., vol. **199**, Springer, Berlin, 1985, 112–127.
103. **Foissy L.** Les algèbres de Hopf des arbres enracinés décorés. – Thèse, Université de Reims, 2002.
104. **Forster O.** Riemannsche Flächen. – Springer, 1977.
105. **Fried M. D.** Exposition of an arithmetic–group theoretic connection via Riemann's existence theorem. – In: "The Santa Cruz Conference on Finite Groups" (B. Cooperstein, G. Mason eds.), Proceedings of Symposia in Pure Mathematics, AMS, 1980, vol. **37**, 571–602.
106. **Fried M. D.** Arithmetic of 3 an 4 branch point covers. A bridge provided by noncongruence subgroups of $SL_2(\mathbb{Z})$. – In: "Séminaire de Théorie des Nombres, Paris, 1987–88", Birkhäuser, "Progress in Mathematics", 1990, vol. **81**, 77–117.
107. **Fried M. D., Jarden M.** Field Arithmetic. – Springer, 1986 ("Ergebnisse der Mathematik und ihrer Grenzgebiete, 3.Folge, Band 11).
108. **Fulton W.** Intersection Theory, 2nd ed. – Springer-Verlag, Berlin, 1998.
109. **Geissinger L.** Hopf algebras of symmetric functions and class functions. – In: "Combinatoire et Représentations du Groupe Symétrique", Springer, Lect. Notes in Math., 1977, vol. **579**, 168–181.
110. **Gelfand I. M., Dikii L. A.** Fractional powers of operators, and Hamiltonian systems. – *Funkcional. Anal. i Prilozhen*, 1976, vol. **10**, no. 4, 13–29.
111. **Getzler E.** Operads and moduli spaces on genus 0 Riemann surfaces. – In: "The Moduli Spaces of Curves", Birkhäuser, Boston, 1995, 199–230.
112. **Goryunov V., Lando S.K.** On enumeration of meromorphic functions on the line. – In: "The Arnoldfest", Fields Inst. Commun., AMS, Providence RI, 1999, vol. **24**, 209–223.
113. **Goulden I. P., Jackson D. M.** Combinatorial Enumeration, John Wiley & Sons, Inc., New York, 1983.
114. **Goulden I. P., Jackson D. M.** The combinatorial relationship between trees, cacti and certain connection coefficients for the symmetric group. – *Europ. J. Comb.*, 1992, vol. **13**, 357–365.
115. **Goulden I. P., Jackson D. M.** Transitive factorization into transpositions, and holomorphic mappings on the sphere. – *Proc. Amer. Math. Soc.*, 1997, vol. **125**, 51–60.
116. **Goulden I. P., Jackson D. M.** A proof of a conjecture for the number of ramified coverings of the sphere by the torus. – *J. Combin. Theory, Ser. A*, 1999, vol. **88**, 246–258.
117. **Goulden I., Jackson D., Vakil R.** The Gromov–Witten potential of a point, Hurwitz numbers, and Hodge integrals. – *Proc. London Math. Soc.*, 2001, vol. **83**, no. 3, 563–581.

118. **Goulden I. P., Jackson D. M., Vainshtein A.** The number of ramified coverings of the sphere by the torus and surfaces of higher genera. – *Annals of Combin.*, 2000, vol. **4**, 27–46.
119. **Goupil A., Schaeffer G.** Factoring n-cycles and counting maps of given ginus. – *Europ. J. Combinat.*, 1998, vol. **19**, no. 7, 819–834.
120. **Graber T., Vakil R.** Hodge integrals and Hurwitz numbers via virtual localization. – *Composito Math.*, 2003, vol. **135**, no. 1, 25–36.
121. **Granboulan L.** Construction d'une extension régulière de $\mathbb{Q}(T)$ de groupe de Galois M_{24}. – *Experimental Math.*, 1996, vol. **5**, no. 1, 3–14.
122. **Granboulan L.** Calcul d'objets géométriques à l'aide de méthodes algébriques et numériques : Dessins d'Enfants. – Thèse, Université Paris-7, 1997.
123. **Griffiths P. A.** Introduction to Algebraic Curves. – AMS, 1989 (Translations of Mathematical Monographs, vol. **76**).
124. **Griffiths P. A., Harris J.** Principles of Algebraic Geometry, 2nd ed. – John Wiley & Sons, New York, 1994.
125. **Gross D. J., Migdal A. A.** A nonperturbative treatment of two-dimensional quantum gravity. – *Nuclear Physics B*, 1990, vol. **340**, 333–365.
126. **Gross D. J., Taylor W.** Two-dimensional QCD is a string theory. – *Nuclear Physics B*, 1993, vol. **400**, 181–210.
127. **Gross D. J., Taylor W.** Twists and Wilson loops in the string theory of two dimensional QCD. – *Nuclear Physics B*, 1993, vol. **403**, 395–452.
128. **Gross J. L., Tucker T. W.** Topological Graph Theory. – John Wiley & Sons, 1987.
129. **Grothendieck A.** Esquisse d'un programme (1984). – In: "Geometric Galois Action" (Schneps L., Lochak P. eds.), vol. 1: "Around Grothendieck's *Esquisse d'un Programme*", London Math. Soc. Lecture Notes Series, Cambridge Univ. Press, 1997, vol. **242**, 5–48. (English translation: "Scetch of a programme", the same volume, p. 243–284.)
130. **Guionnet A.** First order asymptotics of matrix integrals; a rigorous approach towards the understanding of matrix models. – Preprint, November 2002, math.PR/0211131
131. **Guralnick R. M., Thompson J. G.** Finite groups of genus zero. – *J. of Algebra*, 1990, vol. **131**, 303–341.
132. **Hamilton W. R.** Letter to John T. Graves "On the Icosian" (17th October 1856). – In: "W. R. Hamilton, Mathematical Papers, vol. III, Algebra" (H. Halberstam, R. E. Ingram eds.), Cambridge Univ. Press, 1967, 612–625.
133. **Hansen V. L.** Braids and Coverings: Selected Topics. – Cambridge University Press, 1989 (London Math. Soc. Student Texts, vol. **18**).
134. **Hanusse N.** Cartes, constellations et groupes : questions algorithmiques. – Thèse, Université Bordeaux I, 1997.
135. **Hanusse N., Zvonkin A. K.** Cartographic generation of Mathieu groups. – In: "Formal Power Series and Algebraic Combinatorics '99", Barcelona, June 1999, 241–253.
136. **Harary F., Palmer E. M.** Graphical Enumeration. – Academic Press, New York and London, 1973
137. **Harer J.** The cohomology of the moduli space of curves. – Lecture Notes in Math., Springer, 1988, vol. **1337**, 138–221.
138. **Harer J., Zagier D.** The Euler characteristic of the moduli space of curves. – *Invent. Math.*, 1986, vol. **85**, 457–485.

139. **Harris J., Morrison I.** Moduli of Curves. – Springer, 1998.
140. **Heffter L.** Über das Problem der Nachbargebiete. – *Math. Ann.*, 1891, vol. **38**, 477–508.
141. **Heffter L.** Über metacyklische Gruppen und Nachbarconfigurationen. – *Math. Ann.*, 1898, vol. **50**, 261–268.
142. **Herman M. R.** Exemples de fractions rationnelles ayant une orbite dense sur la sphère de Riemann. – *Bull. Soc. Math. France*, 1984, vol. **112**, 93–142.
143. **'t Hooft G.** A planar diagram theory for strong interactions. – *Nuclear Physics B*, 1974, vol. **72**, 461–473.
144. **Hurwitz A.** Über Riemann'sche Fläche mit gegebenen Verzweigungspunkten. – *Math. Ann.*, 1891, vol. **39**, 1–61.
145. **Hurwitz A.** Über die Anzahl Riemannschen Flächen mit gegebenen Verzweigungspunkten. – *Math. Ann.*, 1902, vol. **55**, 53–66.
146. **Itzykson C., Zuber J.-B.** The planar approximation. II. – *J. Math. Phys.*, 1980, vol. **21**, no. 3, 411–421.
147. **Itzykson C., Zuber J.-B.** Matrix integration and combinatorics of modular groups. – *Commun. Math. Phys.*, 1990, vol. **134**, 197–207.
148. **Itzykson C., Zuber J.-B.** Combinatorics of the modular group II. The Kontsevich integrals. – *Intern. J. of Modern Physics A*, 1992, vol. **7**, no. 23, 5661–5705.
149. **Jackson D. M.** Counting cycles in permutations by group characters, with an application to a topological problem. – *Trans. Amer. Math. Soc.*, 1987, vol. **299**, no. 2, 785–801.
150. **Jackson D. M., Visentin T. I.** An Atlas of the Smaller Maps in Orientable and Nonorientable Surfaces. – Chapman and Hall, 2001.
151. **Jacques A.** Constellations et graphes topologiques. – In: "Combinatorial Theory and its Applications" (P. Erdös, A. Renyi, V. Sos eds.), North-Holland, Amsterdam, 1970, vol. II, 657–673.
152. **James G., Kerber A.** The Representation Theory of Symmetric Groups. – Addison–Wesley, 1981.
153. **Jendrol̆ S.** A non-involutory self-duality. – *Discrete Math.*, 1989, vol. **74**, 325–326.
154. **Jensen I.** A transfer matrix approach to the enumeration of plane meanders. – *J. Phys. A*, 2000, vol. **33**, no. 34, 5953–5963.
155. **Jones G. A.** Maps on surfaces and Galois groups. – *Mathematica Slovaca*, 1997, vol. **47**, no. 1, 1–33.
156. **Jones G. A.** Characters and surfaces: a survey. – In: "The Atlas of Finite Groups: Ten Years on (Birmingham, 1995)", Cambridge Univ. Press, London Math. Soc. Lecture Notes Series, 1998, vol. **249**, 90–118.
157. **Jones G. A.** Cyclic regular subgroups of primitive permutation groups. – *J. Group Theory*, 2002, vol. **5**, no. 4, 403–407.
158. **Jones G. A., Singerman D.** Theory of maps on orientable surfaces. – *Proc. London Math. Soc.*, 1978, vol. **37**, 273–307.
159. **Jones G. A., Singerman D.** Maps, hypermaps and triangle groups. – In: "The Grothendieck Theory of Dessins d'Enfants" (L. Schneps ed.), London Math. Soc. Lecture Notes Series, Cambridge Univ. Press, 1994, vol. **200**, 115–145.
160. **Jones G. A., Singerman D.** Belyi functions, hypermaps and Galois groups. – *Bull. London Math. Soc.*, 1996, vol. **28**, no. 6, 561–590.

161. **Jones G. A., Streit M.** Galois groups, monodromy groups and cartographic groups. – In: "Geometric Galois Action" (L. Schneps, P. Lochak eds.), vol. 2: "The Inverse Galois Problem, Moduli Spaces and Mapping Class Groups", London Math. Soc. Lecture Notes Series, Cambridge Univ. Press, 1997, vol. **243**, 25–65.
162. **Jones G. A., Zvonkin A. K.** Orbits of braid groups on cacti. – *Moscow Mathematical Journal*, 2002, vol. **2**, no. 1, 129–162.
163. **Jost J.** Compact Riemann Surfaces. An Introduction to Contemporary Mathematics. – Springer, 1997.
164. **Kassel C., Turaev V.** Chord diagram invariants of tangles and graphs. – *Duke Math. J.*, 1998, vol. **92**, no. 3, 497–552.
165. **Kazakov V. A.** Ising model on a dynamical planar random lattice: exact solution. – *Phys. Lett. A*, 1986, vol. **119**, no. 3, 140–144.
166. **Kazakov V. A.** Bosonic strings and string field theories in one-dimensional target space. – In: "Random Surfaces and Quantum Gravity. Proceedings of the NATO Advanced Workshop, Cargèse, 1990" (O. Alvarez, E. Marinari, P. Windey eds.), Plenum Press, 1991, 269–306.
167. **Kepler J.** Harmonice Mundi. Linz, 1619. – In: M. Caspar "Johannes Kepler Gesammelte Werke", Münich, Beck, 1938, vol. 6.
168. **Kharchev S., Marshakov A., Mironov A., Morozov A., Zabrodin A.** Towards unified theory of 2d gravity. – *Nuclear Phys.*, 1992, vol. **B380**, 181-240.
169. **Khovanskii A. G., Zdravkovska S.** Branched covers of S^2 and braid groups. – *J. of Knot Theory and its Ramifications*, 1996, vol. **5**, no. 1, 55–75.
170. **Klein F.** Vorlesungen über das Ikosaeder und die Aflösung der Gleichungen vom fünften Grade. – Leipzig, 1884. (Reprinted: **Klein F.** The Icosahedron and the Solution of Equations of the Fifth Degree. – Dover Publ., 1956.)
171. **Kluitmann P.** Hurwitz action and finite quotients of braid groups. – In "Braids" (J. Birman, A. Libgober eds.), AMS, "Contemporary Mathematics", 1988, vol. **78**, 299–325.
172. **Kneissler J.** The number of primitive Vassiliev invariants up to degree 12, q-alg/9706022.
173. **Kneser H.** Die Deformationssätze der einfach zusammenhängenden Flächen. – *Mathematische Zeitschrift*, 1926, vol. **25**, 362–372.
174. **Knuth D.** The Art of Computer Programming. Vol. 3: Sorting and Searching. – Addison–Wesley, 1973.
175. **Ko K. H., Smolinsky L.** A combinatorial matrix in 3-manifold theory. – *Pacific J. Math.*, 1991, vol. **149**, no. 2, 319–336.
176. **Kochetkov Yu. Yu.** On non-trivially decomposable types. – *Uspekhi Matem. Nauk*, 1997, vol. **52**, no. 5, 205–206.
177. **Köck B.** Belyi's theorem revisited. – To appear in *Beiträge zur Algebra und Geometrie*.
178. **Kontsevich M.** Intersection theory on the moduli space of curves and matrix Airy function. – *Comm. Math. Phys.*, 1992, vol. **147**, no. 1, 1–23.
179. **Kontsevich M.** Vassiliev knot invariants. – *Adv. in Soviet Math.*, 1993, vol. **16**, part 2, 137–150.
180. **Kontsevich M., Manin Yu. I.** Gromov–Witten classes, quantum cohomology, and enumerative geometry. – *Comm. Math. Phys.*, 1994, vol. **164**, 525–562.

181. **Kostov I. K., Mehta M. L.** Random surfaces of arbitrary genus: exact results for $D = 2$ and -2 dimensions. – *Physics Letters B*, 1987, vol. **189**, no. 1-2, 118–124.
182. **Kwak J. H., Lee J.** Enumeration of graph coverings, surface branched coverings and related group theory. – In: "Combinatorial and Computational Mathematics: Present and Future", Papers from the workshop held at the Pohang University, Pohang, February 2000 (S. Hong, J. H. Kwak, K. H. Kim, F. W. Raush eds.), World Scientific, River Edge NJ, USA, 2001, 97–161.
183. **Labelle G.** Sur la symétrie et asymétrie des structures combinatoires. – *Theoret. Comput. Sci.*, 1993, vol. **117**, 3–22.
184. **Labelle G., Leroux P.** Enumeration of (uni- or bicolored) plane trees according to their degree distribution. – *Discrete Math.*, 1996, vol. **157**, no. 1-3, 227–240.
185. **Lando S. K.** On primitive elements in the bialgebra of chord diagrams. – *Amer. Math. Soc. Transl. Ser.* 2, AMS, Providence RI, 1997, vol. **180**, 167–174.
186. **Lando S. K.** On enumeration of unicursal curves. – In: *Amer. Math. Soc. Transl. Ser.* 2, AMS, Providence RI, 1999, vol. **190**, 77–81.
187. **Lando S. K.** On a Hopf algebra in graph theory. – *J. Comb. Theory, Ser. B*, 2000, vol. **80**, 104–121
188. **Lando S. K., Zvonkin A. K.** Meanders. – *Selecta Mathematica Sovietica*, 1992, vol. **11**, no. 2, 117–144.
189. **Lando S. K., Zvonkin A. K.** Plane and projective meanders. – *Theoret. Comput. Sci.*, 1993, vol. **117**, 227–241.
190. **Lando S. K., Zvonkine D.** On multiplicities of the Lyashko–Looijenga mapping on the discriminant strata. – *Functional Analysis and its Applications*, 1999, vol. **33**, no. 3, 21-34.
191. **Lang S.** Algebraic Number Theory. – Springer, 1994 (2nd edition).
192. **Lass B.** Démonstration combinatoire de la formule de Harer–Zagier. – *C. R. Acad. Sci. Paris*, 2001, vol. **333**, 155-160.
193. **Lass B.** Calcul combinatoire ensembliste. – Thèse, Strasbourg, 2001.
194. **Lattès S.** Sur l'itération des substitutions rationnelles et les fonctions de Poincaré. – *C. R. Acad. Sci. Paris*, 1918, vol. **166**, 26–28.
195. **Le T. Q. T., Murakami J.** The universal Vassiliev–Kontsevich invariant for framed oriented links. – *Compositio Math.*, 1996, vol. **102**, no. 1, 41–64.
196. **Lhuilier S. A. J.** Mémoire sur la polyèdrométrie; contenant une démonstration directe du Théorème d'Euler sur les polyèdres, et un examen des diverses exceptions auxquelles ce théorème est assujetti. **(Extrait) par M. Gergonne.** – *Ann. Math. Pures et Appliquées (Annales de Gergonnes)*, 1812-13, t. **3**, 169–189.
197. **Lickorish W. B. R.** Invarinats for 3-manifolds from the combinatorics of the Jones polynomial. – *Pacific J. Math.*, 1991, vol. **149**, no. 2, 337–347.
198. **Lieberum J.** On Vassiliev invariants not coming from semisimple Lie algebras. – *J. Knot Theory Ramif.*, 1999, **vol. 8**, no. 5, 659–666.
199. **Lieberum J.** Chromatic weight system and the corresponding knot invariants. – *Math. Ann.*, 2000, **vol. 317**, no. 3, 459–482.
200. **Liu Y.** Enumerative Theory of Maps. – Kluwer, 1998.
201. **Looijenga E.** The complement of the bifurcation variety of a simple singularity. – *Inv. Math.*, 1974, vol. **23**, 105–116.

202. **Looijenga E.** Intersection theory on Deligne–Mumford compactifications (after Witten and Kontsevich). – In: Séminaire Bourbaki, 1993 (Astérisque, vol. **216**), exposé 768, 187–212.
203. **Looijenga E.** Cellular decompositions of compactified moduli spaces of pointed curves. – In: "The Moduli Space of Curves", Birkhäuser, Progress in Mathematics, 1995, vol. **129**, 369–400.
204. **Luczak T., Pyber L.** On random generation of the symmetric group. – Combin. Probab. Comput., 1993, vol. **2**, 505–512.
205. **Lüroth J.** Note über Verzweigungsschnitte und Querschnitte in einer Riemann'schen Fläche. – Mathematische Annalen, 1871, vol. **4**, 181–184.
206. **Magot N.** Cartes planaires et fonctions de Belyi : Aspects algorithmiques et expérimentaux. – Thèse, Université Bordeaux I, 1997.
207. **Magot N., Zvonkin A.** Belyi functions for Archimedean solids. – Discrete Math., 2000, vol. **217**, 249–271.
208. **Malle G., Matzat B. H.** Inverse Galois Theory. – Springer, 1999.
209. **Malle G., Saxl J., Weigel Th.** Generation of classical groups. – Geometriae Dedicata, 1994, vol. **49**, 85–116.
210. **Malyshev V. A., Minlos R. A.** Gibbs Random Fields. – Kluwer, Dordrecht, 1991.
211. **Manin Yu.** Frobenius Manifolds, Quantum Cohomology, and Moduli Spaces. – American Mathematical Society, Providence RI, 1999.
212. **Matiyasevich Yu. V.** Calculation of generalized Chebyshev polynomials on computer. – Vestnik Mosk. Universiteta, 1996, no. 6, 59–61 (in Russian).
213. **McCanna J.** Is self-duality always involutory? – Congr. Numer., 1990, vol. **72**, 175–178.
214. **McMullen C. T.** Complex Dynamics and Renormalization. – Princeton University Press, NJ, 1994.
215. **Mednykh A. D.** Nonequivalent coverings of Riemann surfaces with a prescribed ramification type. – Siber. Math. J., 1984, vol. **25**, no. 4, 606–625.
216. **Mehta M. L.** A method of integration over matrix variables. – Comm. Math. Physics, 1981, vol. **79**, 327–340.
217. **Mellor B.** The intersection graph conjecture for loop diagrams. – J. Knot Theory Ramifications, 2000, vol. **9**, 187–211.
218. **Milnor J., Moore J.** On the structure of Hopf algebras.– Ann. Math., 1965, vol. **81**, 211–264.
219. **Mohar B., Thomassen C.** Graphs on Surfaces. – Johns Hopkins Univ. Press, 2001.
220. **Moran G.** Chords in a circle and linear algebra over GF(2). – J. Combinatorial Theory, Ser. A, 1984, vol. **37**, 239–247.
221. **Müller P.** Primitive monodromy groups for polynomials. – In "Recent Developments in the Inverse Galois Problem" (M. Fried ed.), AMS, Contemporary Mathematics, 1995, vol. **186**, 385–401.
222. **Natanzon S. M.** Topology of 2-dimensional coverings and meromorphic functions on real and complex algebraic curves. – Selecta Mathematica Formerly Sovietica, 1993, vol. **12**, no. 3, 1993, 251–291. (First published in Russian: Trudy Seminara po Vektornomu i Tensornomu Analizu, 1988, no. 23, 79–103.)
223. **Nedela R.** Regular maps – combinatorial objects relating different fields of mathematics. – J. Korean Math. Soc., 2001, vol. **38**, no. 5, 1069–1105.
224. **Nitaj A.** An algorithm for finding good abc-examples. – C. R. Acad. Sci. Paris, 1993, vol. **317**, 811–815.

225. **Nitaj A.** Algorithms for finding good examples for the *abc* and Szpiro conjectures. – *Experimental Math.*, 1993, vol. **2**, no. 3, 223–230.
226. **Oesterlé J.** Nouvelles approches du "théorème" de Fermat. – Séminaire Bourbaki, exposé 694, Astérisque, 1988, vol. **161–162**, 165–186.
227. **Okounkov A., Pandharipande R.** Gromov–Witten theory, Hurwitz numbers, and matrix models, I. – Preprint, January 2001, 107 pp. (math.AG/0101147)
228. **Pak I.** What do we know about the product replacement algorithm? – In: "Groups and Computations III", Ohio State Univ. Math. Res. Inst. Publ., 2001, vol. **8**, 301–347.
229. **Pakovitch F. B.** Combinatoire des arbres planaires et arithmétique des courbes hyperelliptiques. – *Ann. Inst. Fourier*, 1998, vol. **48**, no. 2, 323–351.
230. **Patras F.** L'algèbre des descentes d'une bigèbre graduée. – *Journal of Alegbra*, 1994, vol. **170**, 547–566.
231. **Penner R. C.** The moduli space of a punctured surface and perturbative series. – *Bull. Amer. Math. Soc.*, 1986, vol. **15**, no. 1, 73–77.
232. **Penner R. C.** Perturbative series and the moduli space of Riemann surfaces. – *J. Diff. Geometry*, 1988, vol. **27**, 35–53.
233. **Penner R. C., Harer J. L.** Combinatorics of Train Tracks. – Princeton Univ. Press, 1992 (Annals of Math. Studies, vol. **125**).
234. **Peterson M. A.** The geometry of Piero della Francesca. – *The Math. Intelligencer*, 1997, vol. **19**, no. 3, 33–40.
235. **Pilgrim K. M.** Dessins d'enfants and Hubbard trees. – *Ann. Sci. de l'École Normale Sup.*, 2000, vol. **33**, no. 5, 671–693.
236. **Prasolov V. V., Sossinsky A.B.** Knots, Links, Braids and 3-Manifolds. An Introduction to the New Invariants in Low-Dimensional Topology. – Translations of Mathematical Monographs, 154, American Mathematical Society, Providence, RI, 1997.
237. **Protopopov A. N.** Topological classification of branched coverings of a two-dimensional sphere. – *Zap. Nauchn. Sem. Leningrad. Otdel. Mat. Inst. Steklov. (LOMI)*, 1988, vol. **167**, "Issled. Topol.", no. 6, 135–156, 192.
238. **Ree R.** A theorem on permutations. – *J. Combinat. Theory*, 1971, vol. **10**, 174–175.
239. **Reshetikhin N. Yu., Turaev V. G.** Ribbon graphs and their invariants. – *Comm. Math. Physics*, 1990, vol. **127**, 1–26.
240. **Reyssat E.** Quelques Aspects des Surfaces de Riemann. – Birkhäuser, 1989.
241. **Ringel G.** Map Color Theorem. – Springer, 1974 (Grundlehren der Mathematischen Wissenschaften, vol. **209**).
242. **Ringel G., Youngs J. W. T.** Das Geschlecht des symmetrischen vollständingen dreifärbbaren Graphen. – *Comm. Math. Helv.*, 1970, vol. **45**, 152–158.
243. **Ritt J. F.** Prime and composite polynomials. – *Trans. Amer. Math. Soc.*, 1922, vol. **23**, 51–66.
244. **Robertson N., Seymour P. D.** Graph minors. XVII. Taming a vortex. – *J. Combinat. Theory*, ser. B, 1999, vol. **77**, no. 1, 162–210.
245. **Royle G. F.** Transitive groups of degree twelve. – *J. Symbolic Comput.*, 1987, vol. **4**, 255–268.
246. **Schmitt W. R.** Incidence Hopf algebras. – *Journal of Pure and Applied Algebra*, 1994, vol. **96**, 299–330.

247. **Schneps L. (ed.)** The Grothendieck Theory of Dessins d'Enfant. – London Math. Soc. Lecture Notes Series, vol. **200**, Cambridge Univ. Press, 1994.
248. **Schneps L.** Dessins d'enfants on the Riemann sphere. – In: "The Grothendieck Theory of Dessins d'Enfant" (L. Schneps ed.), London Math. Soc. Lecture Notes Series, Cambridge Univ. Press, 1994, vol. **200**, 47–77.
249. **Schneps L., Lochak P. (eds.)** Geometric Galois Action. Vol. 1: Around Grothendieck's *Esquisse d'un Programme*. – London Math. Soc. Lecture Notes Series, vol. **242**, Cambridge Univ. Press, 1997.
250. **Schneps L., Lochak P. (eds.)** Geometric Galois Action. Vol. 2: The Inverse Galois Problem, Moduli Spaces and Mapping Class Groups. – London Math. Soc. Lecture Notes Series, vol. **243**, Cambridge Univ. Press, 1997.
251. **Scott G. P.** Braid groups and the group of homeomorphisms of a surface. – *Proc. Camb. Phil. Soc.*, 1970, vol. **68**, 605–617.
252. **Segal G., Wilson G.** Loop groups and equations of KdV type. – *Inst. Hautes Études Sci., Publ. Math.*, 1985, vol. **61**, 5–65
253. **Serre J.-P.** Topics in Galois Theory. – Jones and Bartlett, Boston, 1992.
254. **Servatius B., Servatius H.** Self-dual maps on the sphere. – *Discr. Math.*, 1994, vol. **134**, 139–150.
255. **Servatius B., Servatius H.** The 24 symmetry pairings of self-dual maps on the sphere. – *Discr. Math.*, 1995, vol. **140**, 167–183.
256. **Shabat G. B.** Notes of the seminar on the "dessins d'enfants". – Moscow, 1990, unpublished (in Russian), 70 pp.
257. **Shabat G. B., Voevodsky V. A.** Drawing curves over number fields. – In "The Grothendieck Festschrift" (P. Cartier, L. Illusie, N. M. Katz, G. Laumon, Y. Manin, K. A. Ribet eds.), Birkhäuser, 1990, vol. **3**, 199–227.
258. **Shabat G. B., Zvonkin A. K.** Plane trees and algebraic numbers. – In "Jerusalem Combinatorics '93" (H. Barcelo, G. Kalai eds.), AMS, Contemporary Mathematics, 1994, vol. **178**, 233–275.
259. **Silverman J. H., Tate J.** Rational Points on Elliptic Curves. – Springer, 1992.
260. **Simon B.** The $P(\varphi)_2$ Euclidian (Quantum) Field Theory. – Princeton Univ. Press, 1974.
261. **Singerman D., Syddall R. I.** Belyi uniformization of elliptic curves. – *Bull. London Math. Soc.*, 1997, vol. **29**, 443–451.
262. **Singerman D., Watson P. D.** Weierstrass points on regular maps. – *Geom. Dedicata*, 1997, vol. **66**, no. 1, 69–88.
263. **Smit D.-J.** Summation over equilaterally triangulated surfaces and the critical string measure. – *Comm. Math. Phys.*, 1992, vol. **143**, no. 2, 253–285.
264. **Soboleva E.** Vassiliev knot invariants coming from Lie algebras and 4-invariants. – *J. Knot Theory and its Ramifications*, 2001, vol. **10**, 161–169.
265. **Stanley R. P.** Enumerative Combinatorics. Vol. I. – Wadsworth & Brooks, Monterey CA, 1986.
266. **Steinitz E.** Polyeder und Raumeinteilungen. – In: "Encyclopädie der Mathematischen Wissenschaften", Band 3 (Geometrie), Teil 3AB12, 1922, 1–139.
267. **Stillwell J.** Classical Topology and Combinatorial Group Theory. – Springer, 1993 (2nd edition), Graduate Texts in Mathematics, vol. **72**.
268. **Stothers W. W.** Polynomial identities and Hauptmoduln. – *Quart. J. Math. Oxford, Ser.* 2, 1981, vol. **32**, no. 127, 349–370.

269. **Strebel K.** Quadratic Differentials. – Springer, 1984.
270. **Streit M.** Homology, Belyi functions and canonical curves. – *Manuscripta Mathematica*, 1996, vol. **90**, 489–509.
271. **Streit M., Wolfart J.** Characters and Galois invariants of regular dessins. – *Rev. Mat. Complut.*, 2000, vol. **13**, no. 1, 49–81.
272. **Sullivan D.** Quasiconformal homeomorphisms and dynamics. I: Solution of the Fatou–Julia problem on wandering domains. – *Ann. Math.*, 1985, vol. **122**, 401–418.
273. **Sweedler M. E.** Hopf Algebras. – W. A. Benjamin, Inc., 1969.
274. **Szegö G.** Orthogonal Polynomials. – AMS, 1939 (American Mathematical Society Colloquium Publications, vol. **23**).
275. **Tchebicheff P. L.** Sur l'intégration de la différentielle

$$(x + A)/\sqrt{x^4 + \alpha x^3 + \beta x^2 + \gamma}\, dx\,.$$

– *Bull. Acad. Impériale de Saint-Petersbourg*, 1861, vol. **3**, 1–12. Reprinted in: *Jurnal des Math. Pures et Appl.*, 1864, vol. **9**, no. 2, 225–246.
276. **Temperley H., Lieb E.** Relations between the percolation and coloring problem and other graph-theoretical problems associated with regular planar lattices: some exact results the for percolation problem. – *Proc. Roy. Soc.*, 1971, vol. **A322**, 251-280.
277. **Thom R.** L'équivalence d'une fonction différentiable et d'un polynôme. – *Topology*, 1965, vol. **3**, 297–307.
278. **Thomassen C.** The graph genus problem is NP-complete. – *J. Algorithms*, 1989, vol. **10**, 568–576.
279. **Tutte W. T.** A ring in graph theory. – *Proc. Cambridge Philos. Soc.*, 1947, vol. **43**, 26–40. Reprinted in: "Selected Papers of W. T. Tutte," vol. **1**, Charles Babbage Research Center, Winnipeg, 1979.
280. **Tutte W. T.** Planted plane trees with a given partition. – *Amer. Math. Monthly*, 1964, vol. **71**, 272–277.
281. **Tutte W. T.** Graph Theory. – Addison-Wesley, 1984.
282. **Vaintrob A.** Vassiliev knot invariants and Lie S-algebras. – *Mathematical Research Letters*, 1994, vol. **1**, 579–595.
283. **Vassiliev V. A.** Cohomology of knot spaces. – In: "Theory of Singularities and Its Applications" (V. I. Arnold ed.), Advances in Soviet Math., vol. **1**, AMS, 1990.
284. **Vassiliev V. A.** Complements of Discriminants of Smooth Maps: Topology and Applications. – AMS, 1992.
285. **Vassiliev V. A.** Invariants of knots and complements of discriminants. – In: "Developments in Mathematics, The Moscow School" (V. I. Arnold, M. Monastyrsky eds.), Chapmann & Hall, 1993, 194–250.
286. **Vassiliev V. A.** Topology of Complements to Discriminants. – Fazis, Moscow, 1997 (in Russian).
287. **Vogel P.** Algebraic structure on modules of diagrams. – Preprint, 1995. (http://www.math.jussieu.fr/~vogel/)
288. **Völklein H.** Groups as Galois Groups. An Introduction. – Cambridge University Press, 1996 ("Cambridge Studies in Advanced Mathematics", vol. **53**).
289. **van der Waerden B. L.** Algebra. – Springer, 1966.
290. **Wajnryb B.** Orbits of Hurwitz action for coverings of a sphere with two special fibers. – *Indag. Math., New Ser.*, 1996, vol. **7**, no. 4, 549–558.

291. **Walsh T. R. S.** Hypermaps versus bipartite maps. – *J. Combinat. Theory B*, 1975, vol. **18**, 155–163.
292. **de Weger B. M. M.** Solving exponential diophantine equations using lattice basis reduction algorithms. – *J. Number Theory*, 1987, vol. **26**, 325–367.
293. **Weil A.** The field of definition of a variety. – *Amer. J. Math.*, 1956, vol. **78**, 509–524.
294. **Weil A.** Number Theory. An Approach through History, from Hammurapi to Legendre. – Birkhäuser, 1984.
295. **Weyl H.** Zur Theorie der Darstellung der einfachen kontinuierlichen Gruppen. (Aus einem Schreiben an Herrn I. Schur) – In: "Sitzungsberichte der Preußischen Akademie der Wissenschaften zu Berlin", 1924, 338-345. – Reprinted in: H. Weyl, "Gesammelte Abhadlungen", Band II, Springer, 1968, 453–460.
296. **Weyl H.** The Classical Groups. – Princeton, 1946.
297. **White A. T.** The genus of the complete tripartite graph $K_{mn,n,n}$. – *J. Combinat. Theory*, 1969, vol. **7**, 283–285.
298. **White A. T.** Graphs of Groups on Surfaces. – Elsevier, Amsterdam, 2001.
299. **Witten E.** Two dimensional gravity and intersection theory on the moduli space. – *Surveys in Diff. Geom.*, 1991, vol. **1**, 243–310.
300. **Wolfart J.** The 'obvious' part of Belyi's theorem and Riemann surfaces with many automorphisms. – In: "Geometric Galois Action. Vol. **1**: Around Grothendieck's *Esquisse d'un Programme*" (Schneps L., Lochak P. eds.), London Math. Soc. Lecture Notes Series, Cambridge Univ. Press, 1997, vol. **242**, 97–112.
301. **Zagier D.** On the distribution of the number of cycles of elements in symmetric groups. – *Nieuw Arch. Wiskd., IV. Ser.*, 1995, vol. **13**, no. 3, 489–495.
302. **Zannier U.** On Davenport's bound for the degree of $f^3 - g^2$ and Riemann's Existence Theorem. – *Acta Arithmetica*, 1995, vol. **71**, no. 2, 107–137.
303. **Zapponi L.** Fleurs, arbres et cellules : un invariant galoisien pour une famille d'arbres. – *Composito Mathematica*, 2000, vol. **122**, 113–133.
304. **Zdravkovska S.** Topological classification of polynomial maps. – *Uspekhi Mat. Nauk*, 1970, vol. **25**, no. 4, 179–180.
305. **Zelevinsky A. V.** Representations of Finite Classical Groups. A Hopf Algebra Approach. – Springer, 1981 (Lecture Notes in Math., vol. **869**).
306. **Zinn-Justin P., Zuber J.-B.** On the counting of colored tangles. – *J. of Knot Theory and its Ramifications*, 2000, vol. **9**, no. 8, 1127–1141.
307. **Zvonkin A. K.** How to draw a group. – *Discrete Mathematics*, 1998, vol. **180**, 403–413.
308. **Zvonkin A. K.** Galois orbits of plane trees: a case study. – Unpublished note, 1996.
309. **Zvonkin A. K.** Matrix integrals and map enumeration: An accessible introduction. – *Computers and Mathematics with Applications: Mathematical and Computer Modelling*, special issue "Combinatorics and Physics" (M. Bousquet-Mélou, D. Loeb eds.), 1997, vol. **26**, no. 8-10, 281–304.
310. **Zvonkin A. K.** Towards topological classification of univariate complex polynomials. – In: "Formal Power Series and Algebraic Combinatorics, 12th International Conference, FPSAC'00, Moscow, Russia, June 2000, Proceedings" (D. Krob, A. A. Mikhalev, A. V. Mikhalev eds.), Springer, 2000, 76–87.
311. **Zvonkin A. K.** Megamaps: construction and examples. – *Discrete Mathematics and Theoretical Computer Science*, Conference edition: "Discrete Models: Combinatorics, Computation, and Geometry", July 2001, 329–339.

312. **Zvonkine D.** Multiplicities of the Lyashko–Looijenga map on its strata. – *C. R. Acad. Sci. Paris, Sér. I*, 1997, vol. **324**, no. 12, 1349–1353.
313. **Zvonkine D.** Énumération des revêtements ramifiés des surfaces de Riemann. – Thèse, Université Paris-Sud Orsay, 2003.

Index

τ-function 207
abc conjecture 137
k-constellation 7
1-factor 374
1-flag 33
1-form
   holomorphic 224
   meromorphic 223
     residue of 224
1-term relations 356
2-term relations 364
   for graphs 375
3-term element
   weighted 379
4-bialgebra of graphs 372
4-invariant 372
   multiplicative 372
   of a graph 370
4-term element 340
4-term relations 356
   for graphs 371

absolute Galois group 115
adjacency matrix of a graph 375
adjoint action 171
affine linear group 49
Airy function
   matrix 266
algebra
   Hopf 347
   of 3-graphs 395
   Temperley–Lieb 395
   universal enveloping 387
algebraic numbers

   conjugate 116
antipode 347
antisymmetry relation 394
arc diagram 339
arc diagrams
   concatenation 342
Archimedean solid 113
automorphism group
   of a constellation 68
   of a map 38
automorphism of Frobenius 50
average value of a function 162

bachelor 98, 125
Baker function 208
base
   hypermap 64
   star 19
Belyi
   function 80
     dynamical 139
     pure 110
   pair 80
   theorem 79
Bernoulli numbers 244, 248, 291, 421
bialgebra 344
   of 4-invariants 372
   of finite order knot invariants 359
   of weighted graphs 380
   polynomial 345
   primitive element of 346
bicolored
   graph 43
   map 43

plane tree   45
bipartite graph   43
block of imprimitivity   8
bound of Davenport-Stothers-Zannier
    126
boundary of a moduli space   235
braid   305
braid group   13, 306
    action on cacti   309
    action on constellations   13
    diagonal   315
    Hurwitz   306
    orbit of   308
    pure   306
    sphere   306
    surface   306
bundle
    Hodge   288

cacti
    family of   310
    isomorphic   57
cactus   57
    non-rooted   58
    rooted   58
    special   318
    symmetric   59
canonical
    geometric form of a plane tree   87
    Jenkins–Strebel quadratic differential
        241
    triangulation   51
canonically labelled tree   46
cartographic group
    of a constellation   8
    of a generalized constellation   65
Catalan numbers   158, 178, 191, 198,
        216, 396
caustic   280
center of a face   37, 80
character
    irreducible   10, 400
Chebyshev polynomial   82
    generalized   82
    of the first kind   132
    of the second kind   132
Chern class of a line bundle
    first   249
Chinese character   393

chord diagram   338
    decomposable   348
    intersection graph of   364, 367
    of a singular knot   354
chord diagrams
    comultiplication   343
    multiplication   342
chord separating operator   361
chromatic polynomial   370
    modified   370
circle diagram   393
class
    Chern first   249
    Segre   293
    top   293
classification of coverings
    flexible   13, 304
    rigid   24, 270, 277
coarse moduli space   233
combinatorial map   37
complete
    graph   33
        genus of   33
    tripartite graph   136
    weight system   390
completed Hurwitz space   298
complex
    curve   70
    node of   224
    equivalence of coverings   75
    projective line   70
    structure   70
composition type   321
comultiplication
    of chord diagrams   343
    of knot invariants   359
concatenation
    of arc diagrams   342
    of knots   359
cone   292
    over a variety   292
conjecture
    A_n 325
    S_n 325
    Witten   256
conjugate
    algebraic numbers   116
    constellations   8
    trees   92

connected
  graph   27
  sum of knots   359
constellation   7
  degree of   7
  generalized   64
  genus of   23
  length of   7
  passport of   9
  quotient   68
  regular   69
constellations
  conjugate   8
  isomorphic   8
contraction of an edge in a weighted
    graph   379
contribution of a coupling   169
convolution product in a bialgebra
    347
Conway polynomial   365
counit   344
covariance matrix   163
covering   13
  n-sheeted   14
  decomposable   65
  degree of   14
  finite-sheeted   14
  Galois   18
  ramified   15
  regular   18
  universal   18
  unramified   13
coverings
  complex equivalence of   75
  flexible classification of   304
  isomorphic   14
  rigid classification of   277
critical
  point   15, 21, 72, 81
    degree of   73, 81
    multiplicity of   15, 73, 81
    order of   73, 81
  value   15, 21, 73, 81
    simple   310
cubic field
  cyclic   104
  purely   105
curve
  complex   70

elliptic   107, 110
  Fermat   135
  modular   229
  nodal   224
    genus of   226
  stable   235
  unicursal   188
  universal   234
cycle   27
cycle structure of permutation   9
cyclic cubic field   104

dart   33
  incident to a face   34
  incident to a vertex   33
Davenport–Stothers bound   127
decomposable
  chord diagram   348
  covering   65
  element of a graded bialgebra   348
  polynomial   320
decorated moduli space   237
defect   310
defined over   72, 79, 91
defining polynomial   92, 116
degeneracy of a partition   278
degenerate partition   278
degree
  of a critical point   73, 81
  of a face   30
  of a vertex   28
  of constellation   7
  of covering   14
Dehn twist   30
deletion of an edge in a weighted graph
    379
Deligne–Mumford compactification of
    the moduli space   235
density of a Gaussian measure   162
dessin d'enfant   80
dessins d'enfants
  theory of   77
diagonal braid group   315
diagram
  arc   339
  chord   338
  circle   393
  open   394
diameter of a tree   99

448   Index

differential
   meromorphic   223
   quadratic   226
dilaton equation   255
discrete
   action of a group   228
   Painlevé I equation   197
discriminant   119, 278
   locus   278
double
   flower   132
   point   224
     normalized   225
     scaling limit   202
dual
   hypermap   53
   map   52
dynamical Belyi function   139

edge   27
   midpoint of   37
   polygon in a graph   374
   quadrangle in a graph   373
elementary
   hypermap   44
   transformation   352
   triangulation   52
elimination of quantifiers   118
elliptic
   curve   107, 110
   integral   143
embedded graph   1
   marked   238
enumeration problem
   inverse   99
enveloping algebra
   universal   387
equation
   dilaton   255
   matrix Airy   266
   Painlevé I   201
     discrete   197
   Pell   142
   Pell–Abel   142
   string   253
equivalence of coverings
   flexible   24
   rigid   24
equivalent Shabat polynomials   84

Euler characteristic   31
   orbifold   230
   virtual   230
eventually periodic point   140
exponential generating function   216
extension
   Galois   116
   normal   116
external vertex   393

face
   center of   37, 80
   degree of   30
   of a map   28
   valency of   30
family
   of cacti   310
   of functions
     normal   139
   of trees   94
fat graph   1
Fatou set   139
Feit–Jones theorem   327
Fermat curve   135
fiber   14
field
   cubic cyclic   104
   of definition   152
   of moduli   79, 97
     of a dessin   117
     of an orbit   117
   purely cubic   105
   splitting   116
fine moduli space   234
finite
   mapping   272
   order invariant   353
first Chern class of a line bundle   249
first orthogonality relation   400
flexible
   classification of coverings   13, 304
   equivalence of coverings   24
flower
   double   132
   Leila's   106
form
   modular   426
   quasimodular   426
formula

Frobenius's 10, 406
  generalized 407
Goulden–Jackson 59, 279, 418
  generalized 418
Hurwitz 290
Riemann-Hurwitz 22
Whitney 293
framed
  graph 376
  knot 357
framing 357
Frobenius
  automorphism of 50
Frobenius's
  formula 10, 406
    generalized 407
function
  Airy matrix 266
  Baker 208
  Belyi 80
    pure 110
  generating 215
  meromorphic 71
    on a nodal curve 225
  stable 299
fundamental group 16

Galois
  covering 18
  extension 116
  group 116
    absolute 115
    universal 115
  orbit 92
Gaussian measure
  in $\mathbb{R}^k$ 162
  on the line 160
  on the space of
    Hermitian matrices 164
    symmetric matrices 167
  standard 163
Gegenbauer polynomial 132
general linear group 49
generalized
  Chebyshev polynomial 82
  constellation 64
  Hermite polynomial 195
generating function 215
  exponential 216

genus
  of a constellation 23
  of a map 28
  of a nodal curve 226
Goulden–Jackson formula 59, 279, 418
  generalized 418
graph 27
  bicolored 43
  bipartite 43
  complete 33
    tripartite 136
  complexity 368
  connected 27
  embedded 1
  marked 238
  fat 1
  framed 376
  intersection 367
  invariant 369
  modular 237
  planar 32
  ribbon 1
  simple 369
  star 18, 23
  weighted 379
  with rotation 1
group
  affine linear 49
  algebra 401
  braid 13, 306
    diagonal 315
    Hurwitz 306
    pure 306
    sphere 306
    surface 306
  cartographic 8
  fundamental 16
  Galois 116
    universal (or absolute) 115
  general linear 49
  imprimitive 8
  Klein modular 229
  mapping class 30
  Mathieu
    M_{11} 123
    M_{12} 42, 123
    M_{23} 12, 48, 123
    M_{24} 11, 40, 42
  monodromy 8, 17

450   Index

orthogonal   176
primitive   8
projective linear   49
   representation of   399
      left   399
      right   399
   special   8
      linear   49
      projective   49
      unitary   171

half-edge   33
Hermite polynomial   193
   generalized   195
Hermitian matrix   165
hierarchy
   KdV   206, 256
   Korteweg–de Vries   206, 256
Hodge
   bundle   288
   integral   290
holomorphic 1-form   224
HOMFLY polynomial   366
homotopic loops   16
Hopf algebra   347
Hurwitz
   braid group   306
   formula   290
   numbers   288
   problem   269
   space   288, 328
      completed   298
hyperdual hypermap   54
hypermap   43
   dual   53
   elementary   44
   hyperdual   54
   reciprocal   54
   rooted   46

icosahedron   113
IHX-relation   394
imprimitive group   8
incidence relation   27
integral
   elliptic   143
   Hodge   290
   quasi-elliptic   143
internal vertex   393

intersection graph of a chord diagram
   364, 367
invariant
   of finite order   353
   Tutte   368
   Vassiliev   353
inverse enumeration problem   99
irreducible
   character   10, 400
   components   225
   representation   400
      character of   400
isomorphic
   cacti   57
   constellations   8
   coverings   14
   Riemann surfaces   70
isthmus   30

Jacobi polynomial   132
Jenkins–Strebel quadratic differential
   239
   canonical   241
Julia set   139

KdV hierarchy   206, 256
Klein modular group   229
knot   350
   framed   357
   invariant   350
      of order $\leq n$ 353
   singular   351
knots
   concatenation of   359
   connected sum of   359
Kontsevich
   model   257
   theorem   256
Korteweg–de Vries hierarchy   206, 256

leading term   295
Legendre polynomial   132
Leila's flower   106
lemma
   Schur's   400
length of constellation   7
Lickorish bilinear form   397
lifting   17
loop   16, 27
loops

Index   451

homotopic   16
  product of   16
Lyashko–Looijenga mapping   271

Müller theorem   318
Mandelbrot set   139
map   28
  bicolored   43
  combinatorial   37
  dual   52
  face of   28
  genus of   28
  plane   32
  self-dual   54
  special   42
  topological   28
mapping
  class group   30
  finite   272
  Lyashko–Looijenga   271
  quasihomogeneous   274
marked embedded graph   238
mass-formula   47
matching polynomial   375
Mathieu group
  M_{11} 123
  M_{12} 42, 123
  M_{23} 12, 48, 123
  M_{24} 11, 40, 42
matrix
  Airy
    equation   266
    function   266
  covariance   163
  Hermitian   165
  orthogonal   176
  symmetric   167
  unitary   171
Maxwell stratum   280
mean value of a function   162
meander   190, 396
megamap   330
meromorphic
  1-form   223
    residue of   224
  differential   223
    quadratic   226
  function   71
    on a nodal curve   225

  pole of   71
  stable   299
  zero of   71
midpoint of an edge   37, 110
mirror symmetry   423
modified chromatic polynomial   370
modular
  curve   229
  form   426
  graph   237
module of chord diagrams   341
moduli space
  coarse   233
  decorated   237
  fine   234
monodromy   16
  group   8, 17
multiple edges   27
multiplication
  of chord diagrams   342
  of knot invariants   359
multiplicative
  4-invariant   372
  weight system   362
multiplicity
  of a critical point   73, 81

negative
  resolution of a singular point   351
  triangle   51
nodal curve   224
  genus of   226
  normalization of   225
node of a complex curve   224
non-rooted cactus   58
normal
  extension   116
  family of functions   139
normalization of a nodal curve   225
normalized double point   225
number field   115
numbers
  Bernoulli   244, 248, 291, 421
  Catalan   158, 178, 191, 198, 216, 396
  Hurwitz   288

open diagram   394
operator
  chord separating   361

452   Index

renormalization   361
orbifold   228
   Euler characteristic   230
orbit
   Galois   92
   of braid group   26, 308
   of diagonal braid group   315
order
   of a critical point   73, 81
   of a pole   224
   of symmetry   59
orthogonal
   group   176
   matrix   176
orthogonality relation
   first   400
   second   402

Painlevé I equation   201
   discrete   197
pair
   Belyi   80
   stable   232
parasitic solution   92
partition   9
   canonical polynomials associated to   409
   degeneracy of   278
   degenerate   278
passport   81
   of constellation   9
     refined   9
   of type A_n 325
   of type S_n 325
   polynomial   59
   valuable
     cactus   59
     tree   45
path   27
Pell equation   142
Pell–Abel equation   142
perfect matching   374
periodic point   140
permutation
   cycle structure of   9
planar graph   32
plane
   map   32
   tree   45

canonical geometric form of   87
Platonic solid   114
point
   critical   15, 21, 72, 81
     degree of   73, 81
     multiplicity of   15, 73, 81
     order of   73, 81
   double   224
     normalized   225
   eventually periodic   140
   periodic   140
   ramification   15, 21
   strictly preperiodic   140
pole
   of a meromorphic function   71
   order of   224
polynomial
   bialgebra   345
   Chebyshev   82
     generalized   82
     of the first kind   132
     of the second kind   132
   Conway   365
   decomposable   320
   defining   92, 116
   Gegenbauer   132
   Hermite   193
   HOMFLY   366
   Jacobi   132
   Legendre   132
   matching   375
   passport   59
   Shabat   82
   square-free form of   86
   trigonometric   277
positive
   resolution of a singular point   351
   triangle   51
preinvariant   362
primitive
   element of a bialgebra   346
   group   8
   stratum   280
principal part   294
product
   semidirect   66
   wreath   67
projection   14
projective

line
  complex 70
  linear group 49
  special group 49
pseudorhombicuboctahedron 114
pure
  Belyi function 110
  braid group 306
purely cubic field 105

quadrangle
  edge 373
  vertex 372
quadratic
  Jenkins–Strebel differential 239
    canonical 241
  residue 227
quasi-elliptic integral 143
quasihomogeneous mapping 274
quasimodular form 426
quotient constellation 68

ramification
  locus 21, 73
  point 15, 21
ramified covering 15
reciprocal hypermap 54
refined passport 9
regular
  constellation 69
  covering 18
renormalization operator 361
representation of a group 399
  irreducible 400
    character of 400
  left 399
  right 399
representations
  isomorphic 399
residue
  of a meromorphic 1-form 224
  quadratic 227
resolution of a singular point
  negative 351
  positive 351
ribbon graph 1
Riemann
  data 75
  sphere 70

surface 70
surfaces
  isomorphic 70
Riemann's existence theorem 74
Riemann–Hurwitz formula 22
rigid
  classification of coverings 24, 270, 277
  equivalence of coverings 24
rigidity 76
Ritt theorem 65, 318
root 46
rooted
  cactus 58
  hypermap 46

Sato
  Grassmanian 206
  subspace 206
Schur's lemma 400
second orthogonality relation 402
Segre class 293
  top 293
self-dual map 54
semidirect product 66
semilinear transformation 50
set
  Fatou 139
  Julia 139
  Mandelbrot 139
Shabat
  polynomial 82
  polynomials
    equivalent 84
sheet 14
sheets of a curve at a double point 224
simple
  critical value 310
  graph 369
singular knot 351
solid
  Archimedean 113
  Platonic 114
solution
  parasitic 92
space
  Hurwitz 288, 328
    completed 298
  moduli

454   Index

  coarse   233
  fine   234
spanning subgraph   381
special
  cactus   318
  group   8
  linear group   49
  map   42
  tree   48
sphere braid group   306
splitting field   116
sporadic tree   99
square free form of a polynomial   86
stabilizer of a point   229
stable
  curve   235
  meromorphic function   299
  pair (g, n) 232
standard Gaussian measure   163
star graph   18, 23
stationary equation   201
stratum   279
  Maxwell   280
  primitive   280
strictly preperiodic point   140
string equation   253
STU-relation   394
surface   22
  braid group   306
symmetric
  cactus   59
  matrix   167
symmetry
  order of   59

Temperley–Lieb algebra   395
theorem
  Belyi   79
  Feit–Jones   327
  Kontsevich   256
  Müller   318
  Riemann's existence   74
  Ritt   65, 318
top Segre class   293
topological map   28
transformation
  elementary   352
  semilinear   50
tree   45

  bicolored plane   45
  canonically labelled   46
  diameter of   99
  plane   45
    canonical geometric form of   87
    special   48
  sporadic   99
  valuable passport   45
trees
  conjugate   92
  family of   94
triangle
  negative   51
  positive   51
triangulation
  canonical   51
  elementary   52
trigonometric polynomial   277
Tutte
  invariant   368
  relation   368
type of a composition   321

unicursal curve   188
unitary
  group   171
  matrix   171
universal
  covering   18
  curve   234
  enveloping algebra   387
  Galois group   115
  one-matrix model   204
unramified covering   13

valency   28
  of a face   30
  of a vertex   28
valuable
  cactus passport   59
  tree passport   45
value
  critical   15, 21, 81
Vassiliev invariant   353
vertex   27
  degree of   28
  external   393
  internal   393
  quadrangle in a graph   372

Index   455

valency of   28
virtual Euler characteristic   230

weight
  of a $\mathbb{C}^*$-action   274
  of a coordinate   274
  of a weighted graph   379
  system   362
    complete   390
    multiplicative   362
weighted
  3-term element   379
  graph   379

graph invariant   381
  multiplicative   381
graphs
  bialgebra of   380
Whitehead collapse   238
Whitney formula   293
Wick coupling   164
Witten conjecture   256
wreath product   67

Zannier bound   129
zero
  of a meromorphic function   71

Printing and Binding: Strauss GmbH, Mörlenbach

Breinigsville, PA USA
27 February 2011
256441BV00005B/61/P